Microprocessors and Application

Microprocessors and Application

Editor:

Nelson Bolívar, Ph.D

ArclerPress

www.arclerpress.com

Microprocessors and Application

Editor: Nelson Bolívar, Ph.D

708 3rd Avenue, 6th Floor
New York
NY 10017
United States of America
www.arclerpress.com

ISBN: 978-1-68094-358-0

Library of Congress Control Number: 2016950959

About the Editor

Nelson Bolívar, Ph.D

Nelson Bolivar has a PhD. in physics from the University of Lorraine in France finished in 2014. His expertise is in quantum systems and condensed matter. His interest includes spintronic devices and correspondences between general relativity and condensed matter. He is currently an associate professor at the Central University of Venezuela

Contents

List of Contributors

Muiris Woulfe
Graphics, Vision and Visualisation Group (GV2), School of Computer Science and Statistics, Trinity College Dublin, Dublin, Ireland

Michael Manzke
Graphics, Vision and Visualisation Group (GV2), School of Computer Science and Statistics, Trinity College Dublin, Dublin, Ireland

Takeshi Morishita
Department of Robotics, Toin University of Yokohama, Kanagawa 225-8502, Japan

Osamu Tojo
Department of Robotics, Toin University of Yokohama, Kanagawa 225-8502, Japan

Xiaobin Li
Enterprise Microprocessor Group, Intel Corporation, Santa Clara, CA, USA

Jean-Luc Gaudiot
Department of Electrical Engineering and Computer Science, University of California, Irvine, CA, USA

Huda S Muhammad
Schlumberger Corp., Dubai Internet City, Bldg. 14, Dubai, UAE

Assim Sagahyroon
Department of Computer Science and Engineering, American University of Sharjah, Sharjah, UAE

Martin Schoeberl
Institute of Computer Engineering, Vienna University of Technology, 1040 Vienna, Austria

Gábor János Tornai
Faculty of Information Technology and Bionics, Pázmány Péter Catholic University, Práter str. 50/a, Budapest 1083, Hungary

György Cserey
Faculty of Information Technology and Bionics, Pázmány Péter Catholic University, Práter str. 50/a, Budapest 1083, Hungary

Robert I. Davis
Real-Time Systems Research Group, Department of Computer Science, University of York, YO10 5DD, York, UK

Alan Burns
Real-Time Systems Research Group, Department of Computer Science, University of York, YO10 5DD, York, UK

Reinder J. Bril
Technische Universiteit Eindhoven (TU/e), Den Dolech 2, 5600 AZ Eindhoven, Netherlands

Johan J. Lukkien
Technische Universiteit Eindhoven (TU/e), Den Dolech 2, 5600 AZ Eindhoven, Netherlands

Daniel J. Bernstein
Department of Computer Science, University of Illinois at Chicago, Chicago, IL 60607-7053, USA

Niels Duif
Department of Mathematics and Computer Science, Technische Universiteit Eindhoven, P.O. Box 513, 5600 MB, Eindhoven, The Netherlands

Tanja Lange
Department of Mathematics and Computer Science, Technische Universiteit Eindhoven, P.O. Box 513, 5600 MB, Eindhoven, The Netherlands

Peter Schwabe
Department of Electrical Engineering, National Taiwan University, 1, Section 4, Roosevelt Road, Taipei 10617, Taiwan

Bo-Yin Yang
Institute of Information Science, Academia Sinica, 128 Section 2 Academia Road, Taipei 115-29, Taiwan

R. Różycki
Institute of Computing Science, Poznan University of Technology, 2 Piotrowo St., 60-965 Poznan,´ Poland

J. Węglarz
Institute of Computing Science, Poznan University of Technology, 2 Piotrowo St., 60-965 Poznan,´ Poland

Eric Pop
Department of Electrical and Computer Engineering, Micro and Nanotechnology Lab and Beckman Institute, University of Illinois Urbana-Champaign, Urbana IL 61801, USA

Dmaithan Almajali
Faculty of Economics and Administrative Sciences, Zarqa University, Zarqa, Jordan

Khalid Mansour
Faculty of Economics and Administrative Sciences, Zarqa University, Zarqa, Jordan

Ra'ed Masa'deh
School of Business, The University of Jordan, Amman, Jordan

Mahmoud Maqableh
School of Business, The University of Jordan, Amman, Jordan

Vahid Karimi
Department of Electrical Engineering, Shiraz University of Technology, Shiraz, Iran

Reza Mohseni
Department of Electrical Engineering, Shiraz University of Technology, Shiraz, Iran

Yaser Norouzi
Department of Electrical Engineering, Amirkabir University of Technology, Tehran, Iran

Mohammad Javad Dehghani
Department of Electrical Engineering, Shiraz University of Technology, Shiraz, Iran

Gerard Rushingabigwi
Applied Electromagnetic fields Group, University of Science and Technology of China (USTC), Hefei, China

Liguo Sun
Applied Electromagnetic fields Group, University of Science and Technology of China (USTC), Hefei, China

Qi Zhu
Applied Electromagnetic fields Group, University of Science and Technology of China (USTC), Hefei, China

Yulong Xia
Applied Electromagnetic fields Group, University of Science and Technology of China (USTC), Hefei, China

Yongxia Yu
Applied Electromagnetic fields Group, University of Science and Technology of China (USTC), Hefei, China

Benjamin G. Jacob
Department of Global Health, University of South Florida, Tampa, USA

Robert J. Novak
Department of Global Health, University of South Florida, Tampa, USA

H. Olkkonen
Department of Physics and Mathematics, University of Eastern Finland, Kuopio, Finland

J. T. Olkkonen
VTT Technical Research Centre of Finland, VTT, Finland

Jean-Marc Laheurte
ESYCOM, Université Paris-Est, UPEMLV, Marne-la-Vallée, France

Aladdin Kabalan
ESYCOM, Université Paris-Est, UPEMLV, Marne-la-Vallée, France

Houssam Retima
ESYCOM, Université Paris-Est, UPEMLV, Marne-la-Vallée, France

Eric Piedallu
ESYCOM, Université Paris-Est, UPEMLV, Marne-la-Vallée, France

Fulvio Michelis
Université Paris-Est, IFSTTAR, COSYS/LISIS, Marne-la-Vallée, France

Bérengère Lebental
Université Paris-Est, IFSTTAR, COSYS/LISIS, Marne-la-Vallée, France

Sara Nouh
Electronics & Communications Engineering Department, American University in Cairo, Cairo, Egypt

Nada Elgaml
Electronics & Communications Engineering Department, Cairo University, Giza, Egypt

Nora Ali
Electronics & Communications Engineering Department, Cairo University, Giza, Egypt

Ahmed Khattab
Electronics & Communications Engineering Department, Cairo University, Giza, Egypt

Ramez Daoud
Electronics & Communications Engineering Department, American University in Cairo, Cairo, Egypt

Hassanein Amer
Electronics & Communications Engineering Department, American University in Cairo, Cairo, Egypt

Jaspreet Kaur Kohli
Department of Computer Engineering, Sandip Foundation Institute of Technology and Research Centre, Nashik, India

Amol Potgantwar
Department of Computer Engineering, Sandip Foundation Institute of Technology and Research Centre, Nashik, India

Jesvin Veancy Boaz
Department of Electronics and Communication Engineering, Easwari Engineering College, Chennai, India

Yogesh Palanichamy
Department of Information Science and Technology, Anna University, Chennai, India

Athikkan Venkatasami
Department of EEE, Einstein College of Engineering, Tirunelveli, India

Pitchai Latha
Department of CSE, Government College of Engineering, Tirunelveli, India

Preface

Microprocessors are a mass storage device. They are the advanced form of computers. They are also called as microcomputers. The impact of microprocessor in different lures of fields is significant. The availability of low cost, low power and small weight, computing capability makes it useful in different applications. The microprocessor is a multipurpose, clock driven, register based, programmable electronic device which accepts digital or binary data as input, processes it according to instructions stored in its memory, and provides results as output. Microprocessors contain both combinational logic and sequential digital logic. Microprocessors operate on numbers and symbols represented in the binary numeral system. A microprocessor incorporates the functions of a CPU on a single integrated circuit or a few integrated circuits. It is a computer processor on a microchip and is a multipurpose, programmable device that uses digital data as input and provides results as an output once it processes the input according to instructions stored in its memory. Microprocessors use sequential digital logic as they have internal memory and operate on numbers and symbols represented in the binary numeral system. They are designed to perform arithmetic and logic operations that make use of data on the chip. General purpose microprocessors in PCs are used for multimedia display, computation, text editing and communication. Several microprocessors are part of embedded systems. These embedded microprocessors provide digital control to several objects including appliances, automobiles, mobile phones and industrial process control.

This book takes applications of microprocessors that goes from academic to practical elements to the industrial and domestic illustrating modern applications. Now a days, a microprocessor based systems are used in instructions, automatic testing product, speed control of motors, traffic light control, light control of furnaces etc. The applications of microprocessors are enormous. Almost every aspect of our lives are affected by a device that works on a microprocessor at some point. In instrumentation devices such as frequency counters, function generators, frequency synthesizers, spectrum analyses and many other instruments. Control and communications, from the everyday devices like microwave oven, washing machine and so on, microprocessors are being used in controlling various parameters like speed, pressure, temperature. It is very useful in the field of instrumentation. Frequency counters, function generators, frequency synthesizers, spectrum analyses and many other instruments are available, when microprocessors are used as controller. It is also used in medical instrumentation. In telephone industry, these are used in digital telephone sets, telephone exchanges and modem. The use of microprocessor in television, satellite communication made teleconferencing possible. LAN and WAN for communication of vertical information through computer networking. Microprocessor based microcomputer with software packages has changed the office environment. Microprocessors based systems are being used for word processing, spread sheet operations, storage etc. The microprocessor has revolutionize the publication technology. The use of microprocessor in toys, entertainment equipment and home applications is making them more entertaining and full of features. The use of microprocessors is more widespread and popular

Editor
Nelson Bolívar, Ph.D

INTRODUCTION

A microprocessor is a computer processor which incorporates the functions of a computer's central processing unit (CPU) on a single integrated circuit (IC), or at most a few integrated circuits. The microprocessor is a multipurpose, clock driven, register based, programmable electronic device which accepts digital or binary data as input, processes it according to instructions stored in its memory, and provides results as output. Microprocessors contain both combinational logic and sequential digital logic. Microprocessors operate on numbers and symbols represented in the binary numeral system. A microprocessor incorporates the functions of a CPU on a single integrated circuit or a few integrated circuits. It is a computer processor on a microchip and is a multipurpose, programmable device that uses digital data as input and provides results as an output once it processes the input according to instructions stored in its memory. Microprocessors use sequential digital logic as they have internal memory and operate on numbers and symbols represented in the binary numeral system. They are designed to perform arithmetic and logic operations that make use of data on the chip. General purpose microprocessors in PCs are used for multimedia display, computation, text editing and communication. Several microprocessors are part of embedded systems. These embedded microprocessors provide digital control to several objects including appliances, automobiles, mobile phones and industrial process control.

Kinds of Microprocessors

There are several different kinds of microprocessors suitable for programming at Future Electronics. We stock many of the most common types categorized by several parameters including clock frequency, data bus width, MMAC/MIPS/FLOPS, Flash size, RAM size, ROM size, I\O voltage, cache memory size and packaging type. Our parametric filters will allow you to refine your search results according to the required specifications.

A *microprocessor* is one of the most central parts of a modern personal computer or, in fact, any advanced computer device. It integrates the functions of a *central processing unit*, the portion of a computer responsible for carrying out programmed instructions, onto a single integrated circuit that couples the important thinking devices of the machine with the electrical infrastructure needed to support them. Microprocessor design is able to incorporate a tremendous amount of processing power in a very small space. Perhaps more than any other component of the modern computer, the microprocessor has a long and storied history, and an almost mythological status thanks to its great importance. Each step of the way to modern microprocessing has spurred more triumphs, innovations, and competition.

History of Microprocessors

Before the development of the microprocessor, there were a variety of early technologies for simulating logic functions in computing devices. Many of these early inventions were spurred by wartime necessity during World War II. These early technologies were extremely expensive, slow, and prone to failure;€“ and by todayâ€™s standards their capabilities were amazingly primitive. Computing technologies based on vacuum tubes and transistors helped make IBM a giant in the large-scale computing industry, but were not realistic for business or home use due to their prohibitive costs and intensive maintenance schedules. Early integrated circuits appeared in calculators, of all things, in the early 1960s; years before Intel began work on the first recognizable microprocessor.

Though IBM was a titan in the early huge-scale computing race, it was Intel, founded in 1968, that provided much of the pioneering work associated with first-generation microprocessor technology. The power of a given microprocessor was, and still is, measured in *bits*. Put simply, this refers to the most basic unit of coded instructions, expressed in a string of binary 1s and 0s, which the computer interprets to carry out tasks. The more powerful the processor, the more instructions it can carry out at one time, leading to faster processing and more effectiveness at complex tasks. Though 4-bit chips were fairly simple, they gave weight to the prophetic prediction known as *Moore's Law*, predicting that the capacity of integrated circuits (and thus, of computers) would double every two years.

Intel remained a forerunner in early microprocessor technology, releasing its first 8-bit microprocessor, the 8008, in 1972. Many of the same engineers involved with the original 4-bit models contributed ideas and design specifications to this new generation of chips, which had taken only about four years to develop. By this time, other companies were becoming increasingly interested in the microprocessor field, thanks largely to Intel's own evangelization of the potential benefits. By the middle of the decade, Intel's new 8080 microprocessor made major inroads in catching the eye of computer developers and engineers and the ascent of microprocessors was only beginning.

In the mid-1970s, National Semiconductor entered the field of microprocessor development, enticed by the fast pace of breakthroughs at Intel. Unfortunately, the era of 16-bit chips was short, even by the standards of such an industry. 16-bit microprocessors including National Semiconductor's PACE were relatively slow, though more advanced iterations like the Motorola 6800 found their way into early Macintosh systems and others. By the time 16-bit chips were living up to their potential, the 32-bit era had already begun to arrive and 16-bit technology was destined to be eclipsed by the next big thing a foretaste of things to come in an industry where research never ends.

By the late 1970s, development of 32-bit microprocessors was in full swing and they began to appear on the mass market in the 1980s, courtesy of National Semiconductor and Hewlett-Packard. Desktop engineering devices entered a new phase with the inception of 32-bit processing. 32-bit microprocessors were relatively advanced and can be looked upon as the forerunner of modern microprocessors.

64-bit chips have been available since 1992, and are now in the mainstream of computer use. Much of the 64-bit microprocessor era has been dominated by the battle between Intel and AMD. The latter company was founded in 1969 and rose to prominence after inking a deal with IBM and Intel by which AMD would become the second source manufacturer of Intel-designed processors. From that humble agreement, prompted by IBM's internal policies at the time, was born legal controversy, corporate intrigue, and ultimately, the development of a vigorous microprocessing market with two major superpowers and the potential for many more as technology develops.

RISC, standing for Reduced Instruction Set Computer, derives from an IBM research project dating back to the second half of the 1970s. RISC chips were intended to streamline computer programming by omitting the most complex instructions and relying more effectively upon the simpler, more common instructions that could be processed and acted upon more quickly. Such chips are still used today and have developed into a spectrum of full-fledged computing architectures in competition with CISC methods of development, which makes full use of complex instructions. RISC architecture is widely used in devices such as cellular phones and certain types of travel PCs.

- ***Microprocessors were categorized into five generations***: first, second, third, fourth, and fifth generations. Their characteristics are described below:
- ***First-generation:*** The microprocessors that were introduced in 1971 to 1972 were referred to as the first generation systems. First-generation microprocessors processed their instructions serially—they fetched the instruction, decoded it, and then executed it. When an instruction was completed, the microprocessor updated the instruction pointer and fetched the next instruction, performing this sequential drill for each instruction in turn.
- ***Second generation:*** By the late 1970s, enough transistors were available on the IC to usher in the second generation of microprocessor sophistication: 16-bit arithmetic and pipelined instruction processing. Motorola's MC68000 microprocessor, introduced in 1979, is an example. Another example is Intel's 8080. This generation is defined by overlapped fetch, decode, and execute steps. As the first instruction is processed in the execution unit, the second instruction is decoded and the third instruction is fetched. The distinction between the first and second generation devices was primarily the use of newer semiconductor technology to fabricate the chips. This new technology resulted in a five-fold increase in instruction, execution, speed, and higher chip densities.
- ***Third generation:*** The third generation, introduced in 1978, was represented by Intel's 8086 and the Zilog Z8000, which were 16-bit processors with minicomputer-like performance. The third generation came about as IC transistor counts approached 250,000.
- Motorola's MC68020, for example, incorporated an on-chip cache for the first time and the depth of the pipeline increased to five or more stages. This generation of microprocessors was different from the previous ones in that all major workstation manufacturers began developing their own RISC-based microprocessor architectures (Computer, 1996).
- ***Fourth generation:*** As the workstation companies converted from commercial microprocessors to in-house designs, microprocessors entered their fourth generation with designs surpassing a million transistors. Leading-edge microprocessors such as Intel's 80960CA and Motorola's 88100 could issue and retire more than one instruction per clock cycle.
- ***Fifth generation:*** Microprocessors in their fifth generation, employed decoupled super scalar processing, and their design soon surpassed 10 million transistors. In this generation, PCs are a low-margin, high-volume-business dominated by a single microprocessor.

Applications of Microprocessor

Microprocessors are a mass storage device. They are the advanced form of computers. They are also called as microcomputers. The impact of microprocessor in different lures of fields is significant. The availability of low cost, low power and small weight, computing capability makes it useful in different applications. Now a days, a microprocessor based systems are used in instructions, automatic testing product, speed control of motors, traffic light control, light control of furnaces etc. The wide range of application area of the microprocessor can be broadly classified into two groups: general purpose application and special purpose application.

1. General purpose application

i) Single board micro computers

Single board microcomputers are simple and cheaper. They have the minimum possible software and hardware configuration. They are used to train the students. They are also used to build small computer based systems.

ii) Personal Computers

The home computer made up of a 8 bit microprocessor are used for playing video games and learning simple programs. The 16 bit computers are used for word processing, payroll, business accounts, etc.

iii) Super Minis and CAD

32 bit processors are used to build powerful microcomputers. Their performance is better than mainframe and mini-computers. These systems are used in engineering side as computer aided design machines.

2. Special purpose application

i) Instrumentation

The microprocessor based instrument can be made intelligent using feature programmability. Microprocessors can be used as controllers in various instruments. They are also used in medical instruments to measure blood pressure and temperature.

ii) Control

Microprocessor controllers are now available in home appliances such as microwave oven, washing machine. In industry, microprocessors are used in controlling various process parameters such as speed, temperature, moisture and pressure.

iii) Communication

In the telephone industry, microprocessors are used in digital telephone sets, telephone exchanges and modems. They are also used in railway reservation system at the national level and air reservation system at international level. Satellite communication systems, mobile phones and televisions are also using microprocessors.

iv) Office Automation and Publication

With the availability of inexpensive and user friendly microcomputers along with wide range of software packages, office works are computerized. Microcomputers are used in office to perform word processing, spreadsheet operations, storage and retrieval of huge information. In publishing houses microprocessor based systems are used for making automatic photo copies. Microprocessor based LASER printers are used to achieve good speed in printing.

How Microprocessors Work

The computer you are using to read this page uses a microprocessor to do its work. The microprocessor is the heart of any normal computer, whether it is a desktop machine, a server or a laptop. The microprocessor you are using might be a Pentium, a K6, a PowerPC, a Sparc or any of the many other brands and types of microprocessors, but they all do approximately the same thing in approximately the same way.

A microprocessor -- also known as a CPU or central processing unit -- is a complete computation engine that is fabricated on a single chip. The first microprocessor was the Intel 4004, introduced in 1971. The 4004 was not very powerful -- all it could do was add and subtract, and it could only do that 4 bits at a time. But it was amazing that everything was on one chip. Prior to the 4004, engineers built computers either from collections of chips or from discrete components (transistors wired one at a time). The 4004 powered one of the first portable electronic calculators.

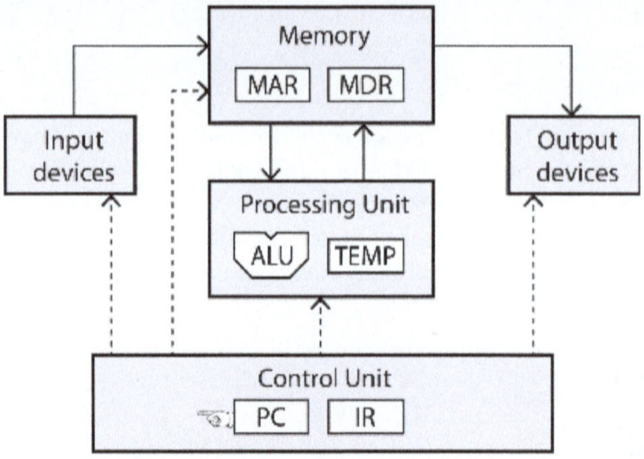

Difference between Microprocessor and Microcontroller

The term microprocessor and microcontroller have always been confused with each other. Both of them have been designed for real time application. They share many common features and at the same time they have significant differences. Both the IC's i.e., the microprocessor and microcontroller cannot be distinguished by looking at them. They are available in different version starting from 6 pin to as high as 80 to 100 pins or even higher depending on the features.

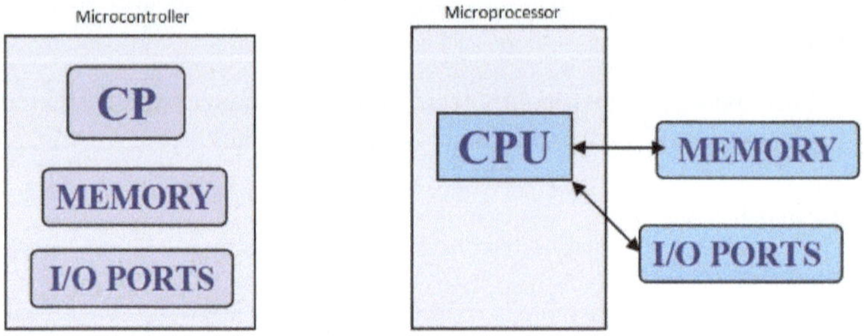

Microprocessor is an IC which has only the CPU inside them i.e. only the processing powers such as Intel's Pentium 1,2,3,4, core 2 duo, i3, i5 etc. These microprocessors don't have RAM, ROM, and other peripheral on the chip. A system designer has to add them externally to make them functional. Application of microprocessor includes Desktop PC's, Laptops, notepads etc.

But this is not the case with Microcontrollers. Microcontroller has a CPU, in addition with a fixed amount of RAM, ROM and other peripherals all embedded on a single chip. At times it is also termed as a mini computer or a computer on a single chip. Today different manufacturers produce microcontrollers with a wide range of features available in different versions. Some manufacturers are ATMEL, Microchip, TI, Freescale, Philips, Motorola etc.

Microcontrollers are designed to perform specific tasks. Specific means applications where the relationship of input and output is defined. Depending on the input, some processing needs to be done and output is delivered. For example, keyboards, mouse, washing machine, digicam, pendrive, remote, microwave, cars, bikes, telephone, mobiles, watches, etc. Since the applications are very specific, they need small resources like RAM, ROM, I/O ports *etc.* and hence can be embedded on a single chip. This in turn reduces the size and the cost.

Microprocessor find applications where tasks are unspecific like developing software, games, websites, photo editing, creating documents etc. In such cases the relationship between input and output is not defined. They need high amount of resources like RAM, ROM, I/O ports etc.

The clock speed of the Microprocessor is quite high as compared to the microcontroller. Whereas the microcontrollers operate from a few MHz to 30 to 50 MHz, today's microprocessor operate above 1GHz as they perform complex tasks. Read more about what is microcontroller.

Comparing microcontroller and microprocessor in terms of cost is not justified. Undoubtedly a microcontroller is far cheaper than a microprocessor. However microcontroller cannot be used in place of microprocessor and using a microprocessor is not advised in place of a microcontroller as it makes the application quite costly. Microprocessor cannot be used stand alone. They need other peripherals like RAM, ROM, buffer, I/O ports *etc.* and hence a system designed around a microprocessor is quite costly.

CHAPTER 1

A HYBRID FIXED-FUNCTION AND MICROPROCESSOR SOLUTION FOR HIGH-THROUGHPUT BROAD-PHASE COLLISION DETECTION

Muiris Woulfe and Michael Manzke

Keywords: Broad phase, Collision detection, Fixed-function microarchitecture, Microprocessor, Hybrid system, Energy efficiency

INTRODUCTION

As technology progresses, increasingly greater realism is demanded by the consumers of real-time graphics applications. Collision detection is an important factor in achieving this realism. It determines if simulated objects are intersecting, and, in cooperation with collision response, it maintains realism by preventing objects from interpenetrating. Collision detection is found in computer games, animation, robotics and computer-aided design (CAD). An improvement in collision detection will benefit myriad applications.

Despite decades of research, collision detection remains a fundamental problem. It can form a computational bottleneck in many applications. Interactive applications are particularly challenging as they demand a frame rate of at least 30 fps to ensure the illusion of visual continuity. Moreover, the inter-frame durations must be sufficient to execute the entire program loop, which potentially comprises input processing, collision detection, collision response, physics, AI, audio and rendering. The classic solution is to trade accuracy for speed. This trade-off is undesirable for most applications, and it is particularly problematic for robotics and CAD. Additional research is necessary to find sufficient throughput enhancements.

Algorithms can be executed on fixed-function microarchitectures on platforms such as application-specific integrated circuits (ASICs) or on general-purpose microprocessors such as CPUs and GPUs. Microarchitectures sacrifice programmability to dissipate less power and exhibit superior throughput. These advantages result from providing the designer with complete control over component layout and from eliminating the overhead of executing instructions. As many graphics applications require the recurrent execution of algorithms at interactive frame rates, these algorithms are good candidates for microarchitectures, providing they are utilised sufficiently and do not require programmability. GPU rasterisation is a good example of an effective microarchitecture.

Recent articles on the topic of integrated circuit (IC) power consumption have demonstrated that future ICs will require additional functionality to be implemented as microarchitectures. A power dissipation problem is evident in current multicore architectures. Native transistor switching speeds continue to double every two process generations, while processor frequencies are not increasing substantially. This serves to reduce the amount of utilisation necessary to justify adding custom microarchitectures. Their addition is further justified through the current desire for mobile and wearable computing devices, which demand energy efficiency to maximise finite battery lifespans.

We identify collision detection as an algorithm that is computationally expensive and satisfies the utilisation requirement for its implementation as a microarchitecture. We specifically select the broad phase due to its parallelisability, its compute-bound nature and its need for minimal control logic. Two alternative microarchitectures are proposed: one focuses on minimising resource consumption while the other supports greater object quantities. Both use pipelines comprising a plurality of memories that rearrange the input into a format maximising parallelism and bandwidth. To increase the object counts supported and to improve the computational

Citation: Woulfe, M. and Manzke, M (2017). A hybrid fixed-function and microprocessor solution for high-throughput broad-phase collision detection, EURASIP Journal on Embedded Systems, 2017; 1. DOI:10.1186/s13639-016-0037-7.

complexity, we propose a hybrid solution that combines these microarchitectures with a spatial-partitioning stage on a CPU or GPU. We further propose reusing the hierarchies created by a ray tracer to minimise construction costs. For 1,024,000 objects, this system achieves an acceleration of 812 × relative to a CPU and an acceleration of 161 × relative to a GPU, while maintaining energy efficiency.

This article makes the following contributions:

- Two fixed-function microarchitectures for performing broad-phase collision detection that offer significant throughput and power advantages relative to CPU and GPU equivalents
- A novel technique for combining these microarchitectures with ray-tracing data structures hosted on a general-purpose microprocessor
- A hybrid system for collision detection comprising the aforementioned

RELATED WORK

Collision Detection

Collision-detection systems check a set of *n* objects for collision. Most are multiphase, but there are many ways to delineate these phases. This article will utilise the following two definitions:

- ***Broad Phase:*** This uses an approximate test to create a potentially colliding set comprising pairs of objects.
- ***Narrow Phase:*** This checks the potentially colliding set using a more accurate algorithm, and it may also compute the distance between objects as well as the point and time of collision.

Multiphase collision detection is based on the hypothesis that the broad phase's approximate test will eliminate the vast majority of objects from consideration. This scheme typically leads to a significant improvement in throughput.

The broad phase is concerned with bounding volumes. These are convex shapes that simplify complex and non-convex environment geometry. A plurality of bounding volumes exist. Spheres [1] are the same as their geometric counterparts, and their advantage is that they are invariant under rotation. Axis-aligned bounding boxes (AABBs) [2, 3] are cuboids whose axes are aligned with those of the environment. Oriented bounding boxes (OBBs) [4] extend AABBs by removing the axis-aligned requirement. Discrete-oriented polytopes (*k*-DOPs) [5] are *k*-sided parallelepipeds where the surfaces consist of hyperplanes whose normals belong to a fixed set of *k* vectors. There exist a number of algorithms to check these bounding volumes for collision. All-pairs checks every object for collision against every other object, resulting in n(n−1)2n(n−1)2comparisons. An alternative is full-sort sweep and prune [2], which sorts axes to determine when a collision begins and ends. Incremental sweep and prune [3] improves on this by using insertion sort to exploit coherence. Spatial partitioning [6] is another alternative that uses grids to divide the environment into cells before placing each object within an appropriate cell. It reduces the number of pairwise collision tests by only checking objects within the same cell.

The narrow phase typically uses bounding-volume hierarchies (BVHs) [7]. BVH algorithms traverse these hierarchies to prune branches where a collision is impossible. Deformable narrow phases [8] attempt to refit BVHs to objects undergoing deformation. Recent research has attempted to improve the accuracy of these algorithms [9]. Continuous collision detection [10] is also a topic of current interest. These algorithms attempt to fit BVHs to the motion of objects so that collisions are not missed within the intervals between cycles. Alternatives to BVH traversal such as Lin-Canny [11] and V-Clip [12] work by tracking the closest features of polyhedra.

There has also been research interest in performing collision detection on GPUs. Originally, this research repurposed rasterisation to find overlapping objects [13]. As GPUs developed fully programmable cores, researchers moved to utilise these. Liu et al. [14] outline a broad phase that represents objects as a collection of spheres processed using spatial partitioning, followed by full-sort sweep and prune along a single axis chosen to minimise the number of overlaps. The narrow phase is avoided using a penalty algorithm for rigid-body dynamics, although this can introduce substantial divergences from expected results. Combining this with the use of sweep and prune along only a single axis is likely to compound these inaccuracies. Significant accelerations are demonstrated, but the throughput starts to deteriorate above 128,000 objects, leading to scalability concerns. Avril, Gouranton and Arnaldi [15] outline an alternative broad phase that uses a hybrid system comprising CPU-based spatial partitioning and GPU-based full-sort sweep and prune. A novel mapping function and square root approximation logic avoid global memory accesses and reduce atomic operations. The authors demonstrate significant accelerations, but the rate declines as the object count increases, leading to the same scalability concerns as for the previous research. A narrow-phase algorithm is proposed by Lauterbach, Mo and Manocha [16]. This algorithm exploits GPU cores using a parallelised front-based traversal method. This method can be specialised for deformable objects [17]. A derivative exploiting GPU texture memory has been proposed by Zhang and Kim [18]. HPCCD [19] increases the level of parallelism by splitting a narrow-phase algorithm across a hybrid system comprising a CPU and GPU operating simultaneously. The CPU performs BVH traversal while the GPU executes elementary collision tests.

Ray tracing

We restrict our review of ray tracing to spatial-partitioning hierarchies, as these are the only element germane to this article. Traditionally, the most common hierarchy was the *k*-d tree [20]. *k*-d trees divide the environment by splitting along an arbitrary plane aligned to the world axes. There has recently been significant interest in BVHs for ray tracing [21, 22], which are not the same generalisable hierarchies used in collision detection but are instead AABB hierarchies constructed in accordance with the surface-area heuristic (SAH) metric. Their advantage is that, unlike *k*-d trees, they can be refitted in dynamic scenarios.

Microarchitectures

There is currently significant research interest in microarchitectures, as multicore architectures are expected to soon encounter a utilisation wall when they reach silicon power-density limits. This wall will limit the fraction of a processor that can run at full speed. It results from increasing transistor counts combined with an inability to reduce the power to switch a transistor. Esmaeilzadeh et al. [23] posit that at 8 nm, over 50 % of a processor will be unutilised. This will result in a 14 % throughput increase per annum, which is substantially less than current trends. The solution proposed by most researchers is specialisation [24–26], which involves offloading parallelisable computations to embedded microarchitectures. It has already been used effectively in a variety of processors. An increasing number of CPUs include specialised primitives [27], and the Apple iPhone 6 A8 comprises approximately 64 % fixed-function logic [28].

An early attempt at a collision-detection microarchitecture is outlined by Atay, Lockwood and Bayazit [29]. This exclusively focuses on the narrow phase and is designed for robotics. The triangle-triangle intersection test employed by the microarchitecture allows it to achieve high accuracy at the expense of interactivity. The CollisionChip [30] is an alternative narrow-phase microarchitecture that uses 24-DOP hierarchies storing triangles in the leaf nodes. It traverses a single hierarchy combining those of the two objects being tested, using an algorithm designed to reduce memory accesses and node transformations. A specialised separating-axis test (SAT) is used to test the *k*-DOPs and triangles for collision. The design is specialised for CAD objects with extremely large quantities of triangles and, like the previous microarchitecture, is not focused on achieving real-time results. An alternative approach is employed by the now discontinued AGEIA PhysX, which is a commercial IC and associated driver designed to accelerate physics including collision detection. A patent [31] outlines two possible designs, which both revolve around a specialised very-long instruction word (VLIW) processor with a plurality of floating-point units. It is ambiguous as to whether this system performs collision detection on the PhysX IC, the CPU or a combination of both.

Previous Research

An earlier revision of our design [32] achieved an acceleration of 1.5 ×, despite being limited to 512 objects due to object duplication in memory. The current article builds on this by providing results for up to 1,024,000 objects, achieved via the integration of the microarchitectures into a complete hybrid system comprising the reuse of ray-tracing spatial-partitioning hierarchies. This article additionally compares and contrasts two alternative microarchitecture designs, and it provides the expected throughput if the microarchitectures were implemented on an ASIC.

DESIGN

The design of our hybrid collision-detection system comprises three stages:

- ***Spatial-Partitioning Broad Phase:*** This executes on a processor and divides the environment into cells.
- ***Cell-Based Broad Phase:*** This utilises one of the two microarchitectures to perform collision detection on the contents of each cell.
- ***Narrow Phase:*** This stage executes on a processor and performs conventional narrow-phase processing.
- We begin by outlining the core element of our system, the cell-based broad phase, before complementing this with a description of the spatial-partitioning broad phase.

Cell-Based Broad Phase

We chose a microarchitecture as the platform for our cell-based broad phase for three primary reasons.

- ***Degree of Parallelism:*** Microarchitectures can accelerate algorithms with a high degree of parallelism. The broad phase offers significant scope for parallelisation with a workload that can be statically balanced to ensure consistently high utilisation throughout the microarchitecture.
- ***Compute and Memory Bound:*** Microarchitectures exploit parallelism to accelerate compute-bound algorithms but offer fewer advantages to memory-bound algorithms. The broad phase involves a high degree of computation with memory accesses that can be aggregated to reduce their impact.
- ***Sequence of Operations:*** Consistent sequences of operations require minimal control logic and facilitate efficient pipelining through the reuse of standardised computation engines, thereby lending themselves to microarchitecture implementation. The broad phase tends to use standard, recurring collision tests performed in a consistent sequence.

The selection of a microarchitecture was also influenced by evidence that the broad phase consumes a considerable portion of the interactive application program loop. Lin and Gottschalk [33] and Fan et al. [34] discovered that collision detection is often a bottleneck. We calculated from the 22 benchmarks of the third experiment in Woulfe and Manzke [35] that a mean of 47 % of the overall collision-detection time is spent in the broad phase. This calculation can be considered a conservative estimate, as only the high throughput dynamic bounding-volume tree (DBVT) algorithm from the Bullet Physics SDK was considered. Finally, it should be noted that even if the broad phase were not to account for a major part of the program loop in a given scenario, the microarchitecture would still provide throughput improvements that would facilitate increased realism.

To design the microarchitectures, we began by investigating the various broad-phase algorithms. The most commonly used is incremental sweep and prune. However, it is difficult to parallelise across more threads of execution than the number of coordinate axes. One solution would be to switch from incremental to full-sort sweep and prune, but this essentially obviates the algorithm's primary advantage of coherence. The GPU sweep-and-prune implementations designed by Liu et al. [14] and Avril, Gouranton and Arnaldi [15] represent an attempt to trade coherence for parallelism. Despite the promise shown, full-sort sweep and prune would not be entirely amenable to microarchitectures as it makes significant use of sorting. Sorting tends to be problematic due to the overhead of memory access latencies and, therefore, tends to either inadequately exploit parallelism [36] or have undesirable throughput-to-area trade-offs [37]. Harkins et al. [38] claim that algorithms utilising sorting are not suited to microarchitecture implementation. For these reasons, sweep and prune would be a suboptimal choice. This corresponds to the findings of Chen et al. [39] that the best serial algorithms can have poor parallel scalability.

In contrast, we discovered that all-pairs is ideal for microarchitecture implementation. It is embarrassingly parallel, and this parallelism can be used to effectively exploit resources. AABBs were selected as the bounding volumes, since they tend to provide a good object fit while requiring a relatively low quantity of arithmetic components, thereby enabling many operations to be performed in parallel. AABBs have also been successfully used by the I-COLLIDE [3] and SOLID [7] libraries. A sequential version of the algorithm is:

function ALLPAIRSAABB
  n: Object count
  $\min_a^b$: Minimum of AABB a along axis b
  $\max_a^b$: Maximum of AABB a along axis b
  for $i \leftarrow 1$ **to** $n-1$ **do**
    for $j \leftarrow i+1$ **to** n **do**
      collision $\leftarrow$

$$\left(\max_i^x \geq \min_j^x\right) \wedge \left(\min_i^x \leq \max_j^x\right) \wedge \left(\max_i^y \geq \min_j^y\right) \wedge \left(\min_i^y \leq \max_j^y\right) \wedge \left(\max_i^z \geq \min_j^z\right) \wedge \left(\min_i^z \leq \max_j^z\right)$$

      if *collision* **then**
        result $\leftarrow$ *result* $\cup \langle i, j \rangle$
      end if
    end for
  end for
  return *result*
end function

Another advantage of all-pairs is that its throughput is deterministic. In contrast, sweep and prune has a computational complexity of $O(n+s)$, where s denotes the number of swapping operations required to maintain the algorithm's sorted object lists. As s cannot be determined a priori, the behaviour of sweep and prune can vary significantly. Scenarios with many moving objects result in significant increases in s that lead to decreases in throughput. Tracy, Buss and Woods [40] demonstrate that scenarios with few moving objects can also perform poorly if the total number of objects is very high. Only all-pairs facilitates the accurate deduction of the most complex scenario that can be executed within a given timeframe, without concern that the frame rate will decrease in certain scenarios. All-pairs unlocks the possibility of a wider range of scenarios through the avoidance of the non-deterministic throughput of sweep and prune.

Area-efficient Microarchitecture

The first design of the cell-based broad-phase microarchitecture is area efficient. In other words, it uses a minimal quantity of resources to exploit available parallelism. However, the trade-off is that it is limited in the quantity of objects supported. The design consists of a pipeline implementing two primary operations—buffer and compare. A schematic is provided in Fig. 1.

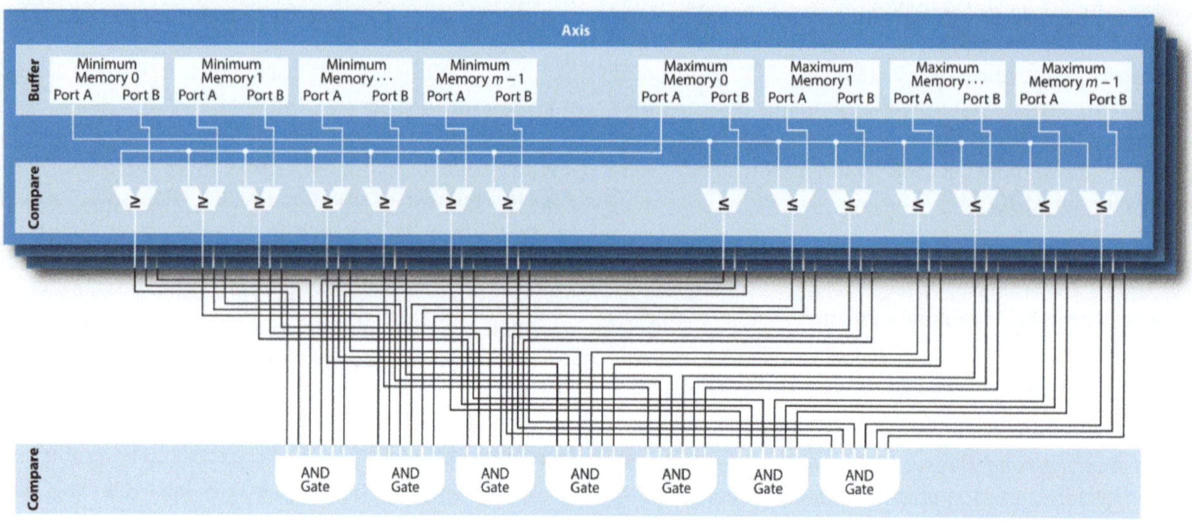

Figure 1. Schematic of the area-efficient microarchitecture. The buffer and the comparators from the compare operation are replicated three times to cover the three axes.

As the availability of resources can vary, the microarchitecture is designed to be extensible via a factor m. When many resources are available, the design can take full advantage of these to gain the maximum achievable acceleration, while it will still fit and execute efficiently when resources are constrained.

The microarchitecture could represent numbers using fixed-point formats, but these have relatively low accuracy. Moreover, their economical use of resources would offer little advantage as the limiting factor tends to be the quantity of memory and not the quantity of logic consumed. In addition, effective use of pipelining almost entirely eliminates the throughput gains that could be achieved. Therefore, the microarchitecture represents numbers using the single-precision IEEE 754 floating-point format. There is a wealth of research highlighting the efficiency of performing floating-point computations on microarchitectures [41]. The number of mainstream libraries defaulting to single precision, such as SOLID [7] and Bullet, indicates that this offers sufficient accuracy. All platforms can use this format, precluding the need to translate when communicating data.

Buffer

The buffer stores each AABB's data in an efficient manner for processing by the subsequent compare operation. During initialisation, the buffer reads each AABB and stores the data in $6m$internal dual-port memories. The $6m$ memories correspond to m memories for each of the minimum x, maximum x, minimum y, maximum y, minimum z and maximum z values. The data are replicated across each set of m memories, so that each of the m memories contains the same data. This results in six logical $2m$-port memories, allowing $12m$ data to be outputted in a single clock cycle.

In the following sections, the first port of each dual-port memory will be referred to as A and the second port will be referred to as B. The six memories that contain each AABB's data and that share an index will be referred to as a *memory group*. For example, memory group 0 contains minimum x memory 0, maximum x memory 0, minimum y memory 0, maximum y memory 0, minimum z memory 0 and maximum z memory 0. In the following sequence, the inputs to each memory belonging to a given memory group remain the same at all times.

To enable the required sequence of object-object comparisons, the AABBs are outputted from the memory groups in a specific sequence. Initially, the address input to memory group 0's A is set to 0, while the remaining $2m-1$ are set to the subsequent addresses. On the subsequent cycle, 0's A retains its value, and the remaining are each incremented by $2m-1$. This sequence continues until any input selects $n-1$. At this stage, 0's A is set to 1, while the remaining are set to the subsequent addresses. The sequence continues until 0's A selects $n-2$. Addresses after $n-1$ may be accessed using this sequence; these must be subsequently removed. The sequence is exemplified in Table 1.

Table 1. Area-efficient microarchitecture sequence

	Address								Comparison						
Cycle	0A	0B	1A	1B	2A	2B	3A	3B	0–1	0–2	0–3	0–4	0–5	0–6	0–7
1	0	1	2	3	4	5	6	7	0–1	0–2	0–3	0–4	0–5	0–6	0–7
2	0	8	9						0–8	0–9					
3	1	2	3	4	5	6	7	8	1–2	1–3	1–4	1–5	1–6	1–7	1–8
4	1	9							1–9						
5	2	3	4	5	6	7	8	9	2–3	2–4	2–5	2–6	2–7	2–8	2–9
6	3	4	5	6	7	8	9		3–4	3–5	3–6	3–7	3–8	3–9	
7	4	5	6	7	8	9			4–5	4–6	4–7	4–8	4–9		
8	5	6	7	8	9				5–6	5–7	5–8	5–9			
9	6	7	8	9					6–7	6–8	6–9				
10	7	8	9						7–8	7–9					
11	8	9							8–9						

An exemplar of the sequencing of the dataflow through the area-efficient microarchitecture with extensibility factor m=4 and object count n=10. On each clock cycle, the microarchitecture requests the specified memory addresses from the memory groups and ports indicated. These data are subsequently compared according to the comparison sequence outlined

In this proposal, the parallelism per cycle varies. It would be preferable to maintain a consistent high level of parallelism throughout execution, and a variety of schemes could be used to achieve this. However, although practical in theory, these schemes become impossible to implement in a microarchitecture, as the complexity of the required control logic would consume large quantities of resources and the design would fail to achieve an adequate clock frequency. Through experimentation, we selected the outlined design as the variable parallelism is compensated for by the ability to maintain a high clock frequency.

In the buffer, the memory bandwidth is $2fmw$ bit/s, where f is the clock frequency of the microarchitecture in hertz and w is the bit-width of a single memory location. The number of cycles required to generate the sequence is

$$\sum_{i=0}^{n-2}\left\lceil \frac{n-i-1}{2m-1}\right\rceil.$$

Compare

The compare operation performs the comparison from all-pairs using the data supplied by the buffer. It compares the data outputted by memory group 0 with the data outputted by all other memory groups. It comprises $6m-3$ greater-than-or-equal-to and $6m-3$ less-than-or-equal-to comparators. The outputs are connected to $2m-1$ logical AND gates. Each gate takes six inputs corresponding to the six comparator results forming an AABB pair. If a collision is detected, the indices of the two colliding objects are written to a single line of memory.

Many-object Microarchitecture

The second design of the cell-based broad-phase microarchitecture supports significantly greater object quantities. It achieves this advantage through the use of additional resources to avoid data replication. It, therefore, exhibits less parallelism. The design consists of a pipeline implementing three primary operations—buffer, reorder and compare. A schematic is provided in Fig. 2. This microarchitecture is also extensible via a factor m and also uses single-precision floating point.

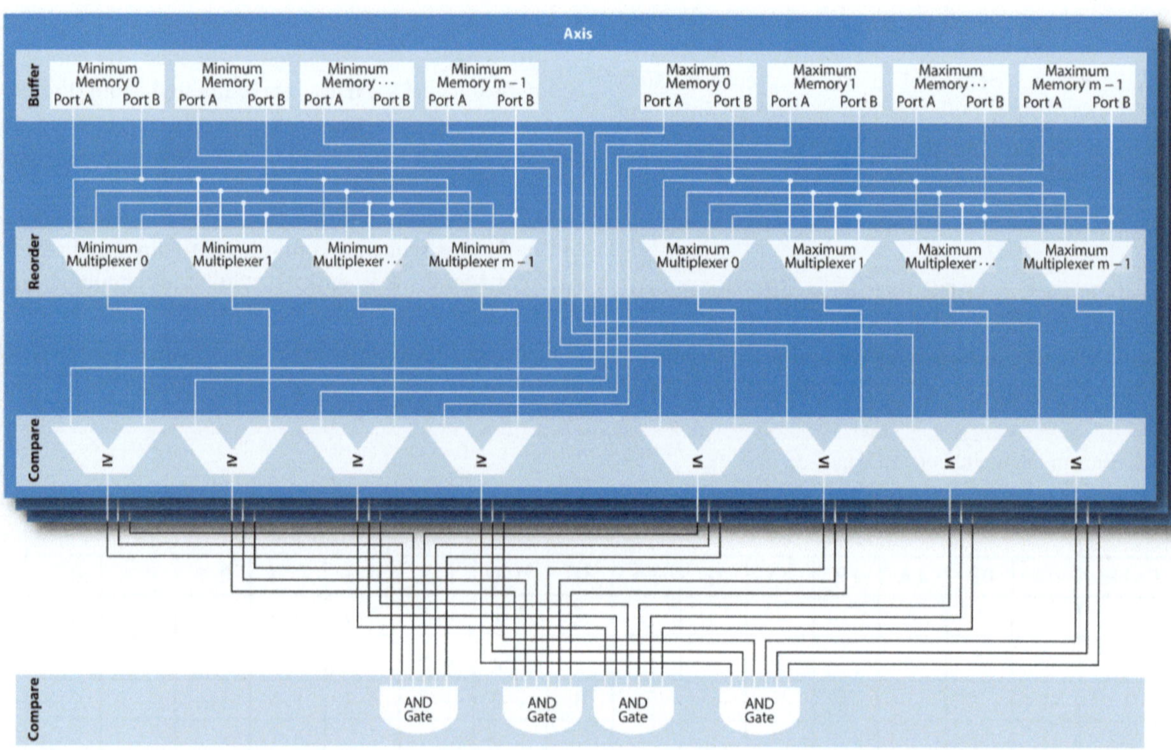

Figure 2. Schematic of the many-object microarchitecture. The buffer, the reorder and the comparators from the compare operation are replicated three times to cover the three axes.

Buffer

During initialisation, the buffer works in the same way as for the area-efficient microarchitecture. It reads each AABB and stores the data in $6m$ dual-port memories. However, unlike the area-efficient microarchitecture, these data are stored across the m memories in a format that precludes data duplication with, for example, the first AABB stored in the first address of the first memory group and the second AABB stored in the first address of the second memory group. This data layout, which is exemplified in Table 2, results in a schism between the memory group address and the index of the AABB being retrieved; the index can be computed using $am+j$ where a is the address and j is the memory group being accessed.

Table 2. Many-object microarchitecture buffer layout

	Object index			
Address	**0**	**1**	**2**	**3**
0	0	1	2	3
1	4	5	6	7
2	8	9		

An exemplar indicating the layout of the objects in the buffer operation of the many-object microarchitecture with extensibility factor m=4 and object count n=10

As for the area-efficient microarchitecture, the AABBs are outputted from the memory groups in a specific sequence. Initially, all memory groups' A address inputs are set to 0 while 0's B is set to 1. The remaining Bs are set to 0. On the subsequent cycle, 1's B is incremented to 1, and the other Bs retain their previous values. This sequence continues until 0's B selects $\left\lceil \frac{n}{m} \right\rceil$, which ensures that all comparisons to AABB $n-1$ have been performed. At this stage, all As are set to 1, 0's B is set to 2, and the remaining Bs are set to 1. The sequence continues until some A selects $\left\lceil \frac{n}{m} \right\rceil - 1$, which is the address corresponding to AABB $n-2$, and 0's B selects $\left\lceil \frac{n}{m} \right\rceil$. The sequence is exemplified in Table 3. As for the area-efficient microarchitecture, this design exhibits a variable degree of parallelism, as addresses after $n-1$ may be accessed; the outlined solution represents a compromise between parallelism and clock frequency.

Table 3. Many-object microarchitecture sequence

	Address				Object index				Address				Object index				Multiplexer			
Cycle	**0A**	**1A**	**2A**	**3A**	**0A**	**1A**	**2A**	**3A**	**0B**	**1B**	**2B**	**3B**	**0B**	**1B**	**2B**	**3B**	**0**	**1**	**2**	**3**
1	0	0	0	0	0 ∗	1 †	2 ‡	3 §	1	0	0	0	4 §	1 ∗	2 †	3 ‡	1	2	3	0
2	0	0	0	0	0 ∗	1 †	2 ‡	3 §	1	1	0	0	4 ‡	5 §	2 ∗	3 †	2	3	0	1
3	0	0	0	0	0 ∗	1 †	2 ‡	3 §	1	1	1	0	4 †	5 ‡	6 §	3 ∗	3	0	1	2
4	0	0	0	0	0 ∗	1 †	2 ‡	3 §	1	1	1	1	4 ∗	5 †	6 ‡	7 §	0	1	2	3
5	0	0	0	0	0 ∗	1 †	2 ‡	3 §	2	1	1	1	8 §	5 ∗	6 †	7 ‡	1	2	3	0
6	0	0	0	0	0 ∗	1 †	2 ‡	3 §	2	2	1	1	8 ‡	9 §	6 ∗	7 †	2	3	0	1
7	0	0	0	0	0 ∗	1 †	2 ‡	3 §	2	2	2	1	8 †	9 ‡	§	7 ∗	3	0	1	2
8	0	0	0	0	0 ∗	1 †	2 ‡	3 §	2	2	2	2	8 ∗	9 †	‡	§	0	1	2	3
9	0	0	0	0	0 ∗	1 †	2 ‡	3 §	3	2	2	2	§	9 ∗	†	‡	1	2	3	0
10	1	1	1	1	4 ∗	5 †	6 ‡	7 §	2	1	1	1	8 §	5 ∗	6 †	7 ‡	1	2	3	0
11	1	1	1	1	4 ∗	5 †	6 ‡	7 §	2	2	1	1	8 ‡	9 §	6 ∗	7 †	2	3	0	1
12	1	1	1	1	4 ∗	5 †	6 ‡	7 §	2	2	2	1	8 †	9 ‡	§	7 ∗	3	0	1	2
13	1	1	1	1	4 ∗	5 †	6 ‡	7 §	2	2	2	2	8 ∗	9 †	‡	§	0	1	2	3
14	1	1	1	1	4 ∗	5 †	6 ‡	7 §	3	2	2	2	§	9 ∗	†	‡	1	2	3	0
15	2	2	2	2	8 ∗	†	‡	§	3	2	2	2	§	9 ∗	†	‡	1	2	3	0

An exemplar of the sequencing of the dataflow through the many-object microarchitecture with extensibility factor m=4 and object count n=10. On each clock cycle, the microarchitecture requests the specified memory addresses from the memory groups and ports indicated. These memory addresses result in the outputting of the specified object indices. The symbols ∗, †, ‡ and § indicate the indices used in each comparison performed within the microarchtecture's compare operation, which are chosen using the multiplexer selectors specified

In the buffer, the memory bandwidth is $2fmw$ bit/s. The number of cycles required to generate the sequence is

$$\sum_{i=0}^{\left\lceil \frac{n}{m} \right\rceil - 1} (n - im - 1).$$

Reorder

If the data emitted from the buffer were immediately sent to the compare operation, only a fraction of the required comparisons would take place and some of these would be repeated. The goal of the reorder operation is to rectify this using $6m$ multiplexers to create the appropriate sequence of comparisons. These multiplexers consume significant resources, which is the reason this microarchitecture is less area efficient than the previous.

Following from the definition of a memory group, we use the term *multiplexer group* to denote a set of six multiplexers sharing the same index. For example, multiplexer group 0 comprises minimum x multiplexer 0, maximum x multiplexer 0, minimum y multiplexer 0, maximum $y$multiplexer 0, minimum z multiplexer 0 and maximum z multiplexer 0.

On initialisation of the reorder operation, multiplexer group 0's selector is set to 1, 1's is set to 2,$m-2$'s is set to $m-1$ and $m-1$'s is set to 0. On each cycle, every selector is incremented by 1 mod$m$. The sequence restarts any time the buffer's A is modified. This is exemplified in Table 3.

Compare

The compare operation comprises $3m$ greater-than-or-equal-to and $3m$ less-than-or-equal-to comparators. The outputs are connected to m logical AND gates. Each gate takes six inputs corresponding to the six comparator results forming an AABB pair. If a collision is detected, the indices of the two colliding objects are written to a single line of memory.

Spatial Partitioning

Despite the microarchitectures' effective exploitation of parallelism and bandwidth, there are two potential concerns. The first concern is that the depth of the memories results in a restriction on the quantity of bounding volumes and, therefore, on the quantity of objects. Although many microarchitecture memories are now of substantial depth, the imposition of any such limit could be considered unsatisfactory. The second concern is that all-pairs suffers from an undesirable computational complexity of $O(n^2)$, resulting from the algorithm's non-exploitation of coherence. Both issue affects scenarios of small or moderate size, and the microarchitectures operating in isolation are sufficient to accelerate these. However, it is desirable to find a solution to these issues in order to unlock the possibility of larger scenarios.

Our solution is to transform the broad phase into a hybrid system combining the microarchitectures with a processor. This processor executes spatial partitioning to divide the list of objects into appropriately sized cells for microarchitecture processing. It has the primary advantages of overcoming object limits and reducing computational complexity. An auxiliary advantage is the possibility of increased parallelism through the overlapping of computations performed by the different stages. Once the potentially colliding set corresponding to a cell is received from the microarchitecture, narrow-phase processing of the cell can proceed while the microarchitecture processes the subsequent cell.

However, this new stage could consume additional computational resources and negatively affect overall system throughput. To ameliorate this issue, we reuse the hierarchies from ray tracing. Reusing an existing data structure offers a significant reduction in construction costs and memory footprint. Moreover, by selecting ray-tracing hierarchies, we benefit from the current high degree of research interest in ray tracing, while aligning with the direction in which graphics applications are ultimately heading.

One significant difference exists in the way ray tracing optimally consumes hierarchies and the way our microarchitectures optimally consume them. For ray tracing's broad phase, it is usually beneficial to subdivide hierarchy branches as far as possible, aiming to achieve approximately one object per leaf. The ray tracer will use the generated leaf nodes to perform ray-object culling. This high degree of subdivision is not beneficial for the microarchitectures; they are designed to efficiently process moderate quantities of objects to effectively exploit parallelism. To reconcile this difference, we retain, during construction of the hierarchy, the leaf nodes with a quantity of objects less than or equal to the selected microarchitecture object limit. Once broad-phase collision-detection processing commences, the contents of the recorded cells are accessed by the microarchitectures.

Most contemporary ray tracers use BVHs, but these are not entirely suitable for collision detection as they permit objects to overlap cell boundaries. These overlapping objects would need to be resolved before performing collision detection, and this resolution would nullify many of the benefits of reuse. One ray-tracing hierarchy that provides a solution is the k-d tree, as this hierarchy places objects that overlap cell boundaries within all overlapped cells. Although a k-d tree would offer excellent throughput for collision detection, it would be less desirable for ray tracing, as k-d trees cannot be easily refitted for dynamic scenarios. To reconcile the throughput of BVHs with the flexibility of k-d trees, we propose a two-level hierarchy. Our proposal consists of a k-d tree that is subdivided until each leaf node contains a quantity of objects less than or equal to the microarchitecture object limit. Within each leaf, a BVH splits the cells until each contains a single object in accordance with ray-tracing practice. The two-level hierarchy is not time consuming to construct, as only relatively few levels of the k-d tree are required, and the throughput degradation is negligible as they can be constructed in $O(n \log n)$ [20]. In some cycles of the interactive application program loop, it may be necessary to perform slight alterations to the k-d tree if the position of objects changes significantly. This could require migration of objects between cells, but the large cell sizes mean migration will occur infrequently and the cost will be negligible. It is, furthermore, unlikely that the quantity of objects in a cell will precisely equal the cell-size limit, thereby allowing cells to accommodate additional objects without rebuilding or refitting in many cases. The underlying BVHs serve to maximise throughput as they can be efficiently refitted on each cycle. Therefore, the proposed two-level hierarchy maximises the throughput of both collision detection and ray tracing.

IMPLEMENTATION

We envisage our cell-based broad-phase microarchitectures fabricated as part of an IC that would also execute the remainder of the interactive application program loop. Using a single platform would allow for the elimination of data transfer overheads. This concept has been successfully adopted to integrate a CPU and GPU within some Intel Core processors [42] as well as AMD accelerated processing units (APUs) [43], such as those in the PlayStation 4. Within the spectrum of platforms readily available today, our microarchitectures could naturally reside within the fixed-function logic of GPUs, as there is already a significant focus on relocating many elements of the interactive application program loop to these platforms [44]. The remainder of the program loop could utilise the programmable elements of the GPU. Adding one of the microarchitectures would not compromise GPU programmability, as all GPUs include some fixed-function logic such as rasterisation. This is unlikely to change due to power-density limits as well as the lacklustre throughput achieved when traditionally fixed-function elements have been reimplemented using the programmable elements of a GPU [45]. Moreover, it is not prohibitively expensive to include one of the microarchitectures, as the large production volumes of commodity platforms amortises the cost [27]. Therefore, there exists sufficient motivation for the fabrication of our logic as part of a future GPU.

These platforms were unavailable to us, and we were limited to prototyping our microarchitectures on a field-programmable gate array (FPGA), to which we translated, mapped, placed and routed a complete design written in hardware-description language (HDL). FPGAs are ICs that are reconfigurable, meaning that a single FPGA can implement different microarchitectures at different times. However, this reconfigurability incurs significant throughput, power and area penalties.

One of the primary characteristics of the microarchitectures is the possibility of their adaptation to the size of the underlying platform. We found that the limiting factor for both microarchitectures was the quantity of internal memories, which constrained both designs to $m=16$. Based on this value, it was possible to process a maximum of 1024 objects using the area-efficient microarchitecture and a maximum of 16,384 objects using the many-object microarchitecture.

When targeting platforms such as GPUs, the FPGA can be used to verify the functionality of the microarchitectures and to analyse their behaviour, but it is insufficient for gaining a true reflection of throughput. To address this, we adapted the throughput metrics from the FPGA implementations to a clock frequency of 500 MHz in accordance with assumptions made in existing research [46–48]. Since 500 MHz is significantly lower than the clock frequencies of modern GPUs, we, therefore, derive a conservative estimate of throughput. We excluded the communication overhead as we intend that our microarchitectures would reside on the same IC as all associated computation. All other elements retained the same values as their FPGA counterparts. In practice, however, it is likely that the constraints on internal memory sizes would be less severe, thereby facilitating the possibility of simulating many more objects without spatial partitioning. In addition, it is likely that m could be increased due to the greater density of resources. Therefore, all throughput metrics are conservative estimates of what would be achieved in a practical system centred around a GPU.

Our CPU-based software utilised the Bullet Physics SDK. We added custom C++ code to adapt the broad phase to gather the relevant data from the ray-tracing hierarchies, before invoking the appropriate microarchitecture operations and reading the resultant potentially colliding sets.

RESULTS AND DISCUSSION

The adapted Bullet code was compiled using G++ with throughput optimisations enabled. The host system consisted of a Quad-Core AMD Opteron 2350 clocked at 2 GHz with 8 GB of RAM. The operating system was 64-bit Ubuntu Linux. Our GPU results were measured using an NVIDIA GeForce GTX 670 with 2 GB of external memory.

Our experiments used an updated version of the framework for benchmarking collision detection [35]. Our benchmarks consisted of 1000 collision-detection cycles of a scenario comprising a cube enclosing n objects, as illustrated in Fig. 3. The dimensions of the cube were $\sqrt[3]{50^3 \times 5n}$ m. The objects were uniformly distributed throughout the environment, and all object properties were determined using the uniform probability distribution with different values possible for each axis. The objects were spheres, cuboids, cylinders and cones, and their sizes lay between 25 and 75 m. The linear velocity spanned from (−25,−25,−25) to (25,25,25) m/s, and the angular velocity spanned from (−2.5,−2.5,−2.5) to (2.5,2.5,2.5) rad/s. In all of the benchmarks, our goal was to generate a large quantity of objects undergoing a high degree of movement and interaction, in order to thoroughly assess throughput under what are traditionally considered the most challenging cases for broad-phase algorithms. In addition, we used relatively simple objects, as the broad phase is only concerned with objects and ignores details such as their complexity.

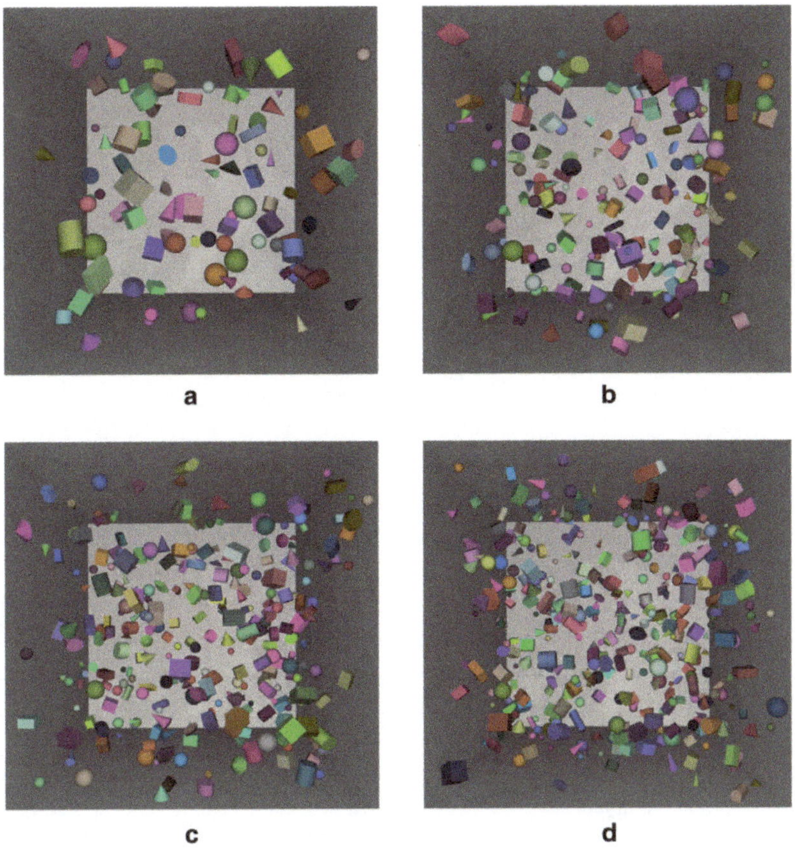

Figure 3. Sample benchmarks. **a** 100 objects. **b** 200 objects. **c** 300 objects. **d** 400 objects.

The throughput was computed using 20 variants of the benchmark. For our system, only the microarchitecture computations were recorded; the spatial-partitioning hierarchy construction was excluded as this would form part of a ray tracer and would not add to the system's execution time. A variety of cell sizes—1024 for the area-efficient microarchitecture and 1024, 2048, 4096, 8192 and 16,384 for the many-object microarchitecture—was tested to determine the optimal. Throughput counters embedded in the microarchitectures were used to record the timing metrics. For comparison purposes, we measured the execution times of broad-phase algorithms executing on a CPU and GPU. For the CPU comparison, we selected Bullet's DBVT algorithm, for which we recorded the execution times using high-resolution throughput counters present in the CPU. We selected this algorithm as it achieves optimal throughput for the vast majority of scenarios. It performs spatial partitioning using two AABB hierarchies whose nodes can be dynamically rearranged. One hierarchy represents static objects and the other represents dynamic objects. The objects are moved between the two hierarchies as their statuses change. For the GPU comparison, we selected Bullet's GPU-based sweep and prune, for which we recorded the time spent checking for collisions between objects, while excluding the transfer of data between the CPU and GPU. It is an implementation of the design proposed by Liu et al. [14]. The execution times of all these systems are tabulated in Table 4, and the acceleration factors of our system relative to the CPU and GPU are plotted in Fig. 4.

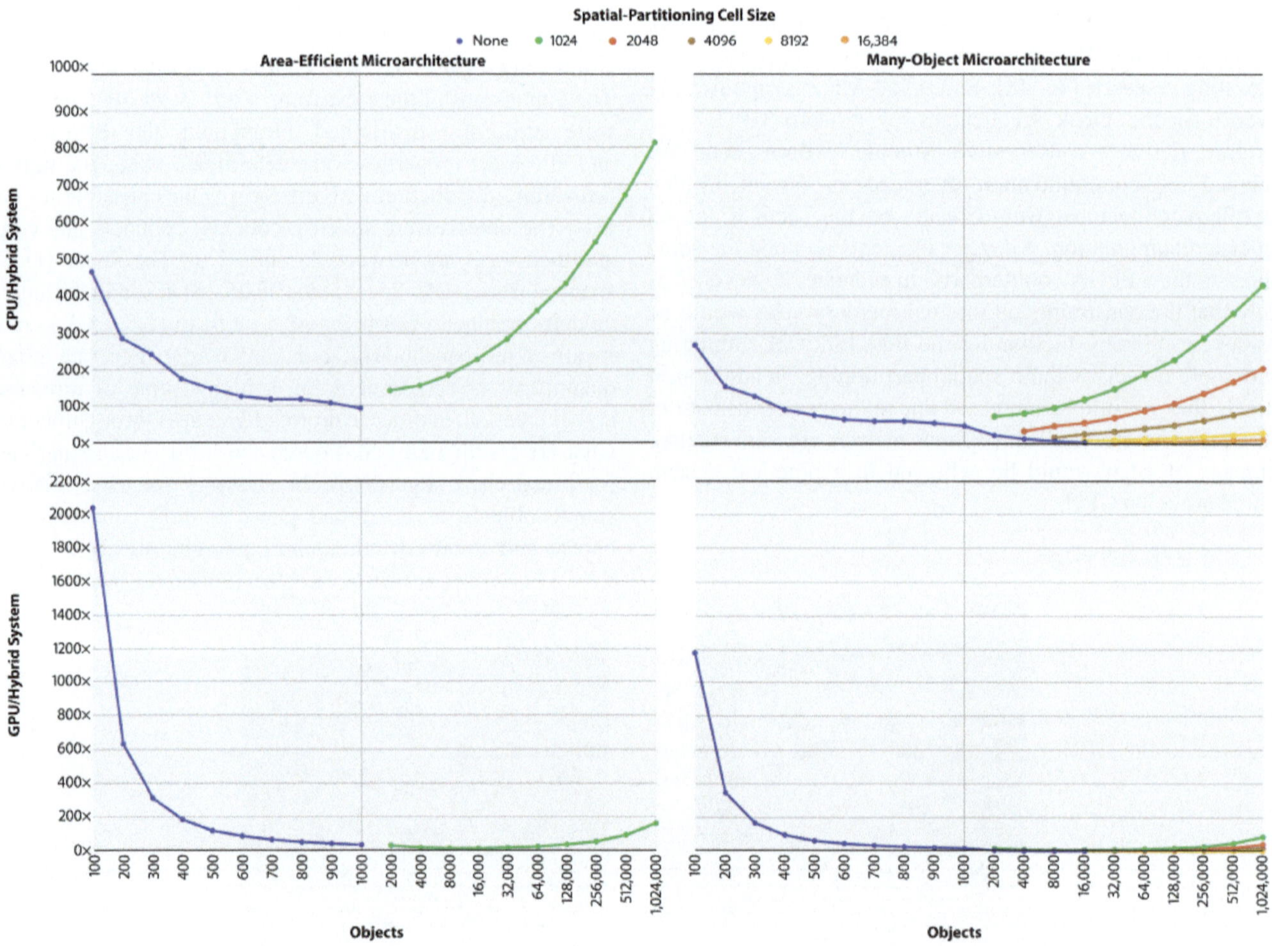

Figure 4. Acceleration factors of the hybrid system relative to the CPU and GPU. The *vertical lines*indicate a change of scale along the *x* axis.

Table 4. Execution times for the CPU, GPU, area-efficient microarchitecture and many-object microarchitecture

			Area-efficient microarchitecture		Many-object microarchitecture					
Objects	**CPU**	**GPU**	**None**	**1024**	**None**	**1024**	**2048**	**4096**	**8192**	**16,384**
100	0.2213	0.9732	**0.0005**		**0.0008**					
200	0.4478	0.9982	**0.0016**		**0.0029**					
300	0.7990	1.0347	**0.0033**		**0.0062**					
400	0.9994	1.0441	**0.0057**		**0.0107**					
500	1.2966	1.0302	**0.0087**		**0.0165**					

600	1.5770	1.0727	**0.0124**		**0.0236**					
700	1.9871	1.0796	**0.0168**		**0.0319**					
800	2.5760	1.0889	**0.0217**		**0.0414**					
900	2.9700	1.1072	**0.0273**		**0.0522**					
1000	3.1694	1.0945	**0.0336**		**0.0643**					
2000	6.0023	1.2040		**0.0423**	0.2535	**0.0804**				
4000	14.2332	1.4039		**0.0913**	1.0070	**0.1727**	0.4036			
8000	30.7177	1.9585		**0.1665**	4.0140	**0.3153**	0.6270	1.7638		
16,000	69.8204	3.6801		**0.3074**	16.0280	**0.5821**	1.2199	2.5936	8.2487	
32,000	159.5770	8.3066		**0.5676**		**1.0749**	2.2525	4.7891	15.2312	36.9132
64,000	375.3380	22.8188		**1.0481**		**1.9848**	4.1592	8.8431	28.1244	68.1603
128,000	834.9720	67.5312		**1.9353**		**3.6649**	7.6800	16.3288	51.9317	125.8579
256,000	1940.2600	187.3310		**3.5735**		**6.7672**	14.1811	30.1512	95.8919	232.3967
512,000	4423.3300	605.7710		**6.5985**		**12.4957**	26.1853	55.6742	177.0645	429.1206
1,024,000	9897.0100	1958.8776		**12.1841**		**23.0732**	48.3512	102.8025	326.9497	792.3714

All execution times are in milliseconds. 'CPU' is Bullet's DBVT, 'GPU' is Bullet's GPU sweep and prune, and the numeric table headers indicate the spatial-partitioning cell size used with the microarchitectures. Bold results highlight the optimal time for each microarchitecture

All of our results indicate that very significant accelerations can be achieved, as all benchmarks have been significantly accelerated relative to both the CPU and GPU. The results also show that the area-efficient microarchitecture outperforms the many-object microarchitecture by a factor of 3.14 ×. The higher degree of acceleration can be explained by the greater quantity of comparisons performed in parallel. Using the results from the area-efficient microarchitecture, it can be observed that for those benchmarks involving no spatial partitioning, the acceleration never decreases below 94 × relative to the CPU or below 33 × relative to the GPU. If the results comprising spatial partitioning are included, the accelerations increase further. Compared to the CPU, our system achieves an acceleration of up to 812 ×, while compared to the GPU, it achieves an acceleration of up to 161 ×. In both cases, these accelerations correspond to the highest object counts. The results indicate that our system provides significant advantages to all benchmarks, both with and without spatial partitioning, and that the most challenging benchmarks for traditional collision detection are those most accelerated by our system.

The results also indicate that spatial partitioning has a beneficial effect on throughput. Without spatial partitioning, an $O(n^2)$ computational complexity can be observed. This complexity can be disregarded for up to 1000 objects, as it is minimal and has little impact due to the significant accelerations provided by parallelism. However, it is apparent that the complexity may affect throughput for larger object quantities. This is rectified by spatial partitioning, which our results indicate lowers the computational complexity to $O(n \log n)$. The positive effect is most evident when comparing against the CPU; it is less evident when comparing against the GPU as the GPU throughput is relatively poor for benchmarks with 1000 objects or fewer. The results also indicate that the optimal cell size is 1024, irrespective of microarchitecture. This is the optimal as the microarchitectures execute efficiently with moderate quantities of objects. Since both microarchitectures accommodate this cell size, we recommend selecting the area-efficient microarchitecture in all cases due to its enhanced throughput.

The results also demonstrate that our system is not significantly affected by object count, even when the count greatly exceeds what can be typically expected. This demonstrates that our designs do not need optimisation or specialisation for specific scenarios. In contrast, there are certain object-count ranges for which the CPU and GPU perform poorly. Higher object counts significantly degrade CPU throughput. The GPU tends to perform poorly with both large object counts which cannot be processed efficiently, as well as small object counts which inadequately exploit parallelism.

We further compare the previously provided throughput figures from the area-efficient microarchitecture with the throughput figures provided within two recent GPU broad-phase collision-detection articles in Table 5 and Fig. 5. The results demonstrate that our system significantly outperforms all of the algorithms for all object counts, with up to 14 × acceleration over Liu et al. [14] and up to 40 × acceleration over Avril, Gouranton and Arnaldi [15]. These results are not as representative of expected throughput as our prior results, since the benchmarks used in the articles differ significantly from those that we used. Our benchmarks included a high degree of object movement and interaction, to thoroughly assess the most challenging scenarios and prove that use of our system facilitates scenarios traditionally avoided due to their potentially negative impact on throughput. In contrast, many of the benchmarks used by the aforementioned articles choose scenarios that are more favourable to algorithms sensitive to object movement and interaction. Irrespective of these differences, the results clearly demonstrate that our system significantly outperforms the state of the art.

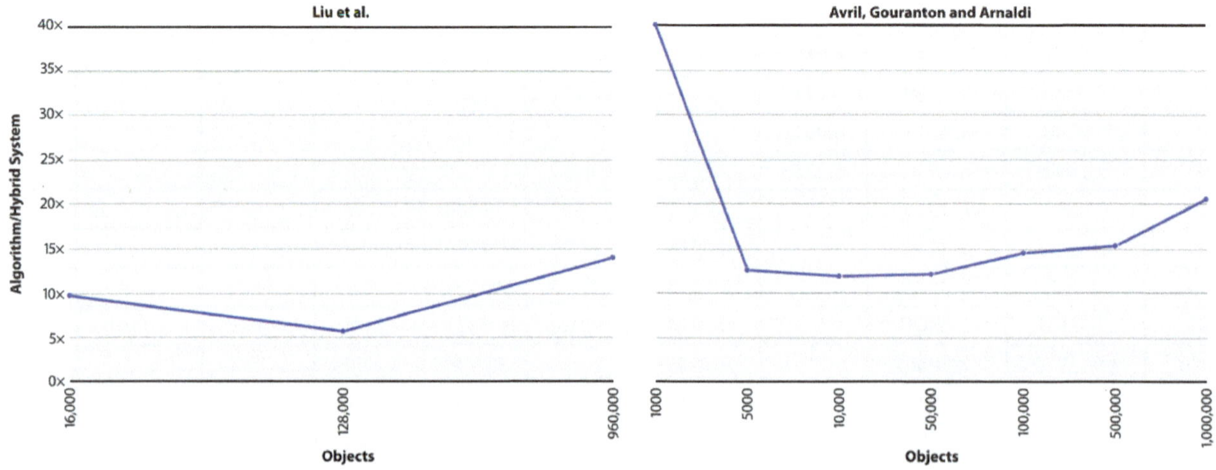

Figure 5. Acceleration factors of the hybrid system using the area-efficient microarchitecture relative to recent GPU broad-phase collision-detection algorithms.

Table 5. Execution time comparisons for recent GPU broad-phase collision-detection algorithms against the area-efficient microarchitecture

Article	Objects	Article execution times	Microarchitecture execution times
Liu et al.	16,000	3.0000	**0.3074**
	128,000	11.0000	**1.9353**
	960,000	161.0000	**11.4859**
Avril, Gouranton and Arnaldi	1000	1.3500	**0.0336**
	5000	1.3900	**0.1101**
	10,000	2.4100	**0.2017**
	50,000	10.2100	**0.8379**
	100,000	22.5200	**1.5471**
	500,000	116.7200	**7.5946**
	1,000,000	245.9500	**11.9223**

All execution times are in milliseconds. Bold results highlight the optimal time for each object count

We previously mentioned that we envisage our microarchitectures implemented as fixed-function logic on a platform such as a GPU. To assess how the microarchitectures would fit, we computed the approximate size if implemented on the NVIDIA GeForce GTX TITAN X, using information on process widths, die sizes and transistor counts for our FPGA and the GPU, as well as research into the difference in area consumption between FPGA and ASIC implementations [49]. We deduced that the area-efficient microarchitecture would consume approximately 0.072 % of the GPU die area, while the many-object microarchitecture would consume approximately 0.086 %. It would also be useful to have assessed the expected power consumption of the microarchitectures when implemented on a GPU. Unfortunately, there were insufficient data available to accurately compute this information. However, a substantive body of research [23–26] indicates that specialised logic consumes significantly less power than programmable logic due to the removal of overheads such as instruction processing, register accesses and inefficient memory layouts. Analyses indicate a 16 × power decrease can be expected, with up to a 500 × decrease possible for some applications [50]. It is, therefore, evident that the microarchitectures would reduce power consumption appreciably.

CONCLUSIONS

We presented a hybrid system comprising one of two fixed-function microarchitectures complemented by a general-purpose microprocessor. The objective was to achieve significant acceleration and energy efficiencies for collision detection through the effective exploitation of parallelism and memory bandwidth. The area-efficient microarchitecture focused on occupying minimum resources, while the many-object microarchitecture focused on supporting greater object quantities. To compensate for potential challenges, such as a limit on the quantity of objects supported, the microarchitectures were combined with a spatial-partitioning phase executing on a processor. We outlined a novel means of combining this spatial-partitioning phase with the hierarchies constructed by ray tracers, thereby reducing computation time and memory consumption.

Overall, we demonstrated a significant enhancement in the throughput of collision detection, with a 1,024,000-object benchmark demonstrating an acceleration of 812 × when compared against a CPU and an acceleration of 161 × when compared against a GPU. Throughput was maintained for large object counts where traditional systems perform inadequately. The area-efficient microarchitecture with a spatial-partitioning cell size of 1024 was found to be the optimal of our designs due to its greater exploitation of parallelism. Moreover, fixed-function microarchitectures significantly reduce power consumption, and this allowed us to mitigate silicon power-density challenges, while facilitating computation on mobile and wearable computing devices.

Although we have demonstrated that our system is highly efficient, there exists some scope for expansion. In particular, it would be interesting to experiment with shifting additional stages of the interactive application program loop from processors to microarchitectures. For instance, a microarchitecture for constructing BVHs [48] could complement ours. This would accept the *k*-d tree cells generated by the processor and construct or refit the BVHs associated with each cell. It would also be constructive to try adding an aforementioned narrow-phase microarchitecture [29, 30]. Overall, it is evident that these proposed ideas would develop our system, but, even without adaptation, our system achieves significant throughput and power efficiency advantages.

ACKNOWLEDGEMENTS

The authors wish to express their gratitude to Michael J. Doyle for providing guidance on the ray-tracing elements of this article.

REFERENCES

1. PM Hubbard, Approximating polyhedra with spheres for time-critical collision detection. ACM Trans. Graph.15(3), 179–210 (1996).
2. D Baraff, *Dynamic simulation of non-penetrating rigid bodies. PhD thesis* (Cornell University, Ithaca, New York, USA, 1992).
3. JD Cohen, MC Lin, D Manocha, MK Ponamgi, in *Proceedings of the 1995 Symposium on Interactive 3D Graphics (i3D 1995)*. I-COLLIDE: An interactive and exact collision detection system for large-scale environments (ACMNew York City, New York, USA, 1995), pp. 189–196.
4. S Gottschalk, MC Lin, D Manocha, in *Proceedings of the 23rd Annual Conference on Computer Graphics and Interactive Techniques (SIGGRAPH 1996)*. OBBTree: A hierarchical structure for rapid interference detection (ACMNew York City, New York, USA, 1996), pp. 171–180.
5. K Zikan, P Konečný, in *Proceedings of the 5th International Conference in Central Europe on Computer Graphics and Visualization (WSCG 1997)*. Lower bound of distance in 3D (University of West BohemiaPlzeň, Czechia, 1997), pp. 640–647.
6. H Jin, Z Liu, T Wu, Y Wang, in *Proceedings of the 2009 International Conference on Computer Engineering and Technology (ICCET 2009)*. The research of collision-detection algorithm based on spatial subdivision (IEEE Computer SocietyLos Alamitos, California, USA, 2009), pp. 452–455.
7. G van den Bergen, Efficient collision detection of complex deformable models using AABB trees. J. Graph. Tools. 2(4), 1–14 (1997).
8. M Teschner, S Kimmerle, B Heidelberger, G Zachmann, L Raghupathi, A Fuhrmann, M-P Cani, F Faure, N Magnenat-Thalmann, W Strasser, P Volino, Collision detection for deformable objects. Comput. Graph. Forum. 24(1), 61–81 (2005).
9. Z Wang, M Tang, R Tong, D Manocha, TightCCD: Efficient and robust continuous collision detection using tight error bounds. Comput. Graph. Forum. 34(7), 289–298 (2015).
10. L He, R Ortiz, A Enquobahrie, D Manocha, in *Proceedings of the 19th Symposium on Interactive 3D Graphics and Games (i3D 2015)*. Interactive continuous collision detection for topology changing models using dynamic clustering (ACMNew York City, New York, USA, 2015), pp. 47–54.
11. MC Lin, JF Canny, in *Proceedings of the 1991 IEEE International Conference on Robotics and Automation (ICRA 1991)*. A fast algorithm for incremental distance calculation (IEEENew York City, New York, USA, 1991), pp. 1008–1014.
12. B Mirtich, V-Clip: Fast and robust polyhedral collision detection. ACM Trans. Graph.17(3), 177–208 (1998).
13. NK Govindaraju, D Knott, N Jain, I Kabul, R Tamstorf, R Gayle, MC Lin, D Manocha, Interactive collision detection between deformable models using chromatic decomposition. ACM Trans. Graph.24(3), 991–999 (2005).
14. F Liu, T Harada, Y Lee, YJ Kim, Real-time collision culling of a million bodies on graphics processing units. ACM Trans. Graph.29(6), 154:1–154:8 (2010).
15. Q Avril, V Gouranton, B Arnaldi, in *Proceedings of the 2012 Eurographics Symposium on Parallel Graphics and Visualization (EGPGV 2012)*. Fast collision culling in large-scale environments using GPU mapping function (EurographicsAire-la-Ville, Switzerland, 2012), pp. 71–80.
16. C Lauterbach, Q Mo, D Manocha, gProximity: Hierarchical GPU-based operations for collision and distance queries. Comput. Graph. Forum. 29(2), 419–428 (2010).
17. M Tang, D Manocha, J Lin, R Tong, in *Proceedings of the 2011 Symposium on Interactive 3D Graphics and Games (i3D 2011)*. Collision-streams: Fast GPU-based collision detection for deformable models (ACMNew York City, New York, USA, 2011), pp. 63–70.
18. X Zhang, YJ Kim, Scalable collision detection using *p*-partition fronts on many-core processors. IEEE Trans. Visual. Comput. Graph.20(3), 447–456 (2014).
19. D Kim, J-P Heo, J Huh, J Kim, S-e Yoon, HPCCD: Hybrid parallel continuous collision detection using CPUs and GPUs. Comput. Graph. Forum. 28(7), 1791–1800 (2009).
20. I Wald, V Havran, in *Proceedings of the 2006 IEEE Symposium on Interactive Ray Tracing (RT 2006)*. On building fast kd-trees for ray tracing, and on doing that

in O(N log N) (IEEE Computer SocietyLos Alamitos, California, USA, 2006), pp. 61–69.

21. I Wald, S Boulos, P Shirley, Ray tracing deformable scenes using dynamic bounding volume hierarchies. ACM Trans. Graph. 26(1), 16:1–16:18 (2007).

22. J Pantaleoni, D Luebke, in *Proceedings of the 2010 Conference on High-Performance Graphics (HPG 2010)*. HLBVH: Hierarchical LBVH construction for real-time ray tracing (EurographicsGoslar, Germany, 2010), pp. 87–95.

23. H Esmaeilzadeh, E Blem, R St. Amant, K Sankaralingam, D Burger, Dark silicon and the end of multicore scaling. ACM Trans. Graph.39(3), 365–376 (2011).

24. G Venkatesh, J Sampson, N Goulding, S Garcia, V Bryksin, J Lugo-Martinez, S Swanson, M Bedford Taylor, in *Proceedings of the 15th International Conference on Architectural Support for Programming Languages and Operating Systems (ASPLOS 2010)*. Conservation cores: reducing the energy of mature computations (ACMNew York City, New York, USA, 2010), pp. 205–218.

25. ES Chung, PA Milder, JC Hoe, K Mai, in *Proceedings of the 43rd Annual IEEE/ACM International Symposium on Microarchitecture (MICRO 2010)*. Single-chip heterogeneous computing: Does the future include custom logic, FPGAs, and GPGPUs? (IEEE Computer SocietyLos Alamitos, California, USA, 2010), pp. 225–236.

26. T Mitra, Heterogeneous multi-core architectures. IPSJ Trans. Syst. LSI Design Methodol.8:, 51–62 (2015).

27. R Rajwar, M Dixon, R Singhal, in *Proceedings of the 7th Biennial Conference on Innovative Data Systems Research (CIDR 2015)*. Specialized evolution of the general-purpose CPU (Asilomar, California, USA, 2015). http://cidrdb.org/cidr2015/.

28. YS Shao, S Xi, V Srinivasan, G-Y Wei, D Brooks, in *Proceedings of the 2015 Sensors to Cloud Architectures Workshop (SCAW 2015)*. Toward cache-friendly hardware accelerators (Bay Area, California, USA, 2015). http://darksilicon.org/hpca/SCAW-2015/.

29. N Atay, JW Lockwood, B Bayazit, A collision detection chip on reconfigurable hardware. Technical Report WUCSE-2005-33, (Washington University, St. Louis, Missouri, USA 2005).

30. A Raabe, F Zavelberg, in *Proceedings of the 16th International Conference in Central Europe on Computer Graphics, Visualization and Computer Vision (WSCG 2008) – Communication Papers*. Defying the memory bottleneck in hardware accelerated collision detection (UNION Agency-SciencePlzeň, Czechia, 2008), pp. 25–31.

31. C Davis, M Hedge, OA Schmid, M Maher, JP Bordes, Physics Processing Unit. International Patent WO/2005/038559 (2005).

32. M Woulfe, J Dingliana, M Manzke, in *Proceedings of the 4th Workshop on Virtual Reality Interaction and Physical Simulation (VRIPHYS 2007)*. Hardware accelerated broad phase collision detection for realtime simulations (EurographicsAire-la-Ville, Switzerland, 2007), pp. 79–88.

33. MC Lin, S Gottschalk, in *Proceedings of the 8th IMA Conference on the Mathematics of Surfaces*. Collision detection between geometric models: a survey (IMASouthend-on-Sea, Essex, UK, 1998), pp. 37–56.

34. W Fan, B Wang, J-C Paul, J Sun, A hierarchical grid based framework for fast collision detection. Comput. Graph. Forum. 30(5), 1451–1459 (2011).

35. M Woulfe, M Manzke, in *Proceedings of the 25th Spring Conference on Computer Graphics (SCCG 2009)*. A framework for benchmarking interactive collision detection (Comenius UniversityBratislava, Slovakia, 2009), pp. 221–228.

36. A Beechick, S Casselman, L Yarbrough, Internal sorting and FPGA. Proc. SPIE. 2914:, 66–71 (1996).

37. S Wong, S Vassiliadis, JY Hur, in *Proceedings of the 16th Annual Workshop on Circuits, Systems and Signal Processing (ProRISC 2005)*. Parallel merge sort on a binary tree on-chip network (Technology Foundation STWUtrecht, The Netherlands, 2005), pp. 365–368.

38. J Harkins, T El-Ghazawi, E El-Araby, M Huang, in *Proceedings of the 2005 IEEE International Conference on Field-Programmable Technology (FPT 2005)*. Performance of sorting algorithms on the SRC 6 reconfigurable computer (IEEENew York City, New York, USA, 2005), pp. 295–296.

39. Y-K Chen, J Chhugani, CJ Hughes, D Kim, S Kumar, V Lee, A Lin, AD Nguyen, E Sifakis, M Smelyanskiy, High-performance physical simulations on next-generation architecture with many cores. Intel Technol. J.11(3), 251–261 (2007).

40. DJ Tracy, SR Buss, BM Woods, in *Proceedings of the 2009 IEEE Virtual Reality Conference (VR 2009)*. Efficient large-scale sweep and prune methods with AABB insertion and removal (IEEE Computer SocietyLos Alamitos, California, USA, 2009), pp. 191–198.

41. R Scrofano, L Zhuo, VK Prasanna, Area-efficient arithmetic expression evaluation using deeply pipelined floating-point cores. IEEE Trans. Very Large Scale Int. Syst.16(2), 167–176 (2008).

42. N Kurd, M Chowdhury, E Burton, TP Thomas, C Mozak, B Boswell, P Mosalikanti, M Neidengard, A Deval, A Khanna, N Chowdhury, R Rajwar, TM Wilson, R Kumar, Haswell: A family of IA 22 nm processors. IEEE J. Solid-State Circuits. 50(1), 49–58 (2015).

43. B Munger, D Akeson, S Arekapudi, T Burd, HR Fair III, J Farrell, D Johnson, G Krishnan, H McIntyre, E McLellan, S Naffziger, R Schreiber, S Sundaram, J White, K Wilcox, Carrizo: A high performance, energy efficient 28 nm APU. IEEE J. Solid-State Circuits. 51(1), 105–116 (2016).

44. M Joselli, J Ricardo da Silva Junior, M Zamith, E Clua, M Pelegrino, E Mendonça, E Soluri, in *Proceedings of the 4th International IEEE Consumer Electronics Society Games Innovation Conference (IGIC 2012)*. Techniques for designing GPGPU games (IEEE Computer SocietyLos Alamitos, California, USA, 2012), pp. 51–55.

45. S Laine, T Karras, in *Proceedings of the 2011 Conference on High-Performance Graphics (HPG 2011)*. High-performance software rasterization on GPUs (ACMNew York City, New York, USA, 2011), pp. 51–55.

46. J Spjut, A Kensler, D Kopta, E Brunvand, TRaX: A multicore hardware architecture for real-time ray tracing. IEEE Trans. Computer-Aided Design Integr. Circuits Syst.28(12), 1802–1815 (2009).

47. J-H Nah, J-S Park, C Park, J-W Kim, Y-H Jung, W-C Park, T-D Han, T&I engine: Traversal and intersection engine for hardware accelerated ray tracing. ACM Trans. Graph.30(6), 160:1–160:10 (2011).

48. MJ Doyle, C Fowler, M Manzke, A hardware unit for fast SAH-optimised BVH construction. ACM Trans. Graph.32(4), 139:1–139:10 (2013).

49. I Kuon, J Rose, Measuring the gap between FPGAs and ASICs. IEEE Trans. Computer-Aided Design Integr. Circuits Syst.26(2), 203–215 (2007).

50. R Hameed, W Qadeer, M Wachs, O Azizi, A Solomatnikov, BC Lee, S Richardson, C Kozyrakis, M Horowitz, in *Proceedings of the 37th Annual International Symposium on Computer Architecture (ISCA 2010)*. Understanding sources of inefficiency in general-purpose chips (ACMNew York City, New York, USA, 2010), pp. 37–46.

CHAPTER 2

INTEGER INVERSE KINEMATICS METHOD USING FUZZY LOGIC

Takeshi Morishita and Osamu Tojo

Keywords: Inverse kinematics, Integer algorithm, Reduction control, Fuzzy model, Multijoint control model

INTRODUCTION

Small robots are gradually becoming a common feature in the daily living environment of modern societies. They are often subject to cost constraints, and this, along with their size, makes the mounting of large-scale devices difficult. As a countermeasure, the control software for small size microcomputers requires a degree of ingenuity and an efficient algorithm.

Many robots actively work with limbs constructed from links and joints. These joints, e.g., robot arms and legs, often have one or more degrees of freedom (DOF). Forward and inverse kinematics for controlling the limb positions of the target machine are used to relate the controlled object to the coordination system. This is a standard approach to robot control. Naturally, the mathematics involved in this level of control usually includes trigonometric functions and matrix techniques based on real numbers. In addition, the higher is the number of joints of the robot, the more complex and nonlinear the inverse kinematics equation. Such mathematical functions are computationally expensive and a burden on small processors.

To solve the problem of multijoint control, many authors have proposed a control approach based on fuzzy logic or other related technology for simplifying trajectory generation and extending the range of applications. In addition, the purposes of a conventional fuzzy control method for robotic trajectories are an increased robustness in spite of noise or sensor failure, the ability to handle non-linearity without degradation or failure, redundancy resolution, and high-accuracy positioning. However, because the natural fuzzy logic and normalized general calculations rely on a conventional algorithm based on real numbers and their operations, float or double-precision floating number type variables are used in the microprocessor. This is significant burden for small microprocessors that can be reduced by an integer method. [1–6].

Fuzzy logic deals with intermediate values via a membership function, in contrast to traditional binary logic. The membership function expands the range of the defined function from {0, 1} to [0, 1]. The function specifies the degree to which an input is included in the set and the numerical value grade. When explaining the theory and mathematical logic using an intermediate value, the grade is often normalized to aid understanding and use. This is natural fuzzy logic and is a general method for quantifying ambiguity. However, normalization is not necessarily needed for computational calculations for fuzzy logic, because it can be calculated by only four arithmetic operations in some defined method of membership function.

In this paper, we propose a simplifying method to implement inverse kinematics using only integer arithmetic. In addition, we show how this approach can be applied to experiments with an actual self-contained multijoint robot and evaluate the reproducibility of the orbit to produce the integer inverse kinematics.

INTEGER INVERSE KINEMATICS MODEL

Extended Range of Fuzzy Set

Fuzzy logic can accommodate an unnormalized definition function. We define a fuzzy set A of the universal set X consisting of integer multiples in the one byte range:

Citation: Morishita, T. and Tojo, O (2013). Integer inverse kinematics method using Fuzzy logic, DOI:10.1007/s11370-012-0115-1.

$$\mu_A : X \rightarrow [0, 255] \quad (1)$$

where $\mu_A(x)$ is the membership function of A with $x \in X$ and is an integer value in the closed interval [0, 255].

FUZZY MODEL FOR MULTIJOINT ROBOT CONTROL

A limb with n joints can be controlled using the following inverse kinematics control model based on fuzzy logic. The ith control rule is given by the fuzzy relation R_i defined by the space (x, y, θ) where x and y are the coordinates of bicycle pedal and θ is the joint angle. Then, the fuzzy relations for the robot mechanism joints are given by

$$R_{ij} = \left(x, y, \theta_j\right) \quad (2)$$

where j denotes a specific joint, i.e., the hip, knee, or ankle joint.

The following four rule bases use the sentence connective "also":

$$R_{HP} = \bigcup_{i=1}^{n} R_{iHP}$$

or

$$R_{HR} = \bigcup_{i=1}^{n} R_{iHR}$$

or

$$R_{KN} = \bigcup_{i=1}^{n} R_{iKN}$$

or

$$R_{AN} = \bigcup_{i=1}^{n} R_{iAN} \quad (3)$$

Here, HP, HR, KN, and AN denote the hip pitch, hip roll, and knee and ankle joints, respectively.

The input membership function $\mu_{i1}(x)$ of the x coordinate and the input membership function $\mu_{i2}(y)$ of the y coordinate are defined as

$$\omega_i = \mu_{i1}(x) \wedge \mu_{i2}(y) \quad (3)$$

where ω_i is the grade of the ith rule. The maximum grade for this system is 255, and with this step, the grade value can be truncated to an integer value by the input x. In general, the fuzzy relation between the input and the output membership function $B_{ij}(\theta_j)$ is

$$B_j^o(\theta_j) = \left[\omega_1 \wedge B_{1j}(\theta_j)\right] \vee \left[\omega_2 \wedge B_{2j}(\theta_j)\right] \vee \ldots \vee \left[\omega_n \wedge B_{nj}(\theta_j)\right] = \bigvee_{i=1}^{n} \left[\omega_i \wedge B_{ij}(\theta_j)\right] \quad (5)$$

Here, the measurement points are set to x^0 and y^0; an input of (x^0, y^0) gives the output joint angle $B_j^0\left(\theta_j\right)$. In Eq. (5), ∧ is minimum criterion logic, ∨ is maximum criterion logic, and the function B^0 overlaps each output membership result, which are reduced to their fuzzy set in Eq. (4).

We aim to increase the joint control performance with a lightweight algorithm, and, therefore, we apply a center of gravity (COG) defuzzification method using a singleton-type output membership function:

$$\theta_j = \frac{\sum_{i=1}^{n} \omega_i B_{ij}}{\sum_{i=1}^{n} \omega_i} \quad (6)$$

This fuzzy logic control model will be applied to multijointed robots. Thus, this logic can derive each joint angle from the tip coordinate of the limb [7]. Moreover, since this algorithm is, by definition, an integer variable system, any decimal value obtained by division is rounded down according to the microprocessor specifications.

ANALYSIS OF HUMAN ACTION

We have analyzed the bicycle pedaling action requiring the cooperative operation of two legs, as shown in Fig. 1, using video images of a human subject. Four points (the hip joint, knee joint, ankle, and tip of the toe) were selected for the analysis, and the movement was recorded from two directions (side-on and from behind). In addition, we confirmed that humans do not consciously consider the angles of their joints. Moreover, we determined the following from the two types of sample images of the bicycle action:

1. There is little change in the yaw and roll of the hip, knee, and ankle joint angle or the ankle joint pitch angle.
2. Pedaling defines a trajectory that is approximately twodimensional. We thus define the controlled object to be a single foot with 3 DOF.

Figure 1. Analysis of bicycle pedaling action using video images of a human subject.

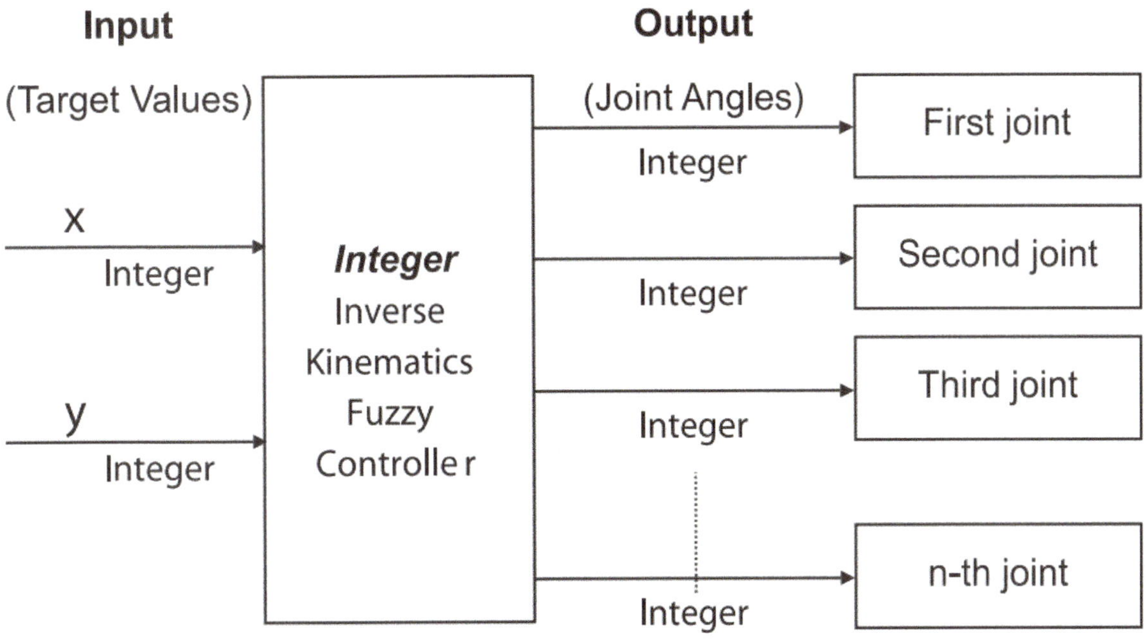

Figure 2. Block diagram outlining the integer inverse kinematics controller.

ROBOT SYSTEM BASED ON HUMAN MODEL

Integer Inverse Kinematics Control Model

The controller utilizes the algorithm defined in Sect. 2 to construct the quantification method for the behavior of the human subject and the formulation of linguistic rules. This algorithm easily describes the behavior of the human subject by the four arithmetic integer operations. In this way, we have realized a lightweight and flexible control process for a small robot.

The advantage of the fuzzy model is that human behavior models can be easily applied by the microprocessors inside small robots [8]. To demonstrate, we applied the fuzzy controller to the legs of a small autonomous multijoint type robot. Figure 2 shows the system block diagram of the control section designed by the fuzzy logic.

Linguistic Quantification

Using the results of sample image analysis (Table 1), we used fuzzy sets to quantify the (x, y) coordinates of the foot position on the bicycle pedal. Figure 3 shows the input membership functions, scaling factors, language labels, and fuzzy sets. The output membership functions are shown in Fig. 4.

Table 1. Data analysis based on human bicycle pedaling

Pedal θ_p (°)	Hip θ_1 (°)	Knee θ_2 (°)	Ankle θ_3 (°)
0	83	62	65
45	83	78	69
90	69	96	66
135	57	106	74
180	51	103	77
225	51	88	72
270	56	74	72
315	74	59	70

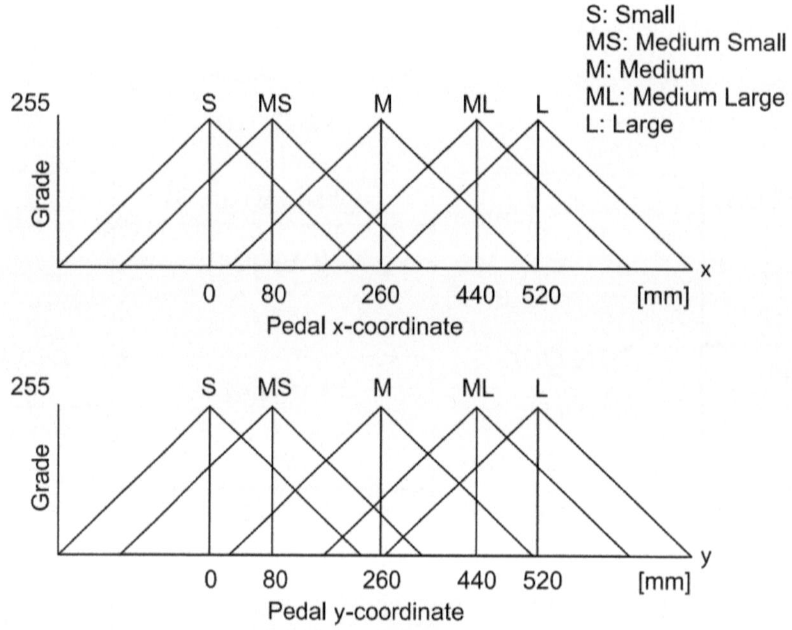

Figure 3. Input membership functions based on human behavior analysis.

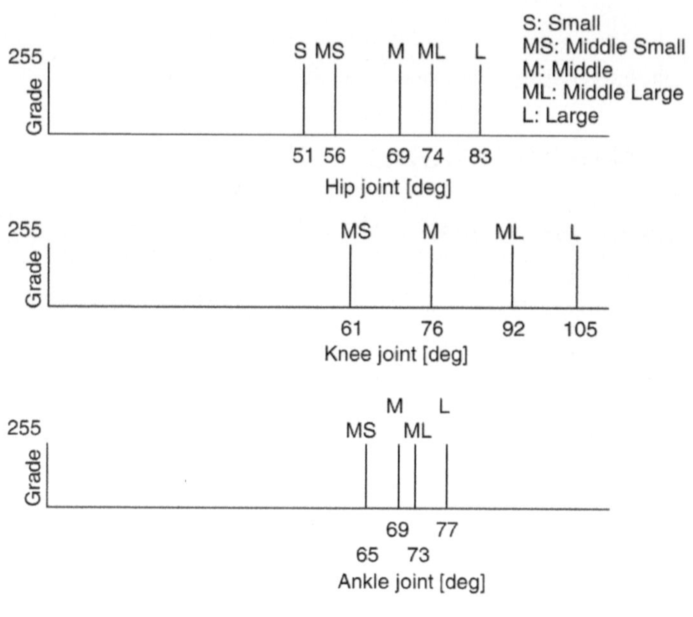

Figure 4. Singleton output membership functions based on human behavior analysis.

Table 2. Fuzzy rules for the 3 DOF based on human behavior

		Hip joint					Knee joint					Ankle joint				
		X					X					X				
		S	MS	M	ML	L	S	MS	M	ML	L	S	MS	M	ML	L
Y	S			S					L					L		
	MS		MS		S			L		ML			ML		ML	
	M	M				MS	ML				M	MS				ML
	ML		L		ML			M		MS			M		M	
	L			L					MS					MS		

The turning radius of the pedals, the input target object, is 250 mm, and we use eight points at 45° intervals on the circumference of this circle to define the center of each input membership function fuzzy set to the x-axis. In a similar manner, the joint angles of the leg are allocated to the center of each output membership function fuzzy set. Here, the output membership functions are defined by singletons (Fig. 4). At the same time, fuzzy labels of the singleton set points on the x-axis used as the results of Table 1 are defined for the fuzzy sets of all membership functions using the dynamic characteristic data from the human analysis. These set points are chosen based on previous experimental experience using a general fuzzy technique and the data from Fig. 1.

Fuzzy Rule for Inverse Kinematics

We consider the hip joint fuzzy-control rules R^i based on Table 1, Figs. 3, and 4 in the following form:

$$R^i : \text{IF (x is M) AND (y is S) THEN } \theta_i \text{ is S} \tag{7}$$

Thus, there is a set of 8 rules for moving each link, and in total, 24 rules describe the inverse kinematic mapping of the 3 hip, knee, and ankle joints in one leg.

Table 2 shows the fuzzy control rule table that was obtained from the human action analysis and the 24 rules. This shows the relationship between the language label and the corresponding joint angle dynamics. The rule constitutes the inverse kinematics model used to calculate the joint angles of the 3 DOF system by using fuzzy reasoning for the target coordinates. The rule table is presented in one table for conciseness, but the hip, knee, and ankle joints of the actual rule table are independent. That is, this table relates the (x, y) coordinates of the bicycle pedal to each joint angle using a fuzzy labels, referring to Fig. 1. In addition, the inference algorithm uses the minimum–maximum composition method and a singleton type membership function.

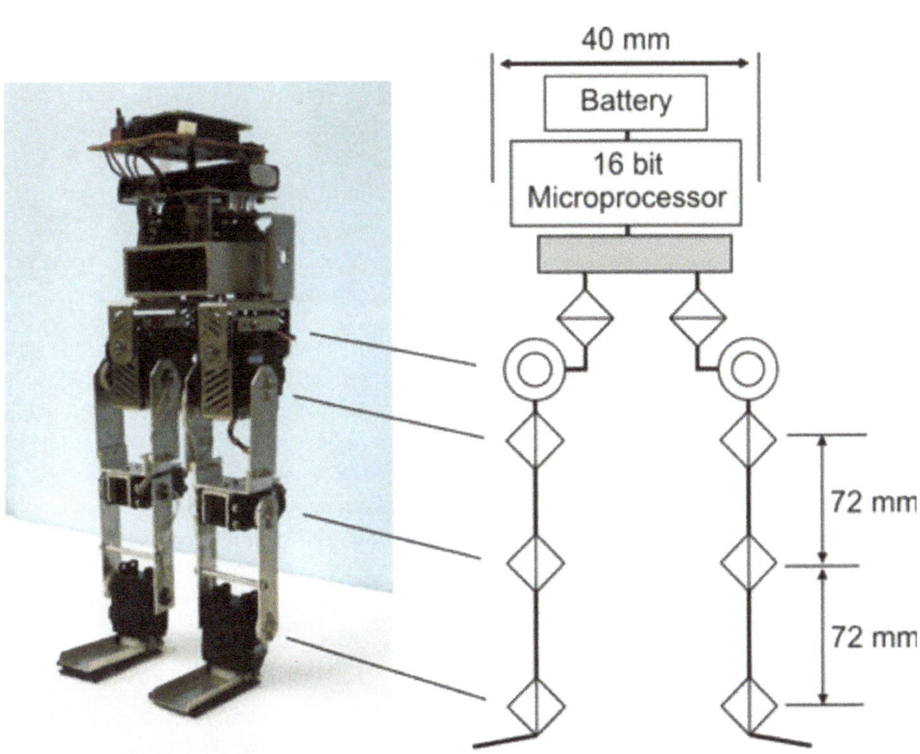

Figure 5. The autonomous humanoid robot system.

Table 3. The specifications of the robot system

Microprocessor:
H8/300H CPU
25 MHz
RAM 8 KB
Battery: 7.4 V Lipo
Actuator: RC Servomotor × 10

Robot Hardware Model

Figure 5 shows a photograph of the robot used in the experiment and the hardware configuration. The lower body is an autonomous 10 DOF humanoid robot with a 16-bit microprocessor [8–12]. However, we have defined the leg to consist of 3 DOF, and, therefore, 6 DOF in total are controlled. The system specifications are shown in Table 3.

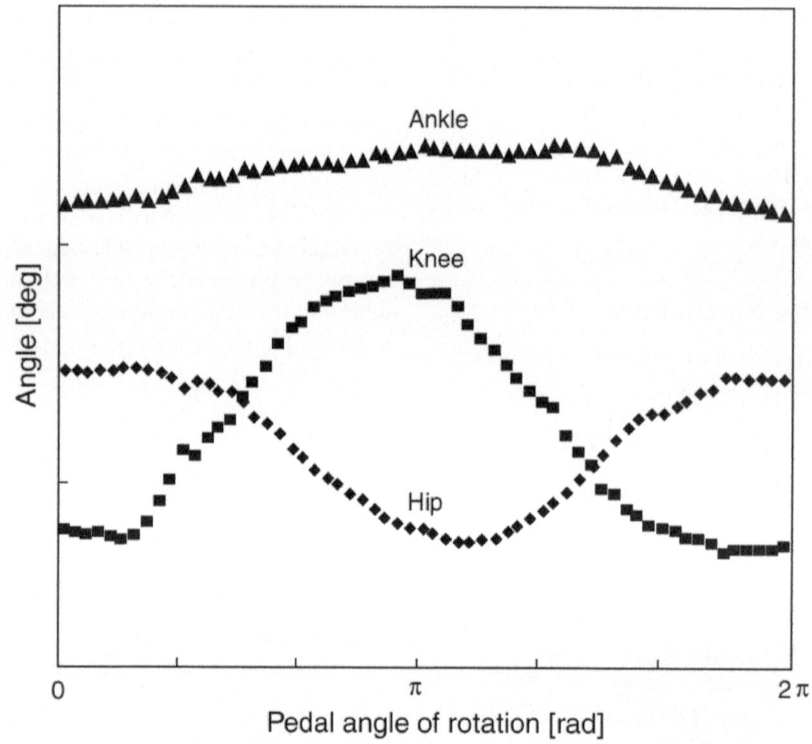

Figure 6. Trajectories of robot joint angles during pedaling action.

REPRODUCING HUMAN ACTION

Bicycle Driving Experiment and Results

System checks were undertaken by initial simulations before the actual experiment was conducted. The coordinates of pedals were entered into the simulated integer inverse kinematics fuzzy control model, and the output results were confirmed to be the appropriate joint angles. The model was then implemented in a small, low performance 16-bit microprocessor, and an experiment was carried out using the robot. Because the two pedals differ by 180° (π rad), this was taken into account in calculating the joint angles of each leg.

Figure 6 shows the trajectories of each actual robot joint angle as a function of the pedal angle over a sampling time of 100 ms. The pedal angle of rotation was used for the input data at intervals of 1° (π/180 rad). These results show that the control model, constructed from eight points, complements the angle between these points continuously and smoothly from the angle of the pedal. Figure 7 shows photographs of the bicycle operation performed by the robot. The robot exhibits a fluid pedaling action and good bicycle control. In addition, this result demonstrates that appropriately interpolated trajectories can be used to realize a collaborative two legged cycling work system.

Integer Inverse Kinematics Processing Capacity

Table 4 lists the calculation times of the joint angles using this algorithm for increasing DOF on a 16-bit microprocessor with a clock frequency of 25 MHz. In the case of a small humanoid robot with 24 DOF, the processing time is about 1 ms.

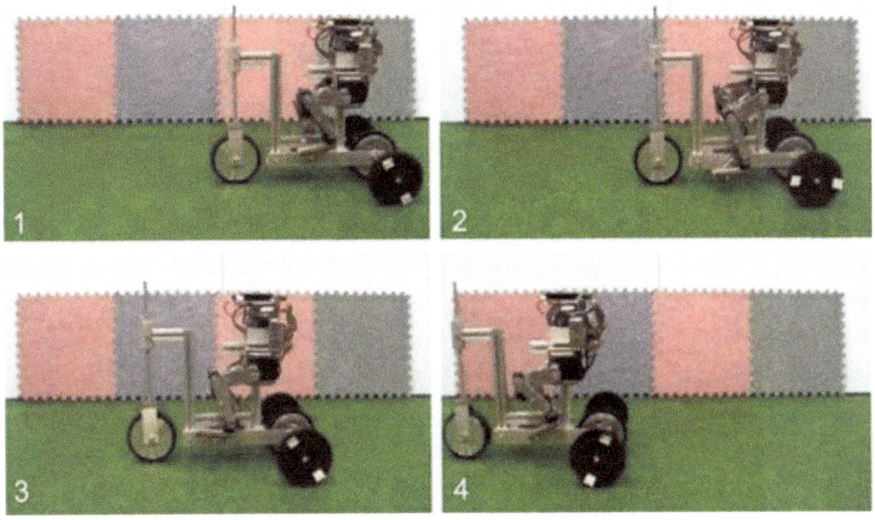

Figure 7. Pedaling action performed by the biped autonomous robot.

Table 4. Joint angle calculation times

4 DOF	8 DOF	12 DOF	24 DOF
175 μs	350 μs	510 μs	1040 μs

This is equal to the pulse-width modulation (PWM) control process time of a standard commercial remote-controlled servo motor. The results in Table 4 show that the microprocessor can control all joints without overloading.

EVALUATION OF INTEGER INVERSE KINEMATICS

The general inverse kinematics model always gives the same result for the same input. The integer inverse kinematics model with fuzzy logic is constructed by the membership functions and control rules based on three joint angles and movement from sampling image data of human behavior. Therefore, if we can construct an integer inverse kinematics model based on the behavior results from the general inverse kinematics model, then we can evaluate the motion trajectory generative capacities of this integer model.

Feature Extraction from General Inverse Kinematics

We first calculated the 3 DOF movement of the leg from a general inverse kinematics model. Figure 8 shows the simulation models of the real small humanoid behavior used in the experiment. Here, the angles θ_1, θ_2, θ_3, and θ_p represent the hip joint, knee joint, ankle joint, and pedal angles, respectively. The angle of the foot is kept constant at 20°, and the posture for eight pedaling positions at 45° intervals is shown. The region of interest for this behavior is the hip, knee, and ankle with the sampling posture being the posture of the leg at pedal positions at 45° intervals (see Fig. 8). The truncated integer joint angle for each posture is recorded in Table 5. Each joint angle at a pedal angle of 90° is shown in Fig. 8.

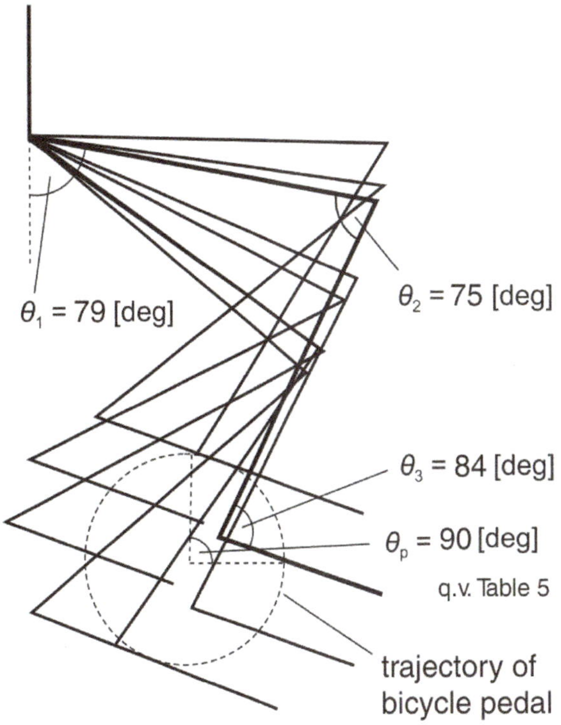

Figure 8. Simulation based on the general inverse kinematic model. The angles θ_1, θ_2, θ_3, and θ_p represent the hip joint, knee joint, ankle joint, and pedal angles, respectively.

Fuzzy Integer Inverse Kinematics Model Derived from General Inverse Kinematics

The membership functions and fuzzy control rules are derived from the behavior and joint angle characteristics. When the (x, y) coordinates of the pedal rotational position is input in 1° steps, the reproducibility of the integer inverse kinematics model can be demonstrated through a comparison with the calculated results of the general inverse kinematics model. In addition, x and y are used as inputs to express rotation in terms of integer coordinates. Figures 9 and 10 show the input and output membership functions constructed using the survey data of Table 5 and the general inverse kinematic model. The joint angles in Table 5 were sorted in descending order, assigned a label, and from this, the output membership functions and fuzzy control rules (Table 6) were developed.

EVALUATION AND CONSIDERATION OF INTEGER INVERSE KINEMATICS MODEL REPRODUCIBILITY

We calculated the hip, knee, and ankle joint angles during the pedaling action using the x and y coordinates of the pedal position for both the general and integer inverse kinematic models. Figures 11 and 12 show the simulation results of the general and integer inverse kinematics models, respectively. The results of the fuzzy inverse kinematics model using real variables are shown in Fig. 13.

Table 5. Data analysis based on the general inverse kinematics

Pedal	Hip	Knee	Ankle
θ_p (°)	θ_1 (°)	θ_2 (°)	θ_3 (°)
0	81	48	57
45	88	59	77
90	79	75	84
135	67	86	82
180	54	89	73
225	50	81	61
270	54	63	46
315	63	53	45

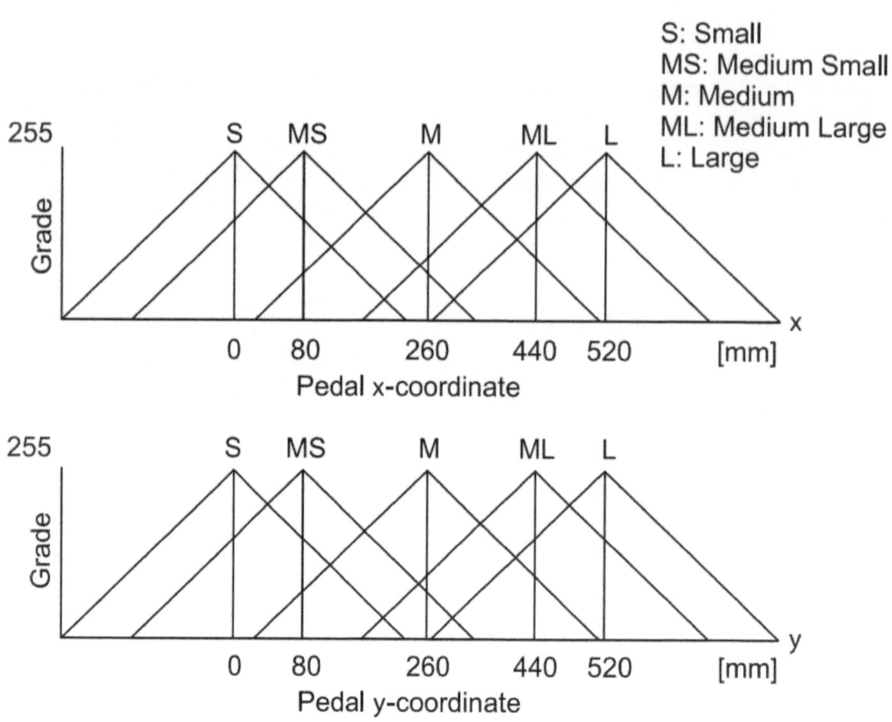

Figure 9. Input membership functions for the general inverse kinematics model.

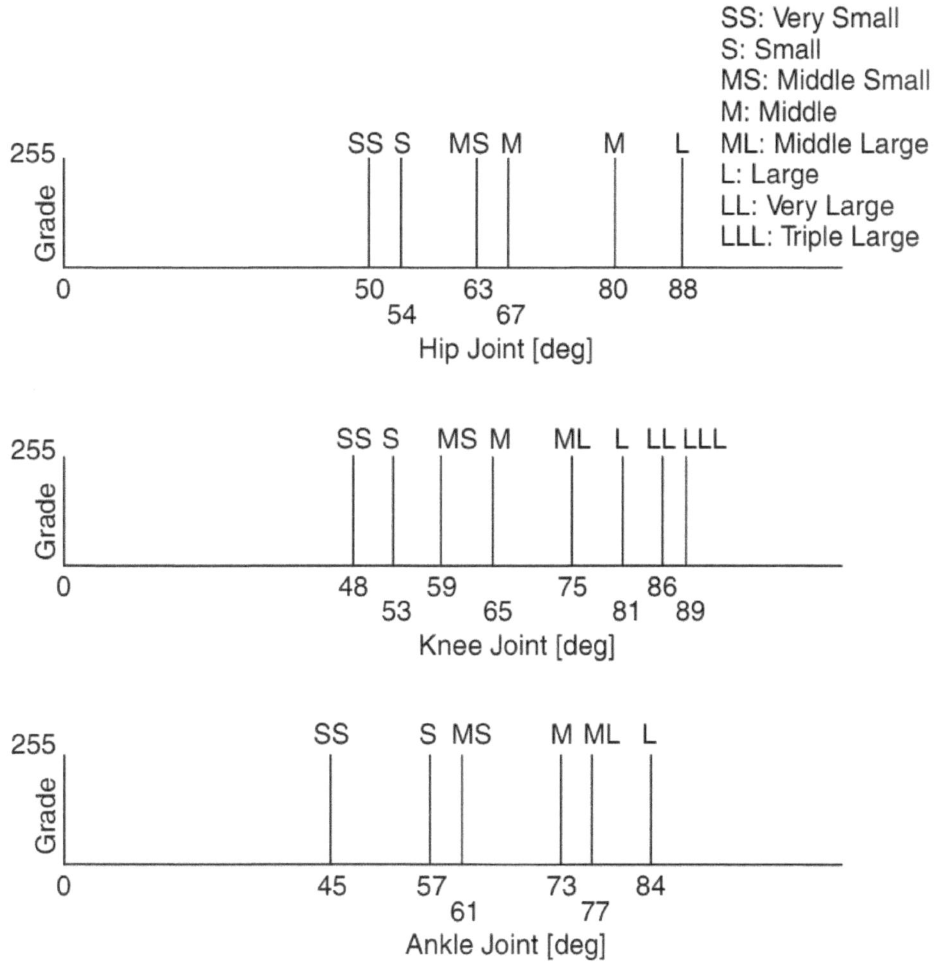

Figure 10. Singleton output membership functions of the general inverse kinematics model.

Table 6. Fuzzy rules for the 3 DOF based on general inverse kinematics

		Hip joint					Knee joint					Ankle joint				
		X					X					X				
		S	MS	M	ML	L	S	MS	M	ML	L	S	MS	M	ML	L
Y	S			S					L					M		
	MS		SS		M			LLL		M			MS		L	
	M	S				ML	LL				S	SS				L
	ML		MS		L			ML		SS			SS		ML	
	L			ML					MS					S		

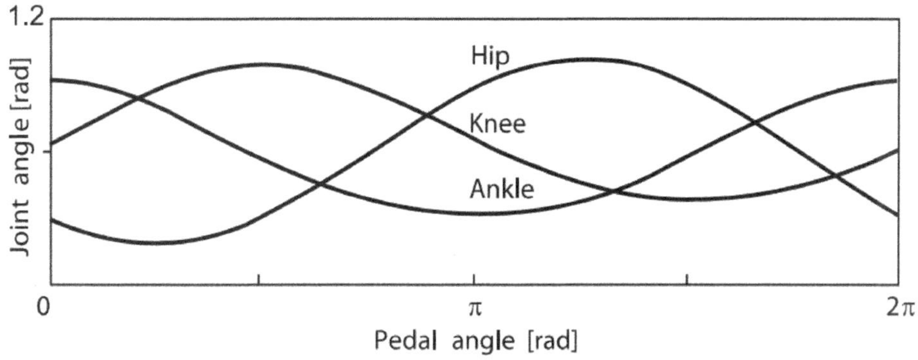

Figure 11. Simulation of joint angles during cycling based on the general inverse kinematics model.

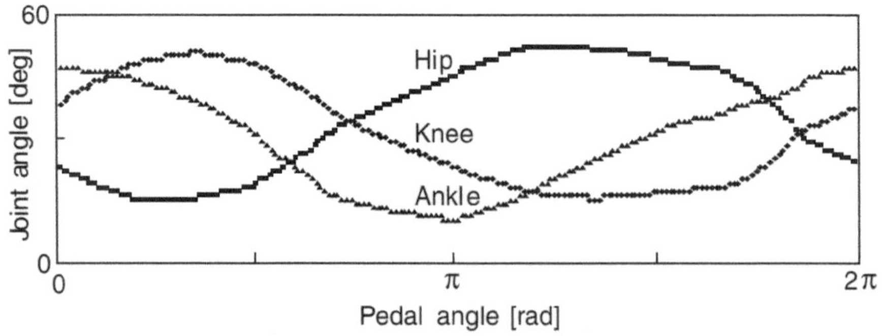

Figure 12. Fuzzy inverse kinematics outputs defined by integer number and based on the general inverse kinematics.

The results of the integer inverse kinematics model are very similar to the output of the general model. This demonstrates that the integer inverse kinematics model can reproduce the original behavior by only performing integer operations. In addition, when comparing Figs. 12 and 13, the integer inverse kinematics model shows discretely controlled results of the model based on real variables. From these results, the integer inverse kinematics model is con- firmed to be a simplified method that can approximate traditional inverse kinematics. Figure 14 shows the simulated behavior using the results of Fig. 12 when a 45° interval and the x, y coordinates of the pedal are inputted. We note that Figs. 8 and 14 show very similar results and, thus, further confirm that this model approximates traditional methods well.

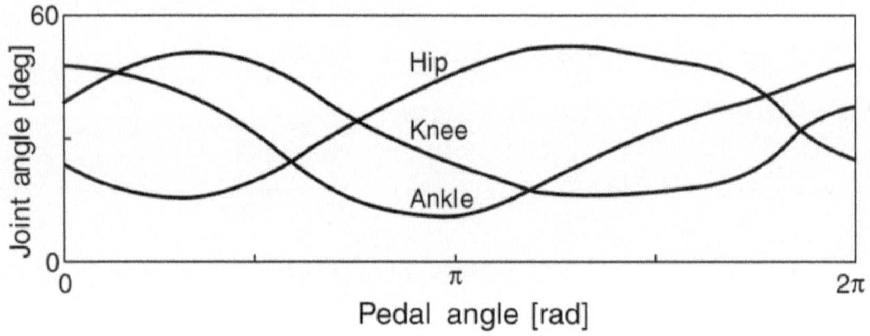

Figure 13. Fuzzy inverse kinematics outputs defined by real numbers and based on the general inverse kinematics.

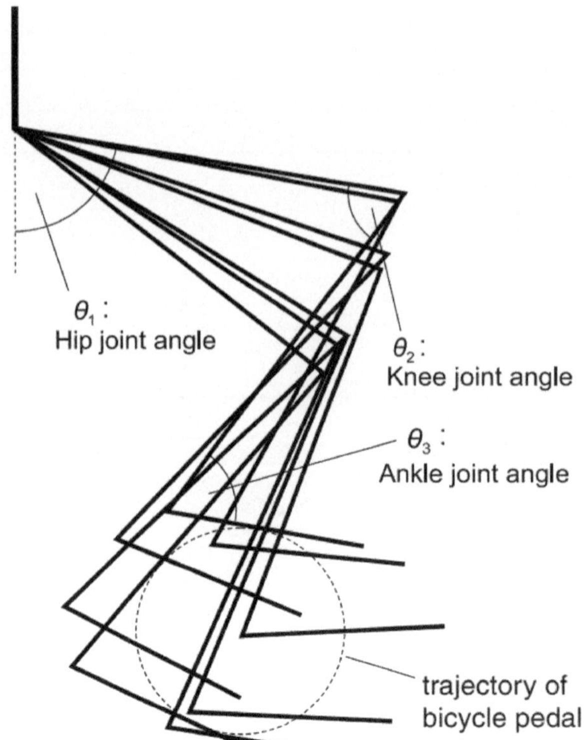

Figure 14. Simulation based on the integer inverse kinematics model.

CONCLUSION

This paper has outlined an inverse kinematics model consisting of integer variables and proposed a method to realize inverse kinematics using integer input and output and integer arithmetic. We have also shown that the application of this periodic trajectory model to an actual humanoid robot system reproduces the cycling behavior of a human very well. Furthermore, we have confirmed the reproducibility of this model based on the original target behavior when deriving this inverse kinematics model. We have also showed that it is possible to generate approximated behavior of the original motion. Moreover, the integer inverse kinematics model is shown to be a simplified calculation method that discretizes the real variables of a conventional robot joint control program.

Although the model in this paper was evaluated through an experiment with the robot performing a periodic controlled motion, future work will involve evaluating other controlled objects or motion using the proposed inverse kinematic model that includes results of our previous research.

ACKNOWLEDGMENTS

This work was partly supported by a Grant-inAid for Scientific Research (C) (21500844).

REFERENCES

1. Kim SW, Lee JJ, Sugisaka M (1993) Inverse kinematics solution based on fuzzy logic for redundant manipulators. Proc IEEE/RSJ Int Confer Intell Robots Syst 2: 904–910. doi:10.1109/IROS.1993.583249
2. Kumbla KK, Jamshidi M (1994) Control of robotic manipulator using fuzzy logic. Proc Third IEEE World Confer Fuzzy Syst 1: 518–523. doi:10.1109/FUZZY.1994.343731
3. Chen C-Y, Her M-G, Hung Y-C, Karkoub M (2002) Approximating a robot inverse kinematics solution using fuzzy logic tuned by genetic algorithms. Int J Adv Manuf Technol 20: 375– 380. doi:10.1007/s001700200166MATHCrossRef
4. Alavandar S, Nigam MJ (2008) Neuro-fuzzy based approach for inverse kinematics solution of industrial robot manipulators. Int J Comput Commun Control 3: 224–234
5. Banga VK, Kumar R, Singh Y (2011) Fuzzy-genetic optimal control for robotic systems. Int J Phys Sci 6: 204–212
6. Piltan F, Sulaiman N, Nasiri H, Allahdadi S, Bairami MA (2011) Novel robot manipulator adaptive artificial control: design a novel siso adaptive fuzzy sliding algorithm inverse dynamic like method. Int J Eng 5: 399–418
7. Sugeno M (1988) Fuzzy control. Nikkan Kougyou Shinnbun, Tokyo (in Japanese)
8. Tojo O, Morishita T (2009) Compact autonomous robot with inverse kinematics algorithm based on fuzzy control model. In: Proceedings of 6th Int Conf Ubiquitous Robots Ambient Intell (URAI 2009), pp 402–405
9. Tojo O, Morishita T (2010) Method for Integer Inverse Kinematics and its Application. In: Proceedings of 7th Int Conf Ubiquitous Robots Ambient Intell (URAI 2010), pp 233–236
10. Umeda K, Asanuma K, Kikuchi T, Ueda R, Osumi H, Arai T (2005) Development of a simulator of environment and measurement for multiple autonomous mobile robots considering camera characteristics. J Robot Soc Jpn 23:112–119 (in Japanese)
11. Ohkubo Y, Okada K, Morishita T, Inaba M, Inoue H (2004) Portable Situation-Reporting System by a Palmtop Humanoid Robot for Daily Life. Proc 2004 IEEE/RSJ Int Conf Intell Robots Sys 4:3553–3558. doi:10.1109/IROS.2004.1389966
12. Urata J, Nakanishi Y, Miyadera A, Mizuuchi I, Yoshikai T, Inaba M (2006) Posture sensor for a spherical joint using cellular phone micro camera. Proc 23rd Annu Conf Robotics Soc Jpn 23:3E31. (in Japanese)

CHAPTER 3

TOLERATING RADIATION-INDUCED TRANSIENT FAULTS IN MODERN PROCESSORS

Xiaobin Li and Jean-Luc Gaudiot

Keywords: Soft-error, Computer architecture, Fault-tolerant, Control flow checking, Multi-threading

INTRODUCTION

As MOS device sizes continuing shrinking, the reliability of the computer systems that are built upon these devices is of increasing concern [1]. For one thing, radiation-induced soft errors will become more significant in the near future [2–4]. In order to handle these inevitable errors, we must integrate in our design fault-tolerant features so that processors can continue to correctly perform their specified tasks despite the occurrence of logic errors [5]. Such designs as the Intel Itanium [6,7], the IBM Power6 [8], the z10 [9], the Fujitsu SPARC64 [10], etc., already include transient fault detection and recovery mechanisms.

Depending on the kind of micro-architectural components where the error occurs, from a program execution perspective, the soft error induced may manifest itself in different ways. For instance, an ALU computation error would lead to a data error whereas if the program counter is corrupted by the incoming particle, the execution would end up an incorrect execution sequence. Therefore, the fault-tolerance design is as usual an ad-hoc process and typically different approaches are needed to protect different parts of the processor against different failure modes.

First, many faults manifest themselves as control flow errors that cause processors to violate the correct sequencing of instructions. Signature checking is a well-known scheme used to detect this type of error. We will present a basic compile-time signature assignment algorithm and describe a novel approach to improve the fault detection coverage of the basic algorithm.

Citation: Li, X. and Gaudiot, J (2010). Tolerating Radiation-Induced Transient Faults in Modern Processors, *International Journal of Parallel Programming*, 2010; 38: 85. DOI:10.1007/s10766-009-0114-9.

As far as data errors are concerned, since the very concept of Simultaneous Multi-Threading (SMT) provides the necessary redundancy, some proposed schemes entail running two copies of the same thread on top of SMT platforms in order to detect and correct soft errors. This allows, upon detection of an error, the rolling back of the processor state to a known safe point, and then a retry of the instructions, thereby effecting a completely error-free execution.

The goal of this paper is thus to describe two approaches which will enhance the robustness of processors in the presence of transient faults such as those caused by cosmic rays. Section 2 defines control flow errors and outlines our signature approach to protect against them. The complete algorithm is described in Sect. 3 and the hardware enhancement is present in Sect. 4. In Sect. 5, we turn our attention to those errors which have not been caught by the signature approach and use the inherent redundancy in Simultaneous Multi-Threaded processors to produce redundant execution of threads and verify the correctness of program execution. Two crucial implementation issues are then addressed: Sect. 6 describes the design trade-off between the fault detection coverage and design costs while Sect. 7 discusses the possible occurrence of deadlock situations. The design overhead evaluation of the proposed design is presented in Sect. 8. We conclude our work in Sect. 9.

CONTROL FLOW ERRORS

From an instruction set architecture standpoint, an abstraction of program execution behavior can be based on various considerations which include control flow, memory and I/O access, and object type and range [11,12]. We first examine how to protect against control flow errors (those which cause a processor to violate the correct sequencing of instructions).

The causes of control flow errors can be traced to the failure of one of many microarchitecture components such as the instruction cache, the program counter unit, the jump execution unit, etc. The instruction cache stores the program content; for instance, if an error causes a stored branch instruction content to change, the pipeline would then be executed in the wrong program order. The program counter unit is merely a pointer to the instructions to be fetched. Conceivably, if its content is corrupted by some radiation-induced error, incorrect instructions would be fed into the execution stage, which in turn would end up as an instruction sequencing violation. Similarly, the jump execution unit computes the target address of the jump instruction. When an error causes an incorrect computation, we would consequently encounter an unexpected program order. Indeed, it has been found that these control flow errors account for between 33% [13] and 77% [14] of all run-time errors

Signature control flow checking techniques are used to monitor the program execution sequence in order to determine if the legal control flow is being followed. Various signature checking techniques have been proposed in the recent past [12,14–22]. Basically, there are two steps in signature checking: compile-time signature generation followed by run-time signature verification.

In the back-end stages, in order to express the program control flow structures, compilers typically build a control flow graph (CFG) in which a node or a basic block is a sequence of instructions with no branch-in except for the entry point and no branch-out except for the exit point and where directed edges are used to represent jumps in the program control flow [23]. Figure 1 illustrates the CFG concept by a simple example. Thus, in the first step of signature checking, which is based upon the CFG, the compiler pre-computes the signatures associated with each node of the CFG, and then either embeds those signatures into the original codes [12,14,16,17,19,22], or provides that information directly to the watchdog [18,21]. It should be noted that at this step, we could have two signature pre-computing techniques: the first, assigned-signature control flow checking [18,19,22], associates with each node an arbitrary signature, for example, a prime number. Conversely, the second technique, derived-signature control flow checking [12,14,16,17,21], derives signatures from the nodes themselves. For example, we derive a checksum from the binary code of the instruction inside a node and then use that checksum as the corresponding node signature.

```
I0:  load R1 = 5
I1:  load R2 = -1
I2:  R1 = R1 + R2
I3:  if (R1 = = 0) goto I5
I4:  goto I2
I5:  halt
```

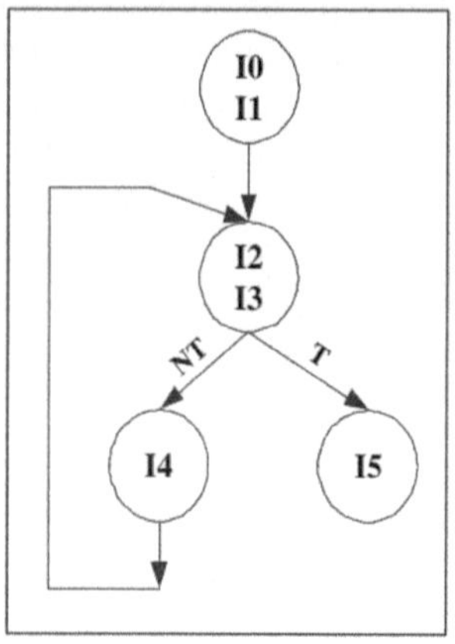

Figure 1. An example program and its control flow graph.

During the second step, the checking engine, which can be either the watchdog or the host microprocessor, computes the run-time signatures and then checks them against the compile time pre-computed signatures. If the signatures differ, it means that an error occurred.

Although the second step of both the assigned-signature checking algorithm and the derived-signature checking algorithm are essentially the same, assigned signature checking techniques have two major drawbacks: the need for registers to hold signatures and the performance overhead due to the need to execute extra instructions related to the assigned-signature checking [12,22]. For example, Oh et al. [22] have shown that the overhead in terms of code size ranges from 26.6 to 61.9% while the overhead in terms of execution time ranges from 16.2 to 58.1%. Conversely, derived signature checking techniques might not guarantee that each node has a unique signature, which might consequently impact the fault coverage. For instance, assume two nodes with identical signatures and a control flow error that is transferring execution into an incorrect node but with the same signature. Such an error cannot be detected. It should be noted that Borin et al. [24] have demonstrated that the performance slowdown of the software-only derived-signature checking would be an average of 30%.

ALGORITHM FOR COMPILER-ASSISTED SIGNATURE CHECKING

As discussed above, the assigned-signature checking technique is based on a comparison between the compiler assigned signature with the one calculated at run time. Any difference between these two signatures indicates that a control flow error has occurred. This section describes in detail a compiler-assisted signature assignment and checking algorithm. As for how to address the inherent performance overhead associated with assigned-signature checking, we propose to use additional hardware to trade this off, as will be explained in Sect. 4.

The Basic Assigned-Signature Control Flow Checking Algorithm

Compiler Time Assigned Reference Signature (S)

As discussed before, the program control flow structure can be expressed as a CFG. We start with a given node Vi of the CFG, and assign to it a unique number which is called the state code of the node. This state code is denoted D(i). Then, we compute the reference signature S(i) of this node by using the following formula:

$$S(i) = D(i) \oplus D(pred(V_i)). \tag{1}$$

where pred(V_i) is the immediate predecessor of node V_i in the CFG (note that ⊕ is an exclusive-OR function). Furthermore, we assume for the moment that each node has only one immediate predecessor. More complex cases will be discussed later.

Run Time Signature (G)

A global register holds the run time signature G of the node currently executing. When the program execution changes the control flow to a new node, e.g., Vi , G is updated by the following formula:

$$G = G \oplus S(i). \tag{2}$$

where S(i) is the reference signature of the current new node V_i.

Then, the core of the control flow checking mechanism consists in the checking of the run time signature G against the static state code D(i) (the one assigned by the compiler) as follows:

Do $G \oplus D(i)$
JNZ exception-handler

Note that this comparison would take place whenever the run time control flow enters a new node of the CFG.

Justifying Signature (J)—Handling Multiple-Branch-In Nodes

Now, we start considering the more complex case when a node has multiple immediate predecessors. Indeed, in normally complex CFGs, a node may have multiple immediate predecessors. We would call such a node a multiple-branch-in (MBI) node: it is a node whose number of immediate predecessors is greater than one. To simplify the discussion, we denote the set of pred(MBI) as:

$$\mathbb{S} = \{V_k \mid V_k \text{ is an immediate predecessor of MBI}\}. \tag{3}$$

When dealing with such MBI nodes, as required in eq. (1), we must choose one of the immediate predecessors as the primary immediate predecessor (or primary node, for short). Also, since there is more than one path up from an MBI node, we associate with each immediate predecessor an additional parameter which we call the justifying signature. The justifying signature is used at run time to verify that all immediate predecessors to the MBI node are legal antecedents to that node.

The following outlines the compile-time MBI node handling algorithm:

1. Arbitrarily select a node from $\mathbb{S}$ as the MBI node primary node (assume V_j for the rest of this discussion). Note that we leave the discussion of primary node selection later.
2. The reference signature of the MBI node is governed by the selected primary node V_j as:

$$S(MBI) = D(MBI) \oplus D(j). \tag{4}$$

3. For every node $V_k \in \mathbb{S}$, associate it with the justifying signature given by the following formula:

Note that now each node $V_k \in \mathbb{S}$ has two signatures: the reference signature S(k) and the justifying signature J(k) as illustrated in Fig. 3 where V_7, V_8 and V_9 are the MBI nodes.

$$J(k) = D(k) \oplus D(j). \tag{5}$$

The run-time MBI node handling algorithm could be described as follows:

4. We denote control flow changes as: $V_k \rightarrow$ MBI. First, the run time signature is updated according to the following formula:

$$G = G \oplus S(MBI) \oplus J(k). \tag{6}$$

5. Finally, the MBI node run-time control flow checking can be applied as discussed before:

Do $G \oplus D(\text{MBI})$
JNZ exception-handler

We use Fig. 2 to illustrate the simple example of an MBI node and how justifying signature works. Node V_3 is an MBI node and its immediate predecessor set is $\mathbb{S} = \{V_1, V_2\}$ whereas V_1 is arbitrarily selected as the primary node. At compile time, after assigning the reference signatures S(1) and S(2) to nodes V_1 and V_2, the justifying signatures J(1) and J(2) are also evaluated. As we can see from Formula 6, the run time signature G carried when entering the MBI node V_3 is updated by two exclusive-OR functions (Refer to Formula 2, only one exclusive-OR function is required there): first, perform an exclusive-OR with the reference signature of node V_3, and then perform

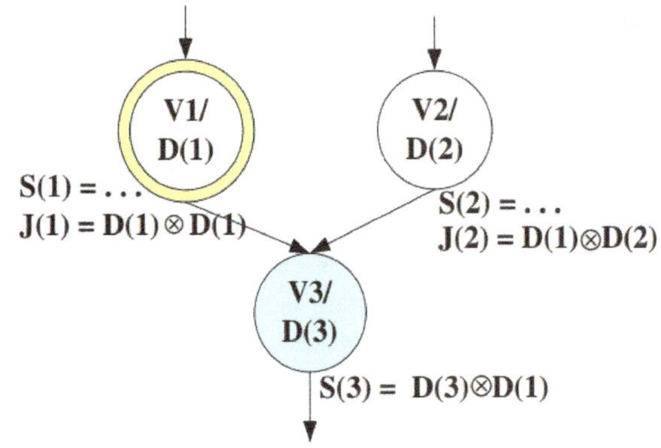

Figure 2. An example of MBI node and justifying signatures.

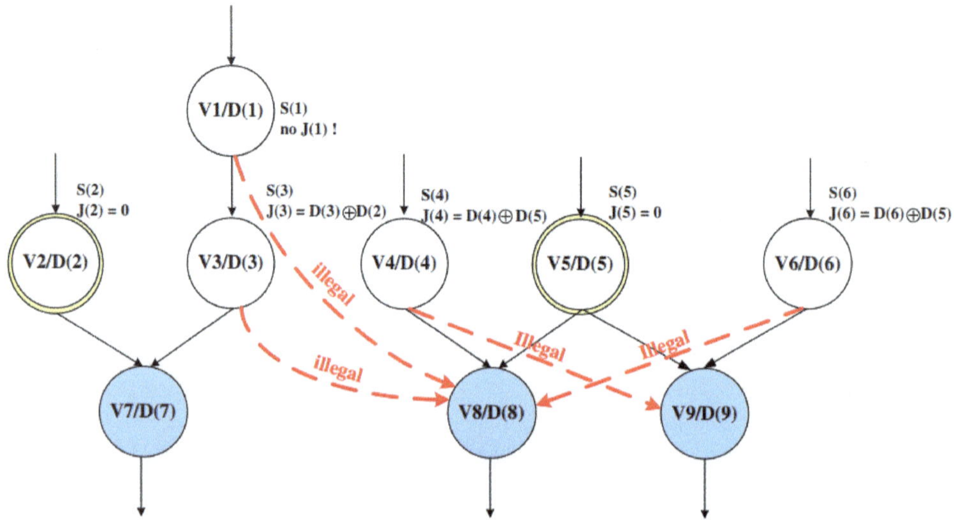

Figure 3. Fault detection coverage analysis for the MBI node handling algorithm.

another exclusive-OR with the justifying signature of the jumping node. For example, when the control flow change is $V_2 \rightarrow V_3$, the run time signature is calculated by:

$$\begin{aligned} G &= G \oplus S(3) \oplus J(2) \\ &= D(2) \oplus D(3) \oplus D(1) \oplus D(1) \oplus D(2) \\ &= D(3) \end{aligned}$$

Analyzing the Fault Detection Coverage of the Basic Algorithm

To the end of improving the fault detection coverage of the above basic algorithm, we start by first analyzing the coverage: consider the general case of an MBI node control flow change: $V_k \rightarrow$ MBI. According to the relationship between the node V_k and the node "MBI," there are three possible cases:

1. If $V_k \in \mathbb{S}$, which means that the control flow change is legal, as discussed before, we can easily prove that the updated run time signature is: G = D(MBI).
2. If $V_k \notin \mathbb{S}$, which means that the control flow change is illegal and that two cases must be separately considered:
   a. V_k is an immediate predecessor of another MBI node, which means that J(K) has been defined;
   b. Vk is not an immediate predecessor of any MBI node. Then, J(k) is null at compile-time and then without loss of generality, J(k) will be a random number at run-time (whatever is left in the corresponding storage cell at that time).

Justifying Signature Has Been Defined

Consider the following control flow change: $V_i \rightarrow V_j$ where V_j is an MBI node and its set of pred(V_j) is $\mathbb{S}_j$. However, $V_i \notin \mathbb{S}_j$, i.e., the control flow change is illegal, V_i is instead an immediate predecessor of another MBI node, say V_m, hence $V_i \in \mathbb{S}_m$. Moreover, V_x has been selected as the primary node of V_j and V_y as the primary node of V_m. Then, when entering V_j, the run time signature is updated by using the following formula:

$$\begin{aligned} G &= G \oplus S(j) \oplus J(i) \\ &= D(i) \oplus D(j) \oplus D(x) \oplus D(i) \oplus D(y) \\ &= D(j) \oplus D(x) \oplus D(y) \end{aligned}$$

Therefore, two cases need to be considered:

1. As long as D(x) ⊕ D(y) = 0, we end up with G = D(j), which means that a control flow error has escaped detection. The faulty condition D(x)⊕ D(y) = 0 is satisfied only if D(x) = D(y), i.e., node V_x is the same as node V_y (remember that the state code of each node is unique). Examples of illegal control flow changes such as $V_6 \rightarrow V_8$ and $V_4 \rightarrow V_9$, are shown in Fig. 3. In both cases, two MBI nodes V_8 and V_9 share node V_5 as their primary node.

   Observation 1

   When two MBI nodes, V_j and V_m, share their primary node, V_x (= V_y), any illegal control flow change: V_i(∈ $\mathbb{S}_m$ and ∉ $\mathbb{S}_j$) → V_j and any illegal control flow change: V_l(∈ $\mathbb{S}_j$ and ∉ $\mathbb{S}_m$) → V_m cannot be detected. Further, we denote the probability of these illegal control flow changes as PND1.

2. On the other hand, if $V_x \neq V_y$, i.e., no primary node sharing, we have: G ≠ D(j) which means that the control flow error can be successfully detected. An example for this case is shown in Fig. 3 as an illegal control flow change, from $V_3 \rightarrow V_8$.

   To summarize the above two cases, we can state the following:

   `Observation 2

   The fault detection coverage may decrease if two MBI nodes share their primary node. In other words, if a node V_x has multiple branch-outs, for example, the exit statement of the node is a conditional branch, and if more than two (including two) branch destination nodes are MBI nodes, the node V_x should not be selected as a primary node.

Random Justifying Signature

Consider the following control flow change: $V_i \rightarrow V_j$ where V_j is an MBI node and its set of pred(V_j) = $\mathbb{S}_j$. Further assume that V_x has been selected as the primary node of V_j. However, $V_i \notin \mathbb{S}_j$, i.e., the control flow change $V_i \rightarrow V_j$ is illegal. Also, V_i is not an immediate predecessor of any MBI node such that J(i) has not been defined and we deal with it as a random number. An example of an illegal control flow change would be $V_1 \rightarrow V_8$, which is shown in Fig. 3. In this case, when entering V_j, the run-time signature is updated by using the following formula:

$$\begin{aligned} G &= G \oplus S(j) \oplus J(i) \\ &= D(i) \oplus D(j) \oplus D(x) \oplus J(i) \end{aligned}$$

Because of the randomness of J(i), two cases must be considered:

1. If D(i) ⊕ D(x) ⊕ J (i) = 0, we have G = D(j) which means that the control flow error escapes detection;
2. If D(i) ⊕ D(x) ⊕ J (i) ≠ 0, we have G ≠ D(j) which means that the algorithm has successfully detected the control flow error.

Fortunately, the probability for D(i) ⊕ D(x) ⊕ J (i) to be zero is very low: it can happen only if J (i) = D(i)⊕ D(x). Given the n-bit size of state codes and signatures, the probability is:

$$P_{ND2} = P\{J(i) = D(i) \oplus D(x)\} = 2^{-n}. \tag{7}$$

In summary, the fault detection coverage of the MBI node handling algorithm is:

$$C \equiv P\{\text{fault detection} | \text{fault existence}\} \tag{8}$$

$$= P\{\text{control flow error detection} | V_k \rightarrow \text{MBI and } V_k \notin \mathbb{S}\} \tag{9}$$

$$= 1 - P_{ND1} - P_{ND2}. \tag{10}$$

Improving the Fault Detection Coverage of the Basic Algorithm

Based on the above discussion, we know that to improve the fault detection coverage, reducing P_{ND1} is the primary approach, which in turn requires selecting the primary node from those not being shared by MBI nodes. In other words, by carefully handing those cases we analyzed in Sect. 3.2, we can improve the fault detection coverage over the previous work, such as described by Oh et al. [22]. Therefore, we propose a two pass algorithm for the purpose of selecting the primary node for MBI nodes, which is specified in Algorithm 1: given an MBI's immediate predecessor list, we choose the one that has the smallest chance of being shared. To this end, we first traverse CFG to count how many children nodes a node has, and then the node with the least number of children nodes is the best candidate to be the primary node.

Algorithm 1: Two-pass algorithm for selecting the primary node

```
Initialization: for every node in CFG do
    pn_cnt = 0;
    pn_selected = FALSE;
end
First-Pass: for every MBI node do
    for each of its primary node do
        pn_cnt ++;
    end
end
Second-Pass: for each MBI node do
    select primary node with min(pn_cnt) and pn_selected = FALSE ;
    mark the selected primary node's pn_selected = TRUE;
    for each of this MBI's immediate predecessor but not being selected do
        pn_cnt - -;
    end
end
```

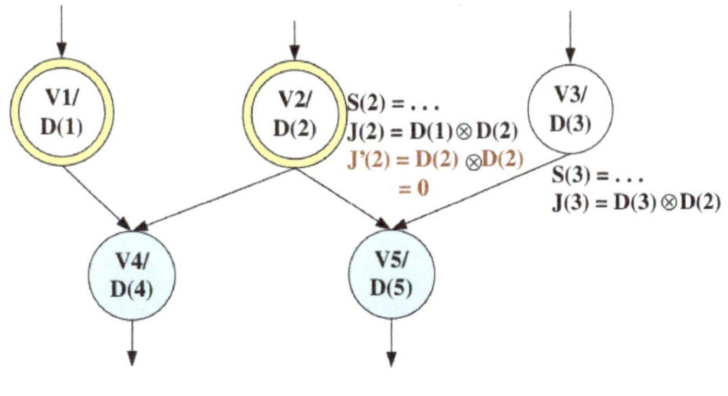

Figure 4. An example of ITE node with two justifying signatures.

Handling the If-Then-Else Node

In addition to the fault detection coverage improvement, we feel it is important to remark that randomly selecting the primary node may result in conflicts as illustrated in Fig. 4: Indeed, if V_1 had been selected as the primary node for V_4 and V_2 for V_5, respectively, we would have to create two justifying signatures for node V_2: as far as MBI node V_4 is concerned, the justifying signature of V_2 is: J (2) = D(1) ⊕ D(2); whereas as far as MBI node V_5 is concerned, the justifying signature of V_2 is: J' (2) = D(2) ⊕ D(2) = 0.

Hence, for the control flow change: $V_2 \rightarrow V_4$, J(2) should be used to update the run-time signature whereas for the control flow change : $V_2 \rightarrow V_5$, only J'(2) is the correct choice. Anything corrupted up to this level could result in faulty control flow error detection. Simply speaking, for the legal control flow change: $V_2 \rightarrow V_4$, if the justifying signature J'(2) had been used to update the run time signature, we would end up with $G \neq D(4)$ such that a control flow error could be flagged, a false alarm. Such situations have not been addressed by Oh et al. [22].

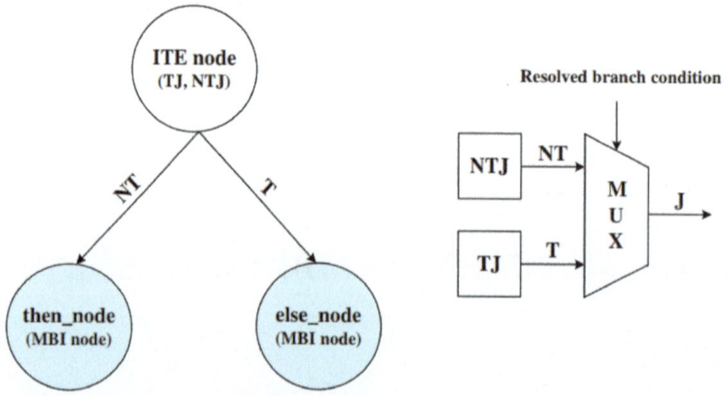

Figure 5. An hardware approach for ITE node with two justifying signatures (T = taken; NT = not-taken; TJ = justifying signature for ITE_node → else_node; NTJ = justifying signature for ITE_node → then_node).

The necessary conditions for a node associated with two justifying signatures are:

1. The exit of the node is a conditional branch, that is to say the node is an if-then-else (ITE) node;
2. Both branch destinations are MBI nodes.

In short, we need to distinguish the two justifying signatures: one for the then branch flow (the resolved branch condition is not-taken), the other for the else-branch flow (the resolved branch condition is taken).

Figure 5 shows a hardware-based algorithm: at compile time, when an MBI node traces back its immediate predecessors for the purpose of justifying signatures, the associated directed edges are checked (directed edges are given by the CFG): if the edge is a "taken" path, the associated justifying signature will be placed into the TJ register; whereas if it is a "not-taken" path, the NTJ register is used for the associated justifying signature. At run time, the resolved branch condition is used to select the appropriate justifying signature for updating the run time signature. More details will be given in Sect. 4.

HARDWARE ENHANCEMENT FOR CONTROL FLOW CHECKING

As discussed before, the assigned-signature checking technique has an inherent performance overhead drawback. However, with advances in CMOS technology, we have an abundance of cheap hardware resources [1]. Moreover, our proposed mechanism can be simply implemented in any modern microprocessor at little additional cost. Hence, in this section, we first introduce three additional instructions dedicated to control flow checking, and then design a simple hardware implementation to execute these instructions.

Table 1. Three additional instructions specification

No.	Mnemonic	Format	Function description
1	`SIC`	`SIC imm1, imm2` where `imm1` = S(i); `imm2` = D(i)	**Signature checking :** 1 Update G as: G = G ⊕ imm1 2 **If** (G == imm2) fault free, **Else** control flow error;
2	`SIJ`	`SIJ imm1, imm2` where `imm1` / `imm2` = D(i) ⊕ D(j) **If** (ITE node) `imm1` for NTJ, `imm2` for TJ, **Else** `imm1` for J	**Signature justifying:** Update J as: J = imm1/imm2 depended on resolved branch condition
3	`SIJC`	`SIJC imm1, imm2` where `imm1` = S(i); `imm2` = D(i)	**MBI node signature checking:** 1 Update G as: G = G ⊕ imm1 ⊕ J 2 **If** (G == imm2) fault free, **Else** control flow error;

We will also provide a comprehensive control-flow checking algorithm based on these hardware enhancements. In the end, we will show the benefit from trading hardware off a reduction in performance overhead.

Additional Instructions

The three additional instructions dedicated to the assigned-signature control flow checking are succinctly described in Table 1. Instruction SIC is used to check for control flow errors in non-MBI nodes. Instruction SIJ is dedicated to signature justification. Instruction SIJC is used to check for control flow errors in MBI nodes. The compiler is responsible for the insertion of these additional instructions into the original program so as to achieve run-time control flow checking. The detailed algorithm will be presented in sect. 4.3.

Implementation of Additional Instructions

The on-chip control flow checker to execute the above three additional instructions is relatively easily designed: assume a simple five-stage pipeline: Fetch → Decode → Execution → Memory access → Write back. Our on-chip control flow checker would be located in the "Decode" and "Execution" stages.

As seen in Fig. 6, a total of five registers are needed to hold the necessary information: register G is used for the run time signature; registers D/S and NTJ receive immediate values from the imm1 field of their instruction words and registers D and TJ receive immediate values from the imm2 field of the instruction words. For each instruction, Table 2 shows the control signals generated by the opcode decoder (also shown in Fig. 6).

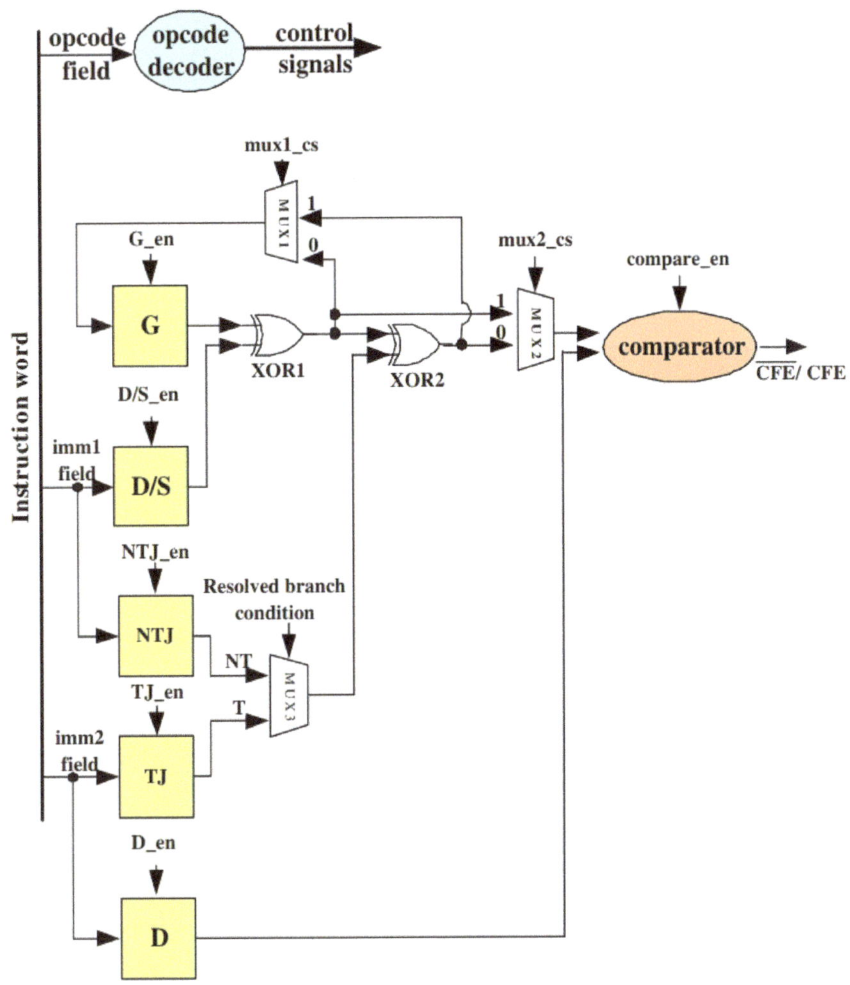

Figure 6. On-chip control flow checker block diagram.

Table 2. On-chip control flow checker control signals

Instruction	G_en	D/S_en	D_en	NTJ_en	TJ_en	mux1_cs	mux2_cs	br. cond.	compare_en
SIC	en	en	en	$\overline{en}$	$\overline{en}$	0	1	$\overline{en}$	en
SIJ	$\overline{en}$	$\overline{en}$	$\overline{en}$	en	en	X	X	$\overline{en}$	$\overline{en}$
SIJC	en	en	en	$\overline{en}$	$\overline{en}$	1	0	from BRU	en

en = enable; $\overline{en}$ = disable; X = don't care; BRU = branch unit

Operation of SIC Instructions

When an instruction word SIC imm1, imm2 is decoded, its opcode field is fed into the opcode decoder. The decoder then generates the control signals specified in Table 2. The imm1 filed is received by the enabled register

$$D/S\left(D/S\_en = enable \text{ and } NTJ\_en = \overline{enable}\right)$$

while the imm2 field is received by the enabled register

$$D\left(D\_en = enable \text{ and } TJ\_en = \overline{enable}\right).$$

The content of register D/S goes into XOR1 along with the content of register G (G_en = enable). The result is selected by mux2 (mux2_cs = 1). Now the enabled comparator compares the two inputs which are received from mux2 and register D. These have performed the run-time control flow checking. Also, we can see the result of XOR1 is sent to modify register G since we have mux1_cs = 0 and G_en = enable.

Operation of SIJ Instructions

When an instruction word SIJ imm1, imm2 is decoded, its opcode field is fed into the opcode decoder. The decoder then generates the control signals specified in Table 2. The imm1 field is received by the enabled register

$$NTJ\left(NTJ\_en = enable \text{ and } D/S\_en = \overline{enable}\right)$$

while the imm2 field is received by the enabled register

$$TJ\left(TJ\_en = enable \text{ and } D\_en = \overline{enable}\right).$$

Operation of SIJC Instructions

When an instruction word SIJC imm1, imm2 is decoded, its opcode field is fed into the opcode decoder. The decoder then generates the control signals specified in Table 2. The imm1 field is received by the enabled register

$$D/S\left(D/S\_en = enable \text{ and } NTJ\_en = \overline{enable}\right)$$

while the imm2 field is received by the enabled register

$$D\left(D\_en = enable \text{ and } TJ\_en = \overline{enable}\right).$$

The content of register D/S goes into XOR1 along with the content of register G (G_en = enable). Based on the resolved branch condition, either the content of register NTJ or that of register TJ XOR2 with the result of XOR1. Once again, if no conditional branch results from the branch unit, i.e., not an ITE node, the default "resolved branch condition = NT" such that the content of register NTJ is selected at this point. MUX2 selects the result of XOR2 (mux2_cs = 0). Now the enabled comparator compares the two inputs which are received from mux2 and register D. These have performed the run-time control flow checking. Furthermore, we can see that the result of XOR2 is sent to modify register G since we have mux1_cs = 1 and G_en = enable.

Using Additional Instructions

To summarize the above discussion, Algorithm 3 is a comprehensive signature assignment algorithm based on our hardware enhancement instructions. Returning to the example of Fig. 1, our compiler algorithm would produce the modified diagram shown in Fig. 7. The left-hand side illustrates the state code assignment results and the primary node selection of the MBI node V_2. The right-hand side shows the CFG after insertion of our control flow checking instructions.

Comparing Code Size Overhead

To compare our algorithm with that of Oh et al. in [22], consider a typical node consisting of 7–8 instructions [1]. In order to check for control flow errors, the algorithm in [22] adds 2–4 instructions to each node. The overhead is between 27 and 53%.

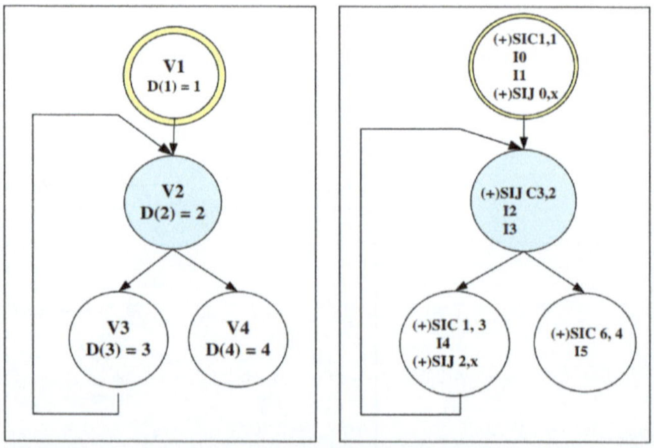

Figure 7. State code and signature assignment example.

For a number of benchmarks, the code size overhead is between 26.6 and 61.9% whereas the execution time overhead is between 16.2 and 58.1%). Conversely, in our hardware-enhanced approach, the additional instructions are a maximum of 1 or 2 for each node. Therefore, the overhead is only 13–27%, which is a significant improvement over [22]. Furthermore, the execution time is given by the following formula [1]:

$$\text{Execution time} = \text{Instr count} \times \text{Clock cycle time} \times \text{Cycles per instr.} \quad (11)$$

With the assistance of our on-chip control flow checker, we could expect a lower execution time than that obtained in [22] when executing the program with the signature checking. This is because we have a smaller instruction count given the same clock cycle time and number of cycles per instructions.

TRANSIENT FAULT-TOLERANT SMT PROCESSORS

The control flow checking targets faults which result in the correct program sequencing being violated. To increase the fault coverage to tolerate those errors manifested as data errors, we propose in this section a transient fault-tolerant Simultaneous MultiThreading processor design.

The multi-threading execution paradigm inherently provides the spatial and temporal redundancy necessary for fault-tolerance: the basic idea is that we run two copies of the same threads on Simultaneous Multi-Threading platforms [25,26,4,27,28] and then the results of the two copied threads are compared to detect transient faults. This allows, upon detection of an error, the rolling back of the processor state to a known safe point, and then a retry of the instructions, thereby effecting a completely error-free execution.

Trade-Off between Fault Detection Coverage and Design Costs

A design trade-off is associated with the above basic idea: Generally speaking, it requires redundant execution of the two identical threads to cover a given processor component. Hence, the higher the fault coverage desired, the more redundant the execution required. However, redundant execution inevitably comes at the cost of performance overhead, hardware overhead, design complexity, etc. Consequently, how to trade off between fault detection coverage and the associated costs is essential for the applicability of the basic idea. Consider a general five-stage pipeline [1]: the instruction fetch stage, the decode stage, the issue stage, the execution stage, and the retire stage can all be exploited to generate the redundant instructions [29,30,26,4,27,28]. Take the fetch stage as an example, we can generate the redundant threads by fetching instructions twice [26,4,27,28]. Since the instruction fetch stage is the first pipeline stage, the redundant execution would then cover all the pipeline stages, thus, the largest possible pipeline fault coverage could be achieved. However, allowing two redundant threads to fetch instructions would mean halving the effective fetch bandwidth. Furthermore, given N/2 effective fetch bandwidth, the maximum pipeline throughput is limited to N/2. Additionally, the redundant thread generated in the fetch stage would then compete not only for the decode bandwidth, the issue bandwidth, and the retire bandwidth, but also for IssueQ and ROB capacity, which are all identified as the key factors that affect the performance of the redundant execution [30]. Conversely, we can re-issue the retired instructions from ROB back to the functional units for redundant execution. In doing so, the bandwidth and spatial occupancy contention at IssueQ and ROB can be reduced, thus the performance overhead is lowered [30]. However, this retire stage-based design comes at the cost of reducing the fault detection coverage: only the EXE stage is covered.

Hence, considering the trade-off between fault detection coverage and design cost, we simply fetch the instructions once and then copy the fetched instructions to generate the redundant thread, In so doing, there is no need for partitioning the fetch bandwidth between the redundant threads. Furthermore, we rely on the dispatch thread scheduling and redundant thread reduction to relieve the contention in the IssueQ and ROB (details can be found in Subsects. 6.2 and 6.3).

Possible Occurrence of Deadlocks

Other than the design trade-off, another issue associated with the basic idea is the need to prevent deadlocks. In a fault-tolerant SMT design, two copies of the same thread are now cooperating with each other. Such cooperation could cause deadlock situations [27]. We thus present a systematic deadlock analysis and conclude that as long as ROB, LQ, and SQ have allocated some dedicated entries for the trailing thread, the deadlock situations identified can be prevented. Based on this observation, we propose two ways of preventing any possible deadlock situations (more details in Sect. 7): (1) static allocation of entries in ROB, LQ, and SQ for the redundant thread copy; (2) dynamic deadlock monitoring.

LOWERING THE PERFORMANCE OVERHEAD OF TRANSIENT FAULT-TOLERANT SMT PROCESSORS

This section presents our proposed fault-tolerant SMT datapath designs. As discussed, to consider the trade-off between the fault detection coverage and the design costs, we simply fetch the instructions once and then copy the fetched instructions to generate the redundant thread. As far as the fault detection coverage is concerned, there are three major components in the fetch stage: I-cache, PC, and BPUs. By copying fetched instructions to generate the redundant threads, any transient faults which would happen inside the I-Cache might not be detected. However, ECC-like mechanisms that are very effective in handling transient faults in memory structures are now widely implemented in modern microprocessors [7,31,10]. Further, the fault which occurs in the BPUs will have no effect on the ultimate correct program execution [32], whereas the critical component PCs must be protected by ECC-like mechanisms.

Copy Fetched Instructions to Generate the Redundant Thread

As shown in Fig. 8, our instruction copy operation is simple: simply buffer in two instruction queues the instructions fetched, hence, the copy operation would not increase the pipeline cycle time, nor would another pipeline stage be added. To be specific, each instruction fetched is bound to a sequential number and a unique thread ID. For instructions that are stored in IFQ, the "leading thread" (LT) is used as the thread ID, whereas for those stored in another IFQ, called trace queue (traceQ), the "trailing thread" (TT) is used. It should be noted that traceQ also serves in the two performance overhead lowering techniques which will be described in detail in the following subsections.

Reducing the Complexity of TT

Focusing on our redundant execution mode, the key factors that affect the performance of redundant execution could be identified as: the bandwidth contention for issue, FUs, and retire operations; and the capacity contention in IssueQ and ROB [33,30]. This subsection addresses those resource contention problems by making TT "lighter" (recall that the purpose of executing TT is just fault detection). In doing so, the contention in IssueQ, ROB and FUs could be reduced accordingly.

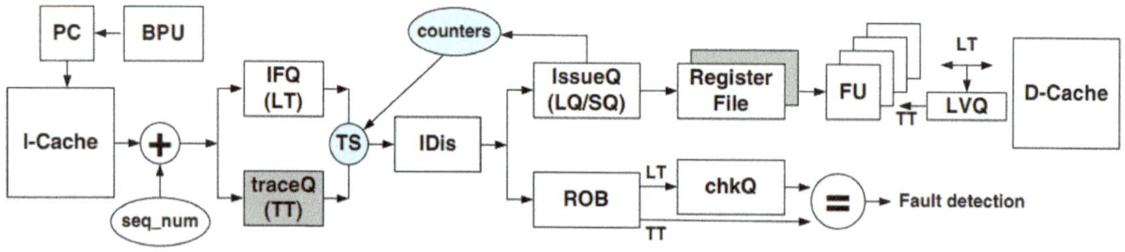

Figure 8. Functional diagram of our proposed fault-tolerant SMT data path.

Preventing the Dispatch of TT Wrong-Path Instructions

The number of dynamic wrong-path instructions might be a significant portion of all the instructions fetched. For example, Kang et al. [34] observed that 16.2–28.8% of the instructions fetched would be discarded from the pipeline even with a high branch prediction accuracy. Hence, if we could prevent TT wrong-path instructions from being dispatched, the effective utilization of IssueQ, ROB, and FUs would be improved accordingly. Based on this observation, we leverage LT branch resolution results to prevent all the wrong-path instructions of TT from being dispatched. It should also be pointed out that in our designs, neither a branch outcome queue [4] nor a branch prediction queue [28] are needed.

If a branch instruction is encountered in traceQ, the dispatch operation will check the status of this TT branch instruction: if its prediction outcome has been resolved by its counterpart from LT, we continue its dispatch operation; otherwise, the TT dispatch operation will be paused (from here, we can see that in order not to pause TT dispatch operation, LT must be executed ahead of TT. The LT ahead of TT execution mode is called the staggered execution mode (more on this in Subsect. 6.3.2).

To set the TT branch instruction status (the initial status is set as “unresolved”), every completed branch instruction from LT will search traceQ to match its TT counterpart. Furthermore, if the branch has been mispredicted, LT will perform its usual branch misprediction recovery, and at the same time it will flush all those instructions inside traceQ that are located behind the matched counterpart branch instruction. In other words, LT performs the branch misprediction recovery for both LT and TT, thus, TT does not recover any branch misprediction by itself. After recovery, the status of the TT branch instruction will be set as “resolved.” On the other hand, if the branch has been correctly predicted, the status of the matched counterpart TT branch instruction will be set as “resolved” as well.

In addition, the sequential numbers provide the mean for matching two redundant threads instructions. As we have seen, each instruction fetched is associated with a sequential number at first, then the fetched instruction is replicated to generate the redundant thread. In doing this, two copied instructions will have the same sequential numbers in different threads.

Load Value Queue (LVQ) Design

We adopt the LVQ design [4] and include it in our design as shown in Fig. 8. Basically, when an LT load fetches data from the cache (or the main memory), the data fetched and the matching tag associated are also buffered into LVQ. Instead of accessing the memory hierarchy, TT loads simply checks and matches the LVQ for the data fetched. In doing so, the TT might reduce the D-cache miss penalties and then improve its performance. It should be noted here that in order to fully benefit from the LT data prefetching, we must guarantee that LT is always ahead of TT, which requires a staggered execution mode.

Dispatch Thread Scheduling

It is well known that there are many idle slots in the execution pipeline. Hence, we must design for the redundant execution to exploit those idle slots as much as possible in order to circumvent the contention which would otherwise degrade performance [33,30].

Modified ICOUNT Policy Applied at the Dispatch Stage

To exploit the idle slots, we must ensure that when one thread is idle for some reason, the execution resources can be promptly allocated to another thread which can utilize those resources more efficiently. The ICOUNT policy was proposed to schedule threads in order to fill IssueQ with issuable instructions, i.e., restrict threads from clogging IssueQ [36]. However, we argue that it is the dispatch stage that directly feeds IssueQ with useful instructions, hence, scheduling threads in the dispatch stage level would cause a prompter reaction to the thread idleness in IssueQ. Therefore, we modify the ICOUNT policy as follows: as illustrated in Fig. 8, at each clock cycle, we count the number of instructions that are still waiting in IssueQ from LT and TT individually. A higher dispatch priority is assigned to the thread with the lower instruction count. To be specific, when the dispatch rate is eight instructions per cycle, the selected thread is allowed to dispatch as many instructions as possible (up to eight). If any dispatch slot is left from the selected thread, the alternative thread would consume the remaining slots. We call the above policy “ICOUNT.2.8.dispatch.”

Staggering Execution

Having shown how to reduce the complexity of TT and made it “light,” a staggered execution mode [4,28] is required. To this end, we developed the “slack dispatch” scheme: in the instruction dispatch stage, if the selected thread is TT, we check the instruction distance between LT and TT. If the distance is less than a predefined threshold, we skip the TT dispatch operation and continue buffering TT in traceQ. This means that the size of traceQ (the number of entries in traceQ) must meet the following requirement:

$$\text{size of (traceQ)} > \text{size of (IFQ)} + \text{predefined distance} \quad (12)$$

Moreover, for purposes of fault-detection, all retired LT instructions and their execution results are buffered into the checking queue (chkQ), as shown in Fig. 8. Hence, TT is responsible for triggering the result comparison. We further assume that the register file of TT is protected by ECC-like mechanisms. This means that, if any fault is detected, the register file state of TT could be used to recover the LT.

DEADLOCK ANALYSIS AND PREVENTION

As pointed out before, the two copies of a thread cooperate with each other for fault checking and recovery. However, if not carefully synchronized, such cooperation could result in deadlocks where neither copy could make any progress [27]. To prevent such situations, a detailed analysis and the introduction of appropriate synchronization mechanisms are necessary.

Deadlock Analysis

In general, resource sharing is one of the underlying causes of deadlocks [37]. In our case, there is much resource sharing between the two thread copies. For example, IssueQ is a shared hardware resource and both thread copies contend for it. The availability of instructions being issued is another type of resource sharing: the issue bandwidth is dynamically partitioned between the two thread copies.

Figure 9 shows the resource allocation graph. TraceQ is a resource which is requested only by LT but is owned only by TT: LT needs entries in traceQ for its instruction fetching and generation of the redundant thread (see Subsect. 6.1). In contrast, chkQ is different in the following sense: only if there is a free entry in chkQ could LT retire its instruction and back up the retiring instructions and execution results there. On the other hand, the entry in chkQ can only be freed by TT: only after an instruction has been retired and compared, can the corresponding entry in chkQ be released. Further, due to the similarity between dispatch and issue operations, we combine them under the term "issue resource" in the discussion which follows.

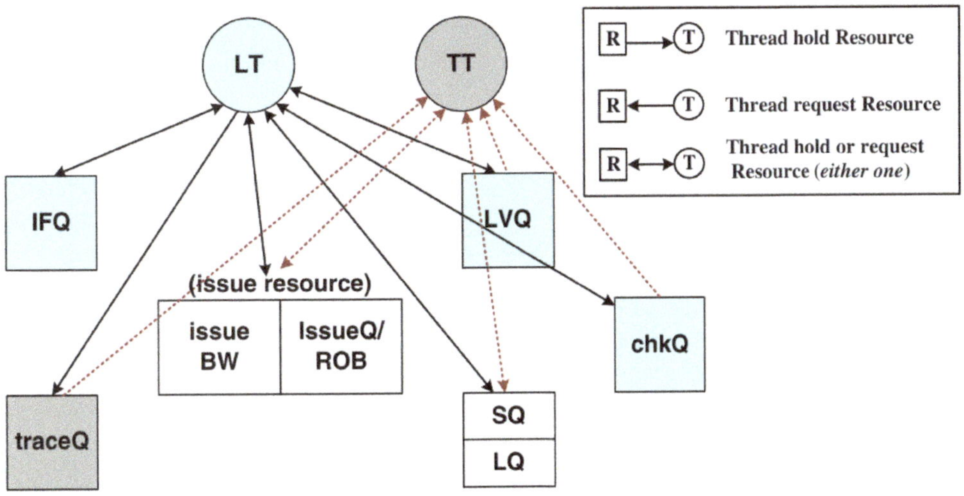

Figure 9. Resource allocation graph for deadlock analysis.

Based on Fig. 9, we can list all possible circular wait conditions. However, some conditions obviously do not end up in a deadlock, for example, "LT → traceQ → TT → SQ → LT." All the possible deadlock scenarios are listed as follows:

LT → chkQ → TT → issue resource → LT

Scenario: When chkQ is full, LT cannot retire its instructions. Then, those instructions which are ready to retire from LT are simply stalled in ROB. If this stalling results in the ROB being full of instructions from LT (note that the situation where the ROB is full of instructions from LT could be exacerbated by the fact that LT is favored by the dispatch thread scheduling policy for the stagger execution mode), the instruction dispatch operation will be blocked, thus, TT will be stalled in traceQ. Consequently, no corresponding instructions from TT can catch up to release the chkQ entries and then a deadlock can happen. In summary, the condition for this deadlock situation is the following:

Observation 1

When chkQ is full and ROB is full of instructions from LT, a deadlock happens.

LT → LVQ → TT → issue resource →LT

Scenario: When LVQ is full, a load instruction from LT cannot proceed with its execution (an LT load instruction must guarantee to successfully buffer loaded data into LVQ to complete its execution). On the one hand, the stalled load instruction could result in the LQ being full of instructions from LT since LT is favored for dispatching. Further, a full LQ will block the instruction dispatch operation, thus, the corresponding load instructions from TT will be unable to catch up to release the LVQ entries and then a deadlock can happen. Hence, a deadlock observation follows:

Observation 2

When LVQ is full and LQ is full of instructions from LT, a deadlock happens. On the other hand, the stalled load instruction could result in a full ROB, thus, the instruction dispatch operation will be blocked. Furthermore, if there is no load instruction from TT in the full ROB, i.e., no load instruction from TT in LQ, no corresponding load instructions from TT will be able to catch up to release the LVQ entries and then a deadlock can happen.

Observation 3

When LVQ is full and ROB is full and there is no load instruction from TT in ROB, a deadlock situation takes place.

LT → SQ → TT → issue resource → LT

Scenario: When SQ is full of instructions from LT, the instruction dispatch operation will be blocked (worse, the stalled store instructions would result in a full ROB which also blocks the dispatch operation). Consequently, no corresponding store instructions from TT can catch up to release the SQ entries and then a deadlock could happen.

Observation 4

When SQ is full of instructions from LT, a deadlock happens.

Deadlock Prevention

Based on the above systematic deadlock analysis, we propose two mechanisms to handle all possible deadlock situations: static hardware resource partitioning and dynamic deadlock monitoring.

Static Hardware Resource Partitioning

In static hardware resource partitioning, i.e., each thread has itself allocated resources, the deadlock conditions identified can be broken such that we can prevent them altogether. For example, we can partition the ROB in order to prevent the possible deadlock situation identified in Observation 1: if some entries of ROB are reserved for TT, TT dispatch operations could continue since when chkQ is full, the partitioned ROB cannot be full of instructions from LT. Subsequently, those dispatched TT instructions will be issued and their execution completed afterwards. After completion, they will trigger the result comparison and then free the corresponding chkQ entries if the operation was found to be fault-free. After some chkQ entries have been freed, LT is allowed to proceed.

Moreover, we find that only three hardware resources (ROB, LQ, and SQ) need to be partitioned in order to prevent all the deadlock situations we identified:

- Partitioning the ROB to break the deadlock situation identified in Observation 1: ROB will never be full of instructions from LT such that TT will be dispatched and then chkQ will be released.
- Similarly, partitioning LQ to break the deadlock situation identified in Observation 2;
- Partitioning SQ to break the deadlock situation identified in Observation 4.

Figure 10 illustrates how the static three-queue partitioning mechanism prevents the deadlock situation identified in Observation 3. When LVQ is full, the load instruction k in LQ of LT cannot be be issued. However, since ROB is now partitioned between LT and TT, the stalled load instruction k in ROB will only block LT from being dispatched. In other words, the TT dispatch operation will not be blocked by the stalled load instruction k, thus, for example, the load instruction i from TT will be dispatched which will then release the LVQ entry occupied by the counterpart load instruction i from LT. Once free LVQ entries are made available, the stalled LT load instruction k can be issued. In summary, we have the following:

Observation 5

For each ROB, LQ, and SQ, allocating some dedicated entries for TT will prevent the deadlock situations identified from occurring.

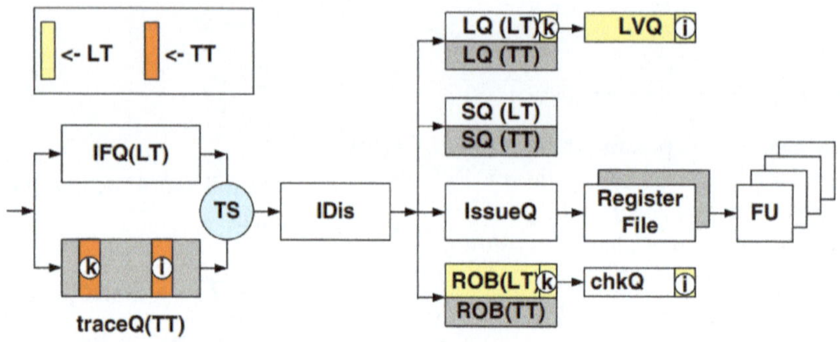

Figure 10. Exemplified static hardware resource partitioning for deadlock prevention.

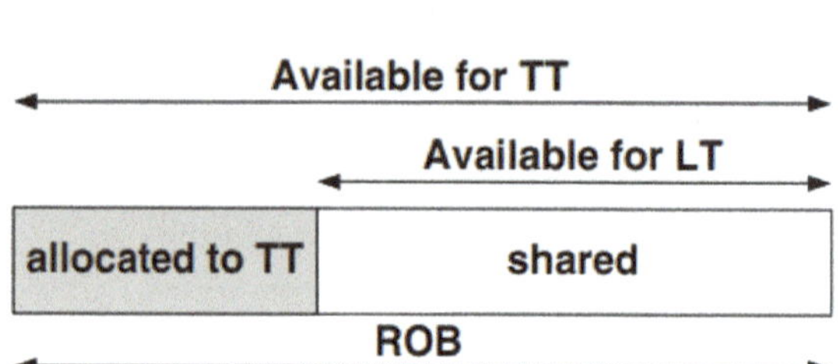

Figure 11. Static ROB partitioning mechanisms.

Algorithm 2: Dispatch thread scheduling policy (with dynamic deadlock monitoring)

Apply ICOUNT.2.8.dispatch policy;
if *selected thread is LT* ***AND*** *IFQ not empty* ***AND*** *no precaution signal* **then**
  Dispatch from IFQ;
else if *distance between LT and TT meets predefined stagger execution mode* ***AND*** *traceQ is not empty* ***AND*** *not an unresolved branch instruction* **then**
  Dispatch from traceQ;
else
  nothing to be dispatched;
end

It should be noted, however, that static hardware resource partitioning has some performance impact on an SMT processor, particularly when partitioning ROB, LQ, and SQ (instruction issue queues for load and store instructions) [38]. Figure 11 shows how the static partitioning is implemented: we allocate a minimum number of entries for TT to prevent deadlocks and the remainder of the queue is shared between LT and TT. Hence, the maximum available entry number for LT is the total number of queue entries minus the number of reserved entries whereas the maximum number of available entries for TT is the total number of queue entries.

Dynamic Deadlock Monitoring

From our deadlock analysis, we also conclude that if we can dynamically regulate the LT progress such that neither ROB nor LQ and SQ could be filled with instructions from LT, the deadlock situations identified can be prevented.

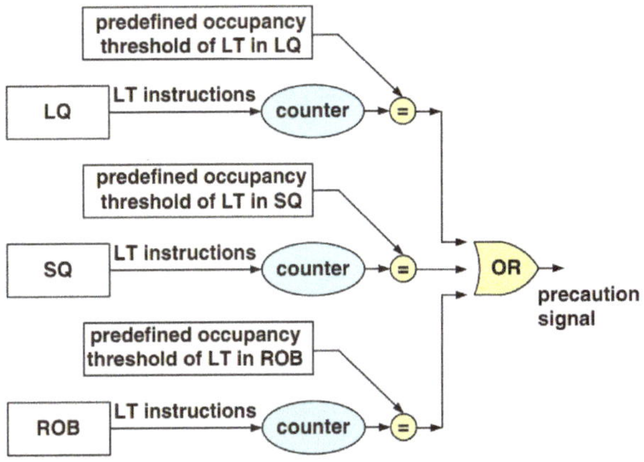

Figure 12. Dynamic monitoring for deadlock prevention.

As illustrated in Fig. 12, instead of statically partitioning hardware resources, we dynamically count the instruction numbers from LT in ROB, LQ, and SQ, respectively, and then a caution signal is generated if at least one of the numbers of instructions counted exceeds the corresponding predefined occupancy threshold. Furthermore, if the caution signal is generated, the dispatch thread scheduling policy will pause LT from being dispatched.

More specifically, the comprehensive dispatch thread scheduling policy we developed is listed in Algorithm 2: first, we apply the ICOUNT.2.8. dispatch policy. If the selected thread is LT, we must then check whether IFQ is empty since no instruction can be dispatched in case of an empty IFQ. Furthermore, we need to make sure no precaution signal has been generated. If there is a such signal, we must stop dispatching from LT. On the other hand, if the selected thread is TT, we check the following conditions before dispatching TT:

- The stagger execution mode requirement;
- Is traceQ not empty;
- Are we not encountering an unresolved branch instruction.

It should be noted that compared with static resource partitioning, the dynamic deadlock monitoring approach comes with a higher design flexibility: by adjusting the predefined occupancy thresholds, we can manipulate the resource allocation between the cooperating threads. However, this flexibility comes at the cost of additional hardware: counters and additional logic, as well as a a more complicated thread scheduling policy

DESIGN OVERHEAD EVALUATION OF TRANSIENT FAULT-TOLERANT SMT PROCESSORS

This section will present our fault-tolerant design overhead evaluation methodology and results. Our main purpose is to present performance numbers and sizing numbers of traceQ and chkQ for the schemes presented in Sect. 6 and Sect. 7.

Table 3. Key design differences between two simulated FT-SMT processors

	SRT-like design	Our design
Redundant thread generation	Fetch twice	Copy fetched instructions
Thread scheduling policy	ICOUNT.2.8	ICOUNT.2.8.dispatch
Deadlock handling	Static hardware partitioning	Dynamic deadlock monitoring
Eliminate TT mispredicted instructions	Using branch outcome queue to prevent being fetched	Using dispatch thread scheduling to prevent being dispatched
Stagger execution mode, chkQ, and LVQ	Implemented	Implemented

Table 4. Simulated FT-SMT processor configuration

Component	Configuration
Processor	2-way SMT, 8-way out-of-order issue, 32-entry IFQ, 256-entry trace queue (256-entry IFQ when simulating superscalar), 64-entry issue queue,128-entry ROB,64-entry load queue, 64-entry store queue, 64-entry LVQ, 256-entry chkQ
Functional unit	6 Int ALU, 2 Int Multiply, 4 FP Add, 2 FP Multiply
Branch predictor	Hybrid: 8K-entry bimodal, 8K-entry gshare, 8K-entry 2-bit selector, 16-entry RAS, 4-way 1K BTB, 10-cycle misprdiction penalty
L1 I- and D-cache	64KB, 32-byte block size, 4-way associative, 1-cycle hit latency
Unified L2 cache	1MB, 64-byte block size, 4-way associative, 6-cycle hit latency
Main memory	100-cycle latency

Performance Overhead Evaluation of Two Fault-Tolerant SMT Processor Designs

We now examine the performance overhead of a fault-tolerant SMT processor. To this end, we have developed an SMT performance simulator based on the SimpleScalar toolset [39]. Moreover, two fault-tolerant SMT processor designs have been simulated. Table 3 shows the key design differences between the two designs: the first one is similar to that discussed by Reinhardt et al. in [4] whereas our design has been presented in Sects. 6 and 7. From a performance overhead viewpoint, the major differences are the way of generating redundant threads and the thread scheduling policy: an SRT-like design fetches a thread twice for redundant execution, adopts an ICOUNT policy [36], and applies the policy in the instruction fetch stage whereas our design copies the fetch instructions to generate the redundant thread, adopts a modified ICOUNT policy, and applies this in the instruction dispatch stage (details in Subsect. 6.3.1). Table 4 summarizes the key parameters of the processors simulated. The others microarchitectural configuration parameters such as fetch bandwidth, function units, L1 I- and D- caches, unified L2-cache, and main memory, are similar to the ones used in [4].

Table 5. SPEC benchmarks, input data sets and IPCs in superscalar execution mode

Benchmark	Version	Input	IPC
099.go	CINT95	test	1.11
132.ijpeg	CINT95	test	2.31
164.gzip	CINT2000	lgred.graphic	1.84
197.parser	CINT2000	lgred.in	1.51
171.swim	CFP2000	test	3.0
179.art	CFP2000	lgred.in	0.72
256.bzip2	CINT2000	lgred.source	2.16
130.li	CINT95	test	1.88
134.perl	CINT95	lgred.makerand.pl	1.87
175.vpr	CINT2000	lgred	1.65
101.tomcatv	CFP95	test	2.64
177.mesa	CFP2000	lgred.in	3.40
183.equake	CFP2000	lgred.in	1.70

We evaluated our designs using 13 SPEC benchmarks shown in Table 5. Those SPEC benchmarks characteristics and their IPCs in the superscalar execution mode are also shown. Furthermore, those input data sets that are not marked as "test" are from MinneSPEC [40], whereas those marked as "test" input data sets have been obtained directly from SPEC. In addition, we used the compiler optimization level "-O2" to compile the SPEC benchmarks executables. Moreover, when simulating, we fast-forwarded past the first 300 million instructions (during that time, the IPC is extremely unstable) for each thread and then simulated all the remaining instructions for each threads.

Moreover, we need to point out that using IPC to evaluate the fault-tolerant SMT performance is appropriate: in the fault-tolerant multi-threading execution mode, two copies of the same thread actually carry only a single computation task, i.e., from the user's viewpoint, they function as if only one thread was executing. On the contrary, in the conventional multi-threading execution mode, two threads are typically independent, hence the executions carry out two tasks. In this sense, the fault-tolerant multi-threading which is running two copies of the same thread is functionally comparable to the superscalar where IPC is conventionally used to evaluate the performance. Furthermore, in our fault-tolerant multi-threading simulations, we computed IPCs as the retired and then result-compared instructions numbers from the leading thread divided by the total simulated clock cycle numbers.

The simulation results are shown in Fig. 13. Compared with the superscalar case (without fault-tolerance features), the performance overhead of our design is lowered by only 12% on average whereas an SRT-like design has 17% overhead. Further, if we choose the baseline as the two-way SMT without any fault-tolerance features (the baseline SMT processor is executing two identical threads), the performance overhead is 34% on average whereas the overhead of SRT-like design is 42%. Also, note that in Fig. 13, there are benchmarks that show that the IPCs of "superscalar" (without FT) is lower than those of "our design." [4] reported similar cases that for some benchmarks, the fault-tolerant SMT implementations displayed higher throughput than the conventional SMT without the fault-tolerant features. The authors suggested that the reason for this apparently odd behavior could be the branch target buffer prefetching that limited the TT branch misprediction penalties.

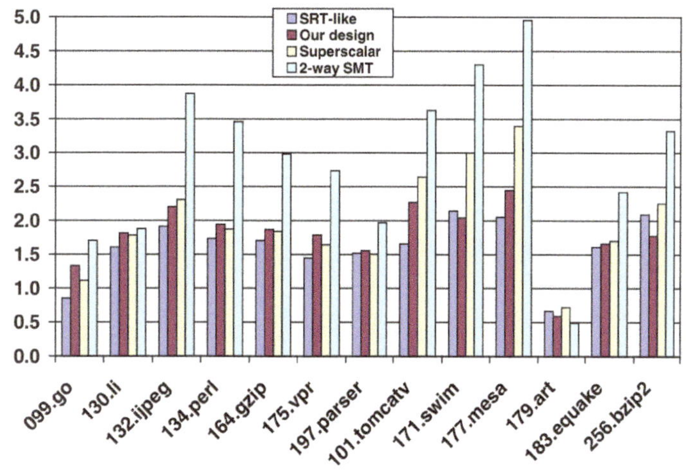

Figure 13. IPC comparison of two FT-SMT processors, superscalar, and two-way SMT

In our designs, there are two reasons: first, because of the LVQ design, the LT prefetches the data for the TT, which improves TT execution. However, the data prefetching has not been supported in "superscalar" and "2-way SMT." Secondly, because of the copying of fetched instructions to generate the redundant thread, only LT needs to access the instruction caches and experience instruction cache misses. However, in conventional designs such as "superscalar" and "2-way SMT," all active threads fetch instructions on their own.

Hardware Overhead Evaluation

This subsection examines two queues which are introduced for the fault-tolerant schemes we have designed: traceQ and chkQ. We want to evaluate how their sizes affect the fault-tolerant performance overhead. To this end, using the performance simulators we developed, we varied the size of the queue and then observed the pipeline throughput to determine the optimal design points.

Sizing TraceQ

TraceQ functions as a buffer holding the TT instructions to support the staggered execution mode. However, from Eq. 12, we know that to support the staggered execution mode, the minimum size of traceQ must be larger than the size of IFQ plus the predefined instruction distance. Hence when choosing the size of traceQ as 32-entry, the staggered execution mode is not supported whereas when traceQ is a 64-entry or 128-entry unit, the predefined instruction distance is 32-instruction.

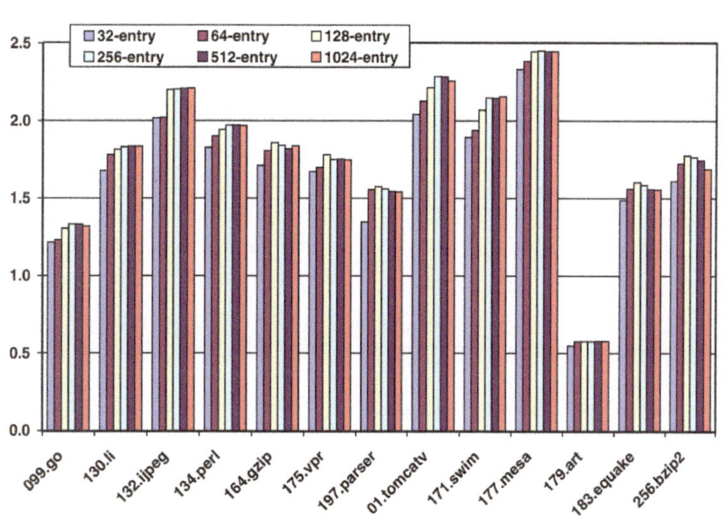

Figure 14. IPC comparison of varying size of traceQ (IFQ: 32-entry, chkQ: 256-entry).

Likewise, when traceQ has more than 256 entries, the predefined instruction distance is 128 instructions. Figure 14 shows the simulation results. Obviously, when traceQ is small, the benefit of the staggered execution mode cannot be fully exploited, and the IPC is degraded: compared with a 256-entry traceQ, IPC in those cases is lowered by 9% on average. On the other hand, if we continued increasing traceQ, the IPC improvement would be limited: on average, there would be no significant IPC improvement. Therefore, we choose a 256-entry traceQ as our design point. Furthermore, each entry of traceQ holds an instruction word (32-bit as typical size), a thread ID (1-bit), and a branch resolve status bit (1-bit). Hence, the storage core of traceQ is : 256 × (32+1+1) = 1088 bytes.

Sizing chkQ

Figure 15 shows the simulation results when we vary the size of chkQ, given the size of traceQ is 256-entry and the predefined instruction distance is 128-instruction. We can see that increasing the size of chkQ has diminishing throughput return: beyond a 64-entry chkQ, no significant improvement can be observed. The reason for that is because chkQ is buffering those retired (but waiting for comparison) LT instructions and because TT is reduced such that TT can catch up with LT more after being dispatched. Hence, chkQ is not frequently full. In other words, chkQ can be made to be smaller than traceQ. Hence, we choose a 64-entry chkQ. In addition, it should be noted that each entry of chkQ holds sequential numbers (10-bit, for instance), instruction word (32-bit), execution results (32-bit, for instance), and status bit (1-bit). Hence, the storage core of chkQ is: 64 × (10 + 32 + 32 + 1) = 600 bytes.

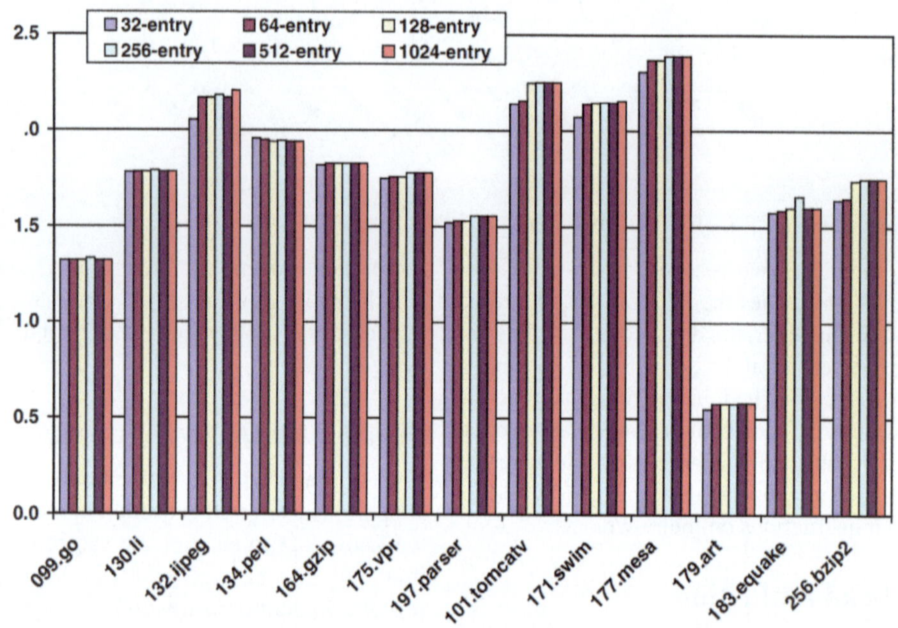

Figure 15. IPC comparison of varying size of chkQ (traceQ: 256-entry, predefined instruction distance: 128-instruction).

CONCLUSIONS

We have presented two architecture mechanisms to protect microprocessors against the effect of radiation. First, we have observed that many faults manifest themselves as control flow errors that cause the processor to violate the correct sequencing of instructions. To detect such control flow errors, we started from a basic compile-time signature assignment algorithm. Then, approaches to improving the fault detection coverage of the basic algorithm were discussed. Moreover, to allow the processor to efficiently check the run-time sequence and detect control flow errors, an on-chip assigned-signature checker was designed: this checker is capable of executing three additional instructions (SIC, SIJ, SIJC). The algorithm of embedding three additional instructions into programs were also described.

Second, we realized that since the very concept of SMT inherently provides redundancy, we could run two copies of the same thread on top of SMT platforms in order to detect and correct soft errors. This allows, upon detection of an error, the rolling back of the processor state to a known safe point, and then a retry of the instructions, thereby effecting a completely error-free execution. We have discussed two crucial implementation issues introduced by this basic scheme: (1) the design trade-off between the fault detection coverage versus design costs; (2) the possible occurrence of deadlock situations.

Algorithm 3: Embedding three additional instructions into programs

```
Derive CFG of the given program;
Assuming we have nodes: V_i (i=1,2,...,N) where N is the total number of nodes in the CFG;
Assign a unique state code D(i) to every node V_i;
Apply Two-Pass primary node selection algorithm;
for every node V_i in the CFG do
    if V_i is not an MBI node then
        Compute its assigned reference signature as: S(i) = D(i) ⊕ D(pred(V_i));
        Place the instruction SIC imm1, imm2 at the beginning of node V_i and before the
        SIJ instruction, if any;
        Assign the values of imm1 and imm2 as: imm1=S(i) and imm2=D(i);
    else
        /* V_i is an MBI node
        (assume V_j is selected as its primary node);
        Assign the reference signature of V_i as: S(i) = D(i) ⊕ D(j);
        Place the instruction SIJC imm1, imm2 at the beginning of node V_i and before the
        SIJ instruction, if any;
        Assign the values of imm1 and imm2 as: imm1=S(i), imm2=D(i);
        for every node V_k ∈ 𝕊 (including V_j) do
            Place instruction SIJ imm1, imm2 into the node V_k and after the SIC and/or
            SIJC instructions;
            Assign the values of imm1 and imm2 as follows;
            if V_k → V_i is a taken path then
                imm1=X, imm2=D(k) ⊕ D(j);
            else if V_k → V_i is a not-taken path then
                imm1=D(k) ⊕ D(j), imm2=X;
            else
                /* V_k → V_i is not a conditional branch path
                imm1=D(k) ⊕ D(j), imm2=X;
            end
        end
    end
end
```

To achieve the largest possible fault detection coverage, we replicate the instructions fetched so as to generate the redundant thread copies. Further, we apply the SMT thread scheduling at the level of the instruction dispatch stage to lower the performance overhead. As a result, when compared to the baseline processor, our simulation results show that by using our two new schemes, the performance overhead can be reduced down to 34% on average, down from 42%. Finally, in the fault-tolerant execution mode, since the two copied threads are cooperating with each other, deadlock situations could be quite common. We thus presented a detailed deadlock analysis and then concluded that allocating some entries of ROB, LQ, and SQ for the trailing thread would be sufficient to prevent such deadlocks.

In terms of future research, we would like to design a simulation tool which would allow the random injection of faults into the microprocessor model to evaluate the effectiveness of our proposed protection schemes.

REFERENCES

1. Hennessy J.L., Patterson D.A.: Computer architecture: a quantitative approach. 3rd edn. Morgan Kaufmann Publishers, Inc. (2002)
2. Borkar, S.: Design challenges of technology scaling. IEEE Micro. (1999)
3. Yang, P. Chern, J.-H.: Design for reliability: the major challenge for VLSI. Proceedings of the IEEE (1993)
4. Reinhardt, S.K., Mukherjee, S.S.: Transient fault detection via simultaneous multithreading. In: 27th international symposium on computer architecture (2000)
5. Hennessy, J.: The future of systems research. IEEE Comput. (1999)
6. Stackhouse, B., Bhimji, S., et al.: A 65 nm 2-billion transistor quad-core itanium processor. IEEE Trans. Solid-State Circuits (2009)
7. Quach, N.: High Availability and reliability in the itanium processor. IEEE Micro. (2000)
8. Sanda, P.N., Kellington, J.W., Kudva, P., Kalla, R., McBeth, R.B., Ackaret, J., Lockwood, R., Schumann, J., Jones, C.R.: Soft-error resilience of the IBM POWER6 processor. IBM J. Res. Dev. (2008)
9. Clarke, W.J., Alves, L.C., Dell, T.J., Elfering, H., Kubala, J.P., Lin, C., Mueller, M.J., Werner, K.: IBM System z10 design for RAS. IBM J. Res. Dev. (2009)

10. Ando, H., Yoshida, Y., Inoue, A., Sugiyama, I., Asakawa, T., Morita, K., Muta, T., Motokurumada, T., Okada, S., Yamashita, H., Satsukawa, Y., Konmoto, A., Yamashita, R., Sugiyama, H.: A 1.3-GHz Fifth-generation SPARC64 Microprocessor. IEEE J. Solid-State Circuits (2003)
11. Intel Corporation, (Santa Clara): IA-32 intel architecture software developer's manuals (2006)
12. Wilken, K., Shen, J.P.: Continuous signature monitoring: low-cost concurrent-detection of processor control errors. IEEE Trans. Comput. Aided Des. (1990)
13. Ohlsson, J., Rimén, M., Gunneflo, U.: A study of the effects of transient fault injection into a 32-bit RISC with built-in watchdog. In: 29th international symposium on fault-tolerant computing (1991)
14. Schuette, M.A., Shen, J.P.: Processor control flow monitoring using signatured instruction streams. IEEE Trans. Comput. (1987)
15. Mohmood, A., McCluskey, E.J.: Concurrent error detection using watchdog processors—a survey. IEEE Trans. Comput. (1988)
16. Schuette, M.A., Shen, J.P.: Exploiting instruction-level parallelism for integrated control-flow checking. IEEE Trans. Comput. (1994)
17. Warter, N.J., Hwu, W.-M.W.: A software based approach to achieving optimal performance for signature control flow checking. 20th international symposium on fault-tolerant computing (1990)
18. Michel, T., Leveugle, R., Saucier, G.: A new approach to control flow checking without program modification. In: 21st international symposium on fault-tolerant computing (1991)
19. Alkhalifa, Z., Nair, S., Krishnamurthy, N., Abraham, J.A.: Design and evaluation of system-level checks for on-line control flow error detection. IEEE Trans. Parallel Distrib. Syst. (1999)
20. Shirvani, P.P., McCluskey, E.J.: Fault-tolerant systems in a space environment: The CRC ARGOS Project. Tech. Rep. CRC-TR 98-2, Stanford University (1998)
21. Bagchi, S., Srinivasan, B., Whisnant, K., Kalbarczyk, Z., Iyer, R.K.: Hierarchical error detection in a software implemented fault tolerance (SIFT) environment. IEEE Trans. Knowl. Data Eng. (2000)
22. Oh, N., Shirvani, P.P., McCluskey, E.J.: Control-flow checking by software signatures. IEEE Trans. Reliab. (2002)
23. Aho A.V., Sethi R., Ullman J.D.: Compilers: Principles, techniques, and tools. Addison-Wesley Publishing Company, Wokingham, UK (1986)
24. Borin, E., Wang, C., Wu, Y., Araujo, G.: Dynamic binary control-flow errors detecttion. ACM SIGARCH Computer Architecture News (2005)
25. Saxena, N.R., McCluskey, E.J. Dependable adaptive computing systems- the ROAR project. In: 1998 IEEE international conference on systems, man and cybernetics (1998)
26. Rotenberg, E.: AR-SMT: A microarchitectural approach to fault tolerance in microprocessors. In: 29th international symposium on fault-tolerant computing (1999)
27. Mukherjee, S.S., Kontz, M., Reinhardt, S.K.: Detailed design and evaluation of redundant multithreading alternatives. In: 29th international symposium on computer architecture (2002)
28. Vijaykumar, T.N., Pomeranz, I., Cheng, K.: Transient-fault recovery using simultaneous multithreading. In: 29th international symposium on computer architecture (2002)
29. Ray, J., Hoe, J.C., Falsafi, B.: Dual use of superscalar datapath for transient-fault detection and recovery. In: 34th international symposium on microarchitecture (2001)
30. Smolens, J.C., Kim, J., Hoe, J.C., Falsafi, B.: Efficient resource sharing in concurrent error detecting superscalar microarchitectures. In: 37th international symposium on microarchitecture (2004)
31. Bossen, D.C., Tendler, J.M., Reick, K.: Power4 system design for high reliability. IEEE Micro. (2002)
32. Mukherjee, S.S., Weaver, C., Emer, J., Reinhardt, S.K., Austin, T.: A systematic methodology to compute the architectural vulnerability factors for a high-performance microprocessor. In: 36th international symposium on microarchitecture (2003)
33. Mendelson, A., Suri, N.: Designing high-performance & reliable superscalar architectures the out of order reliable superscalar (O3RS) approach. In: International conference on dependable systems and networks (2000)
34. Kang, D., Gaudiot, J.-L.: Speculation control for simultaneous multithreading. In: 18th international parallel and distributed processing symposium (2004)
35. Compaq Computer Co., Massachusetts: Alpha 21264/EV68CB and 21264/EV68DC Hardware Reference Manual, 1.1 ed. (2001)
36. Tullsen, D.M., Eggers, S.J., Levy, H.M.: Simultaneous multithreading: maximizing on-chip parallelism. In: 22nd international symposium on computer architecture (1995)
37. Silberschatz A., Galvin P.B., Gagne G.: Applied operating system concepts. John Wiley & Sons, Inc. (2000)
38. Raasch, S.E., Reinhardt, S.K.: The impact of resource partitioning on SMT processors. In: 12th international conference on parallel architectures and compilation techniques (2003)
39. Burger, D., Austin, T.M.: The SimpleScalar Tool Set, Version 2.0. Tech. Rep. 1342, University of Wisconsin-Madison Computer Sciences Department (1997)
40. KleinOsowski, A., Lilja D.J.: MinneSPEC: a new SPEC benchmark workload for simulation-based computer architecture research. Tech. Rep. ARCTiC Lab No. 02–08, University of Minnesota, Minneapolis (2002)

CHAPTER 4

VIRTUAL PROTOTYPING AND PERFORMANCE ANALYSIS OF TWO MEMORY ARCHITECTURES

Huda S Muhammad and Assim Sagahyroon

INTRODUCTION

Due to the rapid advances in circuit integration technology, and to optimize performance while maintaining acceptable levels of energy efficiency and reliability, multicore technology or Chip-Multiprocessor is becoming the technology of choice for microprocessor designers. Multicore processors provide increased total computational capability on a single chip without requiring a complex microarchitecure. As a result, simple multicore processors have better performance per watt and area characteristics than complex single core processors [1].

A multicore architecture has a single processor package that contains two or more processors. All cores can execute instructions independently and simultaneously. The operating system will treat each of the execution cores as a discrete processor. The design and integration of such processors with transistor counts in the millions poses a challenge to designers given the complexity of the task and the time to market constraints. Hence, early virtual system prototyping and performance analysis provides designers with critical information that can be used to evaluate various architectural approaches, functionality, and processing requirements.

In these emerging multicore architecture, the ability to analyze (at an early stage) the performance of the memory subsystem is of extreme importance to designers. The latency resulting by the access of different levels of memory reduces the processing speeds causing more processor stalls while the data/instruction is being fetched from the main memory. Ways in which multiple cores send and receive data to the main memory greatly affect the access time and thus the processing speed. In multicore processors, two approaches to memory subsystem design have emerged in recent years, namely, the AMD DirectConnect architecture and the Intel Shared Bus architecture [2–5]. In the DirectConnect architecture, a processor is directly connected to a pool of memory using an integrated memory controller. A processor can access the other processors' memory pool via a dedicated processor-to-processor interconnect. On the other hand, in Intel's dual-core designs, a single shared pool of memory is at the heart of the memory subsystem. All processors access the pool via an external front-side bus and a memory controller hub.

In this work, virtual system prototyping is used to study the performance of these alternatives. A virtual systems prototype is a software-simulation-based, timing-accurate, electronic systems level (ESL) model, used first at the architectural level and then as an executable golden reference model throughout the design cycle. Virtual systems prototyping enables developers to accurately and efficiently make the painful tradeoffs between that quarrelling family of design siblings functionality, flexibility, performance, power consumption, quality, cost, and so forth.

Virtual prototyping can be used early in the development process to better understand hardware and software partitioning decisions and determine throughput considerations associated with implementations. Early use of functional models to determine microprocessor hardware configurations and architectures, and the architecture of ASIC in development, can aid in capturing requirements, improving functional performance and expectations [6].

In this work, we explore the performance of the two memory architectures introduced earlier using virtual prototyping models built from parameterized library components which are part of the VisualSim Environment [7]. Essentially, VisualSim

Citation: Muhammad, H.S. and Sagahyroon, A (2010). Virtual prototyping and performance analysis of two memory architectures, DOI:10.1155/2009/984891.

is a modeling and simulation CAD tool used to study, analyze, and validate specification and verify implementation at early stages of the design cycle.

This paper is organized as follows: in Section 2 we provide an overview of the two processors and the corresponding memory architectures. Section 3 introduces the VisualSim environment as well as the creation of the platform models for the processors. Simulation Results and the analysis of these results form Section 4 of this paper. Conclusions are summarized in Section 5.

OVERVIEW OF PROCESSORS MEMORY ARCHITECTURE

The AMD Opteron Direct Connect Architecture

The AMD's direct Connect Architecture used in the design of the dual core AMD Opteron consists of three elements:

- An integrated memory controller within each processor, which connects the processor cores to dedicated memory,
- A high-bandwidth Hyper Transport Technology link which goes out the computer's I/O devices, such as PCI controllers,
- Coherent Hyper Transport Technology links which allow one processor to access another processor's memory controller and Hyper Transport Technology links.

The Opteron uses an innovative routing switch and a direct connect architecture that allows "glueless" multiprocessing between the two processor cores. Figure 1 shows an Opteron processor along with the system request queue (SRQ) and host bridge, Crossbar, memory controller, DRAM controller, and HyperTransport ports [3, 8].

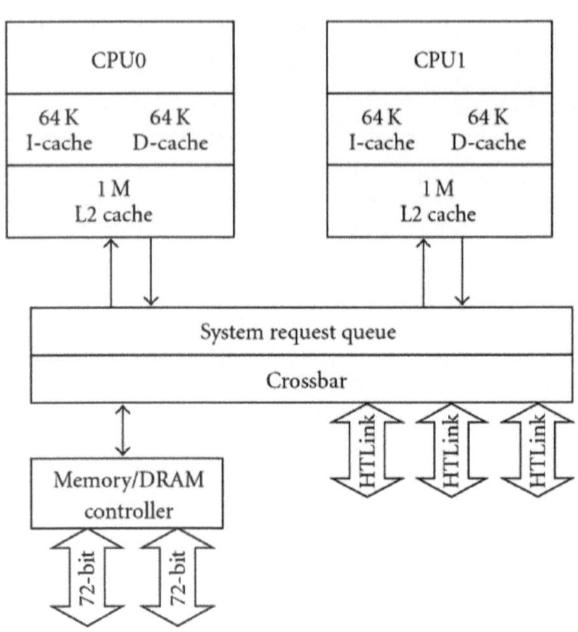

Figure 1. AMD dual core Opteron.

The Crossbar switch and the SRQ are connected to the cores directly and run at the processor core frequency. After an L1 cache miss, the processor core sends a request to the main memory and the L2 cache in parallel. The main memory request is discarded in case of an L2 cache hit. An L2 cache miss results in the request being sent to the main memory via the SRQ and the Crossbar switch. The SRQ maps the request to the nodes that connect the processor to the destination. The Crossbar switch routes the request/data to the destination node or the HyperTransport port in case of an off chip access.

Each Opteron core has a local on-chip L1 and L2 cache and is then connected to the memory controller via the SRQ and the Crossbar switch. Apart from these external components, the core consists of 3 integer and 3 floating point units along with a load/store unit that executes any load or store microinstructions sent to the core [9]. Direct Connect Architecture can improve overall system performance and efficiency by eliminating traditional bottlenecks inherent in legacy architectures. Legacy front-side buses restrict and interrupt the flow of data. Slower data flow means slower system performance. Interrupted data flow means reduced system scalability. With Direct Connect Architecture, there are no front-side buses. Instead, the processors, memory controller, and I/O are directly connected to the CPU and communicate at CPU speed [10].

Intel Xeon Memory Architecture

The Dual-Core Intel Xeon Processor is a 64-bit processor that uses two physical Intel NetBurst microarchitecture cores in one chip [4]. The Intel Xeon dual core processor uses a different memory access technique, by including a Front-Side-Bus (FSB) to the SDRAM and a shared L3 cache instead of having only on-chip caches like the AMD Opteron. The L3 cache and the two cores of the processor are connected to the FSB via the Caching Bus Controller. This controller controls all the accesses to the L3 cache and the SDRAM. Figure 2 below provides an overview of the Intel Dual core Xeon and illustrates the main connections in the processor [5].

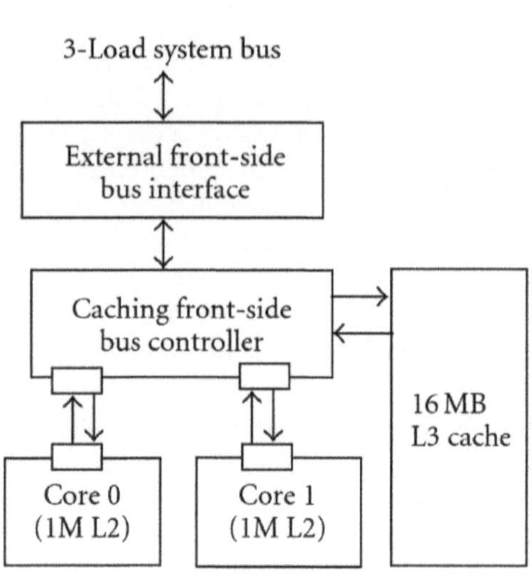

Figure 2. Block diagram of the dual-core Intel Xeon.

Since the L3 cache is shared, each core is able to access almost all of the cache and thus has access to a larger amount of cache memory. The shared L3 cache provides a better efficiency over a split cache since each core can now use more than half of the total cache. It also avoids the coherency traffic between cache in a split approach [11].

VISUALSIM SIMULATION ENVIRONMENT

At the heart of the simulation environment is the VisualSim Architect tool. It is a graphical modeling tool that allows the design and analysis of "digital, embedded, software, imaging, protocols, analog, control-systems, and DSP designs". It has features that allow quick debugging with a GUI and a software library that includes various tools to track the inputs/stimuli and enable a graphical and textual view of the results. It is based on a library of parameterized components including processors, memory controllers, DMA, buses, switches, and I/O's. The blocks included in the library reduce the time spent on designing the minute details of a system and instead provide a user friendly interface where these details can be altered by just changing their values and not the connections. Using this library of building blocks, a designer can for example, construct a specification level model of a system containing multiple processors, memories, sensors, and buses [12].

In VisualSim, a platform model consists of behavior, or pure functionality, mapped to architectural elements of the platform model. A block diagram of a platform model is shown in Figure 3.

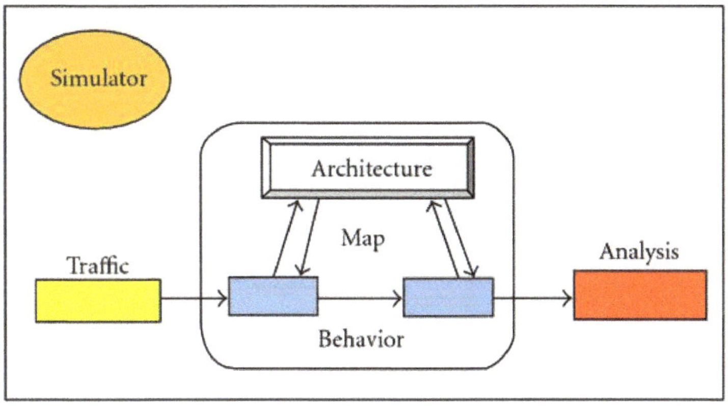

Figure 3. Block diagram of a platform model.

Once a model is constructed, various scenarios can be explored using simulation. Parameters such as inputs, data rates, memory hierarchies, and speed can be varied and by analyzing simulation results engineers can study the various trade-offs until they reach an optimal solution or an optimized design.

The key advantage of the platform model is that the behavior algorithms may be upgraded without affecting the architecture they execute on. In addition, the architecture could be changed to a completely different processor to see the effect on the user's algorithm, simply by changing the mapping of behavior to architecture. The mapping is just a field name (string) in a data structure transiting the model.

Models of computation in VisualSim support block-oriented design. Components called blocks execute and communicate with other blocks in a model. Each block has a well-defined interface. This interface abstracts the internal state and behavior of a block and restricts how a block interacts with its environment. Central to this block-oriented design are the communication channels that pass data from one port to another according to some messaging scheme. The use of channels to mediate communication implies that blocks interact only with the channels they are connected to and not directly with other blocks.

In VisualSim, the simulation flow can be explained as follows: the simulator translates the graphical depiction of the system into a form suitable for simulation execution and executes simulation of the system model, using user specified model parameters for simulation iteration. During simulation, source modules (such as traffic generators) generate data structures. The data structures flow along to various other processing blocks, which may alter the contents of Data Structures and/or modify their path through the block diagram. In VisualSim simulation continues until there are no more data structures in the system or the simulation clock reaches a specified stop time [7]. During a simulation run, VisualSim collects performance data at any point in the model using a variety of prebuilt probes to compute a variety of statistics on performance measures.

This project uses the VisualSim (VS) Architect tool, to carry out all the simulations and run the benchmarks on the modeled architectures. The work presented here utilizes the hardware architecture library of VS that includes the processor cores, which can be configured as per our requirements, as well as bus ports, controllers, and memory blocks.

Models' Construction (Systems' Setup)

The platform models for the two processors are constructed within the VS environment using the parameters specified in Table 1.

Table 1. Simulation models parameters.

	AMD Opteron	Intel Xeon
Core Speed	2 GHz	2 GHz
$Bus_{(Core\text{-}to\text{-}Cache)}$ Speed	N/A	2 GHz
$Bus_{(Core\text{-}to\text{-}SRQ)}$ Speed	2 GHz	N/A
Crossbar Speed	2 GHz	N/A
FSB Speed	N/A	1066 MHz (Width = 4 B)
$Bus_{(SRQ\text{-}to\text{-}RAM)}$ Speed	2 GHz (Width = 4 B)	N/A
L1 Cache Speed	1 GHz	1 GHz
L2 Cache Speed	1 GHz	1 GHz
RAM Speed	1638.4 MHz	1638.4 MHz
I1 Cache Size	64 kB	64 kB
D1 Cache Size	64 kB	64 kB
L2 Cache Size	4 MB (2 MB per core)	4 MB (shared cache)

The Opteron Model

The basic architecture of the simulated AMD dual core Opteron contains two cores with three integer execution units, three floating point units and two loads/stores, and branch units to the data cache [2]. Moreover, the cores contain 2 cache levels with 64 kB of L1 data cache, 64 kB of L1 instruction cache, and 1 MB of L2 cache.

The constructed platform model for the AMD Opteron is shown in Figure 4.

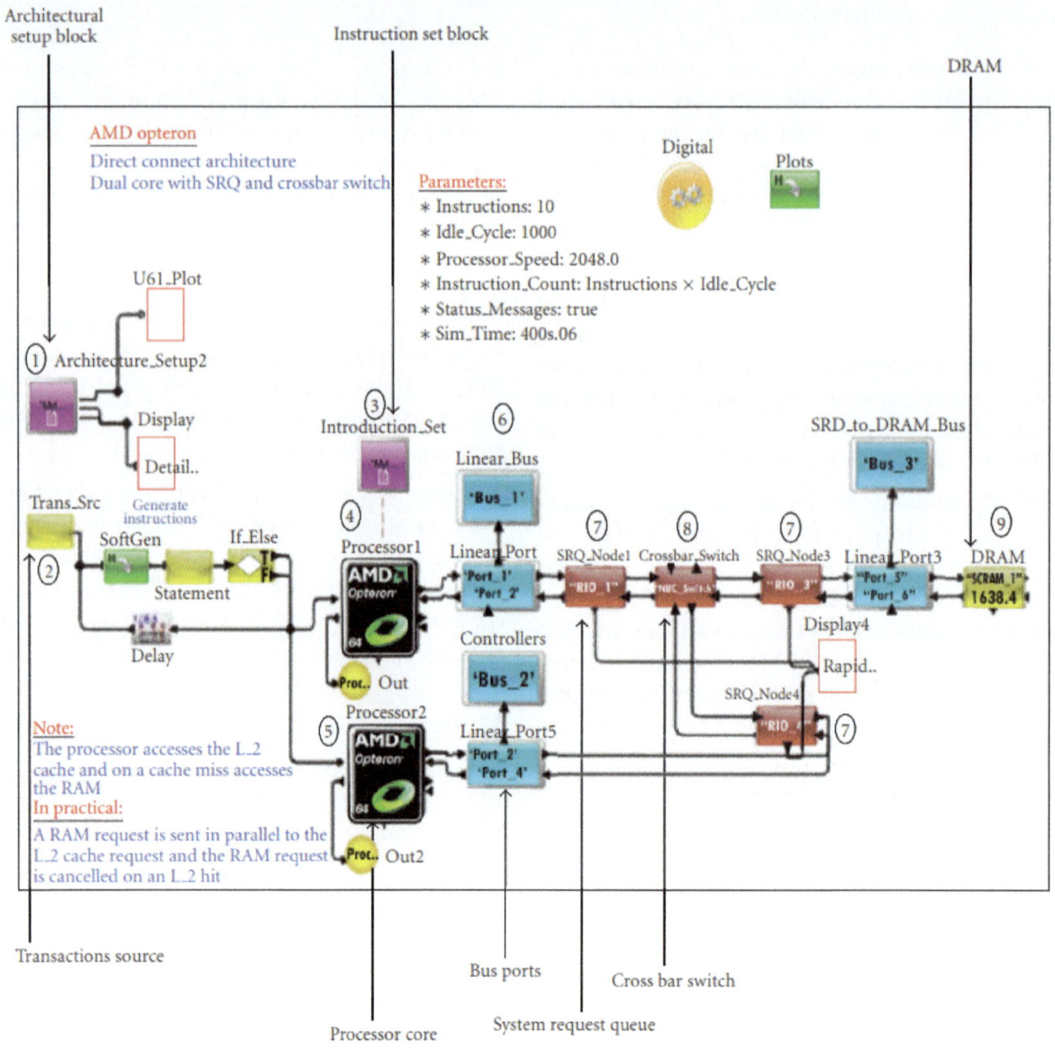

Figure 4. Platform model for the DirectConnect Architecture.

In the above model, the two large blocks numbered 4 and 5, respectively, are the Processor cores connected via bus ports (blocks 6) to the System Request Queue (block 7), and then to the Crossbar switch (block 8). The Crossbar switch connects the cores to the RAM (block 9) and is programmed to route the incoming data structure to the specified destination and then send the reply back to the requesting core.

On the left block 2 components contain the input task to the two cores. These blocks define software tasks (benchmarks represented as a certain mix of floating point, integer and load/store instructions) that are input to both the processors (Opteron and Xeon) in order to test their memory hierarchy performance. The following subsections give a detailed description of each of the blocks, their functionalities, and any simplifying assumptions made to model the memory architecture.

Architecture Setup

The architecture setup block configures the complete set of blocks linked to a single Architecture_Name parameter found in most blocks. The architecture setup block of the model (block 1) contains the details of the connections between the fields mappings of the Data Structure attributes as well as the routing table that contains any of the virtual connections not wired in the model. The architecture setup also keeps track of all the units that are a part of the model and its name has to be entered into each block that is a part of the model.

Core and Cache

Each core of Opteron implemented in the project using VS is configured to a frequency of 2 GHz and has 128 kB of L1 cache (64 kB data and 64 kB instruction cache), 2 MB of L2 cache, and the floating point, integer, and load/store execution units. This 2 MB of L2 cache per core is compatible with the 4 MB of shared cache used in the Intel Xeon memory architecture. The instruction queue length is set to 6 and instructions are included in the instruction set of both the cores, so as to make the memory access comparison void of all other differences in the architectures. These instructions are defined in the instruction block that is further described in a later section.

Certain simplifications have been made to the core of the Opteron in order to focus the analysis entirely on the memory architecture of the processor. These assumptions include the change of the variable length instructions to fixed length micro-ops [9]. Another assumption made is that any L1 cache miss does not result in a simultaneous request being sent to the L2 cache and the RAM. Instead the requests are sent sequentially, where an L1 cache miss results in an L2 cache access and finally an L2 cache miss results in a DRAM access.

Pipeline

The pipeline of the modeled Opteron consists of four stages, mainly the prefetch, decode, execute, and the store. The prefetch of each instruction begins from the L1 cache and ends in a DRAM access in case of L1 and L2 cache misses. The second stage in the pipeline is the decode stage that is implemented by introducing a delay into the entire process. The decode stage does not actually decode the instruction; instead the time required to decode the instruction is added to the time comprising of the delays from the prefetch stage to the end of the execution stage. The third stage, the execution stage, takes place in the five execution units that are present in the cores, and finally after the execution, the write-back stage writes back the specified data to the memory, mainly the L1 cache. The *Pipeline_Stages* (text box in Figure 5) shows the four pipeline stages that have been defined for both cores. It also contains the configuration of one of the cores of the Opteron along with the number of execution units and the instruction queue length. The lower text window depicts the details and actions of the pipeline stages.

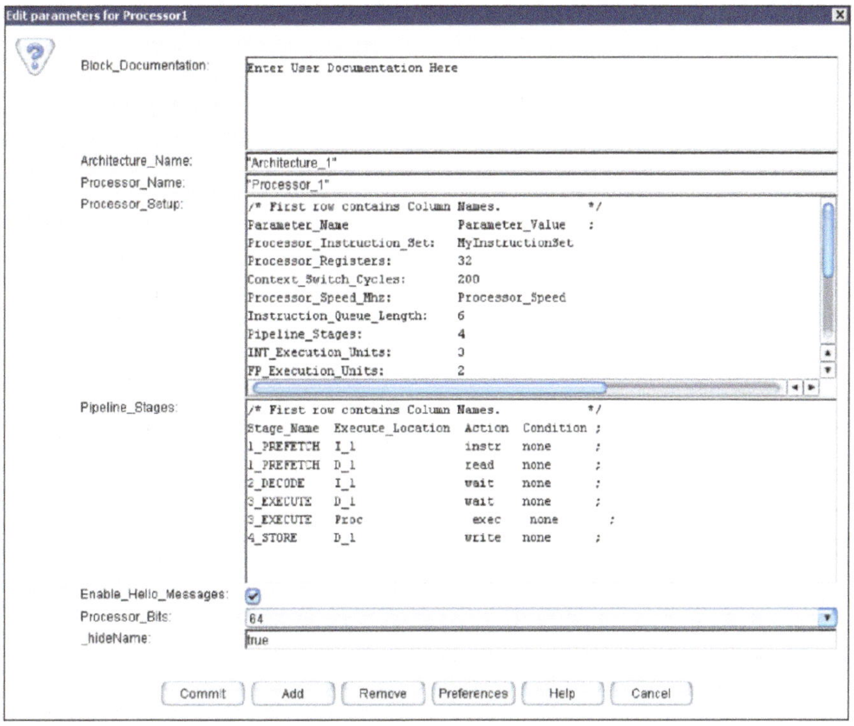

Figure 5. Opteron processor core and pipeline configurations.

Crossbar Switch and the SRQ Blocks

The Crossbar switch configured in the simulation model, is used to route the data packets to the destination specified by the *"A_Destination"* field of the data structure entering the switch. The main memory is connected to the Crossbar switch via the System Request Queue (SRQ) block both of which are implemented using the virtual machine scripting language available in the VisualSim environment. The SRQ accepts only 8 requests in the queue and does not entertain any further requests until there is an empty space in the queue. Each core and the SDRAM are connected to individual SRQ blocks that are in turn connected to the crossbar switch. Figure 6shows the crossbar switch as the *NBC_Switch* and the SRQ blocks as the *RIO_IO_Nodes* which are linked to the bus ports connected to the SDRAM and the processor cores. Figure 4 provides a general overview of the crossbar switch and the SRQ nodes in context of the entire model.

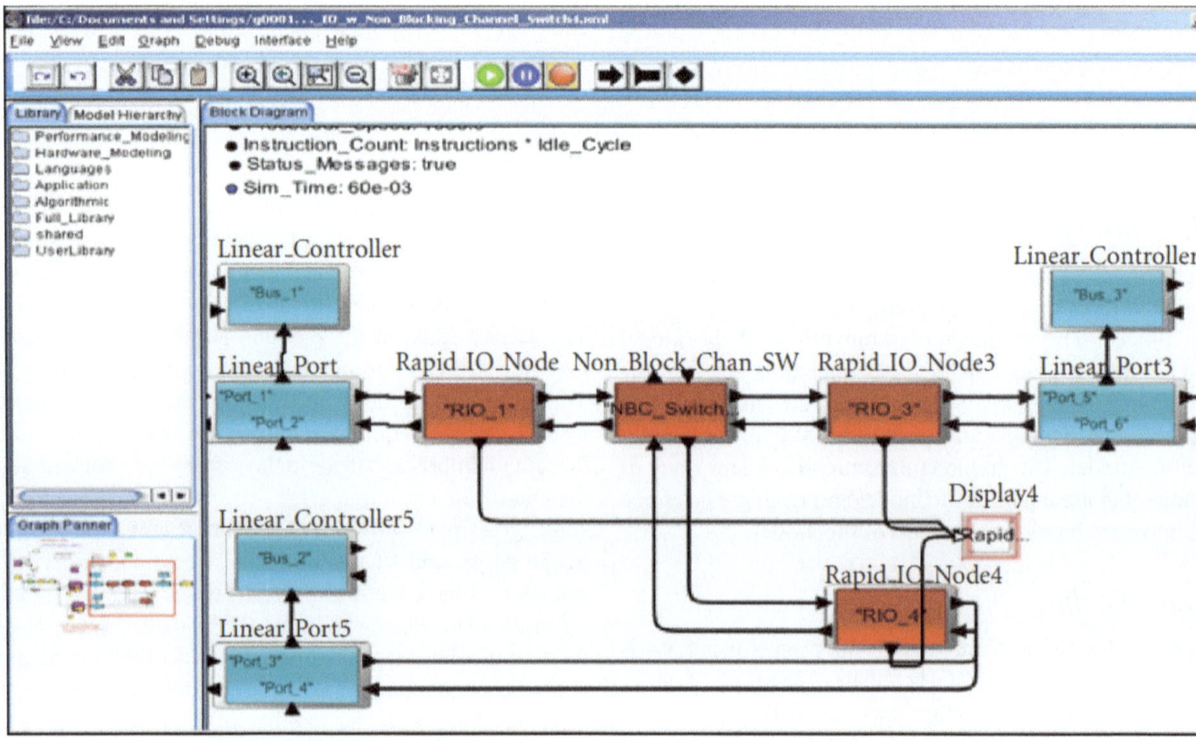

Figure 6. Crossbar switch and SRQ blocks in VisualSim.

Main Memory and the Memory Controller

In the simulation model, the RAM has a capacity of 1 GB and has an in-built memory controller configured to run at a speed of 1638.4 MHz. At this speed and a block width of 4 bytes, the transfer of data from the memory to the cache takes place at a speed of 6.4 GB/s. This rate is commonly used in most of the AMD Opteron processors but can be different depending on the model of the processor. The same rate is also used in the model of the Xeon processor. Each instruction that the RAM executes is translated into delay specified internally by the memory configurations. These configurations are seen in Figure 7 in the *Access_Time* field as the number of clock cycles spent on the corresponding task.

The SDRAM connects to the cores via the SRQ blocks and the Crossbar switch which routes the SDRAM requests from both the cores to the main memory block and then sends a response back to the requesting core, in terms of a text reply. This process requires a certain delay that depends on the type of instruction sent to the SDRAM. In case the SRQ block queue is empty, a single DRAM response time depends on whether the instruction is a memory read, write, read/write, or erase instruction. Each of these instructions takes a fixed number of clock cycles to complete and is determined in the SDRAM configuration as determined by the *Access_Time*field seen in Figure 7. To separate the SDRAM access time from the cache access time, a simplification is made such that the SDRAM access request from the core is not sent in parallel to an L2 cache request as in the actual Opteron; instead, the SDRAM request is issued only after an L2 miss has been encountered.

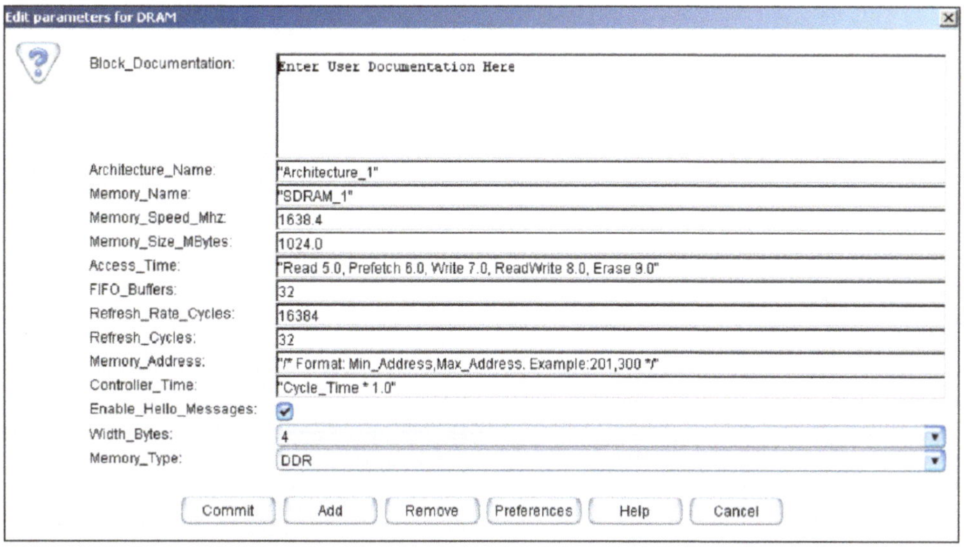

Figure 7. SDRAM and memory controller configurations.

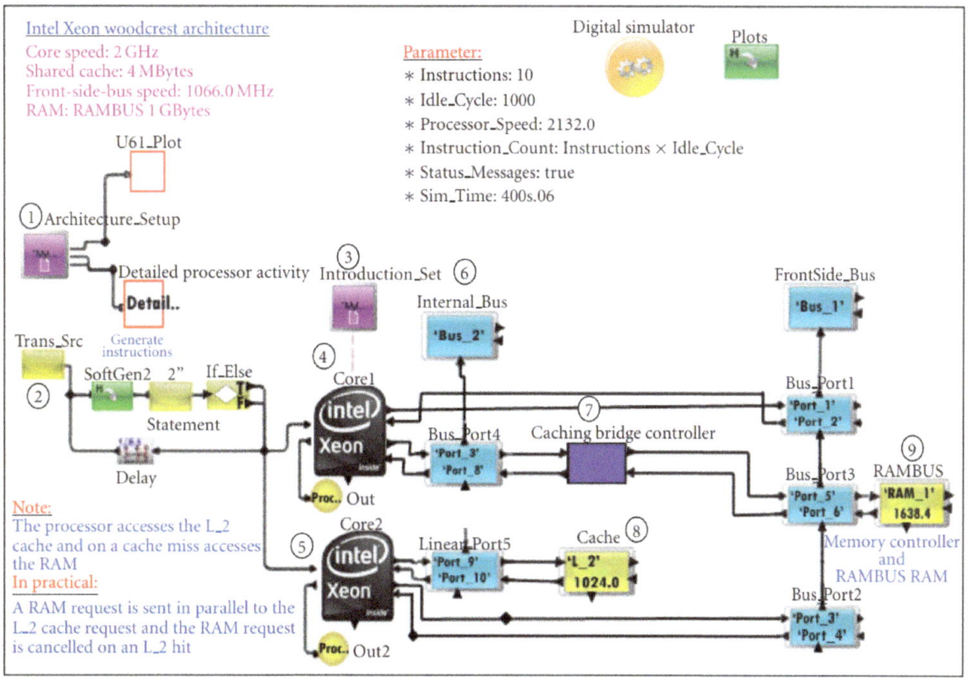

Figure 8. Shared Bus Architecture.

Xeon Model

Basic Architecture

The basic architecture of the Intel Dual Core Xeon is illustrated in Figure 2. The corresponding platform model is depicted in Figure 8. The two cores of the processor are connected to the shared L2 cache and then via the Front-Side-Bus (FSB) interface to the SDRAM. The modeled Intel Xeon processor consists of two cores with three integer execution units, three floating point units, and two loads/stores and branch units to the data cache. The same specifications used to model the Opteron cores in VisualSim are used here as well. Besides, each core is configured with 64 kB of L1 data cache, 64 kB of L1 instruction cache, whereas the L2 cache is a unified cache and is 4 MB in size. The FSB interface, as seen in Figure 8, was constructed using the Virtual Machine block in VS [7] and is connected to the internal bus which links the two cores to the RAM via the FSB. The software generation block (block 2 on the left side of the model) contains the same tasks as the Opteron.

Architecture Setup

The architecture setup block of the model of the Xeon (Figure 8—block 1) is the same as the one implemented in the Opteron and the field mappings of the Data Structure attributes are copied from the Opteron model to ensure that no factors other than the memory architecture affects the results.

Core and Cache

The core implementation of the Xeon is configured using VS to operate at a frequency of 2 GHz and has 128 kB of L1 cache (64 kB data and 64 kB instruction cache), 4 MB of unified and shared L2 cache [5], floating point, integer, and load/store execution units. Here as well, the instruction queue length is set to 6 and instructions are included in the instruction set of both the cores, so as to make the memory access comparison void of all other differences in the architectures. These instructions are defined in the instruction block that is described in a later section.

Certain simplifications have been made to the core of the Xeon in order to focus the analysis entirely to the memory architecture of the processor. The assumption made in accessing the memory is that any L1 cache miss does not result in a simultaneous request being sent to the L2 cache and the RAM. Instead the requests are sent sequentially, where an L1 cache miss results in an L2 cache access and finally an L2 cache miss results in a RAM access.

To simplify the model and the memory access technique, the process of snooping is not implemented in this simulation, and similar to the Opteron, no parallel requests are sent to two memories.

Pipeline

The pipeline of the modeled Xeon consists of four stages (similar to the Opteron model), the prefetch, decode, execute, and the store. The prefetch of each instruction begins from the L1 cache and ends in a RAM access in case of L1 and L2 cache misses. The second stage in the pipeline is the decode stage that is mainly translated into a wait stage. The third stage, the execution stage, takes place in the five execution units that are present in the cores, and finally after the execution, the write-back stage writes back the specified data to the memory, mainly the L1 cache.

Caching Bridge Controller (CBC)

The CBC, block 7 of the model is simply a bridge that connects the L2 shared cache to the FSB [13]. This FSB then continues the link to the RAM (block 9) from which accesses are made and the data/instruction read is sent to the core that requested the data. The CBC model is developed using the VisualSim scripting language and simulates the exact functionality of a typical controller.

Main Memory and the Memory Controller

The dual-core Xeon contains a main memory of type RAMBUS with a speed similar to the memory connected to the Opteron. The size of the RAM is 1 GB and contains a built-in memory controller. This memory controller is configured to run at a speed of 1638.4 MHz. At this speed and a block width of 4 bytes, the transfer of data from the memory to the cache takes place at a speed of 6.4 GB/s. Each instruction that the RAMBUS will carry out implies a certain delay which has been specified internally in the memory configurations. These configurations are seen in Figure 9 in the *Access_Time* field as the number of clock cycles spent executing the corresponding task. The RAM connects to the cores via the CBC and data or instruction requests to the RAM from either core are sent to the main memory block via the FSB. The RAM then sends a response back to the requesting core which can be seen as a text reply on the displays that show the flow of requests and replies. DRAM Access Time is the time taken since a request is made and when the data is made available from the DRAM. It is defined in nanoseconds by the user for each operation like Read, Write, or a Read-Write as an access time parameter in the*Access_Time* field of Figure 9.

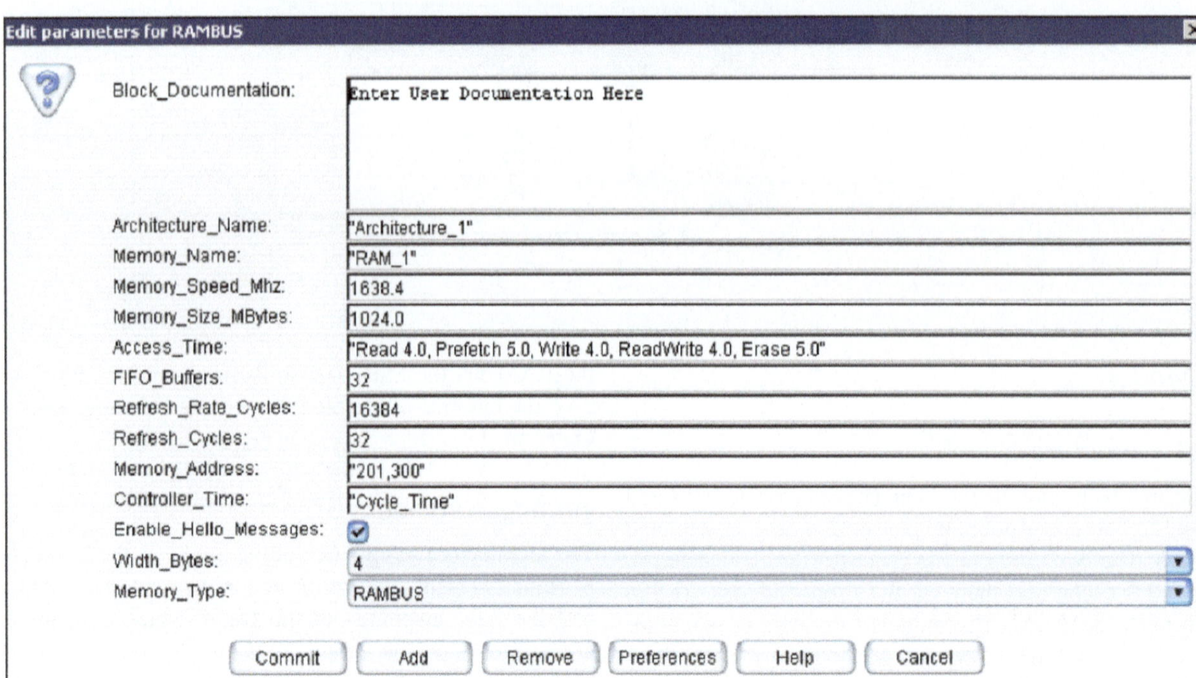

Figure 9. Main memory and controller configurations of the Intel Xeon.

RESULTS AND ANALYSIS

Following a series of experimental tests and numerical measurements using benchmarking software, published literature [14–16] discusses the performance of the AMD Opteron when compared to the Xeon processor using physical test beds comprised of the two processors. These three references provide the reader with a very informative and detailed comparison of the two processors when subjected to various testing scenarios using representative loads.

In this work, we are trying to make the case for an approach that calls for early performance analysis and architectural exploration (at the system level) before committing to hardware. The memory architectures of the above processors were used as a vehicle. We were tempted to use these architectures by the fact that there were published results that clearly show the benefits of the Opteron memory architecture when compared to the Xeon FSB architecture and this would no doubt provide us with a reference against which we can validate the simulation results obtained using VisualSim.

Additionally, and to the best of our knowledge, we could not identify any published work that discusses the performance of the two memory architectures at the system level using an approach similar to the one facilitated by VisualSim.

Using VisualSim, a model of the system can be constructed in few days. All of the system design aspects can be addressed using validated parametric library components. All of the building blocks, simulation platforms, analysis, and debugging required to construct a system are provided in a single framework.

Synopsys integrated Cossap (dynamic data flow) and SystemC (digital) into System Studio while VisualSim combines SystemC (digital), synchronous data flow (DSP), finite state machine (FSM), and continuous time (analog) domains. Previous system level tools typically supported a single modeling specific domain. Furthermore, relative to prior generations of graphical modeling tools, VisualSim integrates as many as thirty bottom-up components functions into a single system level, easy to use, reusable blocks, or modules.

Finally, it is worth mentioning that results obtained using the VisualSim environment in this work are generally in line with results and conclusions found in the literature [14–16].

In the work reported here, Simulation runs are performed using a Dell GX260 machine with a P4 processor running at 3.06 GHz, and a 1 Gbyte RAM.

For simulation purposes and to test the performance of both architectures, traffic sequences are used to trigger the constructed models. These sequences are defined data structures in VisualSim; a traffic generator emulates application-specific traffic. The *Transaction_Source*block in Figures 4 and 8 is used to generate tasks that are applied to the processors as input stimuli. These tasks are benchmarks consisting of a varied percentage mix of integer, floating-point, load/store, and branch instructions. The different percentages are inserted into the software generator's *Instruction_Mix* file and supplied to the processor cores. Thirty three tasks (Table 2) were generated and used to assess the performance of the DirectConnect and the Shared Bus architectures.

Table 2. Benchmark tasks [12].

Model task name	Actual task name
Task_0	DFT
Task_1	DFT
Task_2	DFT
Task_3	CS_Weighting
Task_4	IR
Task_5	Q_Taylor_Weighting
Task_6	CS_Weighting
Task_7	IR
Task_8	Q_Taylor_Weighting
Task_9	CS_Weighting
Task_10	IR
Task_11	Q_Taylor_Weighting
Task_12	CS_Weighting
Task_13	IR
Task_14	Q_Taylor_Weighting
Task_15	CS_Weighting
Task_16	IR
Task_17	Q_Taylor_Weighting
Task_18	CS_Weighting
Task_19	IR
Task_20	DFT
Task_21	DFT
Task_22	DFT
Task_23	DFT
Task_24	DFT
Task_25	DFT
Task_26	DFT
Task_27	DFT
Task_28	DFT
Task_29	DFT
Task_30	DFT
Task_31	DFT
Task_32	DFT

Table 3 shows the "Task Latencies" and "Cycles/Task" measured for both processors. Since the models were constructed by having similar configurations in almost every aspect except the memory architecture, the values obtained are therefore a reflection of the performance of that architecture. Figure 10 is a graph of the "Latency per Task" results for each configuration. This graph is plotted using the total number of clock cycles that were taken by each task. The instructions contained in each of these tasks are defined in the software_generation block. At the end of the execution, a text file is generated by VisualSim that contains the number of cycles that were taken by each of the task, including the stall time and the delays caused by the components linking the RAM to the cores.

Table 3. Task latencies and Cycles/Task.

Task_Name	Latency$_{(Opteron)}$	Latency$_{(Xeon)}$	Cycles/Task$_{(Opteron)}$	Cycles/Task$_{(Xeon)}$
Task_0	1.21E-7	1.27E-7	247	260
Task_1	7.713E-7	7.99E-7	1579	1636
Task_2	7.235E-7	7.299E-7	1481	1494
Task_3	6.538E-7	6.605E-7	1338	1352
Task_4	8.538E-7	1.027E-6	1748	2103
Task_5	7.358E-7	7.415E-7	1506	1518
Task_6	7.453E-7	7.52E-7	1526	1540
Task_7	8.538E-7	1.0222E-6	1748	2093
Task_8	7.362E-7	7.43E-7	1507	1521
Task_9	6.783E-7	7.051E-7	1389	1444
Task_10	8.538E-7	1.0241E-6	1748	2097
Task_11	6.858E-7	7.09E-7	1404	1452
Task_12	7.1629E-7	7.216E-7	1466	1478
Task_13	8.537E-7	1.0179E-6	1748	2084
Task_14	7.298E-7	7.365E-7	1494	1508
Task_15	7.133E-7	7.196E-7	1460	1473
Task_16	8.537E-7	1.0251E-6	1748	2099
Task_17	7.11299E-7	7.169E-7	1456	1468
Task_18	6.862E-7	6.92E-7	1405	1417
Task_19	1.5795E-6	1.9301E-6	3234	3952
Task_20	7.4229E-7	7.482E-7	1520	1532
Task_21	2.6452E-6	3.3117E-6	5417	6782
Task_22	7.4649E-7	7.769E-7	1528	1591
Task_23	2.6454E-6	3.3002E-6	5417	6758
Task_24	7.4529E-7	7.52E-7	1526	1540
Task_25	2.6451E-6	3.3039E-6	5417	6766
Task_26	7.523E-7	7.58E-7	1540	1552
Task_27	2.645E-6	3.3078E-6	5416	6774
Task_28	7.458E-7	7.524E-7	1527	1540
Task_29	2.6453E-6	3.3313E-6	5417	6822
Task_30	7.9129E-7	7.976E-7	1620	1633
Task_31	2.6453E-6	3.2723E-6	5417	6701
Task_32	7.558E-7	7.6189E-7	1547	1560

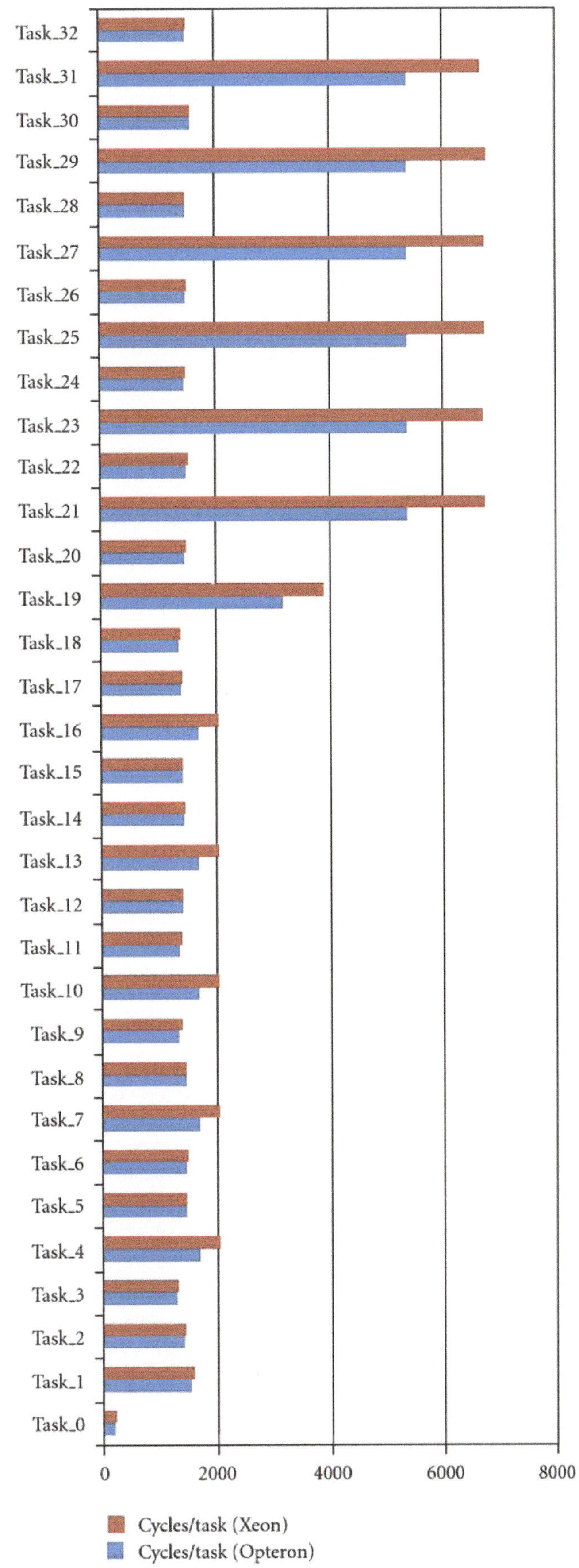

Figure 10. Latency per Task (Cycles per Task).

Cleary, the Xeon has exhibited a higher latency in almost every task. These numbers show that the time taken to execute the same task in the Xeon were more than the Opteron.

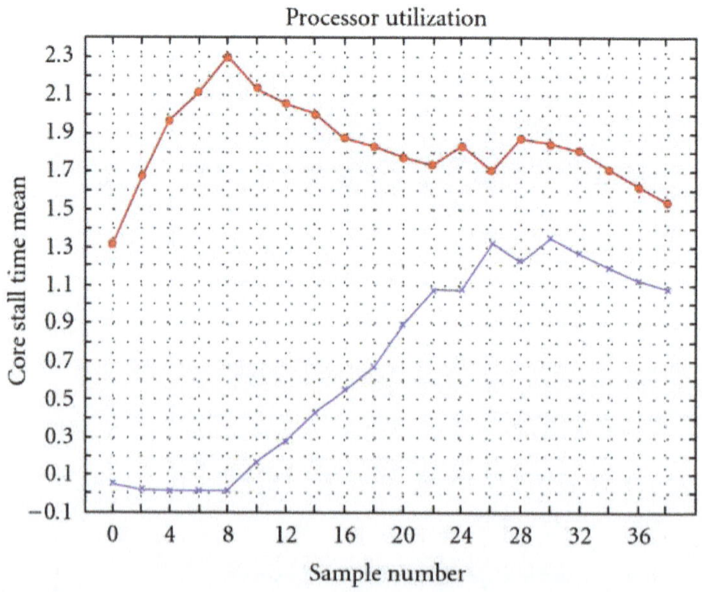

Figure 11. Processor Stall Times (Opteron).

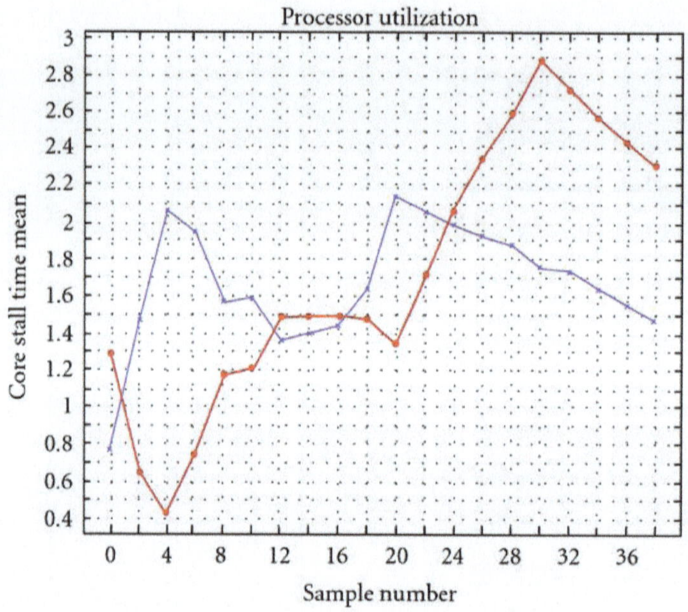

Figure 12. Processor Stall Times (Xeon).

Figures 11 and 12 show a graph of processors' stall times. In both cases, 20 samples are taken during the entire simulation period and the data collected is used in the depicted graphs. During the execution of all the tasks, the maximum time for which the processors stalled was different for each kind of architecture. The maximum stall time for the DirectConnect architecture was 2.3 microseconds whereas for the Shared Bus architecture the maximum stall time was 2.9 microseconds. Due to the shared bus in the Xeon's architecture, delays were greater than the DirectConnect approach of the Opteron, and thus the difference in the stall time.

As the models described earlier suggest, the Opteron follows the split cache approach where each core in the processor has its own L1 and L2 cache; thus no part of the cache is shared between the two cores. On the contrary, the Xeon processor employs the shared cached technique and thus both the cores have access to a larger amount of cache than the ones in the Opteron. Whenever one of the cores in the Xeon is not accessing the shared cache, the other core has complete access to the entire cache which results in a higher hit ratio.

Table 4 shows the hit-ratios of both the architecture models. As the values suggest, the hit-ratios of the Xeon/Shared cache are higher than those of the Opteron/split cache, the reason being discussed above. It is worth mentioning that few simplifying assumptions were made to the memory architectures operation of both processors as discussed in Section 3.

Table 4. Hit ratios.

Hit Ratios (%)	Opteron	Xeon
Processor_1_D_1_Hit_Ratio_Mean	97.23	98.37
Processor_1_I_1_Hit_Ratio_Mean	92.13	95.11
Processor_2_D_1_Hit_Ratio_Mean	98.98	99.92
Processor_2_I_1_Hit_Ratio_Mean	95.21	96.14
L_2_Hit_Ratio_Mean	N/A	96.36

CONCLUSIONS

In this work, we utilized a system modeling methodology above the detailed chip implementation level that allows one to explore different designs without having to write Verilog, VHDL, SystemC, or simply C/C++ code. This approach contributes to a considerable saving in time and allows for the exploration and assessment of different designs prior to implementation.

Since predictability of performance is critical in microprocessors design, simulation models can be used to evaluate architectural alternatives and assist in making informed decisions. Simulation is an acceptable performance modeling technique that can be used to evaluate architectural alternatives and features. In this work, we used Virtual System Prototyping and simulation to investigate the performance of the memory subsystems of both, the Opeteron, and the Xeon dual core processors.

Simulation results indicate that the Opteron has exhibited better latency than the Xeon for the majority of the tasks. In all cases, it either outperformed the Xeon or at least had similar latency. This demonstrates that using an FSB as the only means of communication between cores and memory has resulted in an increase in stalls and latency. On the other hand, in the DirectConnect Architecture, the cores being directly connected to the RAM, via the crossbar switch and the SRQ which were running at processor speed, had minimal delays. Each RAM request from either of the cores was sent individually to the SRQ blocks and they were routed to the RAM that had its memory controller on-chip and the cores did not have to compete for a shared resource.

REFERENCES

1. Balakrishnan S, Rajwar R, Upton M, Lai K: The impact of performance asymmetry in emerging multicore architectures. *Proceedings 32nd International Symposium on Computer Architecture (ISCA '05), June 2005, Madison, Wis, USA* 506-517.
2. Conway P, Hughes B: The AMD Opteron Northbridge architecture. *IEEE Micro*2007,27(2):10-21.
3. Keltcher CN, McGrath KJ, Ahmed A, Conway P: The AMD Opteron processor for multiprocessor servers. *IEEE Micro* 2003,23(2):66-76. 10.1109/MM.2003.1196116.
4. Dual-Core Intel® Xeon® Processor 5000 Series Datasheet May 2006
5. Varada R, Sriram M, Chou K, Guzzo J:Design and integration methods for a multi-threaded dual core 65nm XeonÂ® processor. *Proceedings of IEEE/ACM International Conference on Computer-Aided Design, Digest of Technical Papers (ICCAD '06), November 2006, San Jose, Calif, USA* 607-610.
6. Prowse-Fosler L: Virtual Systems Prototyping Ensures Reusable Design Platforms. http://www.vastsystems.com/solutions-technical-papers.html
7. Mirabilis Design User's Manual vol. 1, http://www.mirabilisdesign.com/
8. HyperTransport I/O Link Specification http://www.hypertransport.org/
9. http://www.chiparchitect.com/
10. http://www.amd.com/us-en/Processors/
11. Rusu S, Tam S, Muljono H, *et al.*:A 65-nm dual-core multithreaded Xeon®processor with 16-MB L3 cache. *IEEE Journal of Solid-State Circuits* 2007,42(1):17-24.
12. Shankar D: Selecting Memory Controllers for DSP Systems. DSP DesignLine, May 2007
13. Tarn S, Rusu S, Chang J, Vora S, Cherkauer B, Ayers D:A 65nm 95W dual-core multi-threaded Xeon® processor with L3 cache. *Proceedings of IEEE Asian Solid-State Circuits Conference (ASSCC '06), November 2006, Hangzhou, China* 15-18.
14. Peng L, Peir J-K, Prakash TK, Chen Y-K, Koppelman D: Memory performance and scalability of Intel's and AMD's dual-core processors: a case study. *Proceedings of the 27th IEEE International Performance, Computing, and Communications Conference (IPCCC '07), April 2007, New Orleans, La, USA* 55-64.
15. Guiang CS, Milfeld KF, Purkayastha A, Boisseau JR: Memory performance of dual-processor nodes: comparison of Intel Xeon and AMD Opteron memory subsystem architectures. *Proceedings for ClusterWorld Conference & Expo, 2003*
16. Advanced Micro Devices: AMD Opteron vs Intel Xeon Processor Performance Comparison http://www.veritest.com/

CHAPTER 5 TIME-PREDICTABLE COMPUTER ARCHITECTURE

Martin Schoeberl

INTRODUCTION

Standard computer architecture is driven by the following paradigm: *Make the common case fast and the uncommon case correct* [1]. However, this design approach leads to architectures where the worst-case execution time (WCET) is high and hard to predict by static analysis. For real-time systems, we have to design architectures with the following paradigm: *Make the worst case fast and the whole system easy to analyze*.

Classic enhancements in computer architectures are pipelining, instruction and data caching, dynamic branch prediction, out-of-order execution, speculative execution, and fine-grained chip multithreading. These features are increasingly harder to model for the low-level WCET analysis. Execution history is the key to performance enhancements, and also the main issue for WCET analysis. Thus, we need techniques to manage the execution history.

Pipelines should be simple, with minimum dependencies between instructions. It is agreed that caches are mandatory to bridge the gap between processor speed and memory access time. Caches in general, and particularly data caches, are usually hard to analyze statically. Therefore, we are introducing caches that are organized to speed up execution time and provide tight WCET bounds. We propose three different caches: (1) an instruction cache for full methods, (2) a stack cache, and (3) a small, fully associative buffer for heap access. Furthermore, the integration of a program—or compiler—managed scratchpad memory can help to tighten bounds for hard-to-analyze memory access patterns.

Out-of-order execution and speculation result in processor models that are too complex for WCET analysis. We discuss that the transistors are better used on chip multiprocessors (CMPs) with simple in-order pipelines. Real-time systems are naturally multithreaded and thus map well to the explicit parallelism of chip multiprocessors.

We propose a multiprocessor model with one processor per thread. Thread switching and schedulability analysis for each individual core disappears, but the access to the shared resource main memory still needs to be scheduled.

We have implemented most of the proposed concepts for evaluation in a Java processor. The Java processor JOP [2] is intended for real-time and safety critical applications written in a modern object-oriented language. It has to be noted that all concepts can also be applied to a standard RISC processor. The following list points out the key arguments for a time-predictable computer architecture.

1. There is a mismatch between performance-oriented computer architectures and worst-case analyzability.
2. Complex features result in increasingly complex models.
3. Caches, a very important feature for high performance, need new organization.
4. Thread level parallelism is natural in embedded systems. Exploration of this parallelism with simple chip multiprocessors is a valuable option.
5. One thread per processor obviates the classic schedulability analysis and introduces scheduling of memory access.

Catching up with WCET analysis of features that enhance the average-case performance is not an option for future real-time systems. We need a sea change and should take the constructive approach by designing computer architectures where predictable timing is a first-order design factor.

Citation: Schoeberl, M (2009). Time-predictable computer architecture, EURASIP Journal on Embedded Systems, 2009; 758480. DOI:10.1155/2009/758480.

The contributions of the paper are twofold: (1) an extensive overview is given of processor features that make WCET estimation difficult, (2) solutions for a time-predictable architecture that can be implemented in RISC, CISC, or VLIW style processors are provided. The implementations of some of the proposed concepts in the context of a Java processor, as described in Section 5, have been previously published in [3, 4].

The paper is organized as follows. Section 2 presents related work on real-time architectures. In Section 3, we describe the main issues that hamper tight WCET estimates of actual processors. We propose solutions for these issues in Section 4. In Section 5, we evaluate the proposed time-predictable computer architecture with an implementation of a Java processor in an FPGA. Section 6 concludes the paper.

RELATED WORK

Bate et al. [5] discuss the usage of modern processors in safety critical applications. They compare commercial off-the-shelf (COTS) processors with a customized processor developed specifically for the safety critical domain. While COTS processors benefit from a large user base and the resulting maturity of the design process, customized processors provide the following advantages:

1. Design in conjunction with the safety argument,
2. Design for good worst-case performance,
3. 1Using only features that can be easily analyzed,
4. .The processor can be treated as a *white box* during verification and testing.

Despite these advantages, few research projects exist in the field of WCET-optimized hardware. Thiele and Wilhelm [6] discuss that a new research discipline is needed for time-predictable embedded systems to "match implementation concepts with techniques to improve analyzability."

Similarly, Edwards and Lee discussed as "It is time for a new era of processors whose temporal behavior is as easily controlled as their logical function" [7]. A first simulation of their PRET architecture is presented in [8]. PRET implements the SPARC V8 instruction set architecture (ISA) in a six-stage pipeline and performs chip level multithreading for six threads to eliminate data forwarding and branch prediction. Scratchpad memories are used instead of instruction and data caches. The shared main memory is accessed via a TDMA scheme, called memory wheel, similar to the TDMA-based arbiter used in the JOP CMP system [9]. The SPARC ISA is extended with a *deadline* instruction that stalls the current thread until the deadline is reached. This instruction is used to perform time-based, instead of lock-based, synchronization for access to shared data.

Berg et al. identify the following design principles for a time-predictable processor: "recoverability from information loss in the analysis, minimal variation of the instruction timing, noninterference between processor components, deterministic processor behavior, and comprehensive documentation" [10]. The authors propose a processor architecture that meets these design principles. The processor is a classic five-stage RISC pipeline with minimal changes in the instruction set: it handles function calls with an explicit instruction for simpler reconstruction of the control flow graph and construction of 32-bit immediate values with two instructions to avoid immediate values in the code segment. The memory system has to be organized in Harvard-style with dedicated busses to the FLASH memory for the code and the SRAM memory for the data. The replacement strategy of caches has to be least-recently used (LRU).

Heckmann et al. provide examples of problematic processor features in [11]. The most problematic features found are the replacement strategies for set-associative caches. A pseudo-round-robin replacement strategy of the 4-way set-associative cache in the ColdFire MCF 5307 effectively renders the associativity useless for WCET analysis. The use of a single 2-bit counter for the whole cache destroys age information within the cache sets. The analysis of that cache results in effectively modeling only a quarter of the cache as a direct-mapped cache. Similarly, a pseudo-LRU replacement strategy for an 8-way set-associative cache of the PowerPC 750/755 uses an age counter for each set. Here, only half of the cache is modeled by the analysis. Slightly more complex pipelines, with branch prediction and out-of-order execution, need an integrated pipeline and cache analysis to provide useful WCET bounds. Such integrated analysis is complex and also demanding with respect to the computational effort. In conclusion, Heckmann et al. suggest the following restrictions for time-predictable processors: (1) separate data and instruction caches, (2) locally deterministic update strategies for caches, (3) static branch prediction, and (4) limited out-of-order execution. The authors discuss for restriction of processor features of actual processors (of the time) for embedded systems, but do not provide suggestions for *additional* or *alternative* features for a time-predictable processor.

The VISA approach [12] adapts a complex simultaneous multithreading processor that can be reconfigured to a simple single-issue pipeline. The complexity of the processor can be dynamically disabled at runtime. WCET analysis is performed for the simple pipeline. A task is divided into subtasks and each subtask is assigned a checkpoint. The task is executed on the complex pipeline and only if the checkpoint is missed, the processor is switched to the simple mode. The checkpoint is inserted early enough to complete the subtask on the simple pipeline before the deadline. The available slack time, when the task is executed on the fast, complex pipeline, is utilized for energy saving.

Puschner and Burns discuss for a single-path programming style [13] that results in a constant execution time. In that case, WCET can easily be measured. However, this programming paradigm is quite uncommon and restrictive. Single-path programming can be inefficient when the control flow is data-dependent. A processor, called SPEAR [14], was especially designed to evaluate the single-path programming paradigm. A single predicate bit can be set with a compare instruction whereby several instructions (e.g., move, arithmetic operations) can be predicated. The SPEAR implements a three-stage in-order pipeline and uses onchip memories for instruction and data instead of caches.

Complex hardware and software architectures hinder hierarchical timing analysis [15]. A radical simplification of the whole system to avoid unwanted timing interactions is proposed—single path programming, execution of a single task/thread per core, simple in-order pipelines, and statically scheduled access to shared memory in CMPs.

Whitham discusses that the execution time of a basic block has to be independent of the execution history [16]. As a consequence, his MCGREP architecture reduces pipelining to two stages (fetch and execute) and avoids caches all together. To reduce the WCET, Whitham proposes to implement the time-critical functions in microcode on a reconfigurable function unit (RFU). The main processor implements an RISC ISA as a microprogrammed, sequential processor. The interesting approach in MCGREP is that the RFUs implement the same architecture and microcode as the main CPU. Therefore, mapping a sequence of RISC instructions to microcode for one or several RFUs is straightforward. With several RFUs, it is possible to explicitly extract instruction level parallelism (ILP) from the original RISC code in a similar way to VLIW architectures.

Whitham and Audsley extend the MCGREP architecture with a trace scratchpad [17]. The trace scratchpad caches microcode and is placed after the decode stage. It is similar to a trace cache found in newer Pentium CPUs to cache the translated micro-operations. The differences from a cache are that the execution from the trace scratchpad has to be explicitly started and the scratchpad has to be loaded under program control. The authors extract ILP at the microcode level and schedule the instructions statically—similar to a VLIW architecture.

WCET ANALYSIS ISSUES

The WCET of tasks is the necessary input for schedulability analysis. Measuring the WCET is not a safe option. Only static WCET analysis can provide safe upper bounds of execution times.

WCET analysis can be separated into high-level and low-level analysis. The high-level analysis is a mature research topic [18–20]. An overview of WCET-related research can be found in [21,22]. The main issues that need to be solved are in the low-level analysis. The processors that can be analyzed are usually several generations behind actual architectures [11, 23, 24] (e.g., Thesing models, in his Ph.D. thesis [25], the MPC755 variant of the PowerPC 750). The PowerPC 750 was introduced in 1997 and the MPC755 was *not recommended for new designs*in 2006.

The main issues in low-level analysis are features that increase average performance. All these features, such as multilevel caches, branch target buffer, out-of-order execution, and speculation, include a state that heavily depends on a large execution history. This caching of the execution history is actually fundamental for performance enhancements. However, it is the history which is hard to model for WCET analysis. A long history leads to a state explosion for the final WCET calculation. Low-level WCET analysis thus usually performs simplifications and uses conservative estimates. One example of this conservative estimate is to classify a cache access as a miss, if the outcome of the cache access is unknown.

Lundqvist and Stenström have shown that this intuitive assumption can be wrong on dynamically scheduled microprocessors [26]. They provide an example of such a timing anomaly in which a cache hit can cause a longer execution time than a cache miss. The principles behind these timing anomalies are further elaborated in [27].

Pipeline Dependencies

Simple pipelines, similar to the original Berkeley/Stanford RISC design [28], are easy to model for WCET analysis. In a nonstalled pipeline, the execution time latency corresponds to the length of the pipeline. The effective execution time itself is only a single cycle. What makes pipeline analysis necessary are stalls introduced by dependencies within the pipeline. Those stalls are introduced by

1. Data dependencies between instructions,
2. Control dependencies between instructions.

In one of the first RISC designs, the MIPS [29], these dependency hazards are explicitly exposed to the compiler. They have to be resolved by the compiler with instruction scheduling for delayed branches and for the single cycle delay between a memory load and the data use. Therefore, these effects are also recognized by the WCET tool. More advanced pipelines avoid exposing stalls from the ISA in order to avoid too many (compiler) target variations and retain binary compatibility between processor versions. Nevertheless, this information is needed for WCET analysis.

Dependencies within a basic block can be easily modeled. The challenge is to merge the effects from different basic blocks and across function boundaries. In [30], the *timing schema* [31] is extended to include the pipeline information. Timing schema is a tree-based WCET analysis. After the determination of basic block execution times, the control flow graph is processed in a bottom-up manner until a final WCET bound is available. Branches are merged with the higher WCET bound as result. For the extension, the pipeline is represented by reservation stations, and the state at the head and tail of a basic block is considered when basic blocks are merged.

Pipelines with timing dependencies can result in an unbounded effect, called *long timing effect* (LTE) [32]. This means that an instruction far back in the history (longer than the pipeline length) influences the execution time of the current instruction. These LTEs can be negative or positive. A positive LTE means longer execution time. An instruction with a possible positive LTE needs a safe approximation of that effect for the pipeline analysis.

More complex pipelines can be analyzed with abstract interpretation, but the analysis time can become impractical. Berg et al. [10] report that up to 1000 states per instruction are needed for the model of the PowerPC 755. This processor was introduced in 1998 and we expect a considerable growth of the states that need to be tracked by abstract interpretation for newer processors.

Instruction Fetch

The instruction fetching is often decoupled from the main memory or the instruction cache by a prefetch unit. This unit fills the prefetch queue with instructions independently of the main pipeline. This form of prefetching is especially important for a variable length instruction set as the x86 ISA or the bytecode instructions of the Java virtual machine (JVM). The fill status of the prefetch queue depends on the history of the instruction stream. The possible length of this history is unbounded. To model this queue for a WCET tool, we need to cut the maximum history and assume an empty queue at such a cut point.

In [33], the authors model the 4-byte-long prefetch queue of an Intel 80188. Even for this simple prefetch queue, the authors have to perform some simplifications in their approach to handle the resulting complexity due to the interaction between the instruction execution and the instruction prefetch (the consuming and the producing ends of the queue).

Caches

Between the middle of the 1980s and 2002, CPU performance increased by around 52% per year, but memory latency decreased only by 9% [1]. To bridge this growing gap between CPU and main memory performance, a memory hierarchy is used. Several layers with different tradeoffs between size, speed, and cost form that memory hierarchy. A typical hierarchy consists of

1. Register file,
2. Per-processor level 1 instruction and data cache,
3. Onchip, shared unified level 2 cache,
4. Offchip level 3 cache,
5. Main memory,
6. Hard disc for virtual memory.

The only layer under the control of the compiler is the register file. The rest of the memory hierarchy is usually not visible—it is not part of the ISA abstraction. Placement of data in the different layers is performed automatically by the hardware for caches and by the OS for virtual memory management. The access time for a word located in a memory block paged out by the OS is several orders of magnitude higher than a level 1 cache hit. Even the access times to the level 1 cache and to the main memory differ by two orders of magnitudes.

Cache memories for the instructions and data are classic examples of the *make the common case fast* paradigm. Avoiding or ignoring this feature in real-time systems, due to its unpredictable behavior, results in a very pessimistic WCET bound. Much effort has been expended on research to integrate the instruction cache into the timing analysis of tasks [34,35], on the cache's influence on task preemption [36, 37], and on integration of the cache analysis with the pipeline analysis [38]. The influence of different cache architectures on WCET analysis is described in [11].

A unified cache for data and instructions can easily destroy all the information on abstract cache states. Access to n unknown addresses in an n-way set-associative cache results in the state *unknown* for all cache lines. Modern processors usually have separate instruction and data caches for the level 1 cache. However, the level 2 cache is usually shared. Most CMP systems also share the level 2 cache between the different cores. The possible interactions between concurrent threads running on different cores are practically impossible to model.

Data caches are considerably harder to analyze than instruction caches. For some data accesses, especially for data allocated on the heap, the addresses cannot be predicted. However, access to the stack can be predicted statically. A data cache that caches heap and stack content suffers from the same problem as a unified instruction and data cache: an unknown address for a heap access will evict one block from all sets in the abstract cache state and will increase the age of all cache blocks.

In a recent paper, Reineke et al. analyzed the predictability of different cache replacement policies [39]. It is shown that LRU performs best with respect to predictability. Pseudo-LRU and FIFO perform similarly, as both perform considerably worse than LRU. In an 8-way set-associative setting, pseudo-LRU and FIFO take more than twice as long as LRU to recover from lost information.

Branch Prediction

Accurate branch prediction is of utmost importance to keep long pipelines filled. The penalty of a wrongly predicted conditional branch is typically almost as long as the pipeline length. Modern branch predictors guess the outcome primarily from results of earlier branches. They heavily rely on the execution history, an effect we want to avoid for a tight worst-case prediction. Global branch predictors and caches have a similar issue: as soon as a single index into the branch history is unknown, the whole information of branch prediction is lost for the analysis at that point.

Two-level branch predictors are not suitable for time-predictable architectures [40]; for example, on the Pentium III, Pentium 4, and UltraSparc III, a decrease in the number of loop iterations can actually result in an increase of the execution time. This is another form of timing anomaly [26].

Branch prediction also interferes with cache contents. When the analysis cannot anticipate the outcome of the prediction, both branch directions need to be considered for cache analysis.

Instruction Level Parallelism

Some microprocessors try to extract ILP from the instruction stream, that is, execute more than one instruction per clock cycle. ILP extractions can be done either statically by the compiler or dynamically by the hardware.

Processors with static-scheduled ILP are known as very long instruction word (VLIW) processors. The main issue of VLIW processors is that the pipeline details are exposed at the ISA. The compiler has to group parallel instructions and needs to consider pipeline constraints. Some processors rely on the compiler to resolve data dependencies and do not stall the pipeline. Therefore, each new generation of VLIW processors needs a new compiler back end. However, this issue is actually an advantage for low-level WCET analysis, as these details are needed for the pipeline analysis.

Dynamically scheduled, superscalar microprocessors combine several parallel execution units with an out-of-order execution to extract the ILP from the instruction stream. In current processors, about hundred instructions (e.g., 128 in the Pentium 4 [1]) can be in flight at each cycle. Analysis of a realistically sized application with an accurate processor model is thus (almost) impossible. Even modeling the pipeline states for basic blocks leads to a state space explosion; and modeling only basic blocks would result in very long penalties for the branches—on a later version of the Pentium 4, a simple instruction takes at least 31 clock cycles from fetch to retire [1].

Despite this complexity, in [41], a hypothetical out-of-order executing microprocessor is modeled for WCET analysis. Verification of the proposed approach on a real processor is missing. We think modeling out-of-order processors is practically not feasible.

Chip Multithreading

Dynamic extraction of ILP is limited to about two instructions per cycle on current processors, such as Pentium 4 and AMD Opteron [1]. Another path to speed up multithreaded workloads is the extraction of thread-level parallelism (TLP). The concept of TLP in a single processor is quite old—it was used in the CDC 6600, a supercomputer from the 1960s—but is now being reconsidered in all desktop and server processors. Fine-grained multithreading can hide the latency of load/use hazards or a cache miss for one thread by the execution of other threads.

The main issue with multithreading in real-time systems arises when the execution time of one thread depends on the state of a different thread. The main source of timing interactions in a CMP comes from shared caches and shared main memory. In the worst case, all latency hiding has to be ignored by the analysis and the sum of the execution times of several threads is the same as the serial execution on a single-threaded CPU. In addition, multithreaded processors usually share the level 1 caches. Therefore, each thread invalidates the abstract cache state of the other threads.

Dynamic ILP and TLP can be combined for simultaneous multithreading (SMT). With this technique, independent threads can be active in the same pipeline stage. This results in a higher utilization of processor resources that are already available for the ILP extraction. Modeling the fine-grained interaction of different SMT threads for WCET analysis seems, at least to the author, an intractable problem.

Chip Multiprocessors

Due to the power wall [1], CMP systems are now state-of-the-art in desktop and server processors. There are three different CMP systems: (1) multicore versions of superscalar architectures (Intel/AMD), (2) multicore chips with simple RISC processors (Sun Niagara), and (3) the cell architecture.

Mainstream desktop processors from Intel and AMD include two or four out-of-order executing processors. These processors are replications of the original, complex cores that share a level 2 cache and the memory bus. Cache coherence protocols on the chip keep the level 1 caches coherent and consistent. Furthermore, these cores also support SMT, sometimes also called hyperthreading.

Sun took a completely different approach with their Niagara T1 [42] by abandoning their superscalar architecture. The T1 contains 8 simple RISC cores, each supporting 4 threads, scheduled round-robin. When a thread stalls due to a cache miss or a load-use dependency, it is skipped in the schedule. The first version of the chip contains a single floating point unit which is shared by all 8 processors. Each core implements a six-stage, single-issue pipeline similar to the original five-stage RISC pipeline. Such a simple pipeline brings WCET analysis back into consideration.

The cell multiprocessor [43–45] takes an approach similar to a distributed memory multiprocessor. The cell contains, beside a PowerPC microprocessor, 8 synergistic processors (SPs). The SPs contain 256 KB onchip memory that is incoherent with the main memory. The PowerPC, the 8 SPs, and the memory interface are connected via a network consisting of four independent rings. Communication between the cores in the network has to be set up explicitly. All memory management, for example, transfer between SPs or between onchip memory and main memory, is under program control, resulting in a new programming model. The time-predictable memory access to the onchip memory and the in-order pipeline of the SPs should be a reasonable target for WCET analysis. The challenge is to include the explicit memory transfers between the cores and the main memory into the analysis.

Intel recently announced a CMP system named Larrabee [46]. Larrabee is intended as a replacement for graphic processing units from other vendors. It is notable that Intel uses several dual issue, in-order x86 cores. They discuss that for some workloads, in-order pipelines are more power efficient than out-of-order cores. The design is based on the first Pentium processor, enhanced with multithreading support and vector instructions.

The main source of timing interactions in a CMP comes from the shared level 2 (and probably level 3) cache and the shared main memory. The shared memory provides an easy-to-use programming model at the cost of unpredictable access time to the data. A shared level 2 cache is practically not analyzable due to the interthread interference. This is the same issue as with multithreading with a shared level 1 cache.

Cache coherence protocols (bus snooping or directory based) enforce a coherent and consistent view of the main memory. These protocols exchange the cache information between all cores on each memory access and introduce a high variability of the cache access time even when the access is a cache hit.

Yan and Zhang analyze a shared instruction cache on a dual core system that executes two threads [47]. To restrict the set of conflicting cache blocks, they introduce the category *always-except one hit* for level 2 cache blocks. Assuming threads A and B, a cache block c is classified as *always-except one hit* for thread A when c is part of a loop in thread A, c conflicts with a block used by thread B, and the conflicting block in thread B is not part of a loop in thread B. However, the approach has two drawbacks: (1) for n threads/cores, several categories (up to $n - 1$) need to be introduced; (2) *not in a loop* is not a proper model for real-time threads as these are usually periodic.

The memory arbitration algorithm determines the worst-case access time to the main memory. Any fairness-based arbitration is, at least, difficult to integrate into WCET analysis. Priority-based arbitration can only guarantee access time for the core with the highest priority because lower priority cores can be blocked indefinitely.

Documentation

To model the processor for the low-level analysis, an accurate documentation of the processor internals is needed. However, this information is often not available or sometimes simply wrong [32]. For actual processors, the documentation of the internals is usually not disclosed. Over time, due to reverse engineering and less competition with other processors, more information becomes available. This is probably another reason why WCET analysis is about 10 years behind the processor technology.

Summary

While conventional techniques in designing processor architectures increase average throughput, they are not feasible for real-time systems. The influence of these architectural enhancements is at best hardly WCET analyzable. From a survey of the literature, we found that modeling a new version of a microprocessor and finding all undocumented details is usually worth a full Ph.D. thesis.

We discuss that trying to catch up on the analysis side with the growing complexity of modern computer architectures is not feasible. A paradigm shift is necessary. Computer architecture has to be redefined or adapted for real-time systems. Predictable and *analyzable* execution time is of primary importance.

TIME-PREDICTABLE ARCHITECTURE

We propose a computer architecture designed especially for real-time applications. We do not want to restrict features only, but we also want to actively add features that enhance performance and are time-predictable.

Figure 1 illustrates the aim of a time-predictable architecture, showing the distribution of the different execution times for a task: they are best-case execution time (BCET), average-case execution time (ACET), worst-case execution time (WCET), and the bound of the WCET that an analysis tool can provide. The difference between the actual WCET and the bound is caused by the pessimism of the analysis resulting from two factors: (a) certain information, for example, infeasible execution paths, not being known statically and (b) the simplifications to make the analysis computationally practical. For example, infeasible execution paths may significantly impact the WCET bound because the static analysis cannot prove that these paths may never be executed. Similarly, dynamic features such as speculative execution and pipelining often need to be modeled conservatively to prevent an explosion of the analysis complexity.

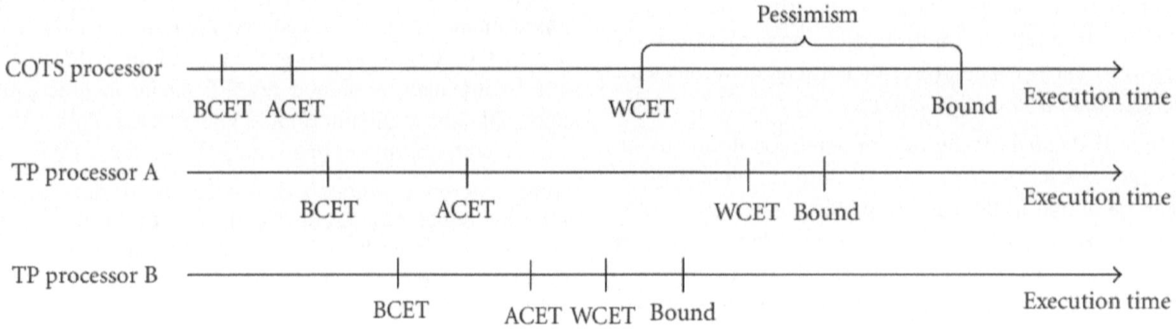

Figure 1. Distribution of the best-case, average-case, and worst-case execution times and the WCET bound of a task on different architectures. A time-predictable processor has less pessimism and the average-case execution time is not important.

The first time line shows the distribution of the execution times for a commercial off-the-shelf (COTS) processor. The other two time lines show the distribution for two different time-predictable processors.

Variant A depicts a time-predictable processor with a higher BCET, ACET, and WCET than a standard processor. Although the WCET is higher than the WCET of the standard processor, the pessimism of the analysis is lower and the resulting WCET bound is lower as well. Even this type of processor is a better fit for hard real-time systems than today's standard processors.

Processor B shows an architecture where the BCET and ACET are further increased, but the WCET and the WCET bound are decreased. Our goal is to design an architecture with a low WCET bound. For hard real-time systems, the likely increase in the ACET and BCET is acceptable because the complete system needs to be designed to reduce the WCET. It should be noted that a processor designed for low WCET will never be as fast in the average case as a processor optimized for ACET. Those are two different design optimizations. We define a time-predictable processor as "under the assumption that only feasible execution paths are analyzed, a time-predictable processor's WCET bound is close to the real WCET."

In the following, we propose time-predictable solutions or replacements, if possible, for the issues we identified in the Section 6. Table 1 summarizes the issues of standard processors for WCET analysis and the proposed architectural solutions.

Table 1. Architectural issues for WCET analysis of standard processors and proposed architectural solutions.

	Standard processor WCET issues	Time-predictable processor
Pipeline	Dependencies and shared state	Simple pipeline
Instruction fetch	Unbounded timing effects	Avoid prefetch queue, use double buffer
Caches	Replacement policy, abstract cache state destruction by unknown addresses	Method cache, stack cache, and a highly associative, small heap cache
Branch prediction	Long history in dynamic predictors	Static branch prediction
Superscalar architectures	Timing anomalies	Avoid, instead use CMP and/or VLIW
Chip-multithreading	Interthread interference (cache, pipeline)	Avoid, instead use CMP
Chip-multiprocessors	Intercore interference via shared memory, cache	TDMA scheduled memory access

Pipeline Dependencies

Pipelining is a major architectural feature to speed up program execution. Different stages of an instruction are overlapped and, therefore, executed in parallel. The theoretical throughput of a scalar pipeline is one instruction per clock cycle.

In contrast to Whitham [16], we think that a time-predictable architecture should be pipelined. The pipeline should be simple and dependencies between instructions avoided, or at least minimized, to avoid unbounded timing effects.

Instruction Fetch

To avoid a prefetch queue, with probably unbounded execution-time dependencies over a stream of instructions, a fixed-length instruction set is recommended. Variable length instructions can complicate instruction cache analysis because an instruction can cross a block boundary. The method cache, as proposed in the following section, avoids this issue. Either all instructions of a function, independent of their length, are in the cache, or none of them.

Fetching variable-sized instructions from the method cache can be performed in a single cycle. The method cache is split into two interleaved memories banks. Each of the two cache memories needs a read port wide enough for a maximum-sized instruction. Accessing both memories concurrently with a clever address calculation overcomes the boundary issue for variable-sized instruction access.

Caches

To reduce the growing gap between the clock frequency of the processor and memory access times, multilevel cache architectures are commonly used. Since even a single-level cache is problematic for WCET analysis, more levels in the memory architecture are practically not analyzable. The additional levels also increase the latency of the memory access on a cache miss.

For the cache analysis, the addresses of the memory accesses need to be predicted. The addresses for the instruction fetch are easy to determine, and access to stack allocated data, for example, function arguments and local variables, is also quite regular. The addresses can be predicted when the call tree is known.

The addresses for heap-allocated data are very hard to predict statically—the addresses are only known during runtime (we found no publication that describes analysis of the data cache for heap-allocated data). Without knowing the address, a single access influences all sets in the cache.

To avoid corruption of the abstract cache state in the analysis by data accesses, separate instruction and data caches are mandatory [11]. Furthermore, we propose to split the data cache into a cache for stack-allocated data and a cache for global- or heap-allocated data. As stack allocated data is guaranteed thread local, the stack cache can be further simplified for CMP systems.

For the instruction cache, we propose a new form of organization where whole functions are loaded on a miss on call or return. Figure 2 shows the proposed organization of the three caches.

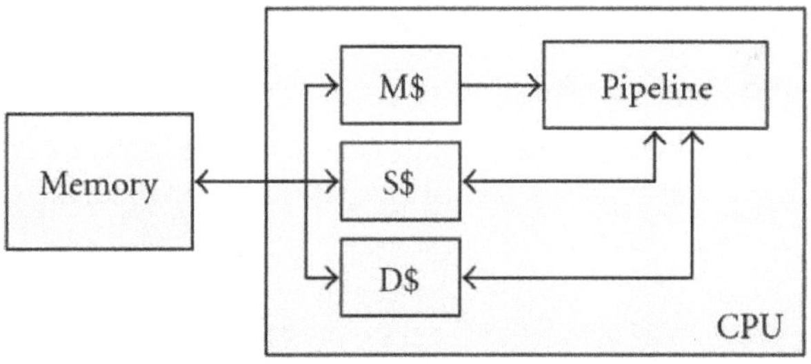

Figure 2. Cache configuration for the time-predictable architecture. The method cache M$ caches instructions of a full method/function. The data cache D$ is augmented by a stack cache S$ to avoid cache trashing of stack allocated data with heap allocated data.

The Instruction Cache

We propose a new form of organization for the instruction cache: the *method cache* [3], which has a novel replacement policy. A whole function or method is loaded into the cache on a call or return. This cache fill strategy pools all the cache misses of a function. All instructions except call and return are guaranteed cache hits. Only the call tree needs to be analyzed during the cache analysis. With the proposed cache organization, the cache analysis can be performed independently of the pipeline analysis.

Filling the cache on call and return only removes another source of interference: there is no competition for the main memory access between instruction cache and data cache. In traditional architectures, there is a subtle dependency between the instruction cache and memory access for a load or store instruction. For example, a load or store at the end of the processor pipeline competes with an instruction fetch that results in a cache miss. One of the two instructions is stalled for additional cycles by the other instruction. With a data cache, this situation can be even worse. The worst-case scenario for the memory stall time for an instruction fetch or a data load is two miss-penalties when both cache reads are a miss.

The main restriction of the method cache is that a whole method needs to fit into the cache. For larger methods, software- and hardware-based options are possible to resolve this issue. The compiler can split large methods into several shorter methods. At the hardware level, there are two options for methods that are too large: the cache can be disabled or the method cache can be switched into a direct-mapped mode.

If we avoid absolute jumps within a method, we can use a relative program counter within the method and place a method at each position within the cache. This property is fulfilled with Java bytecode, but can also be enforced by the compiler for C code.

For a full method load into the cache, we need to know the length of the method. This information is available in the Java class file. For compiled C code, this information can be provided in the executable. A simple convention, implemented in the linker, is to store the method length one word before the actual method starts. In order to use the method cache in an RISC processor, the ISA is extended with a prefetch instruction to force the cache load. The prefetch instruction can be placed immediately before the call or return instruction. It can also be scheduled earlier to hide the cache load latency.

The Stack Cache

Access patterns to stack allocated data are different from heap- or static-allocated data. Addresses into the stack are easy to predict statically because the allocation addresses of stack frames can be predicted by the analysis of the call tree. Furthermore, a new stack frame for a function call does not need to be cache-consistent with the main memory. The involved cache blocks need no cache fill from the main memory.

To benefit from these properties for WCET analysis, we propose to split the data cache into a stack cache and a cache for static- and heap-allocated data (it is possible to further split the data cache into a cache for static data and heap data). The organization of the cache for static- and heap-allocated data, further referred to as data cache, will be proposed in the following section.

The regular access pattern to the stack cache will not benefit from set associativity. Therefore, the stack cache is a simple direct-mapped cache. The stack contains local variables and the write frequency is higher than for other memory areas. The high frequency mandates a write back organization.

A stack cache is similar to a windowed register file as implemented in the Berkeley RISC processor [28]. A stack cache can be organized to exchange data with the main memory on a stack frame basis. When the cache overflows, which happens only during a call, the oldest frame or frames have to be moved to the memory. A frame needs to be loaded from the memory only when a function returns. Exchange with the main memory can be implemented in hardware, microcode, or with compiler visible machine instructions.

If the maximum call depth results in a stack that is smaller than the stack cache, all accesses will be a cache hit. A write back occurs first when the program reaches a call depth resulting in a wrap around within the cache. A cache miss can occur only when the program goes up in the call tree and needs access to a cache block that was evicted by a call down in the call tree.

Figure 3 shows the call and return behavior of a program over time and the changing stack cache window. The stack grows downwards in the figure. The dashed box shows a possibility to enforce a write back at some program point. The following stack changes fit into the enforced stack window and no memory transactions are necessary.

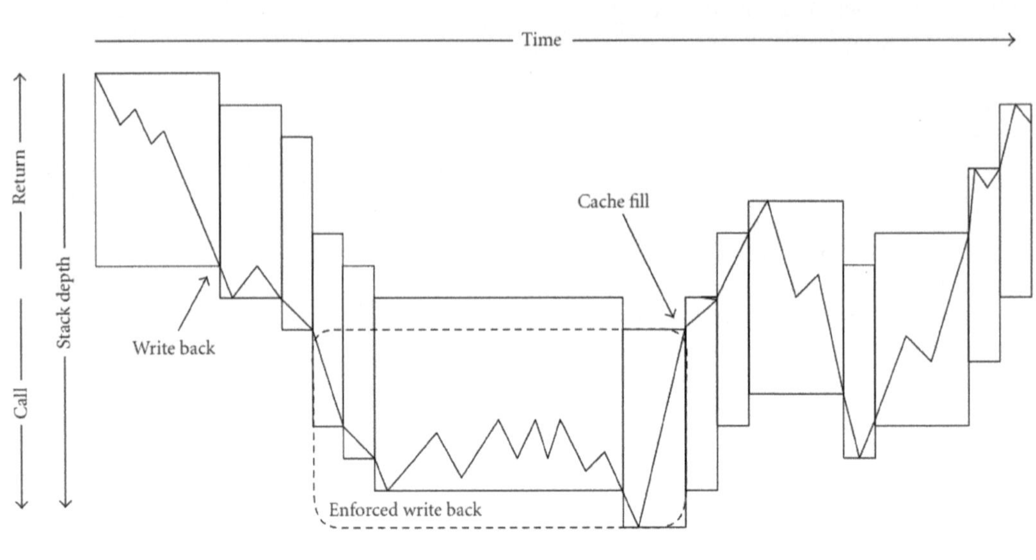

Figure 3. Stack usage for call and return and the resulting stack cache window. When the window overflows on a call, a write back of old frames is necessary. The stack cache fill is caused by an underflow after a return. Enforcing a write back of the whole stack cache can guarantee hits for subsequent, more deeply nested functions.

On a return, the previously used cache blocks can be marked empty because function local data is not accessible after the return (it could be accessed in C by returning a pointer to the stack data; however, this is undefined and considered poor programming practic). As a result, cache lines will never need to be written back on a cache wrap around after return. The stack cache activity can be summarized in the following way.

1. A cache miss can only occur after a return. The first miss is at least *one cache size* away from a leaf in the call tree.
2. Cache write back can only occur after a function call. The first write back is *one cache size* away from the root of the call tree.

We can make the misses and write backs more predictable by forcing them to occur at explicit points in the call tree. At these points, the cached stack frames are written back to the main memory and the whole stack cache is marked empty. If we place the flush points at function calls in the call tree that are within *one cache size* from the leaf functions, all cache accesses into that area are guaranteed hits. This algorithm can actually improve WCET because most of the execution time of a program is spent in inner loops further down the call tree.

Stack data is usually not shared between threads and no cache coherence and consistence protocol—the major bottleneck for CMP scaling—needs to be implemented for a CMP system.

The Data Cache

For conservatively written programs with statically allocated data, the address of the data is known after program linking. Value analysis results in a good prediction of read and write addresses. The addresses are the input for the cache analysis. In [48], control tasks, from a real-time benchmark provided by Airbus, were analyzed. For this benchmark, 90% of the memory accesses were predicted precisely.

In a modern object-oriented language, data is usually allocated on the heap. The address for these objects is only known at runtime. Even when using such a language in a conservative style, where all data is allocated during an initialization phase, it is not easy to predict the resulting addresses. The order of the allocations determines the addresses of the objects. When the order becomes unknown at one point in the initialization phase, the addresses for all following allocations cannot be determined precisely.

It is possible to analyze local cache effects with unknown addresses for an LRU set-associative cache. For an *n*-way associative cache, the history for *n* different addresses can be tracked. Because the addresses are unknown, a single access influences *all* sets in the cache. The analysis reduces the effective cache size to a single set.

The local analysis for the LRU-based cache is illustrated by a small example with a four-word cache. The example cache allocates a cache block on a write. Table 2 shows a code fragment with access to heap-allocated data (objects a, b, c, and d). The cache state after the load or store instruction is shown in the right section of the table. The leftmost column of the cache state represents the youngest element, the rightmost column the oldest (the LRU element). We assume a 4-way set-associative cache for the example. Therefore, we can locally track four different and *unknown* addresses. After the first two constant assignments, we know that a.v and b.v are in the cache. The following load of b.v is trivially a hit and the store into c.v changes the cache content and the age of a.v and b.v. All following loads are hits and only change the age ordering of the cache elements. In this small example we dealt with four different and unknown addresses, but could classify all read accesses as hits for a four-word cache.

Table 2. An example of analyzable accesses to three heap-allocated objects with a four-word LRU cache. The cache content after the execution of a statement is depicted in the right section of the table.

		Cache state			
Instruction	**Memory**	**Youngest**		**Oldest**	
a.v = 123;	store a.v	a.v	—	—	—
b.v = 456;	store b.v	b.v	a.v	—	—
c.v = b.v;	load b.v	b.v	a.v	—	—
	store c.v	c.v	b.v	a.v	—
d.v = b.v;	load b.v	b.v	c.v	a.v	—
	store d.v	d.v	b.v	c.v	a.v
b.v = a.v;	load a.v	a.v	d.v	b.v	c.v
	store b.v	b.v	a.v	d.v	c.v

We propose to implement the cache architecture exactly as it results from this analysis—a small, fully associative cache with an LRU replacement policy. This cache organization is similar to the victim cache [49], which adds associativity to a direct-mapped cache. A small, fully associative buffer holds discarded cache blocks. The replacement policy is LRU.

LRU is difficult to calculate in hardware and only possible for very small sets. Replacement of the oldest block gives an approximation of LRU. The resulting FIFO strategy can be used for larger caches. To offset the less predictable behavior of the FIFO replacement [39], the cache has to be much larger than an LRU-based cache.

The Scratchpad Memory

A common method for avoiding data caches is an onchip memory called scratchpad memory, which is under program control. This program managed memory entails a more complicated programming model although it can be automatically partitioned [50, 51]. A similar approach for time-predictable caching is to lock cache blocks. The control of the cache locking [52] and the allocation of data in the scratchpad memory [53, 54] can be optimized for the WCET. A comparison between locked cache blocks and a scratchpad memory with respect to the WCET can be found in [55].

Exposing the scratchpad memory at the language level can further help to optimize the time-critical path of the application.

Branch Prediction

As the pipelines of current general-purpose processors become longer to support higher clock rates, the penalty of branches also increases. This is compensated by branch prediction logic with branch target buffers. However, the upper bound of the branch execution time is the same as without this feature.

Simple static branch prediction (e.g., backward branches are assumed taken, forward branches not taken) or compiler-generated branch predictions are WCET-analyzable options. One-level dynamic branch predictors can be analyzed [56]. The branch history table has to be separate from the instruction cache to allow independent modeling for the analysis.

Instruction Level Parallelism

Statically scheduled VLIW processors are an option for a time-predictable architecture. The balance between the VLIW width and the number of cores in a CMP system depends on the application domain. For control-oriented applications, we assume that a dual-issue VLIW is a practical architecture. DSP-related applications can probably fill more instruction slots with useful instructions.

Dynamically scheduled superscalar architectures are not considered as an option for a time-predictable architecture. The amount of hardware that is needed to extract ILP from a single thread is better spent on a (VLIW-based) CMP system.

Chip Multithreading

Fine-grained multithreading within the pipeline is in principle not an issue for WCET analysis. The scheduling algorithm of the threads needs to be known and must not depend on the state of the threads. Round-robin scheduling is a time-predictable option. The execution time for simple instructions simply increases by a factor equal to the number of threads. The benefit of hiding pipeline stalls due to data dependencies or branches results in a lower factor for these instructions. Execution of n tasks on an n-way multithreading pipeline takes less (predictable) time than executing these tasks serially on a single threaded processor. However, cache misses, even if a single cache miss could be hidden, result in interference between the different threads because the memory interface is a shared resource.

Fine-grained multithreading resolves the data dependencies for a thread within the pipeline: the thread is only active in a single pipeline stage. Therefore, the forwarding network can be completely removed from the processor. This is an important simplification of the pipeline because the forwarding multiplexer is often part of the critical path that restricts the maximum clock frequency.

To avoid cache thrashing, each thread needs—in addition to its own register file—its own instruction and data cache, which reduces the effectively shared transistors to the pipeline itself. We think that the cost is too high for the small performance enhancement. Therefore, also duplicating the pipeline—resulting in a CMP solution—will result in a better performance/cost factor.

SMT is not an option as the interaction between the threads is too complex to model.

Chip Multiprocessors

Embedded applications need to control and interact with the *real*-world, a task that is inherently parallel. Therefore, these systems are good candidates for CMPs. We discuss that the transistors required to implement superscalar architectures are better used on complete replication of simple cores.

CMP systems share the access bandwidth to the main memory. To build a time-predictable CMP system, we need to schedule the access to the main memory in a predictable way. A predictable scheduling can only be time-based, where each core receives a fixed time slice. This scheduling scheme is called time division multiple access (TDMA). The time slices do not need to be of equal size. The execution time of un-cached loads and stores and the cache miss-penalty depend on this schedule and, therefore, for accurate WCET analysis, the complete schedule needs to be known.

Assuming that enough cores are available, we propose a CMP model with a single thread per processor. In that case, thread switching and schedulability analysis for each individual core disappears. Since each processor executes only a single thread, the WCET of that thread can be as long as its deadline. When the period of a thread is equal to its deadline, 100% utilization of that core is feasible. For threads that have enough slack time left, we can increase the WCET by decreasing their share of the bandwidth on the memory bus. Other threads with tighter deadlines can, in turn, use the freed bandwidth and run faster. The usage of the shared resource main memory is adjusted by the TDMA schedule. The TDMA schedule itself is the input for WCET analysis for all threads. Finding a TDMA schedule, where all tasks meet their deadlines, is thus an iterative optimization problem.

Figure 4 shows the analysis tool flow for the proposed time-predictable CMP with three tasks. First, an initial arbiter schedule is generated, for example, one with equal time slices. That schedule and the tasks are the input of WCET analysis performed for each task individually. If all tasks meet their deadline with the resulting WCETs, the system is schedulable. If some tasks do not meet their deadline and other tasks have some slack time available, the arbiter scheduler is adapted accordingly. WCET analysis is repeated, with the new arbiter schedule until all tasks meet their deadlines or no slack time for an adaption of the arbiter schedule is available. In the latter case, no schedule for the system is found.

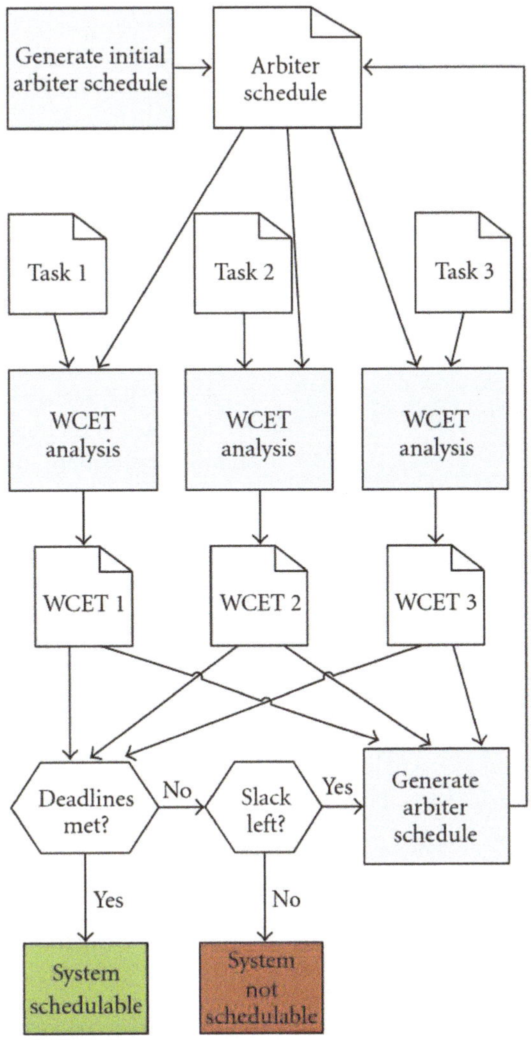

Figure 4. Tool flow for a CMP-based real-time system with one task per core and a static arbiter schedule. If the deadlines are not met, the arbiter schedule is adapted according to the WCETs and deadlines of the tasks. After the update of the arbiter schedule, the WCET of all tasks needs to be recalculated.

Documentation

The hardware description language VHDL was originally developed to document the behavior of digital circuits. Today, digital hardware can be synthesized from a VHDL description. Therefore, the VHDL code for the processor is the ideal form of documentation. VHDL code can also be simulated and all interactions between different components are observable.

An open-source design enables the WCET tool provider to check the real processor when the documentation is missing; documentation errors are also easier to find. Sun provides the Verilog source of their Niagra T1 [42] as open-source under the GNU GPL (http://www.opensparc.net/opensparc-t1/downloads.html).

EVALUATION

In this section, we evaluate some of the proposed time-predictable architectural features with JOP [2], an implementation of a Java processor. We have chosen to natively support Java as it is the language which will be used for future safety critical systems [57, 58]. Java's intermediate representation, the Java class file, is analysis friendly and the type information can be reconstructed from the class file. Executing bytecodes—the instruction set of the Java virtual machine (JVM)—directly in the hardware allows WCET analysis at the bytecode level. The translation step from bytecode to machine code, which introduces timing inaccuracies, can be avoided.

The Java Processor JOP

The major design goal of JOP is the time-predictable execution of Java bytecodes [59]. All functional units, and especially the interactions between them, are carefully designed to avoid any timing dependency between bytecodes.

JOP dynamically translates the Java bytecodes to a stack-based microcode that can be executed in a three-stage pipeline. The translation takes exactly one cycle per bytecode. Compared to other forms of dynamic code translation, the scheme used in JOP does not add any variable latency to the execution time and is, therefore, time-predictable.

JOP contains a simple execution stage with the two topmost stack elements as discrete registers. No write back stage or forwarding logic is needed. The short pipeline (four stages) results in short conditional branch delays; a difficult to analyze branch prediction logic or a branch target buffer can be avoided.

All microcode instructions have a constant execution time of one cycle. No stalls are possible in the microcode pipeline.

Loads and stores of object fields are handled explicitly. The absence of timing dependencies between bytecodes results in a simple processor model for the low-level WCET analysis.

The proposed architecture is open-source and all design files are available (http://www.jopdesign.com/). The instruction timing of the bytecodes is documented.

Method Cache

JOP contains the proposed method cache. The default configuration is 4 KB, divided into 16 blocks of 256 Bytes. The replacement strategy is FIFO.

WCET analysis of the method cache and of standard instruction caches is currently under development. Therefore, we perform only average-case measurements for a comparison between a time-predictable cache organization and a standard cache organization. With a simulation of JOP, we measure the cache misses and miss-penalties for different configurations of the method cache and a direct-mapped cache. The miss-penalty and the resulting effect on the execution time depend on the main memory system. Therefore, we simulate three different memory technologies: static RAM (SRAM), synchronous DRAM (SDRAM), and double data rate (DDR) SDRAM. For the SRAM, a latency of 1 clock cycle and an access time of 2 clock cycles per 32 bit word are assumed. For the SDRAM, a latency of 5 cycles (3 cycles for the row address and 2 cycles for the CAS latency) is assumed. The SDRAM delivers one word (4 bytes) per cycle. The DDR SDRAM has a shorter latency of 4.5 cycles and transfers data on both the rising and falling edges of the clock signal.

The resulting miss-cycles are scaled to the bandwidth consumed by the instruction fetch unit. The result is the number of cache fill cycles per fetched instruction byte: in other words, the average main memory access time in cycles per instruction byte. A value of 0.1 means that for every 10 fetched instruction bytes, one clock cycle is spent to fill the cache.

Table 3 shows the result for different configurations of a direct-mapped cache. For the evaluation, we used an adapted version of the real-time application Kfl (the Kfl benchmark is also used in Section 5.4), which is a node in a distributed control application. As the embedded application is quite small (1366 LOC), we simulated small instruction caches. The best performing configuration depends on the relationship between memory bandwidth and memory latency. The data in bold emphasize the best block size for the different memory technologies. As expected, memories with a higher latency and bandwidth perform better with larger block sizes. For small block sizes, the latency clearly dominates the access time. Although the SRAM has half the bandwidth of the SDRAM and a quarter of the DDR SDRAM, it is faster than the SDRAM memories with a block size of 8 byte. In most cases, a block size of 16 bytes is fastest.

Table 3. Direct-mapped cache, average memory access time.

Cache size	Block size	SRAM	SDRAM	DDR
1 KB	8 B	**0.18**	0.25	0.19
1 KB	16 B	0.22	**0.22**	0.16
1 KB	32 B	0.31	0.24	**0.15**
2 KB	8 B	**0.11**	0.15	0.12
2 KB	16 B	0.14	**0.14**	**0.10**
2 KB	32 B	0.22	0.17	0.11

Table 4 shows the average memory access time per instruction byte for the method cache. Because we load full methods, we have chosen larger block sizes than for a standard cache. All configurations benefit from a memory system with a higher bandwidth. The method cache is less latency sensitive than the direct-mapped instruction cache. For the small 1 KB cache, the access time is almost independent of the block size. The capacity misses dominate. From the 2 KB configuration, we see that smaller block sizes result in less cache misses. However, smaller block sizes result in more hardware for the hit detection since the method cache is in effect fully associative. Therefore, we need a balance between the number of blocks and the performance.

Table 4. Method cache, average memory access time.

Cache size	Block size	SRAM	SDRAM	DDR
1 KB	16 B	0.36	0.21	0.12
1 KB	32 B	0.36	0.21	0.12
1 KB	64 B	0.36	0.22	0.12
1 KB	128 B	0.41	0.24	0.14
2 KB	32 B	0.06	0.04	0.02
2 KB	64 B	0.12	0.08	0.04
2 KB	128 B	0.19	0.11	0.06
2 KB	256 B	0.37	0.22	0.13

The cache conflict is high for the small configuration with 1 KB cache. The direct-mapped cache, backed up with a low-latency main memory, performs better than the method cache. When high-latency memories are used, the method cache performs better than the direct-mapped cache. This is expected as the long latency for a transfer is amortized when more data (the whole method) is filled in one request.

A small block size of 32 bytes is needed in the 2 KB method cache to outperform the direct-mapped cache with the low-latency main memory as represented by the SRAM. For higher latency memories (SDRAM and DDR), a method cache with a block size of 128 bytes outperforms the direct-mapped instruction cache.

The comparison does not show if the method cache is more easily predictable than other cache solutions. It shows that caching full methods performs similarly to standard caching techniques.

Stack Cache

In JOP, a simplified version of the proposed stack cache is implemented. The JVM uses the stack not only for the activation frame and for local variables but also for operands. Therefore, the two top elements of the stack are implemented as registers [4]. With this configuration, we can avoid the write-back pipeline stage.

The fill and spill between the stack cache and the main memory is simplified. The cache content is exchanged only on a thread switch. Therefore, the maximum call depth is restricted by the onchip cache size. In a future version of JOP, we intend to relax this limitation. The cache fill will be performed on a return and the write back on invoke when necessary. A stack analysis tool will add a marker to the methods where a full cache write back will be performed and the stack access in methods deeper in the call tree will be guaranteed hits. Heap-allocated data and static fields are not cached in the current implementation of JOP.

Branch Prediction

In JOP, branch prediction is avoided. This results in pressure on the pipeline length. The microprogrammed core processor has a pipeline length of as little as three stages resulting in a branch execution time of three cycles in microcode. The two slots in the branch delay can be filled with instructions or *nop*. With the additional bytecode fetch and translation stage, the overall pipeline is four stages and results in a four-cycle execution time for a bytecode branch.

WCET Analysis

Bytecode instructions that do not access memory have a constant execution time. Most simple bytecodes are executed in a single cycle. Table 5 shows example instructions and their timing. The access time to object, array, and class fields depends on the timing of the main memory. With a memory with r_{ws} wait states for a read access, the execution time for, for example, getfield, is

$$t_{\text{getfield}} = 11 + 2r_{ws}. \quad (1)$$

Table 5. Execution time of simple bytecodes in cycles.

Instruction	Cycles	Funtion
iconst_0	1	load constant 0 on TOS
bipush	2	load a byte constant on TOS
iload_0	1	load local variable 0 on TOS
iload	2	load a local variable on TOS
dup	1	duplicate TOS
iadd	1	integer addition
isub	1	integer subtraction
ifeq	4	conditional branch

To demonstrate that JOP is amenable to WCET analysis, we have built an IPET-based WCET analyzer [60]. While loop bounds are annotated at the source level, the analysis is performed at the bytecode level. Without dependencies between bytecodes, the pipeline analysis can be omitted. The execution time of basic blocks is calculated simply by adding the execution time of individual bytecodes. For the method cache, we have implemented a simplified analysis where only leaf nodes in the call tree are considered. A return from such a leaf node is a guaranteed hit. (The maximum method size is restricted to half of the cache size.) Invocation of a leaf node in a tight loop (without invocations of other methods) is classified as a miss for the first iteration and a hit for the following iterations. For small benchmarks, the overestimation of the WCET is around 5%. For two real applications (Lift and Kfl) the analysis resulted in an overestimation of 56% and 116%. It should be noted that the overestimation is calculated by comparison with measurement-based WCET estimation, which is not a safe approach.

Another indication that JOP is a WCET *friendly* design is that other real-time analysis projects use JOP as the primary target platform. Harmon has developed a tree-based WCET analyzer for interactive back-annotation of WCET estimates into the program source [61]. Bøgholm et al. have developed an integrated WCET and scheduling analysis tool based on model checking [62].

Comparison with Picojava

We compare the time-predictable JOP design with picoJava [63, 64], a Java processor designed for average-case performance. Simple bytecodes are directly supported by the processor. Most of them execute in a single cycle. More complex bytecodes trap to a software routine. However, the invocation time of the trap depends on the cache state and is between 6 cycles in the best case and 426 cycles in the worst case—a factor in the order of two magnitudes. Some of the trapped instructions (e.g., invokevirtual) can be replaced at runtime by a *quick* version (e.g., invokevirtual quick). This replacement results in different execution times for the first execution of some code and following executions.

To speed up sequences of stack operations, picoJava can fold several instructions into an RISC style three register operation, for example, the sequence: load, load, add, store. This feature compensates for the inefficiency of a stack machine. However, the folding unit depends on a 16-byte instruction buffer with all the resulting unbounded timing effects of a prefetch queue.

However, picoJava implements a 64-word stack buffer as discrete registers. Spill and fill of that stack buffer is performed in background by the hardware. Therefore, the stack buffer closely interacts with the data cache. The interference between the folding unit, the instruction buffer, the instruction cache, the stack buffer, the data cache, and the memory interface causes complications in modeling picoJava for WCET analysis.

Also, picoJava is about 8 times larger than JOP and can be clocked at less than half of the frequency of JOP in the same technology [65]. Therefore, the small size of a time-predictable architecture naturally leads to a CMP system.

Performance

One important question remains: is a time-predictable processor slow? We evaluate the average-case performance of JOP by comparing it with other embedded Java systems: Java processors from industry and academia as well as two just-in-time (JIT) compiler-based systems. For the comparison, we use JavaBenchEmbedded, (available athttp://www.jopwiki.com/JavaBenchEmbedded) a set of open-source Java benchmarks for embedded systems. However, Kfl and Lift are two real-world applications adapted with a simulation of the environment to run as stand-alone benchmarks. UdpIp is a simple client/server test program that uses a TCP/IP stack written in Java.

Table 6 shows the raw data of the performance measurements of different embedded Java systems for the three benchmarks. The numbers are iterations per second whereby a higher value represents better performance. Figure 5 shows the results scaled to the performance of JOP.

Table 6. Application benchmark performance on different Java systems. The table shows the benchmark results in iterations per second—a higher value means higher performance.

	Kfl	UdpIp	Lift
Cjip	176	91	
Jamuth	3400	1500	
EJC	9893	2882	
SHAP	11570	5764	12226
aJ100	14148	6415	
JOP	18275	8467	18649
picoJava	23813	11950	25444
CACAO/YARI	39742	17702	38437

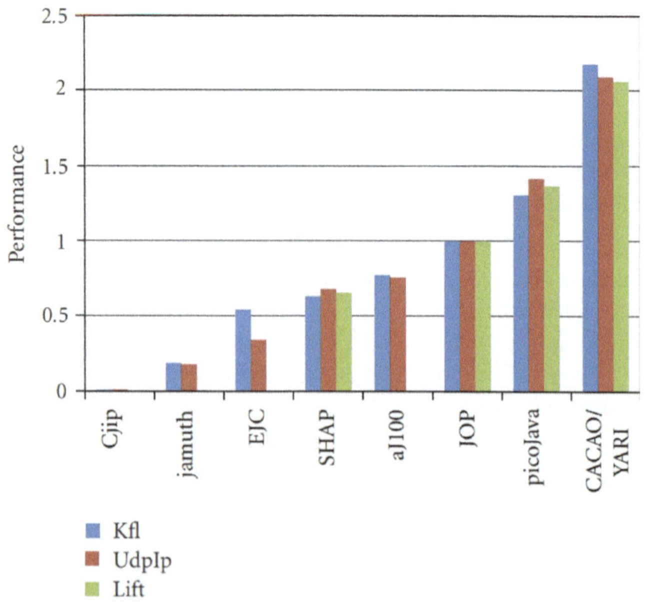

Figure 5. Performance comparison of different Java systems with embedded application benchmarks. The results are scaled to the performance of JOP.

The numbers for JOP are taken from an implementation in the Altera Cyclone FPGA [66], running at 100 MHz. JOP is configured with a 4 KB method cache and a 1 KB stack cache.

Cjip [67] and aJ100 [68] are commercial Java processors, which are implemented in an ASIC and clocked at 80 and 100 MHz, respectively. Both cores do not cache instructions. However, aj100 contains a 32 KB onchip stack memory. jamuth [69] and SHAP [70] are Java processors that are implemented in an FPGA. Also, jamuth is the commercial version of the Java processor Komodo [71], a research project for real-time chip multithreading; and it is configured with a 4 KB direct-mapped instruction cache for the measurements. The architecture of SHAP is based on JOP and enhanced with a hardware object manager. Also, SHAP implements the method cache [72]. The benchmark results for SHAP are taken from the SHAP website (http://shap.inf.tu-dresden.de/, accessed July, 2008); and it is configured with a 2 KB method cache and 2 KB stack cache.

Also, picoJava [73] is a Java processor developed by Sun. picoJava is no longer produced and the second version (picoJava-II) was available as open-source Verilog code. Puffitsch implemented picoJava-II in an FPGA (Altera Cyclone-II) and the performance numbers are obtained from that implementation [65]. picoJava is configured with a direct-mapped instruction cache and a 2-way set-associative data cache. Both caches are 16 KB.

EJC [74] is an example of a JIT system on an RISC processor (32-bit ARM720T at 74 MHz). The ARM720T contains an 8 KB unified cache. To compare JOP with a JIT-based system in exactly the same hardware, we use the research JVM CACAO [75] on top of the MIPS compatible soft-core YARI [76]. YARI is configured with a 4-way set-associative instruction cache and a 4-way set-associative write-through data cache. Both caches are 8 KB.

The measurements do not provide a clear answer to the question of whether a time-predictable architecture is slow. JOP is about 33% faster than the commercial Java processor aJ100. However, picoJava is 36% faster than JOP and the JIT/RISC combination is about 111% faster than JOP. (The numbers of CACAO/YARI are from [76]. In the mean time, YARI has been enhanced and outperforms JOP by a factor of 2.8.) We conclude that a time-predictable solution will never be as fast in the average case as a solution optimized for the average case.

Hardware Area and Clock Frequency

Table 7 compares the resource consumption and maximum clock frequency of a time-predictable processor (JOP), a standard MIPS architecture (YARI), and a complex Java processor (picoJava), when implemented in the same FPGA. The streamlined architecture of JOP results in a small design: JOP is half the size of the MIPS core YARI, and compared to picoJava consumes about 12% of the resources. JOP's size allows implementing a CMP version of JOP even in a low-cost FPGA. The simple pipeline of JOP achieves the highest clock frequency of the three designs. From the frequency comparison, we can estimate that the maximum clock frequency of JOP in an ASIC will also be higher than a standard RISC pipeline in an ASIC.

Table 7. Resource consumption and maximum operating frequency of JOP, YARI, and picoJava.

Soft-core	Logic cells	Memory	Frequency
JOP	3,300	7.6 KB	100 MHz
YARI	6,668	18.9 KB	75 MHz
picoJava	27,560	47.6 KB	40 MHz

Jop CMP System

We have implemented a CMP version of JOP with a fairness-based arbiter [77]. All cores are allotted an equal share of the memory bandwidth. Each core has its own method cache and stack cache. Heap-allocated data is not cached in this design.

When comparing a JOP CMP system against the complex Java processor picoJava, a dual core version of JOP is about 5% slower than a single picoJava core, but consumes only 22% of the chip resources. With four cores, JOP outperforms picoJava by 30% with size of 43% of picoJava.

A configurable TDMA arbiter for a time-predictable CMP system and the integration of the arbitration schedule into the WCET tool [60] is presented in [9].

Summary

A model of a processor with accurate timing information is essential for tight WCET analysis. The architecture of JOP and the microcode are designed with this in mind. Execution time of bytecodes is known cycle accurately [59]. It is possible to analyze the WCET at the bytecode level [78] without the uncertainties of an interpreting JVM [79] or generated native code from ahead-of-time compilers for Java.

CONCLUSION

In this paper, we discuss a time-predictable computer architecture for embedded real-time systems that supports WCET analysis. We have identified the problematic microarchitecture features of standard processors and provided alternative solutions when possible.

Dynamic features, which model a large execution history, are problematic for WCET analysis. Especially interferences between different features result in a state space explosion for the analysis. The proposed architecture is an in-order pipeline with minimized instruction dependencies. The cache memory consists of a method cache containing whole methods and a data cache that is split for stack-allocated data and heap-allocated data. The pipeline can be extended to a dual-issue pipeline when the instructions are compiler scheduled. For further performance enhancements, we propose a CMP system with time-sliced arbitration of the main memory access. Running each task on its own core in a CMP system eliminates scheduling, and the related cache thrashing, from the analysis. The schedule of the memory access becomes an input for WCET analysis. With nonuniform time slices, the arbiter schedule can be adapted to balance the utilization of the individual cores.

The concept of the proposed architecture is evaluated by a real-time Java processor, called JOP. We have presented a brief overview of the architecture. A simple four-stage pipeline and microcoded implementation of JVM bytecodes result in a time-predictable architecture. The proposed method and stack caches are implemented in JOP. The resulting design makes JOP an easy target for the low-level WCET analysis of Java applications.

We compared JOP against several embedded Java systems. The result shows that a time-predictable computer architecture does not need to be slow. A streamlined, time-predictable processor design is quite small. Therefore, we can regain performance by the exploration of thread level parallelism in embedded applications with a replication of the processor in a CMP architecture.

The proposed processor has been used with success to implement several commercial real-time applications [80]. JOP is open-source under the GNU GPL, and all design files and the documentation are available at http://www.jopdesign.com/.

We plan to implement some of the suggested architectural enhancements in an RISC-based system in the future. We will implement the proposed stack cache and the method cache in YARI [76], an open-source, MIPS ISA-compatible RISC implementation in an FPGA.

A scratchpad memory for JOP is implemented and the integration into the programming model is under investigation. We will add a small fully associative data cache to JOP. This cache will also serve as a buffer for a real-time transactional memory for the JOP CMP system. We will investigate whether a standard cache for static data is a practical solution for Java.

ACKNOWLEDGMENT

The author thanks Wolfgang Puffitsch and Florian Brandner for the productive discussions on the topic and suggestions for improving the paper.

REFERENCES

1. Hennessy J, Patterson D: *Computer Architecture: A Quantitative Approach*. 4th edition. Morgan Kaufmann, San Francisco, Calif, USA; 2006.
2. Schoeberl M: A Java processor architecture for embedded real-time systems. *Journal of Systems Architecture* 2008, 54(1-2):265-286. 10.1016/j.sysarc.2007.06.001
3. Schoeberl M: A time predictable instruction cache for a Java processor. In *Proceedings of the On the Move to Meaningful Internet Systems: Workshop on Java Technologies for Real-Time and Embedded Systems (JTRES '04), October 2004, Agia Napa, Cyprus, Lecture Notes in Computer Science. Volume 3292*. Springer; 371-382.
4. Schoeberl M: Design and implementation of an efficient stack machine. *Proceedings of the 12th IEEE Reconfigurable Architecture Workshop (RAW '05), April 2005, Denver, Colo, USA*
5. Bate I, Conmy P, Kelly T, McDermid J: Use of modern processors in safety-critical applications. *The Computer Journal* 2001, 44(6):531-543. 10.1093/comjnl/44.6.531
6. Thiele L, Wilhelm R: Design for timing predictability. *Real-Time Systems* 2004, 28(2-3):157-177.
7. Edwards SA, Lee EA: The case for the precision timed (PRET) machine. In *Proceedings of the 44th ACM/IEEE Annual Conference on Design Automation (DAC '07), June 2007, San Diego, Calif, USA*. ACM Press; 264-265.
8. Lickly B, Liu I, Kim S, Patel HD, Edwards SA, Lee EA: Predictable programming on a precision timed architecture. In *Proceedings of the International Conference on Compilers, Architectures and Synthesis for Embedded Systems (CASES '08), October 2008, Atlanta, Ga, USA*. Edited by: Altman ER. ACM Press; 137-146.
9. Pitter C: Time-predictable memory arbitration for a Java chip-multiprocessor. *Proceedings of the 6th International Workshop on Java Technologies for Real-Time and Embedded Systems (JTRES '08), September 2008, Santa Clara, Calif, USA* 115-122.
10. Berg C, Engblom J, Wilhelm R: Requirements for and design of a processor with predictable timing. In *Perspectives Workshop: Design of Systems with Predictable Behaviour, 2004, Schloss Dagstuhl, Germany, Dagstuhl Seminar Proceedings. Volume 03471*. Edited by: Thiele L, Wilhelm R. Internationales Begegnungsund Forschungszentrum für Informatik (IBFI);
11. Heckmann R, Langenbach M, Thesing S, Wilhelm R: The influence of processor architecture on the design and the results of WCET tools. *Proceedings of the IEEE*2003, 91(7):1038-1054. 10.1109/JPROC.2003.814618
12. Anantaraman A, Seth K, Patil K, Rotenberg E, Mueller F: Virtual simple architecture (VISA): exceeding the complexity limit in safe real-time systems. In *Proceedings of the 30th Annual International Symposium on Computer Architecture (ISCA '03), June 2003, San Diego, Calif, USA*. ACM Press; 350-361.
13. Puschner P, Burns A: Writing temporally predictable code. In *Proceedings of the 7th IEEE International Workshop on Object-Oriented Real-Time Dependable Systems (WORDS '02), January 2002, San Diego, Calif, USA*. IEEE Computer Society; 85-94.
14. Delvai M, Huber W, Puschner P, Steininger A: Processor support for temporal predictability—the SPEAR design example. *Proceedings of the 15th Euromicro Conference on Real-Time Systems (ECRTS '03), July 2003, Porto, Portugal* 169-176.
15. Puschner P, Schoeberl M: On composable system timing, task timing, and WCET analysis. *Proceedings of the 8th International Workshop on Worst-Case Execution Time (WCET) Analysis, July 2008, Prague, Czech Republic*
16. Whitham J: *Real-time processor architectures for worst case execution time reduction, Ph.D. thesis*. University of York, York, UK; 2008.
17. Whitham J, Audsley N: Using trace scratchpads to reduce execution times in predictable real-time architectures. *Proceedings of the 14th IEEE Real-Time and Embedded Technology and Applications Symposium (RTAS '08), April 2008, St. Louis, Mo, USA* 305-316.
18. Li Y-TS, Malik S: Performance analysis of embedded software using implicit path enumeration. In *Proceedings of the ACM SIGPLAN Workshop on Languages, Compilers, & Tools for Real-Time Systems (LCTES '95), November 1995, La Jolla, Calif, USA*. ACM Press; 88-98.
19. Puschner PP, Schedl AV: Computing maximum task execution times—a graph-based approach. *Real-Time Systems* 1997, 13(1):67-91. 10.1023/A:1007905003094
20. Gustafsson J: *Analyzing execution-time of object-oriented programs using abstract interpretation, Ph.D. thesis*. Uppsala University, Uppsala, Sweden; 2000.
21. Puschner P, Burns A: Guest editorial: a review of worst-case execution-time analysis. *Real-Time Systems* 2000, 18(2-3):115-128.
22. Wilhelm R, Engblom J, Ermedahl A, *et al.*: The worst-case execution-time problem-overview of methods and survey of tools. *Transactions on Embedded Computing Systems* 2008, 7(3):1-53.
23. Engblom J, Ermedahl A, Södin M, Gustafsson J, Hansson H: Worst-case execution-time analysis for embedded real-time systems. *International Journal on Software Tools for Technology Transfer* 2003, 4(4):437-455. 10.1007/s100090100054
24. Nilsen KD, Rygg B: Worst-case execution time analysis on modern processors.*ACM SIGPLAN Notices* 1995, 30(11):20-30. 10.1145/216633.216650
25. Thesing S: *Safe and precise worst-case executiontime prediction by abstract interpretation of pipeline models, Ph.D. thesis*. University of Saarland, Saarland, Germany; 2004.
26. Lundqvist T, Stenström P: Timing anomalies in dynamically scheduled microprocessors. In *Proceedings of the 20th IEEE Real-Time Systems Symposium (RTSS '99), December 1999, Phoenix, Ariz, USA*. IEEE Computer Society; 12-21.

27. Wenzel I, Kirner R, Puschner P, Rieder B: Principles of timing anomalies in superscalar processors. In *Proceedings of the 5th International Conference on Quality Software (QSIC '05), September 2005, Melbourne, Australia*. IEEE Computer Society Press; 295-303.

28. Patterson DA: Reduced instruction set computers. *Communications of the ACM*1985, 28(1):8-21. 10.1145/2465.214917

29. Hennessy JL: VLSI processor architecture. *IEEE Transactions on Computers*1984, 33(12):1221-1246.

30. Lim S-S, Bae YH, Jang GT, *et al.*: An accurate worst case timing analysis for RISC processors. *IEEE Transactions on Software Engineering* 1995, 21(7):593-604. 10.1109/32.392980

31. Shaw AC: Reasoning about time in higher-level language software. *IEEE Transactions on Software Engineering* 1989, 15(7):875-889. 10.1109/32.29487

32. Engblom J: *Processor pipelines and static worst-case execution time analysis, Ph.D. thesis*. Uppsala University, Uppsala, Sweden; 2002.

33. Zhang N, Burns A, Nicholson M: Pipelined processors and worst case execution times. *Real-Time Systems* 1993, 5(4):319-343. 10.1007/BF01088834

34. Arnold R, Mueller F, Whalley D, Harmon M: Bounding worst-case instruction cache performance. *Proceedings of the Real-Time Systems Symposium (RTSS '94), December 1994, San Juan, Puerto Rico, USA* 172-181.

35. Healy CA, Whalley DB, Harmon MG: Integrating the timing analysis of pipelining and instruction caching. *Proceedings of the 16th Real-Time Systems Symposium, December 1995, Pisa, Italy* 288-297.

36. Lee C-G, Hahn J, Seo Y-M, *et al.*: Analysis of cache-related preemption delay in fixed-priority preemptive scheduling. *IEEE Transactions on Computers*1998, 47(6):700-713. 10.1109/12.689649

37. Busquets-Mataix JV, Serrano JJ, Ors R, Gil P, Wellings A: Adding instruction cache effect to schedulability analysis of preemptive real-time systems. In *Proceedings of the 2nd IEEE Real-Time Technology and Applications Symposium (RTAS '96), June 1996, Brookline, Mass, USA*. IEEE Computer Society Press; 204-212.

38. Healy CA, Arnold RD, Mueller F, Whalley DB, Harmon MG: Bounding pipeline and instruction cache performance. *IEEE Transactions on Computers* 1999, 48(1):53-70. 10.1109/12.743411

39. Reineke J, Grund D, Berg C, Wilhelm R: Timing predictability of cache replacement policies. *Real-Time Systems* 2007, 37(2):99-122. 10.1007/s11241-007-9032-3

40. Engblom J: Analysis of the execution time unpredictability caused by dynamic branch prediction. In *Proceedings of the 9th IEEE Real-Time and Embedded Technology and Applications Symposium (RTAS '03), May 2003, Toronto, Canada*. IEEE Computer Society; 152-159.

41. Li X, Roychoudhury A, Mitra T: Modeling out-of-order processors for WCET analysis. *Real-Time Systems* 2006, 34(3):195-227. 10.1007/s11241-006-9205-5

42. Kongetira P, Aingaran K, Olukotun K: Niagara: a 32-way multithreaded sparc processor. *IEEE Micro* 2005, 25(2):21-29. 10.1109/MM.2005.35

43. Hofstee HP: Power efficient processor architecture and the cell processor.*Proceedings of the 11th International Symposium on High-Performance Computer Architecture (HPCA '05), February 2005, San Francisco, Calif, USA* 258-262.

44. Kahle JA, Day MN, Hofstee HP, Johns CR, Maeurer TR, Shippy D: Introduction to the cell multiprocessor. *IBM Journal of Research and Development* 2005, 49(4-5):589-604.

45. Kistler M, Perrone M, Petrini F: Cell multiprocessor communication network: built for speed. *IEEE Micro* 2006, 26(3):10-23.

46. Seiler L, Carmean D, Sprangle E, *et al.*:Larrabee: a many-core ×86 architecture for visual computing. *ACM Transactions on Graphics* 2008, 27(3):1-15.

47. Yan J, Zhang W: WCET analysis for multi-core processors with shared L2 instruction caches. *Proceedings of the 14th IEEE Real-Time and Embedded Technology and Applications Symposium (RTAS '08), April 2008, St. Louis, Mo, USA* 80-89.

48. Ferdinand C, Heckmann R, Langenbach M, *et al.*: Reliable and precise WCET determination for a real-life processor. In *Proceedings of the 1st International Workshop on Embedded Software (EMSOFT '01), October 2001, Tahoe City, Calif, USA, Lecture Notes in Computer Science. Volume 2211*. Edited by: Henzinger TA, Kirsch CM. Springer; 469-485.

49. Jouppi NP: Improving direct-mapped cache performance by the addition of a small fully-associative cache and prefetch buffers. *Proceedings of the 17th Annual International Symposium on Computer Architecture (ISCA '90), May 1990, Seattle, Wash, USA* 364-373.

50. Angiolini F, Benini L, Caprara A: Polynomial-time algorithm for on-chip scratchpad memory partitioning. In *Proceedings of the International Conference on Compilers, Architecture, and Synthesis for Embedded Systems (CASES '03), October-November 2003, San Jose, Calif, USA*. ACM Press; 318-326.

51. Verma M, Marwedel P: Overlay techniques for scratchpad memories in low power embedded processors. *IEEE Transactions on Very Large Scale Integration (VLSI) Systems* 2006, 14(8):802-815.

52. Puaut I: WCET-centric software-controlled instruction caches for hard real-time systems. *Proceedings of the 18th Euromicro Conference on Real-Time Systems (ECRTS '06), July 2006, Dresden, Germany* 2006: 217-226.

53. Wehmeyer L, Marwedel P: Influence of memory hierarchies on predictability for time constrained embedded software. *Proceedings of the Conference on Design, Automation and Test in Europe (DATE '05), March 2005, Munich, Germany* 1: 600-605.

54. Suhendra V, Mitra T, Roychoudhury A, Chen T: WCET centric data allocation to scratchpad memory. In *Proceedings of the 26th IEEE International Real-Time*

Systems Symposium (RTSS '05), December 2005, Miami, Fla, USA. IEEE Computer Society; 223-232.

55. Puaut I, Pais C: Scratchpad memories vs locked caches in hard real-time systems: a quantitative comparison. In *Proceedings of the Conference on Design, Automation and Test in Europe (DATE '07), April 2007, Nice, France*. EDA Consortium; 1484-1489.
56. Colin A, Puaut I: Worst case execution time analysis for a processor with branch prediction. *Real-Time Systems* 2000, 18(2-3):249-274.
57. Wellings A: Is Java augmented with the RTSJ a better realtime systems implementation technology than Ada 95. *Ada Letters* 2003, 23(4):16-21.
58. Java Expert Group : Java specification request JSR 302: Safety critical Java technology. http://jcp.org/en/jsr/detail?id=302
59. Schoeberl M: A time predictable Java processor. *Proceedings of the Conference on Design, Automation and Test in Europe (DATE '06), March 2006, Munich, Germany* 1:800-805.
60. Schoeberl M, Pedersen R: WCET analysis for a Java processor. In *Proceedings of the 4th International Workshop on Java Technologies for Real-Time and Embedded Systems (JTRES '06), October 2006, Paris, France. Volume 177*. ACM Press; 202-211.
61. Harmon T, Klefstad R: Interactive back-annotation of worst-case execution time analysis for Java microprocessors. *Proceedings of the 13th IEEE International Conference on Embedded and Real-Time Computing Systems and Applications (RTCSA '07), August 2007, Daegu, Korea* 209-216.
62. Bøgholm T, Kragh-Hansen H, Olsen P, Thomsen B, Larsen KG: Model-based schedulability analysis of safety critical hard real-time Java programs. In*Proceedings of the 6th International Workshop on Java Technologies for Real-Time and Embedded Systems (JTRES '08), September 2008, Santa Clara, Calif, USA*. ACM Press; 106-114.
63. Sun : picoJava-II Microarchitecture Guide. Sun Microsystems, March 1999
64. Sun : picoJava-II Programmer's Reference Manual. Sun Microsystems, March 1999
65. Puffitsch W: *picoJava-II in an FPGA, M.S. thesis*. University of Technology, Vienna, Austria; 2007.
66. Altera : Cyclone FPGA Family Data Sheet. ver. 1.2, April 2003
67. Imsys Im1101c (the Cjip) technical reference manual, v0.25, 2004
68. aJile : aj-100 real-time low power Java processor. preliminary data sheet, 2000
69. Uhrig S, Wiese J: Jamuth: an IP processor core for embedded Java real-time systems. In *Proceedings of the 5th International Workshop on Java Technologies for Real-Time and Embedded Systems (JTRES '07), September 2007, Vienna, Austria*. ACM Press; 230-237.
70. Zabel M, Preußer TB, Reichel P, Spallek RG: Secure, real-time and multi-threaded general-purpose embedded Java microarchitecture. *Proceedings of the 10th Euromicro Conference on Digital System Design Architectures, Methods and Tools (DSD '07), August 2007, Lübeck, Germany* 59-62.
71. Kreuzinger J, Brinkschulte U, Pfeffer M, Uhrig S, Ungerer Th: Real-time event-handling and scheduling on a multithreaded Java microcontroller.*Microprocessors and Microsystems* 2003, 27(1):19-31. 10.1016/S0141-9331(02)00082-0
72. Preußer TB, Zabel M, Spallek RG: Bump-pointer method caching for embedded Java processors. In *Proceedings of the 5th International Workshop on Java Technologies for Real-Time and Embedded Systems (JTRES '07), September 2007, Vienna, Austria*. ACM Press; 206-210.
73. O'Connor JM, Tremblay M: picoJava-I: the Java virtual machine in hardware.*IEEE Micro* 1997, 17(2):45-53. 10.1109/40.592314
74. EJC : The ejc (embedded Java controller) platform. http://www.embedded-web.com/index.html
75. Krall A, Grafl R: CACAO—a 64-bit JavaVM just-in-time compiler. In *Proceedings of the Workshop on Java for Science and Engineering Computation (PPoPP '97), June 1997, Las Vegas, Nev, USA*. Edited by: Fox GC, Li W. ACM Press; Las Vegas, Nev, USA, June 1997
76. Brandner F, Thorn T, Schoeberl M: *Embedded JIT compilation with CACAO on YARI.* Institute of Computer Engineering, Vienna University of Technology, Vienna, Austria; June 2008.
77. Pitter C, Schoeberl M: Performance evaluation of a Java chip-multiprocessor.*Proceedings of the 3rd International Symposium on Industrial Embedded Systems (SIES '08), June 2008, La Grande Motte, France* 34-42.
78. Bernat G, Burns A, Wellings A: Portable worst-case execution time analysis using Java byte code. *Proceedings of the 12th Euromicro Conference on Real-Time Systems (ECRTS '00), June 2000, Stockholm, Sweden* 81-88.
79. Bate I, Bernat G, Murphy G, Puschner P: Low-level analysis of a portable Java byte code WCET analysis framework. *Proceedings of the 7th International Conference on Real-Time Computing and Applications (RTCSA '00), December 2000, Cheju Island, Korea* 39-48.
80. Schoeberl M: Application experiences with a real-time Java processor.*Proceedings of the 17th IFAC World Congress, July 2008, Seoul, Korea*

CHAPTER 6

INITIAL CONDITION FOR EFFICIENT MAPPING OF LEVEL SET ALGORITHMS ON MANY-CORE ARCHITECTURES

Gábor János Tornai and György Cserey

INTRODUCTION

The use of level-set (LS)-based curve evolution has become an interesting research topic due to its versatility and accuracy. These flows are widely used in various fields like computational geometry, fluid mechanics, image processing, computer vision, and materials science [1]. In general, the method entails that one evolves a curve, surface, or image with a partial differential equation (PDE) and obtains the result at a point in the evolution. There is a subset of problems where only the steady state of the LS evolution is of practical interest like segmentation and detection. In this article, only this subset is considered. In addition, we do not form any operator or speedfield (F) for driving the evolution of the LSs. However, two theoretical worst-case bounds of the required number of iterations are proposed to reach the steady state for a well-defined class of LS-based evolution. These bounds depend only on the initial condition. Furthermore, the bounds only allow an extremely small number of iterations if the evolution is calculated with a properly chosen initial condition. These kinds of evolutions are calculated very quickly on many-core devices.

The subject of this paper is both theoretical and practical. The theoretical side is clearly the two new theorems on the worst case of the required number of iterations of the Shi LS evolution [2]. This evolution omits the numerical solution of the underlying PDE and successfully approximates it with a rule-based evolution. It is based on the sign of the driving forces (F) normal to the curves to be changed. The first theorem gives a general bound, and the second one assumes a special kind of discrete convexity defined in subsection 'Basic definitions'. The practical side is presented through two case studies, namely, the LS evolution of Shi is mapped in a straightforward way on two completely different many-core architectures. With a lot of small curves in the initial condition, which would be infeasible on a conventional single-core processor, the proposed theorems ensure small number of iterations. Additionally, with the change in the initial condition (instead of one curve, a lot of small curves are used), the computing width of the many-core platform are utilized.

The first successful method to speed up the LS evolution was introduced in [3]. A local method was proposed in [4] with better characteristics. These methods are labeled as narrow banding methods. However, we are not presenting any PDE operators and do not design any speedfield. Instead, we direct the reader to the work of Sapiro [5] who gives a detailed picture from the art of PDE operator design for a given purpose. Furthermore, there are several results [6–9] regarding region- and model-based evolutions. In [10] the authors used Sobolev norm instead of the standard inner-product-based L_2 norm and showed that this norm allows new energies to implement otherwise considered infeasible.

There are multiple results reporting successful mapping of LS evolution to many-core platforms. An interactive 3D solver for GPU is presented in [11]. This realization uses a graphics application programming interfaces (API) (OpenGL) and the rendering pipeline. Two later works [12, 13] applied the computing-unified device architecture (CUDA) of NVIDIA (NVIDIA Corporation, Santa Clara, CA, USA). Both papers worked with 3D volumes. In [12] the authors mapped a sparse solver, while in [13] a higher-order scheme was used to evolve the LSs. Cellular neural networks (CNN) [14] proved to be an inspiring construct. There have been results regarding the mapping of LS-like evolutions to CNN [15–17]. Cserey et al. [15] successfully mapped the nonlinear histogram modification

Citation: Tornai, G.J. and Cserey, G (2014). Initial condition for efficient mapping of level set algorithms on many-core architectures, EURASIP Journal on Advances in Signal Processing, 2014: 30. DOI:10.1186/1687-6180-2014-30.

to CNN. A later work [17] realized an online boundary detection algorithm based on LS curve evolution to extract the volume of the right atrium. These papers and results indicate that various LS evolutions can be mapped and used on different many-core platforms. In this paper, not a new numerical variant or many core, parallel implementation of the narrow band algorithm is presented. Instead, we are focusing on the given type of evolution on many-core devices and for this evolution, we give two theorems upperbounding the required number of iterations of the evolution process.

This paper is organized as follows. The next section summarizes the theory for curve evolution realized as LS motion, then some definitions follow and the two theorems on worst-case bound of the required number of iterations with their proofs close the theoretical part. Section 'Many-core hardware platforms' describes the hardware platforms used in the two case studies. Section 'Experiments' presents the experimental results. In section 'Validation,' we validate our method against three numerically solved PDE of LS flows. It is followed by section 'Discussion' in and the last section concludes our paper.

THEORY

Basic Curve Evolution

Let us consider a family of closed curves $\gamma(s,t)$ (or surfaces) in R^k, $(k=2,3)$ where sparameterizes the curve and t the (artificial) time. Our aim is to trace $\gamma(s)$, which moves normal to itself, with a given speed function F. If there are no restrictions on the sign of F, then the motion of γ can be complex. Furthermore, it is hardly trackable by curve-based (Lagrangian) methods since these methods cannot handle topological changes and become unstable near singularities. However, the initial curve $\gamma(s,t = 0)$ can be embedded as the zero LS of a higher dimensional function so the evolution of this function is linked to the propagation of the front through the time evolution of this initial value problem. This family of methods is called level set methods (LSM). The evolution equation of these methods is as follows:

$$\frac{\partial \phi}{\partial t} = F|\nabla \phi| \tag{1}$$

γ is embedded into a function that is called LS function. It is denoted by ϕ. The uniformly sampled domain of ϕ is denoted by D and a point $x \in D$ is characterized by its coordinates $(x = (x_1,..x_k))$. The region enclosed by γ is referred to as the object region and is denoted by Ω. Similarly, the region outside γ is referred to as the background region and is denoted by Γ. LetΩ^* and Γ^* denote the true object region and true background region, respectively. ϕ is chosen to be negative inside Ω and positive in Γ. The zero LS is represented implicitly by ϕ. The LS evolution process of [2] is used throughout in this work. In the neighborhood of the zero LS, two sets are defined uniquely with respect to the selected discrete neighborhood:

$$L_{\text{in}} = \{\mathbf{x}|\phi(\mathbf{x}) < 0 \text{ and } \exists \mathbf{y} \in N(\mathbf{x}) \text{ that } \phi(\mathbf{y}) > 0\} \tag{2}$$

$$L_{\text{out}} = \{\mathbf{x}|\phi(\mathbf{x}) > 0 \text{ and } \exists \mathbf{y} \in N(\mathbf{x}) \text{ that } \phi(\mathbf{y}) < 0\} \tag{3}$$

where $N(\mathbf{x})$ is the selected neighborhood defined by Equation 4, but can be any discrete neighborhood known from discrete topology.

$$N(\mathbf{x}) = \{\mathbf{y} \in D| \sum_{k=1}^{K} |y_k - x_k| = 1\} \ \forall \mathbf{x} \in D\} \tag{4}$$

The sets L_{in} and L_{out} are referred to as active front, and ϕ is defined as an approximated signed distance function:

$$\phi(x) = \begin{cases} -3, \text{ if } \mathbf{x} \in \Omega, \text{ and } \mathbf{x} \notin L_{\text{in}} \text{ inner points} \\ -1, \text{ if } \mathbf{x} \in L_{\text{in}} \\ 1, \text{ if } \mathbf{x} \in L_{\text{out}} \\ 3, \text{ if } \mathbf{x} \in \Gamma, \text{ and } \mathbf{x} \notin L_{\text{out}} \text{ outer points} \end{cases} \tag{5}$$

The LSM of Shi uses only the sign of the speed function F to determine the motion of the active front at a given point. The speed function itself is chosen according to the requirements and can be arbitrary (Chan-Vese, geodesic active contour, geodesic active region (GAR) - gaussian mixture model (GMM), nonparametric mixture model, etc.). It is (re) calculated in each cycle for the active front, but its sign at every point of the active front determines the action on these points. Furthermore, from an implementation point of view, both sets of the active front are realized as two separate lists. The motion of the active front is realized by two local switch operators, one for outward and one for inward motion. By applying switch_in() to any points in L_{out} having positive sign F at that point, the active front is moved outward one grid point from that location. Namely, the point is moved from L_{out} to L_{in}, its exterior neighbors are put to L_{out}, and its interior neighbors are deleted from L_{in}. The switch_out() procedure operates similarly in the other direction.

One iteration of the evolution of the algorithm performs four scans. Firstly, it scans L_{out} for switch_in() then it scans L_{in} for elements that are not in the active front any more (all neighbors reside in Ω), afterwards, L_{in} is scanned for switch_out(). Lastly, L_{out} is scanned for elements that are not in the active front any more. Basically, the algorithm makes one step inward or outward according to the sign of the speed function. If it contains a curvature-dependent smoothing term, then sharpened gaussian filtering (shock-filtered heat diffusion) can be applied just on the active front itself in a different cycle after a given number of evolution cycles. For further details see [2, 18, 19].

Basic Definitions

A path p between x and y is a sequence of points $x_l (l = 0,1,\ldots,L) \in D$ subject to $x_l \in N(x_{l+1})$ and $x = x_0$ and $y = x_L$. A set of points $\mathcal{A}$ forms a connected region if and only if there exists a path pbetween every $x, y \in \mathcal{A}$ subject to $\forall x_l \in p$ is an element of $\mathcal{A}$.

The length of a path is a non-negative integer (L) and $L = |p|-1$, where |.| denotes the number of points in the path. A minimum path p_{min} is a shortest path meaning there are no shorter p'paths between x and y. Minimum path is usually not unique and can depend on the chosen discrete neighborhood. The distance between x and y is a non-negative integer that is exactly the length of a minimum path between the two points. This is a real metric and is going to be referred to as d_d.

Within a connected region $\mathcal{A}$, a minimal path p between x and y is minimal if and only if $\mathcal{A} \cap p = p$ and there are no shorter p' paths within $\mathcal{A}$ between **x** and **y**. Like the minimum path, the minimal path may not be unique and may depend on the chosen neighborhood. The diameter B of a connected region is the longest minimum path having at least its endpoints within the connected region. A connected region is considered as convex if all minimal paths are minimum paths at the same time. A configuration $C = \{D \times \phi\}$ is the actual state of the LS function, namely, the shape of the zero LS and the connected regions (Ω_p, Γ_q) composing the object and the background region.

Now we have all the necessary tools to establish proper worst-case bounds on the number of iterations required by the Shi LSM to converge.

Theoretical Results: Worst-case Bounds

Theorem 1 (General bound)

Let the true object region Ω^* be composed of P-connected regions Ω_p^* (where $p = 1 \ldots P$) and the true background region Γ^* be composed of q connected regions Γ_q^* (where $q = 1 \ldots Q$). Assume that F>0 in Ω^* and F<0 in Γ^*. At initialization, C is chosen such that $\Omega_p^* \cap \Omega \neq \emptyset, \forall p = 1 \ldots P$ and $(D \setminus \Omega) \cap \Gamma_q^* \neq \emptyset, \forall q = 1 \ldots Q$. Then, the Shi LSM converges to Ω^* in $N_{it} \leq \max(\max_i(|\Omega_i|), \max_j(|\Gamma_j|))$ iterations, where |.| denotes the number of elements in the region.

Theorem 2 (Convex bound)

If either Ω^* or Γ^* is convex, then the Shi LSM converges to Ω^* in $N_{it} \leq \max(\max_i(B\Omega_i), \max_j(B\Gamma_j))$ iterations, where B denotes the diameter of the given region.

Proof of Theorem 1on General Bound

Let $\Omega^a = \Omega^* \cap \Omega = \bigcup_{p=1}^{P} \Omega_p^a$. These are fixed sets and will not change during the evolution process. Furthermore, $F(x_k) > 0, \forall x_k \in \Omega^a$ which ensures that $\Omega^a \subseteq \Omega$ as Ω evolves.

At initialization for each Ω_i two cases are possible. First case: $\Omega_i \cap \Omega^* = \emptyset$. Therefore, $\Omega_i \subseteq \Gamma^*$ so, $F(x) < 0$. On the boundary of Ω_i, $L_{in,i}$, a switch_out operation is applied so the diameter of Ω_i becomes smaller with two in every iteration. Second case: $\Omega_i \cap \Omega^* \neq \emptyset$. Therefore, the longest possible path in Ω_i gives the upperbound of the number of iterations that is obviously upperbounded by the number of points in Ω_i. Following similar arguments, we can also show this for Γ_j. Taking the maximum of the upperbounds completes the proof of Theorem 1.

Proof of Theorem 2 on Convex Bound

Obviously, the first case of the proof of Theorem 1 obeys the desired bound. The second case is as follows. Since Ω^* is convex, the length of the longest path is bounded by the diameter of Ω_i. In worst case, $\Omega_i \cap \Omega^*$ is one of the endpoints of the diameter. Following similar arguments, we can also show this for Γ_j. Taking the maximum of the diameters in each initial and background region completes the proof of Theorem 2.

Theorem 1 gives a general upperbound on N_{it}, and the iteration cycle checking the stopping condition is not necessary if the number of iteration has reached this upper bound. This worst-case bound is approached if Ω^* or Γ^* are degenerated in some sense (see Figure 1D and Table 1for example). However, in many cases the stricter bound can be applied. We shall emphasize that the worst-case bound is a quantity derived from the initial condition.

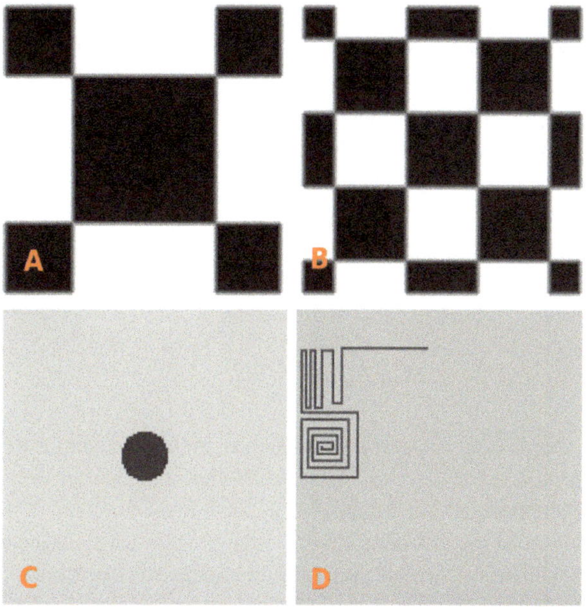

Figure 1. Example of initial condition and test objects. (A and B) Initial conditions belonging to n = 1,2. This chessboard-like initial pattern partitions the domain to several convex pieces. It ensures the theoretical bounds to decrease as n increases. (C) Convex test object. (D) Concave test object. Required number of iterations are presented for (C) and (D) in Table 1.

Table 1. Experimental validation of the theorems

	Values							
Number of circles (n)	1	2^2	4^2	8^2	16^2	24^2	32^2	64^2
Bound according to Theorem 1	64^2	32^2	256	64	16	9	4	1
Bound according to Theorem 2	127	63	31	15	7	5	3	1
Circle N_{it}	26	16	9	6	4	3	3	1
Degenerate N_{it}	145	68	18	7	6	3	3	1

The image was 128^2 pixels. Configuration C was set as squares arranged into n rows and n columns in a chessboard-like pattern (see Figure 1A, B). Two different objects were tested: a circle in the center with radius 11 pixels and a snake-like degenerate object. Configuration and objects are presented in Figure 1.

The possibility of choosing the initial shape of the regions Ω_i and Γ_j is essential to minimize the required number of iterations. It shall be noted that according to the Shi LSM, all calculations are done in the active front that have direct connection with the initial shape of the aforementioned regions. Making both Ω_i and Γ_q smaller, the smaller the worst-case bounds can be. This statement leads us to section 'Experiments', namely, how to construct initial conditions that are minimal in the sense of worst-case bounds and can be mapped and processed efficiently on a many-core architecture. It should be noted that the presentation above does not depend on the dimensionality of the data so the theorems are general from this point of view and the dimension can be arbitrary.

MANY-CORE HARDWARE PLATFORMS

CNN Universal Machine

The CNN universal machine (CNN-UM) [20] is a virtual machine construct like the Turing machine. It is Turing complete and it is universal in the sense that a CNN-UM can present all the behaviors that a predefined CNN dynamics can show. A CNN consists of nonlinear dynamical systems called cells. Each cell has a state, an input, and an output that is a nonlinear function of the state. The cells are arranged in a rectangular grid and each cell is connected to its (nearest) neighbors within a given range (N_r). The connections are weighted and these weights are called templates. The templates define the operator that is applied on the input and the state, hence, they define the output dynamics.

$$\frac{d}{dt}x_{ij}(t) = -x_{ij}(t)+\sum_{kl \in N_r} A_{kl,ij}y_{kl}(t)+\sum_{kl \in N_r} B_{kl,ij}u_{kl}(t)+z_{ij} \quad (6)$$

Here x_{ij}, u_{ij}, y_{ij} stand for the state, input, and output of the cell ij, respectively; A, B, z are the template parameters: A for feedback, B for input, and z_{ij} for bias. The CNN-UM, in addition to the standard CNN, contains memories and control units to allow performing series of template operations and branching.

The experiments were done on an Eye-RIS v1.3 vision system (VS) (Anafocus Ltd., Seville, Spain). It consists of a Q-Eye, Altera NIOS-II 32-b RISC microprocessor and on chip RAM. The Q-Eye is a quarter common intermediate format (QCIF) monochrome image sensor processor with 7- to 8-b de facto accuracy. It is a fine-grain CNN-UM implementation with nearest neighborhood capable operations. The microprocessor handles the memory, the I/O ports and organizes the execution. The consumption of the complete VS is below 750 mW.

GPU

Recent GPUs are feasible for nongraphic operations as well and programmable through general purpose APIs like C for CUDA [21] or OpenCL [22]. In this paper, OpenCL nomenclature is used. The description below is a brief overview of GPUs and, in addition to the basics, gives only those details that have great influence on the LS evolution.

A function that can be executed on the GPU is called a kernel. Any call to a kernel must specify an NDRange for that call. This defines not only the number of work-items to be launched, but also the arrangement of groups of work-items to work-groups and work-groups to the NDRange. The dimensionality of a work-group can be one, two, or three.

Physically, the elementary computing element is the computing element. A few computing element together with a given amount of SDRAM, scheduling unit, and special function unit forms a computing unit (CU). A device consists of several CUs and a global memory (off-chip).

The experiments were done on an NVIDIA 780 GTX GPU. It has 12 CUs, 192 computing elements, and 48KB shared memory in each CU, and 3 GB global memory. The hosting PC runs on Intel core i7-2600 CPU @3.4 GHz with 8 GB system memory, the operating system is Debian with Linux kernel the GPU driver version is 325.15.

EXPERIMENTS

Theorems 1 and 2 give upperbounds on the required number of iterations (N_{it}). A practical proposal of this paper is to construct configurations that have as low worst-case bounds on N_{it} as feasible and can be computed efficiently on many-core architectures. This scenario is presented through two case studies. The first one is on an Eye-RIS v1.3 VS that is a hardware implementation of the CNN-UM and the second one is on a GPU.

The whole image is covered with many-many small active fronts, and as a consequence, the intersection condition ($\Omega_p^* \cap \Omega \neq \emptyset$) is automatically fulfilled. During the case studies, the speedfield F has been very simple, +1 for the object region and −1 for the background regions.

A case study on CNN-UM

In subsection 'CNN universal machine,' a short overview was given on the CNN-UM. Now the details of the mapped algorithm are described. The perspective in this scenario is the precedence of locality which becomes increasingly important as the technology feature size decreases, and the delay together with power consumption of global communication increases.

The mapped algorithm is based on the set theoretic description of the LS function. In addition to L_{in} and L_{out} two other sets are defined.

$$F_{\text{in}} = \{\mathbf{x} \in D |\, \phi(\mathbf{x}) < 0 \wedge \mathbf{x} \notin L_{\text{in}}\} \quad (7)$$

$$F_{\text{out}} = \{\mathbf{x} \in D |\, \phi(\mathbf{x}) > 0 \wedge \mathbf{x} \notin L_{\text{out}}\} \quad (8)$$

In other words, the neighbors of each point in these sets are in the same region as the point itself. Figure 2 shows the universal machine on flows (UMF) diagram of the algorithm. It is a special flowchart description of the algorithms executable on CNN-UM so that it is complete and unique [20].

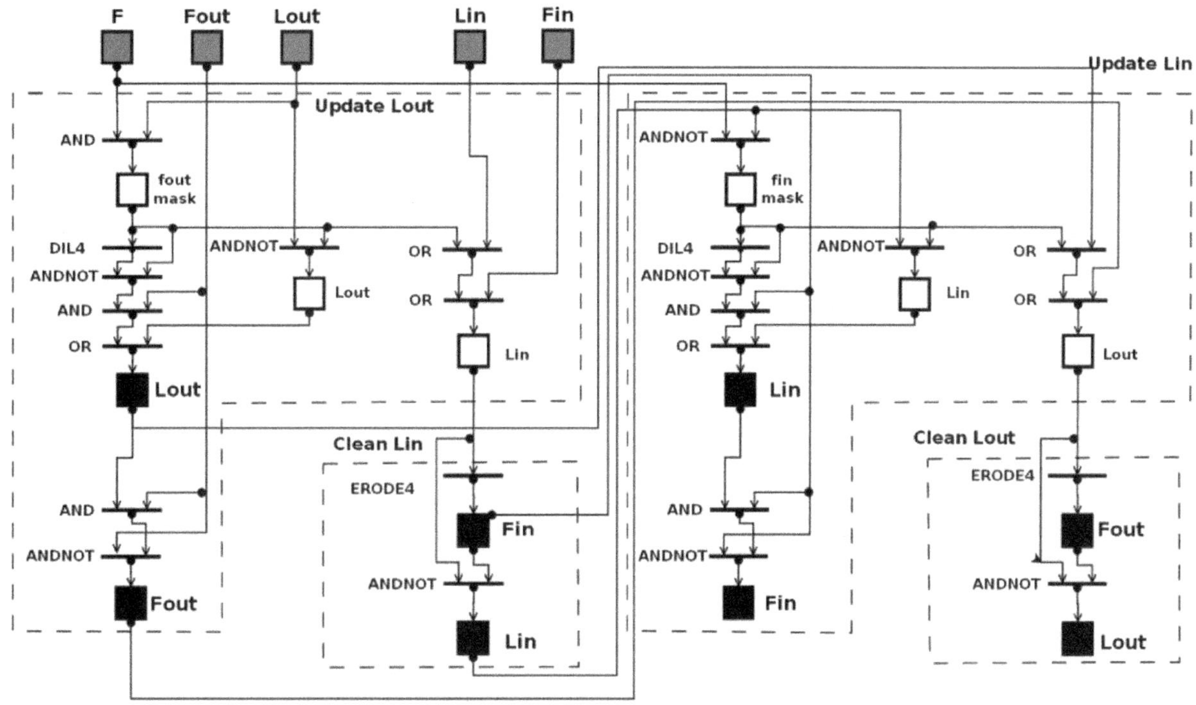

Figure 2. UMF diagram of LS evolution on CNN. Rectangles denote memories, bold short horizontal lines with capital operator names on the left denote template operations. Dashed lines indicate the phases corresponding to the four cycles of the Shi LSM. Black rectangles denote final forms of sets in that phase. Thin lines ending with arrows denote dataflow from memory to an operation, from an operation to an operation, or from an operation to a memory.

Templates AND, OR denote elementary logic, ANDNOT performs logic subtraction (Op 1∧¬Op2), DIL4 and ERODE4 are the four connected dilatation and erosion (spatial logic). All templates are of the nearest neighbor kind. In the 'Update L_{out}' phase, foutmask is computed first. It contains the points that are going to move outward. foutmask is used in three different ways. It is subtracted (ANDNOT) from L_{out}, added (OR) to L_{in} and dilated (DIL4, ANDNOT, AND) to generate its own outer neighbors. This is the new stepped L_{out} part and the unchanged parts are added with an OR operation. The resulting set is finalized as the new L_{out} (black rectangle in Figure 2). From the old F_{out} the new L_{out} is subtracted (ANDNOT) to get the newF_{out} (again, black rectangle in Update L_{out} phase). Finally, the modified L_{in} is added to F_{in}. In the 'Clean L_{in}' phase, the merged foutmask, L_{in}, and F_{in} is the only input. The new L_{in} is the outer pixel layer of this merged input. The new F_{in} is obtained by a simple four connected erosion while L_{in} is the result of a subtraction. 'Update L_{in}' and 'Clean L_{out}' are nearly identical, only the input of the operations are switched, and another mask is used (finmask).

The algorithm is implemented on an Eye-RIS 1.3 VSoC. One step of the algorithm is performed in 400 to 440 μs on a QCIF image. It must be noted that the actual computing is finished within 60 to 70 μs and the remaining time (340 to 370 μs) is required for the data movement from the main memory of the Eye-RIS (on the Altea NIOS-II microprocessor) to the Q-Eye chip memory.

A case study on GPU

The evolution process is divided into two steps. The first one is the planner step and the second is the evolution step. The planner creates the so-called plan. It contains the position offsets of the 16×16 tiles that are calculated actually in the iteration step. The planner works on the indicator image. The indicator is a tiny image and each pixel of the indicator is true if the corresponding tile on the input image shall be processed in this iteration and false otherwise. The size of the plan is calculated by local prefix-sum work-group wise, and global atomic addition is used to correctly determine the offset of the work-group within the plan. The source of the planner kernel is provided as Additional file 1.

The evolution kernel processes only those tiles of the LS function that are inserted in the plan. The evolution kernel makes a step either inward or outward direction depending on the sign of the force field on the LS function. This is done simultaneously unlike in the sequential algorithm. Each work-group processes a 16×16 tile provided in the plan and writes the complete tile back to the global memory. First, each work-item calculates the new value of the pixel of the LS function. Then the neighbors of each pixel are updated as the switch_out() and switch_in()operations require, and the active front is cleaned to maintain the two pixel width. If there is no activity inside the tile, then set the corresponding pixel of the indicator image to false. The boundary of the tile requires special care, namely, to properly update the corresponding neighboring pixels of the LS function and the indicator. The source of the evolution kernel is provided as Additional file 2.

Table 2 shows execution time measurements of the work-efficient parallel algorithm on NVIDIA 780 GTX GPU compared to a baseline single-threaded implementation on Intel core i7-2600 CPU. The execution time was measured by the gettimeofday() C-function which has microsecond resolution. The table specifies the image resolution, the initial condition configuration, and presents the mean of the execution time of an iteration on GPU, on CPU, and the speedup. The iteration time on the GPU contains the execution time of both kernel functions (planner, iteration). The two kernels evenly share the execution time in the case of conventional, sparse initial condition; however, in the case of dense iteration step, the ratio of the evolution kernel can shift to 30:1 with respect to the planner.

Table 2. Time measurements on NVIDIA GTX 780 GPU compared to Intel core i7 CPU

Data size	Initial condition	$\bar{T}_{iteration}$ on GPU (μs)	$\bar{T}_{iteration}$ on CPU (μs)	N_{it}	Speedup
256 × 256	1 × 1	129	1,610	32	12.5
256 × 256	2 × 2	126	2,242	59	17
256 × 256	8 × 8	140	3,164	20	22
256 × 256	32 × 32	143	8,874	8	62
512 × 512	1 × 1	317	3,190	64	10
512 × 512	4 × 4	167	8,724	40	52
512 × 512	16 × 16	157	12,897	25	82
512 × 512	64 × 64	123	16,246	18	132
1,024 × 1,024	1 × 1	534	6,431	129	12
1,024 × 1,024	8 × 8	548	27,461	55	50
1,024 × 1,024	32 × 32	590	43,739	32	74
1,024 × 1,024	128 × 128	490	84,078	12	171
2,048 × 2,048	1 × 1	560	14,972	210	26
2,048 × 2,048	16 × 16	703	79,920	79	113
2,048 × 2,048	64 × 64	830	198,980	28	239
2,048 × 2,048	256 × 256	684	327,541	7	478

Presented results are the mean value of 100 runs.

Number of Iterations

In the experiments more initial configurations were tested. In each configuration, regions of Ω and Γ were placed in a chessboard like pattern as it is showed in Figure 1A,B. Two sample objects are presented in Figure 1C,D that shall be detected. Additionally, the two objects represent the two object families: the degenerate and convex ones having worst-case bounds stated in the theorem 1 and 2.

Iteration results are presented in Table 1 together with the two different bounds of the given configuration. The number of iterations (N_{it}) was measured on the original sequential algorithm of Shi and these values are presented in the table. It is below or equal to the worst-case bounds in every cases.

In the case of CNN-UM, N_{it} coincides with the values presented in the table, while in the case of GPU implementation, N_{it} is consistently higher with one iteration. This means that it exceeded the bounds in the case of n = 32 and n = 64. However, the reason is as follows: the boundary pixels of the subregion have one iteration delay in the cleaning process. This causes the additional iteration so it is not a violation of the theorems.

VALIDATION

In this section, we compare the result of the exact numerical implementation and the Shi LSM for three different speed functions: mean curvature motion, Chan-Vese, and GAR. The quantitative comparison is made by the dice coefficient. Given the state of the two LS functionsΩ_1 and Ω_2 of the two different methods, the coefficient is defined as

$$d(\Omega_1, \Omega_2) = \frac{2\text{Area}(\Omega_1 \cap \Omega_2)}{\text{Area}(\Omega_1) + \text{Area}(\Omega_2)} \tag{9}$$

Its value is in the range of 0 and 1; 0 means complete difference and 1 means complete agreement.

Mean curvature flow

In this case, the speed function is defined as

$$F = -\kappa \tag{10}$$

where κ is the (euclidean) curvature of the LS. It is the norm of the second derivative of γ with respect to the (euclidean) arc length ($\kappa = \|\gamma_{ss}(s)\|$, s is the arc length parametrization of the curve). Another possible, precise, and more easy way to calculate the curvature of an LS from ϕ is as follows:

$$\kappa = \text{div}\,\text{grad}\frac{\nabla\phi}{\|\nabla\phi\|} \tag{11}$$

This force term appears in almost every LS flow as a smoothing and regularizing term. The steady-state solution is a circle with infinitesimal diameter. In practice, the object region vanishes. In this case, not only the steady state but the evolution itself is also investigated. This is an autonomous motion and does not have any control term from an external image.

The details of the numerical approximation are as follows. The LS function ϕ is a signed distance function. It was recalculated after every 30 iterations. The time ($T_{maximum}$) run to 800 units. The time step (Δt) size has been set to 0.4. The curvature has been calculated from the LS function from Equation 11.

In the case of the fast LS evolution, the curvature was calculated according to the work Merriman, Bence, and Osher (MBO) [23, 24], namely, by G⊗ϕ, where G is a 2D gaussian of a given variance.

Figure 3 shows the test initial condition for mean curvature motion and the state of the evolution after 20, 40, 60, and 80 iterations of the fast LS evolution.

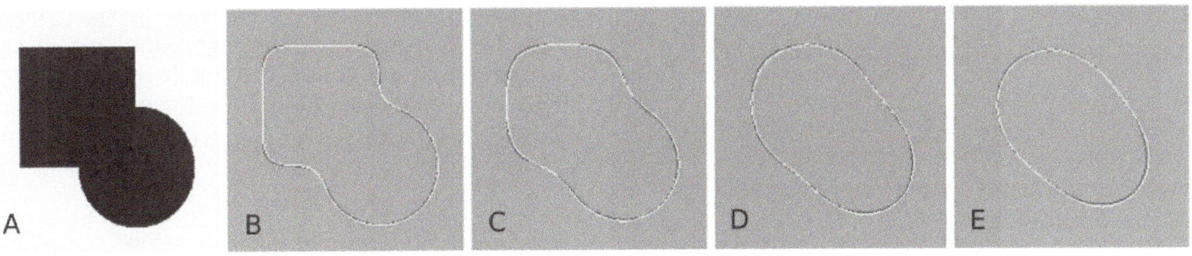

Figure 3. Comparison of mean curvature evolution of PDE approximation and fast LS evolution. This shows the initial condition and evolution of fast LS (white line) and numerical PDE approximation (black line). (A) Test initial condition for validation of mean curvature motion fast LS evolution against numerical PDE approximation. The test region contains positive, negative, and zero curvature regions and singularities as well.(B) State of evolution fast LF at N_{it} = 20 and PDE approximation at T = 56.8. (C) State of evolution fast LF at N_{it} = 40 and PDE approximation at T = 190.8. (D) State of evolution fast LF at N_{it} = 60 and PDE approximation at T = 405.6. (E) State of evolution fast LF at N_{it} = 80 and PDE approximation at T = 706.8.

Figure 4 shows the dice coefficient between the first 80 steps of the fast LS evolution and the corresponding state of the numerical approximation.

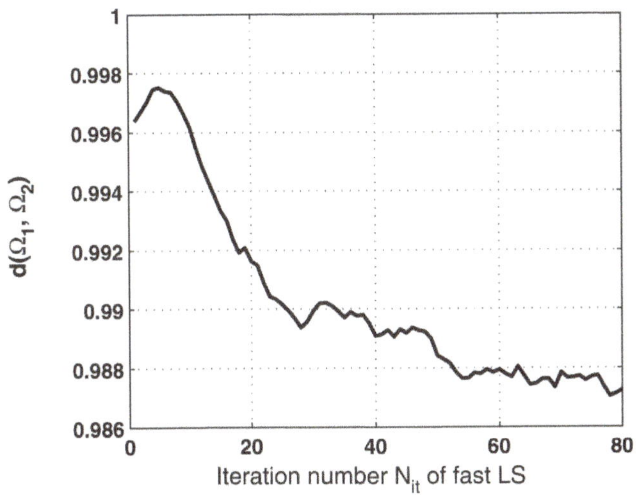

Figure 4. Dice index of mean curvature evolution.Ω_1 is the state of the fast LS evolution, andΩ_2 is the state of the numerical solution. The similarity between the two states is very high.

Chan-Vese Flow

This method was proposed in [6] and its speed term is defined as

$$F = \mu\kappa - \lambda_1(c_1 - I)^2 + \lambda_2(c_2 - I)^2 \tag{12}$$

The parameters are set as follows: $\mu = 1, \lambda_1 = 0.8, \lambda_2 = 0.8$. I represents the input image intensities, the constants $c_1 = 0.5$ and $c_2 = 0$ are simply the means of pixel intensities inside and outside the zero LS. The artificial time parameter runs to 180 units, the time step is 0.5 units. The total number of iterations are 360. The initial condition is 25 circles arranged uniformly in five rows and five columns each with diameter 27 pixels. The LS function (signed distance) is recalculated in each 30 iterations for the numerical solution. The initial condition is 5×5 circles each with diameter 27 pixels. The steady states of the two Cahn-Vese evolutions are shown in Figure 5A. The dice index of the two states is 0.998.

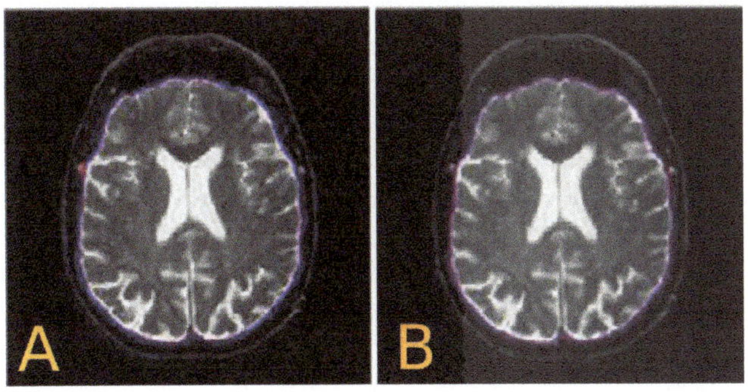

Figure 5. Validation of fast LS evolution.(A) CV and (B) GAR flow. Red corresponds to the numerical PDE solution while blue corresponds to the fast LSM. The two curves are nearly the same and the dice index is 0.998 in both cases.

Geodesic Active Regions Flow

This method was proposed in [7]. This method combines boundary and region-based information to segment an image. In this method, the pixel intensities are modeled with a GMM. The speedfield is as follows:

$$F = -\alpha \log\left(\frac{P(I|R_1)}{P(I|R_2)}\right) + (1-\alpha)\left(b\kappa + \nabla b \frac{\nabla\phi}{|\nabla\phi|}\right) \tag{13}$$

where R_1 and R_2 are the regions to be separated, b is a strictly decreasing function of boundary probability, and α is a balancing constant. In our case α = 0.3, and b is defined as follows:

$$b = \frac{1}{1 + \|\nabla G \otimes I\|} \tag{14}$$

Here G is a 2D gaussian with σ = 3. The GMM parameters are calculated from the image histogram with a recursive expectation maximization algorithm. The artificial time runs to 6 units, the time step is 0.02 units. The total number of iterations are 300. The LS function (signed distance) is recalculated in each 30 iterations for the numerical solution. The initial condition is the same as in the case of Chan-Vese evolution, 5×5 circles each with diameter 27 pixels. Steady states are shown in Figure 5B. The dice index of the two states is 0.998.

DISCUSSION

In this paper, given our investigation of the initial condition and the required number of iterations as a function of it, we presented two bounds on the required number of iterations of an LS evolution. The bounds were proven theoretically and validated experimentally with the original algorithm and also with two different mappings of the algorithm on many-core machines (GPU, CNN-UM). The bounds depend only on the initial configuration of the LS function. The many-core realizations required not only a very small number of iterations less than or equal to the bounds, but the execution of an iteration was also fast (see Table 2 for detailed measurement data).

In addition to the drastic decrease of the required number of iterations, the total execution time decreases as well if dense initial condition is used for the evolution. The total execution time on CPU with sparse initial condition is comparable to the total execution time with dense initial condition. For the smaller images, the dense initial condition was less effective by 30% to 15%; but in the case of the biggest image, the dense iteration was the faster by 35%. In the case of the dense initial condition on GPU, there is a significant speedup compared to the sparse initial condition in all cases since our proposed dense initial condition together with the algorithm utilizes the properties of the underlying architecture. Therefore, greater performance gain can be achieved on GPU if dense initial condition is used.

A great property of the results is the scalability. This is true for the performance as a function of cores and for the number of iterations as a function of size of the disjoint active fronts. Considering the chessboard-like initial configuration with increasingly finer regions, the general bound is proportional to the area of the regions and the convex bound is proportional to the half perimeter of the regions. This is changed in three dimensions to the volume of region in the case of general bound and half of the longest perimeter of the volume in the case of a convex bound.

The assumption on F is stronger in Theorem 1 than the one that was given in the convergence analysis in [2]. In the examples presented there, our stronger assumption stands for at least one of the regions Ω^*,Γ^*. However, there may be cases when for a short period of iterations the sign of F changes. Typically, this is the case when inside the true object region, the actual state of the LS function contains a concave background region with high negative curvature. In these cases, the curvature-based term can be greater than the region term (the pixel-intensity-based terms), but this is a temporary effect. As soon as the local concavity is vanished, the region term becomes again greater and the sign of F changes back. Furthermore, as it was declared in the introduction, the construction of the speedfield and its components is out of the scope of this paper. Additionally, the validations indicate that the method converges de facto to the same state as the exact numerical solutions.

The fact that the active front of the initial condition covers the whole image has a special consequence, namely, separate, disjoint regions of the same object or multiple target objects can be found automatically without user interaction. For example, the grey matter of the brain on an MRI slice is disconnected and may be composed of 8 to 20 disjointed regions on the given slice. The problem of detecting all regions is greatly simplified with our dense initial condition. Similarly, the selected group of cells on a histology image shows this property as well. Additionally, histology images can be extremely large (2 to 30 Mpixel), and the performance gain of our proposed method (initial condition together with the parallel algorithm) becomes more expressed on larger images. A conventional sparse initialization can easily fail this task, with wrongly chosen initial condition, see for instance the initialization and evolution of a gold standard LS implementation of [25], which is a widely used framework for medical image segmentation and analysis.

Figures 6 and 7 show an example. The evolution from a single-circle initial condition is presented on Figure 6B, while our result is presented in Figure 6C,D. It demonstrates its potential and it may be an initial condition for fine-tuning the segmentation with another method. Of course, the dense iteration may have the drawback of increased false-positive rate, for example see Figure 6D where the evolution runs with slightly different parameters, but this could be handled with more sophisticated speed functions or building a priori information into the initial condition. Figure 7 shows a histology image from the skin, where one class of cells is to be segmented.

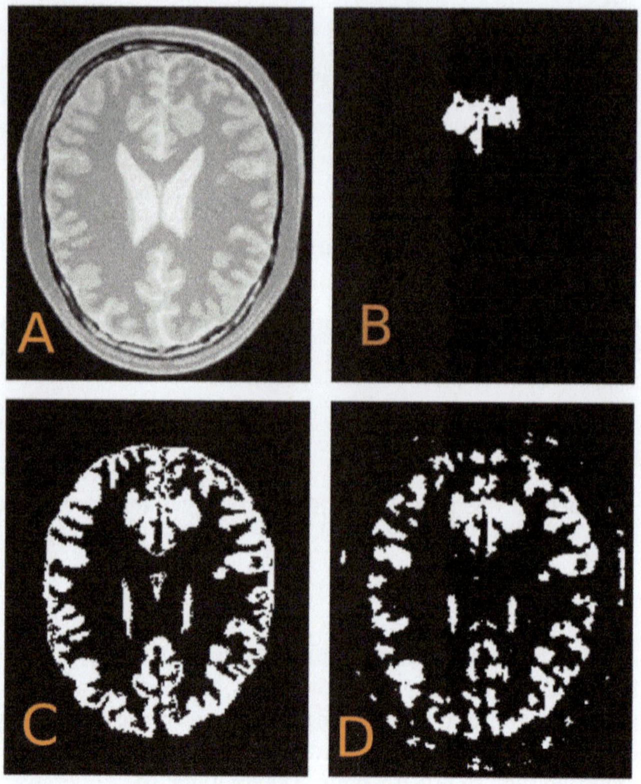

Figure 6. Initial condition dependence of evolution.(A) The original image to be segmented (grey matter of the brain). (Figure 6A is reproduced from [25]). (B) The reached solution of evolution started from a single circle initial condition. (C) The reached solution with our proposed initial condition (32×24 curves with diameter 3 pixels). (D) The reached solution of evolution with slightly modified parameters compared to the evolution shown on Figure 6C.

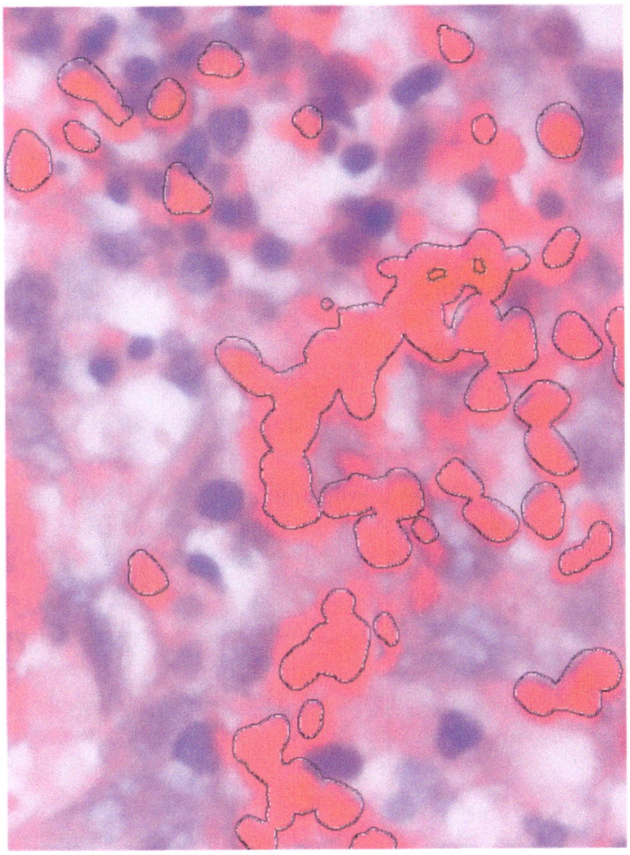

Figure 7. Histology image from the skin.

It must be emphasized that the case studies presented here are not necessarily optimal mappings of the Shi LS evolution by any means. The purpose of presenting them is twofold: (1) to highlight the advantage of the proposed initial condition concept especially on those machines and (2) to give a proof of concept mapping of this fast evolution on two totally differently organized (virtual and physical) many-core machines.

CONCLUSIONS

To automatically detect segment object on an image or on a region of it, the LS-based algorithms are feasible tools. In this paper, it was shown theoretically and experimentally through two case studies that the initial condition plays an essential role in decreasing the execution time. It must be emphasized that this is only validated on many-core architectures where the computations can be distributed among the cores. Furthermore, based on the initial condition configuration, two worst-case bounds were given on the required number of iterations depending on the convexity of the object to be found. The bounds are proven theoretically and validated experimentally. Additionally, the execution time of one iteration was measured on both architectures. It was below 70+370 μs on the Eye-RIS system handling a QCIF image (where 70 μs is the processing time and 370 μs is the outer memory delay). The timing results of the GPU is presented in Table 2 in details. In the case of the proposed dense initial condition on GPU, there is a significant speedup compared to the sparse initial condition in all cases since our dense initial condition together with the algorithm utilizes the properties of the underlying architecture. Therefore, greater performance gain can be achieved (up to 18 times speedup compared to the sparse initial condition on GPU).

The results and tools presented in this paper provide a method to efficiently map LS algorithms on many-core architectures and ensure bounds on the execution time through the two theorems.

ACKNOWLEDGEMENTS

This research was supported by the European Union and the State of Hungary, cofinanced by the European Social Fund in the framework of TÁMOP 4.2.4.A/1-11-1-2012-0001 (National Excellence Program). The support grants TÁMOP-4.2.1.B-11/2/KMR-2011-0002 and TÁMOP-4.2.2/B-10/1-2010-0014 are also gratefully acknowledged. The authors would like to thank Ádám Rák for his help and suggestions.

REFERENCES

1. Sethian JA: Level Set Methods and Fast Marching Methods: Evolving Interfaces in Computational Geometry, Fluid Mechanics, Computer Vision, and Materials Science. New York: Cambridge University; 2000.
2. Shi Y, Karl W: A real-time algorithm for the approximation of level-set-based curve evolution. IEEE Trans. Image Process. 2008, 17(5):645-656. [http://ieeexplore.ieee.org/lpdocs/epic03/wrapper.htm?arnumber = 4480128]
3. Adalsteinsson D, Sethian JA: A fast level set method for propagating interfaces. J. Comput. Phys. 1995, 118(2):269-277. [http://www.sciencedirect.com/science/article/pii/S0021999185710984] 10.1006/jcph.1995.1098
4. Peng D, Merriman B, Osher S, Zhao H, Kang M: A PDE-based fast local level set method. J. Comput. Phys. 1999, 155(2):410-438. [http://www.sciencedirect.com/science/article/pii/S0021999199963453] 10.1006/jcph.1999.6345
5. Sapiro G: Geometric Partial Differential Equations and Image Analysis. New York: Cambridge University; 2001.
6. Chan T, Vese L: Active contours without edges. Image Process. IEEE Trans 2001, 10(2):266-277. 10.1109/83.902291
7. Paragios N, Deriche R: Geodesic active regions: a new framework to deal with frame partition problems in computer vision. J. Vis. Commun. Imag. Rep. 2002, 13(1–2):249-268. [http://www.sciencedirect.com/science/article/pii/S1047320301904754]
8. Joshi N, Brady M: Non-parametric mixture model based evolution of level sets and application to medical images. Int. J. Comput. Vis. 2010, 88: 52-68. [http://dx.doi.org/10.1007/s11263-009-0290-5] 10.1007/s11263-009-0290-5
9. Bertelli L, Chandrasekaran S, Gibou F, Manjunath BS: On the length and area regularization for multiphase level set segmentation. Int. J. Comput. Vis. 2010, 90(3):267-282. [http://www.springerlink.com/index/10.1007/s11263-010-0348-4] 10.1007/s11263-010-0348-4
10. Sundaramoorthi G, Yezzi A, Mennucci A, Sapiro G: New possibilities with Sobolev active contours. Int. J. Comput. Vis. 2009, 84: 113-129. [http://dx.doi.

org/10.1007/s11263-008-0133-9] 10.1007/s11263-008-0133-9

11. Lefohn AE, Cates JE, Whitaker RT: Interactive, GPU-based level sets for 3D segmentation. ed. by Ellis RE, Peters TM, Medical Image Computing and Computer-Assisted Intervention - MICCAI 2003, Lecture Notes in Computer Science. vol. 2878 (Springer, Berlin Heidelberg, 2003) pp. 564–572, [http://dx.doi.org/10.1007/978-3-540-39899-8_70]
12. Roberts M, Packer J, Sousa MC, Mitchell JR: A work-efficient GPU algorithm for level set segmentation. In Proceedings of the Conference on High Performance Graphics. ACM, New York; 2010:123-132.
13. Sharma O, Zhang Q, Anton Q, Bajaj C: Multi-domain, higher order level set scheme for 3D image segmentation on the GPU. In 2010 IEEE Conference on, Computer Vision and Pattern Recognition (CVPR). San Francisco, 13–18 June 2010); 2211-2216.
14. Chua LO, Yang L: Cellular neural networks: applications. Circuits Syst. IEEE Trans. 1988, 35(10):1273-1290. 10.1109/31.7601
15. Cserey G, Rekeczky C, Földesy P: PDE based histogram modification with embedded morphological processing of the level-sets. J. Circuits, Syst Comput 2003, 12(04):519-538. 10.1142/S021812660300101X
16. Rekeczky C, Roska T: Calculating local and global PDEs by analogic diffusion and wave algorithms. In Proceedings of the European Conference on Circuit Theory and Design, Volume 2. (Helsinky University of Technology, Espoo; 2001:17-20.
17. Hillier D, Czeilinger Z, Vobornik A, Rekeczky C: Online 3-D reconstruction of the right atrium from echocardiography data via a topographic cellular contour extraction algorithm. Biomed. Eng. IEEE Trans. 2010, 57(2):384-396.
18. Shi Y, Karl WC: A fast level set method without solving PDEs. In Proceedings on IEEE International Conference on Acoustics, Speech, and Signal Processing (ICASSP'05). Philadelphia, 18–23 March 2005; 2005:97-100.
19. Shi Y: Object based dynamic imaging with level set methods. PhD Thesis, Boston University College of Engineering 2005
20. Chua LO, Roska T, Venetianer PL: The CNN is universal as the Turing machine. Circuits Syst. I: Fundam. Theory Appl. IEEE Trans 1993, 40(4):289-291. 10.1109/81.224308
21. NVIDIA: CUDA C Programming Guide 2011 . Accessed 23 January 2012 [https://developer.nvidia.com/cuda-toolkit-archive]. Accessed 23 January 2012
22. OpenCL specification 1.1 2011. . Accessed 1 April 2012 [http://www.khronos.org/opencl/]. Accessed 1 April 2012
23. Merriman B, Bene J, Osher S: Diffusion generated motion by mean curvature.Computational Crystal Growers Workshop Edited by Taylor J, (Providence, RI 1992) pp.73–83.
24. Merriman B, Bence JK, Osher SJ: Motion of multiple junctions: a level set approach. J.Comput. Phys. 1994, 112(2):334-363. [http://www.sciencedirect.com/science/article/pii/S0021999184711053] 10.1006/jcph.1994.1105
25. The insight toolkit 2012 . Accessed 20 November 2012 [http://www.itk.org]. Accessed 20 November 2012

CHAPTER 7

CONTROLLER AREA NETWORK (CAN) SCHEDULABILITY ANALYSIS

Robert I. Davis, Alan Burns, Reinder J. Bril, and Johan J. Lukkien

Keywords: Controller Area Network (CAN), Fixed priority scheduling, Non-pre-emptive scheduling, Schedulability analysis, Response time analysis, Priority assignment policies

INTRODUCTION

Background

Controller Area Network (CAN) is a serial communications bus designed to provide simple, efficient and robust communications for in-vehicle networks. CAN was developed by Robert Bosch GmbH, beginning in 1983, and presented to a wider audience at the Society of Automotive Engineers (SAE) Congress in 1986—effectively the "birth of CAN". In 1987, the first CAN controller chips were released by Intel (82526) and Philips (82C200). In the early 1990s, Bosch submitted the CAN specification (Bosch, 1991) for standardisation, leading to publication of the first ISO standard for CAN (11898) in 1993 (ISO, 1993).

Mercedes was the first automotive manufacturer to deploy CAN in a production car, the 1991 S-class. By the mid 1990s, the complexity of automotive electronics was increasing rapidly. The number of networked Electronic Control Units (ECUs) in Mercedes, BMW, Audi and VW cars went from 5 or less at the beginning of the 1990s to around 40 at the turn of the millennium. With this explosion in complexity traditional point-to-point wiring became increasingly expensive to manufacture, install, and maintain due to the hundreds of separate connections and tens of kilograms of copper wire required. As a result CAN was rapidly adopted by the cost-conscious automotive industry, providing an effective solution to the problems posed by increasing vehicle electronics content. Following on from Mercedes, other manufacturers including Volvo, Saab, BMW, Volkswagen, Ford, Renault, PSA, Fiat and others all adopted CAN technology

As a result of the wholesale adoption of CAN by the automotive industry, sales of CAN nodes (8, 16 and 32-bit microcontrollers with on-chip CAN peripherals) grew from just under 50 million in 1999 to over 340 million in 2003—see Fig. 1.

By 2004, there were at least 15 silicon vendors manufacturing, in total, over 50 different microprocessor families with on-chip CAN capability.

Today almost all of the cars manufactured in Europe are equipped with at least one CAN bus. In the United States, the Environmental Protection Agency has mandated the use of CAN, for On Board Diagnostics, in all cars and light trucks sold in the US from model year 2008 onwards.

Automotive Applications

In automotive applications, CAN is typically used to provide high speed networks (500 Kbits/s) connecting chassis and power-train ECUs, for example engine management and transmission control. It is also used for low speed networks (100 or 125 Kbits/s) connecting body and comfort electronics, for example door modules, seat modules and climate control. Data required by ECUs on different networks is typically gatewayed between the different CAN buses by a powerful ECU connected to both.

Citation: Davis, R.I., Burns, A., Bril, R.J. and J. J. Lukkien (2007). Controller Area Network (CAN) schedulability analysis, Real-Time Systems, April 2007; Volume 35, Issue 3, pp 239–272. DOI:10.1007/s11241-007-9012-7.

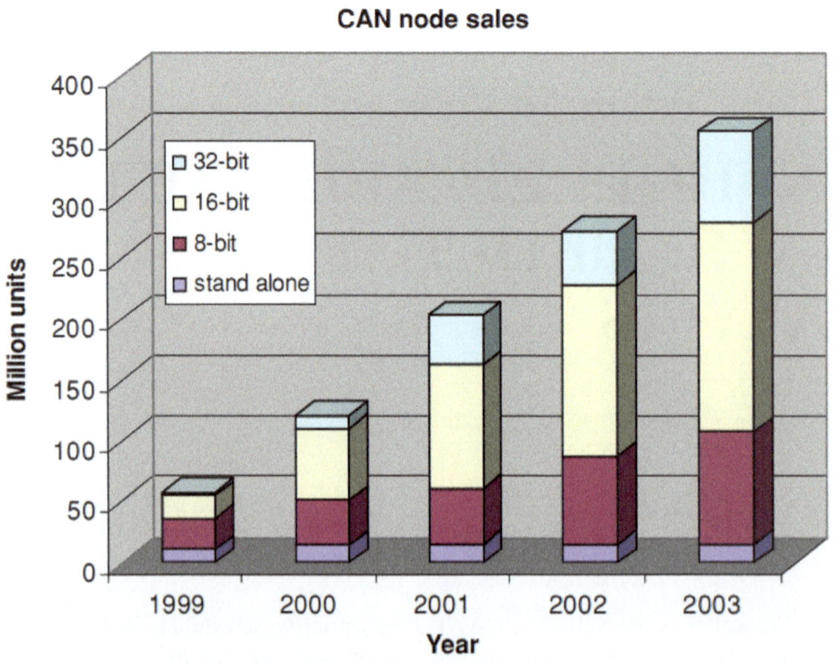

Figure 1. Sales of microcontrollers with onchip CAN peripherals.

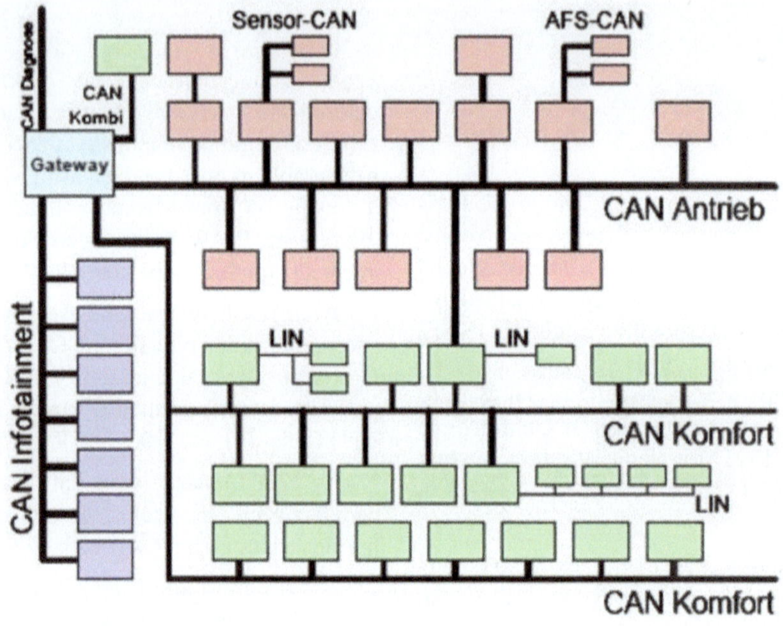

Figure 2. VW Passat network architecture.

The network architecture of the VW Passat (Leohold, 2005) shown in Fig. 2, reproduced with permission from Nolte (2006), illustrates how a number of CAN buses are used to connect around 45 ECUs in that vehicle. Also shown in Fig. 2 are three Local Interconnect Networks (LIN). LIN is a complementary technology to CAN, and is used to provide inexpensive, low speed (20 Kbits/s) connectivity (LIN Consortium, 2003).

Table 1 summarises the requirements placed on in-vehicle networks for the BMW 7 Series. This is typical of automotive applications, where individual CAN buses are used to connect between 2 and 32 ECUs at bandwidths ranging from 100 to 500 Kbits/s (Frischkorn, 2005).

In automotive applications the messages sent on CAN are used to communicate state information, referred to as signals, between different ECUs. Examples of signals include: wheel speeds, oil and water temperature, engine rpm, gear selection, accelerator position, dashboard switch positions, climate control settings, window switch positions, fault codes, diagnostic information and so on. In a high-end vehicle, such as the VW Phaeton, there can be more than 2500 distinct signals (Leohold, 2004), each effectively replacing what would, in a traditional point-to-point wiring loom, have been a separate wire.

Table 1. BMW 7 Series network requirements

	No. of ECUs	Bandwidth	No. of Messages	Cycle times
Body	14–30	100 Kbits/s	300	50 ms-2 s
Chassis	6–10	500 Kbits/s	180	10 ms-1 s
Power-train	3–6	500 Kbits/s	36	10 ms-10 s

Many of these signals have real-time constraints associated with them. For example, an ECU reads the position of a switch attached to the brake pedal. This ECU must send a message over the CAN network, carrying information (a signal) that the brakes have been applied. The ECU responsible for the rear light clusters receives the message, recognises the change in the value of the signal, and switches the brake lights on. All within a few tens of milliseconds of the brake pedal being pressed. Engine, transmission, and stability control systems typically place even tighter time constraints on signals, which may need to be sent as frequently as once every 5 milliseconds to meet their time constraints (Society of Automotive Engineers, 1993).

Research and Real-time Analysis

CAN is a serial data bus that supports priority based message arbitration and non-preemptive message transmission. In the early 1990s, a common misconception about CAN was that although the protocol was very good at transmitting the highest priority message with low latency, it was not possible to guarantee that less urgent signals, carried in lower priority messages, would meet their deadlines.

Tindell and Burns (1994) and Tindell et al. (1994b, 1995) showed how research into fixed priority pre-emptive scheduling for single processor systems could be adapted and applied to the scheduling of messages on CAN. This analysis provided a method of calculating the worst-case response times of all CAN messages. Using this analysis it became possible to engineer CAN based systems for timing correctness, providing guarantees that all messages, and the signals that they carry, would meet their deadlines.

Tindell's seminal research heavily influenced the design of on-chip CAN peripherals such as Motorola msCAN (Motorola, 1998) and has lead to a large body of work into schedulability theory and error models for CAN (Punnekkat et al., 2000; Nolte et al., 2002, 2003; Broster et al., 2002, 2005; Hansson et al., 2002; Broster and Burns, 2003), including at least two PhD theses (Broster, 2003; Nolte, 2006). Overall, this research into CAN scheduling has been cited in over 200 subsequent papers.

In 1995, Tindell's research was recognised by Volvo Car Corporation and successfully used in the configuration and analysis of the CAN buses for the forthcoming Volvo S80 (P23) (Casparsson et al., 1998). Following the success of this project, Volcano Communications Technologies AB used Tindell's analysis as the basis of a commercial CAN schedulability analysis tool, called Volcano Network Architect.

Table 2. CAN messages highlighting flawed analysis

Message	Priority	Period	Deadline	TX time
A	1	2.5 ms	2.5 ms	1 ms
B	2	3.5 ms	3.25 ms	1 ms
C	3	3.5 ms	3.25 ms	1 ms

Since 1998, these tools and the Volcano concept (Casparsson et al., 1998) have been used in the design and development of CAN networks and electronics systems for the Volvo XC90, S80, S/V/XC70, S60, V50 and S40 as well as many other cars from different manufacturers.

Prior to Tindell's work, low levels of bus utilization, up to 30 or 40%, were typical in automotive applications, with extensive testing required to obtain confidence that CAN messages would meet their deadlines. With the advent of a systematic approach based on schedulability analysis, CAN bus utilization could be increased to around 80% (DeMeis, 2005) whilst still guaranteeing that deadlines would be met.

Motivation

The design and development of many in vehicle Controller Area Networks relies on the schedulability analysis of CAN given by Tindell and Burns (1994) and Tindell et al. (1994b, 1995). In this section we show that this analysis is flawed. It may result in computed worst-case response times for messages that are optimistic, i.e. less than the response times that may actually occur. The set of CAN messages listed in Table 2 serve to highlight the problem with the existing schedulability analysis of CAN. As a simple example, we have assumed a 125 Kbit/s network with 3 messages, each of which carries 7 bytes of signal data. Assuming 11-bit identifiers and worst-case bitstuffing, the maximum length of each message is 125 bits. The maximum transmission time of each message is therefore 1 ms.

The analysis method given by Tindell and Burns (1994) and Tindell et al. (1994b, 1995) calculates the worst-case response times of messages A, B and C as 2 ms, 3 ms and 3 ms respectively. Hence the system is deemed to be schedulable—the analysis supposedly guarantees that all of the messages will meet their deadlines in the worst case, despite the high bus utilisation of 97%.

Figure 3 illustrates the worst-case scenario for transmission of message C. We note that the first instance of this message is delayed by higher priority messages A and B, leading to a response time of 3 ms—this is the "worst-case response time" calculated using existing CAN schedulability analysis methods. However, as message transmission is non-pre-emptive, the first transmission of message C has a knock on effect, delaying subsequent transmissions of higher priority messages A and B. Some of this higher priority interference is pushed through into the next period of message C leading to a longer response time for the second instance of message C.

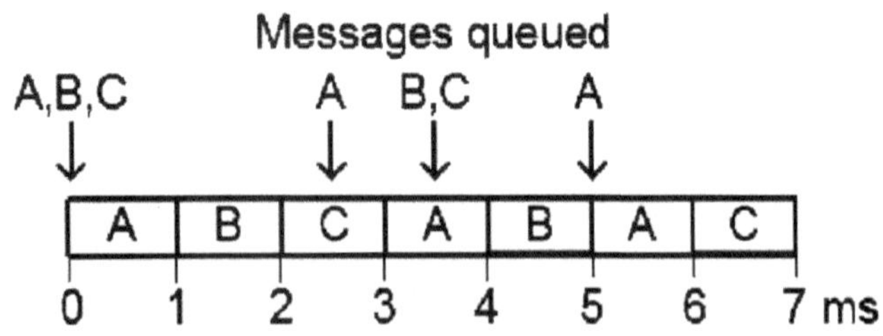

Figure 3. Worst-case scenario for message C.

At time t = 7 ms, the second instance of message C completes transmission with a response time of 3.5 ms. (Note at time t = 7 ms, there are no higher priority messages awaiting transmission and so there is no further push through interference that could delay subsequent instances of message C).

The actual worst-case response time for message C is 3.5 ms, which is greater than its deadline of 3.25 ms. The system is therefore unschedulable; contrary to the guarantees given by Tindell and Burns (1994) and Tindell et al. (1994b, 1995).

In fact, if the periods of messages B and C are shortened from 3.5 ms to 3.25 ms then the existing analysis results in unchanged worst-case response times, implying that the messages will still meet their deadlines. However, with these shorter periods the overall bus utilisation exceeds 100% and so the system cannot possibly be schedulable!

Related Work

The schedulability analysis for CAN builds on previous research into fixed priority scheduling of tasks on single processor systems.

Lehoczky (1990) introduced the concept of a busy period and showed that if tasks have deadlines greater than their periods, referred to as arbitrary deadlines, then it is necessary to examine the response times of all invocations of a task falling within a busy period in order to determine the worst-case response time. Harbour et al. (1991) showed that if deadlines are less than or equal to periods, but priorities vary during execution, then again multiple invocations must be inspected to determine the worst-case response time. We note that non-pre-emptive scheduling is effectively a special case of pre-emptive scheduling with varying execution priority—as soon as a task starts to execute its priority is raised to the highest level. Tindell et al. (1994a) improved upon the work of Lehoczky (1990) providing a formulation for arbitrary deadline analysis based on a recurrence relation.

Building upon these earlier results, George et al. (1996) provided comprehensive schedulability analysis of non-pre-emptive fixed priority scheduling of single processor systems.

Bril (2006) refuted the analysis of fixed priority systems with deferred pre-emption given by Burns (1994), showing that this analysis may result in computed worst-case response times that are optimistic. The schedulability analysis for CAN given by Tindell and Burns (1994) and Tindell et al. (1994b, 1995) builds upon Burns (1994) and suffers from essentially the same flaw. A similar issue with work on pre-emption thresholds (Wang and Saksena, 1999) was first identified and corrected by Regehr (2002).

The revised schedulability analysis presented in this paper aims to provide an evolutionary improvement upon the analysis of CAN given by Tindell and Burns (1994) and Tindell et al. (1994b, 1995). To do so, it draws upon the analysis of Tindell et al. (1994a) for fixed priority pre-emptive scheduling of systems with arbitrary deadlines and the analysis of George et al. (1996) for fixed priority non-pre-emptive systems.

A technical report (Bril et al., 2006a) and a workshop paper (Bril et al., 2006b) highlighted the problem for CAN but did not provide a specific in-depth solution. That is the purpose of this paper. A further technical report (Bril et al., 2006c) complements this paper, providing formal proofs of the worst-case response time of tasks under fixed priority scheduling with deferred pre-emption.

Organisation

The remainder of this paper is organised as follows: Section 2 describes the CAN protocol and terminology before outlining a suitable scheduling model and notation on which to base revised schedulability analysis. Section 3 provides new schedulability analysis for CAN, correcting the flaws in the existing approach. Section 4 discusses the system and message parameters needed for the flaws in the existing analysis to result in incorrect worst-case response times and hence misleading guarantees. Section 5 discusses the issue of optimal priority assignment for CAN messages. Section 6 summarises the implications of flaws in the existing analysis for commercial CAN applications. Finally, Section 7 concludes with a summary of the main contributions of this paper and recommendations for further research.

CONTROLLER AREA NETWORK

This section describes elements of the CAN protocol and characteristics of a system model that are needed to formulate a schedulability test. For a complete description of the CAN protocol, see the CAN specification version 2.0 (Bosch, 1991).

CAN Protocol and Terminology

CAN is an asynchronous multi-master serial data bus that uses Carrier Sense Multiple Access/Collision Resolution (CSMA/CR) to determine access.

CAN was designed as a simple and robust broadcast bus capable of operating at speeds of up to 1 Mbit/s. Message transfer over CAN is controlled by 4 different types of frame: Data frames, Remote Transmit Request (RTR) frames, Overload frames and Error frames.

The layout of a standard format data frame is shown in Fig. 4. Each CAN data frame is required to have a unique identifier. Identifiers may be 11-bit (standard format) or 29-bit (extended format).

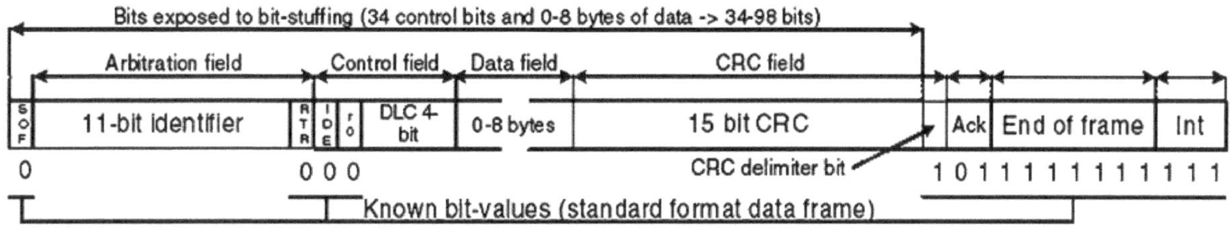

Figure 4. Standard format data frame.

The identifier serves two purposes beyond simply identifying the message. First, the identifier is used as a priority to determine which message, among those contending for the bus, will be transmitted next. Second, the identifier may be used by receivers to filter out messages they are not interested in, and so reduce the load on the receiver's host microprocessor.

In this paper we are interested in the schedulability of data frames, with error frames also considered in Section 3.5. The schedulability analysis can however easily be extended to include RTR frames using the approach given by Tindell et al. (1995).

Priority Based Arbitration

The CAN physical layer supports two states termed dominant ('0') and recessive ('1'). If two or more CAN controllers are transmitting at the same time and at least one of them transmits a '0' then the value on the bus will be a '0'. This mechanism is used to control access to the bus and also to signal errors.

The CAN protocol calls for nodes to wait until a bus idle period is detected before attempting to transmit. If two or more nodes start to transmit at the same time, then by monitoring each bit on the bus, each node can determine if it is transmitting the highest priority message (with a numerically lower identifier) and should continue or if it should stop transmitting and wait for the next bus idle period before trying again. As the message identifiers are unique, a node transmitting the last bit of the identifier field, without detecting a '0' bit that it did not transmit, must be transmitting the highest priority message that was ready for transmission at the start of arbitration. This node then continues to transmit the remainder of its message, all other nodes having backed off.

The requirement for a node to be able to overwrite a recessive bit, and the transmitting node detect this change, limits the combination of physical length and speed of a CAN bus. The duration of each bit must be sufficient for the signal to propagate the length of the network. This limits the maximum data rate to 1 Mbit/s for a network up to 40 m in length or to 125 Kbit/s for a 500 m long network.

The arbitration mechanism employed by CAN means that messages are sent as if all the nodes on the network shared a single global priority based queue. In effect messages are sent on the bus according to fixed priority non-pre-emptive scheduling.

The above high level description is a somewhat simplified view of the timing behaviour of CAN. CAN does not have a global concept of time, rather each CAN controller typically has its own clock which, within a tolerance specified by the protocol, may drift with respect to the clocks of other nodes. The CAN protocol therefore requires that nodes re-synchronise on each message transmission. Specifically, every node must synchronise to the leading edge of the start of frame bit caused by whichever node starts to transmit first.

Normally, CAN nodes are only allowed to start transmitting when the bus is idle. Thus, when the bus is idle beyond the 3-bit inter-frame space and a node starts to transmit a message beginning with the dominant start of frame bit ('0'), then all the other nodes synchronise on the leading edge of this bit and become receivers—i.e. they are not permitted to transmit until the bus next becomes idle. In this case any message that becomes ready for transmission after the leading edge of the start of frame bit has to wait for the next bus idle period before it can enter into arbitration.

However, to avoid problems due to clock drift, the CAN protocol also specifies that, if a CAN node has a message ready for transmission and detects a dominant bit at the 3rd bit of the inter-frame space, it will interpret this as a start of frame bit, and with the next bit, start transmitting its own message with the first bit of the identifier without first transmitting a start of frame bit and without becoming a receiver. Again the leading edge of the start of frame bit causes a synchronisation. This behaviour ensures that any messages that become ready for transmission, whilst another message is being sent on the bus, are entered into the next round of arbitration, irrespective of any within tolerance clock drift.

Error Detection

CAN was designed as a robust and reliable form of communication for short messages. Each data frame carries between 0 and 8 bytes of payload data and has a 15-bit Cyclic Redundancy Check (CRC). The CRC is used by receiving nodes to check for errors in the transmitted message. If a node detects an error in the transmitted message, which may be a bit-stuffing error (see Section 2.1.3), a CRC error, a form error in the fixed part of the message, or an acknowledgement error, then it transmits an error flag. The error flag consists of 6 bits of the same polarity: '000000' if the node is in the error active state and '111111' if it is error passive. Transmission of an error flag typically causes other nodes to also detect an error, leading to transmission of further error flags.

Figure 5 illustrates CAN error frames, for further details see (Bosch, 1991; Punnekkat et al., 2000). The length of an error frame is between 17 and 31 bits. Hence each message transmission that is signalled as an error can lead to a maximum of 31 additional bits of error recovery overhead plus re-transmission of the message itself.

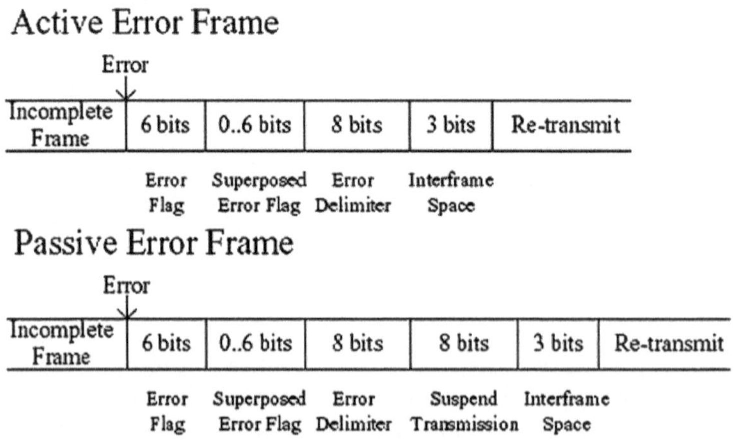

Figure 5. CAN error frames.

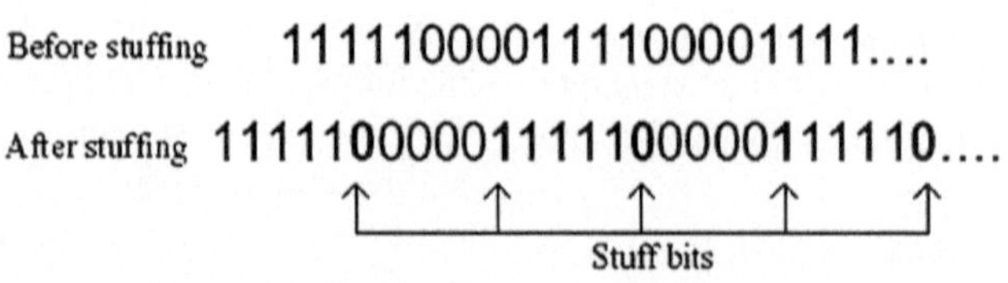

Figure 6. Worst-case bit stuffing.

Bit Stuffing

As the bit patterns '000000' and '111111' are used to signal errors, it is essential that these bit patterns are avoided in the variable part of a transmitted message—see Fig. 4. The CAN protocol therefore requires that a bit of the opposite polarity is inserted by the transmitter whenever 5 bits of the same polarity are transmitted. This process is referred to as bit-stuffing, and is reversed by the receiver.

The worst-case scenario for bit-stuffing is shown in Fig. 6. Note that each stuff bit begins a sequence of 5 bits that is itself subject to bit stuffing.

Stuff bits increase the maximum transmission time of CAN messages. Including stuff bits and the inter-frame space, the maximum transmission time C_m, of a CAN message m containing s_m data bytes is given by:

$$C_m = \left(g + 8s_m + 13 + \left\lfloor \frac{g + 8s_m - 1}{4} \right\rfloor \right) \tau_{\text{bit}} \quad (1)$$

where g is 34 for standard format (11-bit identifiers) or 54 for extended format (29-bit identifiers), $\lfloor a/b \rfloor$ is notation for the floor function, which returns the largest integer less than or equal to a/b, and τ_{bit} is the transmission time for a single bit.

The formula given in Eq. (1) simplifies to:

$$C_m = (55 + 10s_m)\tau_{\text{bit}} \quad (2)$$

for 11-bit identifiers and

$$C_m = (80 + 10s_m)\tau_{\text{bit}} \quad (3)$$

for 29-bit identifiers.

Scheduling Model

In this section we describe an appropriate system model and notation that can be used to analyse worst-case response times of messages on CAN and hence determine system schedulability.

The system is assumed to comprise a number of nodes (microprocessors) connected via CAN. Each node is assumed to be capable of ensuring that, at any given time when arbitration starts, the highest priority message queued at that node is entered into arbitration.

The system is assumed to contain a static set of hard real-time messages, each statically assigned to a node on the network. Each message m has a fixed identifier and hence a unique priority. As priority uniquely identifies each message, in the remainder of this paper we will overload m to mean either message m or priority m as appropriate. Each message has a maximum number of data bytes s_m, and a maximum transmission time C_m, given by Eq. (1).

Each message is assumed to be queued by a software task, process or interrupt handler executing on the host microprocessor. This task is either invoked by, or polls for, the event and takes a bounded amount of time, between 0 and J_m, to queue the message ready for transmission. J_m is referred to as the queuing jitter of the message and is inherited from the overall response time of the task, including any polling delay.

The event that triggers queuing of the message is assumed to occur with a minimum inter-arrival time of T_m, referred to as the message period. This model supports events that occur strictly periodically with a period of T_m, events that occur sporadically with a minimum separation of T_m, and events that occur only once before the system is reset, in which case T_m is infinite.

Each message has a hard deadline D_m, corresponding to the maximum permitted time from occurrence of the initiating event to the end of successful transmission of the message, at which time the message data is assumed to be available on the receiving nodes that require it. Tasks on the receiving nodes may place different timing requirements on the data, however in such cases we assume that D_m is the tightest such time constraint.

The worst-case response time R_m, of a message is defined as the longest time from the initiating event occurring to the message being received by the nodes that require it.

A message is said to be schedulable if and only if its worst-case response time is less than or equal to its deadline ($R_m \le D_m$). The system is schedulable if and only if all of the messages in the system are schedulable.

Practical Implications of the Model

Engineers wanting to use the analysis given in Section 3 to analyse CAN based systems must be careful to ensure that all of the assumptions of the above model hold for their system.

In particular, it is important that each CAN controller and device driver is capable of ensuring that, at any given time when arbitration starts, the highest priority message queued at that node is entered into arbitration. This behaviour is essential if message transmission is to take place as if there were a single global priority queue and for the analysis given in Section 3 to be applicable. As noted by Tindell and Burns (1994), the Philips 82C500 CAN controller cannot in general support this behaviour. Also the Intel 82527 CAN controller has a feature where messages are entered into arbitration in slot order rather than identifier order. In this case it is important that messages are allocated to slots in identifier order to preserve the correct priority based behaviour.

Many on-chip CAN controllers have multiple slots that can be allocated to either transmit or receive a specific message. For example some Motorola, National Semiconductor, Fujitsu and Hitachi on-chip CAN peripherals have 14, 15 or 16 such slots.

These slots typically have only a single buffer and therefore it is necessary to ensure that the previous instance of a message has been transmitted before any new data is written into the buffer, otherwise the previous message will be overwritten and lost. This behaviour provides an additional constraint on message transmission: the deadline of each message must be less than or equal to its period ($D_m \le T_m$).

Recall that the worst-case response time of a message is from the occurrence of the initiating event to the end of successful message reception at the receiving nodes. As noted by Broster (2003), receiving nodes can access the message following the end of frame marker and before the 3-bit inter-frame space—see Fig. 4. The analysis given in the remainder of this paper is slightly pessimistic in that it includes the 3-bit inter-frame space in the computed worst-case response times. To remove this small degree of pessimism, it is valid to simply subtract $3\tau_{bit}$ from the computed response time values.

The time constraint of interest to engineers is the overall end-to-end response time from an initiating event occurring, such as the brake pedal switch closing, to the corresponding output response, for example the brake lights illuminating. The worstcase response time of a message represents only part of this overall end-to-end response time. There is an additional delay to consider, corresponding to the worst-case time between the message becoming available at the receiving node and the output being made. Processing of the signal information contained in a message is typically done either by a task that polls for the message or by an interrupt handler that is triggered by message reception. The worst-case response time of the receiving task or interrupt handler, including any polling delay, needs to be added to the worst-case response time of the message to determine the overall end-to-end response time.

The scheduling model assumed in this paper uses only one time domain, whilst CAN typically has a separate clock source for each node on the network. To ensure that the schedulability analysis for a real network does not produce optimistic results, it is necessary to take clock tolerances into account. This can be achieved by converting to real-time as follows: for message jitters and bit times on the bus the conversion to real-time should assume that the node clocks run as slowly as their tolerance allows. Similarly, message periods and deadlines derived from node clocks should be converted to real-time assuming that the node clocks run as quickly as their tolerance allows.

RESPONSE TIME ANALYSIS

Response time analysis for CAN aims to provide a method of calculating the worst-case response time of each message. These values can then be compared to the message deadlines to determine if the system is schedulable. Initially we provide analysis assuming no errors on the CAN bus. This analysis is then extended, in Section 3.5, to account for errors on the bus.

For systems complying with the scheduling model given in Section 2.2, CAN effectively implements fixed priority non-pre-emptive scheduling of messages. Following the analysis given by Tindell and Burns (1994) and Tindell et al. (1994b, 1995) the worst-case response time of a message can be viewed as being made up of three elements:

1. The queuing jitter J_m, corresponding to the longest time between the initiating event and the message being queued, ready to be transmitted on the bus.
2. The queuing delay w_m, corresponding to the longest time that the message can remain in the CAN controller slot or device driver queue, before commencing successful transmission on the bus.
3. The transmission time C_m, corresponding to the longest time that the message can take to be transmitted.

The worst-case response time of message m is given by:

$$R_m = J_m + w_m + C_m \tag{4}$$

The queuing delay w_m, comprises two elements:

1. Blocking B_m, due to lower priority messages which may be in the process of being transmitted when message m is queued, and
2. Interference due to higher priority messages which may win arbitration and be transmitted in preference to message m.

Given the behaviour of CAN described in the final two paragraphs of Section 2.1.1, the maximum amount of blocking occurs when a lower priority message starts transmission immediately before message m is queued and hence ready to be transmitted on the bus. Message m must wait until the bus is idle before it can be entered into arbitration. The maximum blocking time B_m, is given by:

$$B_m = \max_{k \in lp(m)} (C_k) \tag{5}$$

where lp(m) is the set of messages with lower priority than m.

The concept of a busy period, introduced by Lehoczky (1990), is fundamental in analysing worst-case response times. Modifying the definition of a busy period given by Harbour et al. (1991) to apply to CAN messages, a priority level-m busy period is defined as follows:

1. It starts at some time t^s when a message of priority m or higher is queued ready for transmission and there are no messages of priority m or higher waiting to be transmitted that were queued strictly before time t^s.
2. It is a contiguous interval of time during which any message of priority lower than m is unable to start transmission and win arbitration.
3. It ends at the earliest time t^e when the bus becomes idle, ready for the next round of transmission and arbitration, yet there are no messages of priority m or higher waiting to be transmitted that were queued strictly before time t^e.

The key characteristic of a busy period is that all messages of priority m or higher, queued strictly before the end of the busy period, are transmitted during the busy period. These messages cannot therefore cause any interference on a subsequent instance of message m queued at or after the end of the busy period.

In mathematical terminology, busy periods can be viewed as right half-open intervals: $[t^s, t^e)$ where t^s is the start of the busy period and t^e the end of the busy period. Thus the end of one busy period may correspond to the start of another separate busy period. This is in contrast to the simpler definition given by Lehoczky (1990), which unifies two adjacent busy periods, as we have defined them, and therefore sometimes results in analysis of more message instances than is strictly necessary.

The worst-case queuing delay for message m occurs for some instance of message m queued within a priority level-m busy period that starts immediately after the longest lower priority message begins transmission. This maximal busy period begins with a so-called critical instant (Liu and Layland, 1973) where message m is queued simultaneously with all higher priority messages and then each of these higher priority messages is subsequently queued again after the shortest possible time interval. In the remainder of this paper whenever we refer to a busy period we mean this maximum length busy period.

If more than one instance of message m is transmitted during a priority level-m busy period, then it is necessary to determine the response time of each instance, in order to find the overall worst-case response time of the message.

Basic Analysis and Stopping Condition

Tindell and Burns (1994) and Tindell et al. (1994b, 1995) give the following equation for the worst-case queuing delay:

$$w_m = B_m + \sum_{\forall k \in hp(m)} \left\lceil \frac{w_m + J_k + \tau_{bit}}{T_k} \right\rceil C_k \tag{6}$$

where hp(m) is the set of messages with priorities higher than m, and ⌊a/b⌋ is notation for the ceiling function which returns the smallest integer greater than or equal to a/b.

Although w_m appears on both sides of Eq. (6), as the right hand side is a monotonic non-decreasing function of w_m, the equation may be solved using the recurrence relation given below.

$$w_m^{n+1} = B_m + \sum_{\forall k \in hp(m)} \left\lceil \frac{w_m^n + J_k + \tau_{bit}}{T_k} \right\rceil C_k \tag{7}$$

A suitable starting value is $w_m^0 = B_m$. The recurrence relation iterates until, either $J_m + w_m^{n+1} + C_m > D_m$ in which case the message is not schedulable, or $w_m^{n+1} = w_m^n$ in which case the worst-case response time of the first instance of the message in the busy period is given by: $J_m + w_m^{n+1} + C_m$.

The flaw in the above analysis is that, given the constraint $D_m \le T_m$, it implicitly assumes that if message m is schedulable, then the priority level-m busy period will end at or before T_m. We observe that with fixed priority pre-emptive scheduling this would always be the case, as on completion of transmission of message m no higher priority message could be awaiting transmission. However, with fixed priority nonpre-emptive scheduling, a higher priority message can be awaiting transmission when message m completes transmission, and thus the busy period can extend beyond T_m, as shown by the example in Section 1.4.

The length t_m, of the priority level-m busy period is given by the following recurrence relation, starting with an initial value of $t_m^0 = C_m$, and finishing when $t_m^{n+1} = t_m^n$:

$$t_m^{n+1} = B_m + \sum_{\forall k \in hep(m)} \left\lceil \frac{t_m^n + J_k}{T_k} \right\rceil C_k \tag{8}$$

where hep is the set of messages with priority higher than or equal to m. As the right hand side is a monotonic non-decreasing function of t_m, then the recurrence relation is guaranteed to converge provided that the bus utilisation U_m, for messages of priority m and higher, is less than 1:

$$U_m = \sum_{\forall k \in hep(m)} \frac{C_k}{T_k} \tag{9}$$

If $t_m \le T_m - J_m$, then the busy period ends at or before the second instance of message m is queued. This means that only the first instance of the message is transmitted during the busy period. The existing analysis calculates the worst-case queuing time for this instance, via Eq. (7), and hence provides the correct worst-case response time in this case.

If $t_m > T_m - J_m$, then the existing analysis may give an optimistic worst-case response time, depending on whether the first, or a subsequent instance of message m has the longest response time.

We observe that the analysis presented in appendix A.2 of George et al. (1996) suggests that t_m is the smallest value that is a solution to Eq. (8), however this is not strictly correct. For the lowest priority message $B_m = 0$, and so $t_m = 0$ is trivially the smallest solution when all of the messages have zero jitter. We avoid this problem by using an initial value of $t_m^0 = C_m$.

Checking Multiple Instances

The number of instances Q_m, of message m that become ready for transmission before the end of the busy period is given by:

$$Q_m = \left\lceil \frac{t_m + J_m}{T_m} \right\rceil \qquad (10)$$

To determine the worst-case response time of message m, it is necessary to calculate the response time of each of the Q_m instances. The maximum of these values then gives the worst-case response time.

In the following analysis, we use the index variable q to represent an instance of message m. The first instance in the busy period corresponds to q = 0, and the final instance to $q = Q_m - 1$. The longest time from the start of the busy period to instance q beginning successful transmission is given by:

$$w_m^{n+1}(q) = B_m + qC_m + \sum_{\forall k \in hp(m)} \left\lceil \frac{w_m^n + J_k + \tau_{\text{bit}}}{T_k} \right\rceil C_k \qquad (11)$$

Table 3. CAN messages

Message	Priority	Period	Deadline	TX time
A	1	2.5 ms	2.5 ms	1 ms
B	2	3.5 ms	3.25 ms	1 ms
C	3	3.5 ms	3.25 ms	1 ms

The recurrence relation starts with a value of $w_m^0(q) = B_m + qC_m$, and ends when $w_m^{n+1}(q) = w_m^n(q)$, or when $J_m + w_m^{n+1}(q) - qT_m + C_m > D_m$ in which case the message is unschedulable. For values of q>0 an efficient starting value is given by $w_m^0(q) = w_m(q-1) + C_m$.

The event initiating instance q of the message occurs at time $qT_m - J_m$ relative to the start of the busy period, so the response time of instance q is given by:

$$R_m(q) = J_m + w_m(q) - qT_m + C_m \qquad (12)$$

The worst-case response time of message m is therefore:

$$R_m = \max_{q=0..Q_m-1} (R_m(q)) \qquad (13)$$

We note that the analysis presented above is also applicable when messages have deadlines that are greater than their periods, so called arbitrary deadlines. However, if such timing characteristics are specified then the software device drivers or CAN controller hardware may need to be capable of buffering more than one instance of a message. N_m, the number of instances of each message that need to be buffered is bounded by:

$$N_m = \left\lceil \frac{R_m}{T_m} \right\rceil \qquad (14)$$

We observe that the analysis presented by George et al. (1996) effectively uses $Q_m = \lfloor t_m/T_m \rfloor + 1$ rather than $Q_m = \lceil t_m/T_m \rceil$. This yields a value which is one too large when the length of the busy period plus jitter is an integer multiple of the message period. Although this does not give rise to problems, we prefer the more efficient formulation given by Eq. (10).

Example

In Section 1.4 we showed, with the aid of a simple example, how the existing analysis can provide optimistic worst-case response times and hence flawed guarantees that messages will meet their deadlines. We return to this example to illustrate how the analysis presented in this paper computes the correct worst-case response times. For ease of reference, the table of message parameters is repeated below.

Using the new analysis, the worst-case response time of message C (m = 3) is calculated as follows: As there are no lower priority messages, $B_3 = 0$. Starting with a value of $t_3^0 = C_3 = 1$, the recurrence relation given by Eq. (8) iterates as follows: $t_3^1 = 3,\ t_3^2 = 4,\ t_3^3 = 6,\ t_3^4 = 7$, converging as $t_3^5 = t_3^4 = 7$. The length of the busy period is therefore 7.0 ms, and the number of instances of message C that need to be examined is given by Eq. (10):

$$Q_3 = \left\lceil \frac{7.0}{3.5} \right\rceil = 2$$

This tells us that there is the possibility that the existing analysis will calculate an optimistic worst-case response time. The value could, however, still be correct if the first instance of the message has the longest response time.

Calculation of the response time of the first instance proceeds using Eq. (11): $w_3^0(0) = 0,\ w_3^1(0) = 2$, converging as $w_3^2(0) = w_3^1(0) = 2$. Using Eq. (12), we have $R_3(0) = 3$, the same response time calculated by the existing analysis.

Moving on to the second instance, $w_3^0(1) = w_3(0) + C_m = 3,\ w_3^1(1) = 4,\ w_3^2(1) = 5,\ w_3^3(1) = 6$. At this point computation would normally stop as the response time, given by $J_3 + w_3(q) - qT_3 +$

C_3, has reached 3.5 ms which is greater than the message deadline. However, if we continue iterating, assuming a longer deadline, then the recurrence relation converges on $w_3^4(1) = w_3^3(1) = 6$ and hence $R_3(1) = 3.5$ ms. The worst-case response time of message C is in fact 3.5 ms, as previously illustrated by Fig. 3 in Section 1.4.

Sufficient Schedulability Tests

The analysis given in Sections 3.1 and 3.2 corrects a significant flaw in the existing schedulability analysis for CAN. However, the schedulability test presented is more complex, potentially requiring the computation of multiple response times.

In this section we present two simpler but more pessimistic schedulability tests. These tests are "sufficient but not necessary". By "sufficient" we mean that all systems deemed to be schedulable by the tests are in fact schedulable, and by "not necessary" we mean that not all systems deemed to be unschedulable by the tests are in fact unschedulable.

The schedulability tests given in this section are only applicable given the constraint that message deadlines do not exceed their periods.

The response time of the first instance of a message in the busy period is given by Eq. (7). Assuming that this first instance completes transmission before its deadline and hence before the end of its period, then we have two possibilities to consider.

1. If the busy period ends before the next instance of message m is queued, then Eq. (7) gives the correct worst-case response time.
2. Alternatively, if the busy period continues beyond the time at which the next instance of message m is queued, then we must also consider the response time of the second and any subsequent instances of message m, queued before the end of the busy period.

First, we derive an upper bound on the maximum length of the interval $[s_{m,q}, s_{m,q+1})$, between the times, $s_{m,q}$ and $s_{m,q+1}$, at which two arbitrary but consecutive instances, q and q + 1, of message m start transmission. We then show that this upper bound is also an upper bound on the queuing delay for instance q + 1, and can therefore be used as the basis for a sufficient schedulability test.

We assume that:

1. All q + 1 instances fall within the same busy period,
2. The first q instances are schedulable—we will return to this point later.

We observe that at time $s_{m,q}$, when instance q starts transmission, there can be no other messages currently being transmitted, and no messages of higher priority than m awaiting transmission. Thus, an upper bound on the length of the time interval $[s_{m,q}, s_{m,q+1})$ can be found by making the potentially pessimistic assumption that all higher priority messages are queued just as instance q starts transmission. The smallest solution to Eq. (15) provides an upper bound on the length of this interval.

$$w_m^{n+1} = C_m + \sum_{\forall k \in hp(m)} \left\lceil \frac{w_m^n + J_k + \tau_{\text{bit}}}{T_k} \right\rceil C_k \tag{15}$$

Here, Cm is the transmission time of the qth instance of message m, which is transmitted first in the interval. The summation term represents the interference from higher priority messages, released during the interval, and sent before the (q + 1)th instance of message m can start to be transmitted.

Given the assumption that the first q instances in the busy period are schedulable, and the constraint that $D_m \le T_m$, then the start (and end) of transmission of the qth instance must happen before the end of its period, and hence before the (q + 1)th instance is queued. This means that the queuing delay for the (q + 1)th instance, as measured from the time at which it is queued to the start of its transmission, is less than the length of the interval $[s_{m,q}, s_{m,q+1})$. The queuing delay for the (q + 1)th instance is therefore also bounded by the solution to Eq. (15).

We now return to the assumption that the first q instances are schedulable. Schedulability of the q = 0 instance can be determined using Eq. (7); whilst the schedulability of the second and all subsequent instances within the busy period can be determined, by induction, using Eq. (15).

To determine an upper bound on the queuing delay, a suitable starting value, for use in using Eq. (15), is $w_m^0 = C_m$. The recurrence relation iterates until, either $J_m + w_m^{n+1} + C_m > D_m$ in which case message m is deemed unschedulable, or $w_m^{n+1} = w_m^n$ in which case the second and subsequent instances of message m are schedulable, with an upper bound on their response times of $J_m + w_m^{n+1} + C_m$. Intuitively, we might say that the second and subsequence instances of message m in the busy period are subject to blocking, of at most C_m, due to the previous instance of the same message.

This result suggests a simple sufficient but not necessary schedulability test, formed by combining Eqs. (7) and (15) into a single equation—Eq. (16).

$$w_m^{n+1} = \max(B_m, C_m) + \sum_{\forall k \in hp(m)} \left\lceil \frac{w_m^n + J_k + \tau_{\text{bit}}}{T_k} \right\rceil C_k \tag{16}$$

We observe that the schedulability analysis embodied in Eq. (16) equates to assuming that an instance of message m can be subject to blocking; either of B_m, due to non-pre-emptive transmission of lower priority messages; or of C_m, due to the non-pre-emptive transmission of the previous instance of message m itself.

A further simplification is to assume that the blocking factor always takes its maximum possible value. This leads to a further sufficient but not necessary schedulability test:

$$w_m^{n+1} = B^{\text{MAX}} + \sum_{\forall k \in hp(m)} \left\lceil \frac{w_m^n + J_k + \tau_{\text{bit}}}{T_k} \right\rceil C_k \tag{17}$$

where B^{MAX} corresponds to the transmission time of the longest possible CAN message (8 data bytes), irrespective of the characteristics and priorities of the messages in the system.

Error Model

So far we have assumed that no errors occur on the CAN bus; however, as originally shown by Tindell and Burns (1994) and Tindell et al. (1994b, 1995), schedulability analysis of CAN may be extended to include an appropriate error model.

In this paper we consider only a very simple and general error model. We assume that the maximum number of errors present on the bus in some time interval t is given by the function F(t). We assume no specific details about this function; save that it is a monotonic non-decreasing function of t. For a more detailed discussion of appropriate error models for CAN, see Punnekkat et al. (2000) and Broster et al. (2002, 2005).

We now modify the schedulability equations to account for the error recovery overhead. The worst-case impact of a single bit error is to cause transmission of an additional 31 bits of error recovery overhead plus re-transmission of the affected message. Only errors affecting message m or higher priority messages can delay message m from being successfully transmitted. The maximum additional delay caused by the error recovery mechanism is therefore given by:

$$E_m(t) = \left(31\tau_{\text{bit}} + \max_{k \in hep(m)} (C_k)\right) F(t) \tag{18}$$

Revising Eq. (8) to compute the length of the busy period we have:

$$t_m^{n+1} = E_m\left(t_m^n\right) + B_m + \sum_{\forall k \in hep(m)} \left\lceil \frac{t_m^n + J_k}{T_k} \right\rceil C_k \tag{19}$$

Again an appropriate initial value is $t_m^0 = C_m$. Eq. (19) is guaranteed to converge on a solution provided that the utilisation U_m, including error recovery overhead, is less than 1.

As before, Eq. (10) can be used to compute the number of message instances that need to be examined to find the worst-case response time.

$$w_m^{n+1}(q) = E_m\left(w_m^n + C_m\right) + B_m + qC_m + \sum_{\forall k \in hp(m)} \left\lceil \frac{w_m^n + J_k + \tau_{\text{bit}}}{T_k} \right\rceil C_k \tag{20}$$

Eq. (20) extends Eq. (11) to account for the error recovery overhead. Note that as errors can impact the transmission of message m itself, the time interval considered in calculating the error recovery overhead includes the transmission time of message m as well as its queuing delay. Eqs. (20), (12) and (13) can be used together to compute the response time of each message instance q, and hence find the worst-case response time of each message m in the presence of errors at the maximum rate specified by the error model.

The sufficient schedulability tests given in Section 3.4 can be similarly modified via the addition of the term $E_m(w_m^n + C_m)$ to account for the error recovery overhead.

DISCUSSION

In this section we consider various characteristics of CAN systems and discuss whether flaws in the existing analysis can result in erroneous guarantees under specific circumstances that are relevant to real-world systems.

We seek to answer the following questions.

1. Can the existing analysis give faulty guarantees to messages of any priority?
2. If the bus utilization is low, can the existing analysis still result in optimistic response times?
3. Do error models give sufficient engineering margin for error to account for the flaw in the analysis?
4. Does the omission of diagnostic messages during normal operation reduce interference/blocking enough to ensure that the deadlines of the remaining messages will be met?
5. Which message guarantees can we be sure are not at risk?

Priorities of Messages at Risk

We have found that the existing analysis gives the correct worst-case response times for the highest priority and the 2nd highest priority message. However; it can compute incorrect worst-case response times for messages from the 3rd highest priority to the lowest priority.

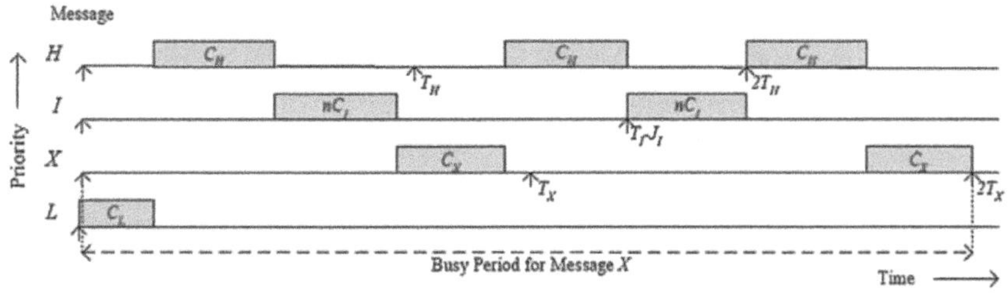

Figure 7. Busy period for message X.

This is illustrated by the example message set constructed below and depicted in Fig. 7. The example message set consists of;

1. A high priority message H;
2. A group of n (where $n \geq 1$) intermediate priority messages, represented by I, which all have the same periods and transmission times;
3. A message X with a priority below those messages in group I, which highlights the flaw in the analysis and
4. A group of k (where $k \geq 0$) low priority messages represented by L, which all have the same transmission times.

The transmission times of the messages are C_H, C_I, C_X and C_L respectively, with the constraint that $C_X > C_L$.

The low priority messages L, are assumed to have very large periods and no jitter. These messages contribute only blocking to the response time of message X. (Note that if there are no lower priority messages, i.e. k = 0, then the example still holds with $C_L = 0$).

The period of message H is:

$$T_H = (C_L + 2C_H + 2nC_I + C_X)/2$$

The period of message X is:

$$T_X = (C_L + 3C_H + 2nC_I + 2C_X)/2$$

The period of the intermediate messages I, is assumed to be large ($T_I \gg 2T_X$); however the period less jitter for each intermediate message is:

$$T_I - J_I = C_L + 2C_H + nC_I + C_X$$

By contrast, messages H and X are assumed to have no jitter.

The busy period for message X is shown in Fig. 7. For simplicity, there is only one intermediate priority message shown in the diagram; however, the transmission time of this message is given as nC_I, representing the arbitrary number of intermediate messages that are considered.

We now show that under certain conditions message X exhibits the problem with the existing analysis. The length of the busy period for message X, given by Eq. (8), is:

$$t_X = C_L + 3C_H + 2nC_I + 2C_X = 2T_X$$

Hence, according to Eq. (10), there are two instances of message X in the busy period. We now compute the response times of these two instances. According to Eq. (11), and as $C_X > C_L$, the queuing delay of the first instance of message X is:

$$w_X(0) = C_L + C_H + nC_I$$

Similarly for the second instance:

$$w_X(1) = C_L + 3C_H + 2nC_I + C_X$$

According to Eq. (12), the response times of the two instances are:

$$R_X(0) = C_L + C_H + nC_I + C_X$$

and

$$R_X(1) = (C_L + 3C_H + 2nC_I + 2C_X)/2$$

Comparing the formulas for R_X (0) and R_X (1), then, provided that $C_H > C_L$, the response time of the second instance is greater than that of the first. Meaning that message X exposes the flaw in the existing analysis. (In fact, assuming that $D_X = T_X$, the second instance of message X is only just schedulable with $R_X = T_X$).

As we can choose an arbitrary number ($n \geq 1$) of intermediate priority messages, and similarly an arbitrary number ($k \geq 0$) of lower priority messages, message X may lie anywhere from the 3rd highest to the lowest priority in a set of messages with cardinality greater than or equal to 3. We conclude that any message from the lowest priority to the 3rd highest priority in a set of 3 or more messages can be given an optimistic response time and therefore a faulty guarantee by the existing analysis.

Breakdown Utilization

The example in Section 1.4 has a bus utilisation of 97%. It is interesting to ask if the existing analysis can yield optimistic worst-case response times for systems with much lower utilisation.

Returning to the example message set, constructed in Section 4.1, we now consider how low the utilisation of that message set can be.

Table 4. Utilisation of message sets breaking the existing analysis

Number of messages	Utilisation
3	45.5%
5	21.4%
10	9.2%
25	3.4%
100	0.82%

To achieve the lowest possible utilisation, we need only consider the contribution from messages H and X; as the utilisation of both the intermediate messages I, and the low priority messages L, tends to zero when their periods are increased to an arbitrarily large value. We therefore have:

$$U = \frac{2C_X}{C_L + 3C_H + 2nC_I + 2C_X} + \frac{2C_H}{C_L + 2C_H + 2nC_I + C_X}$$

with the constraints that $C_H > C_L$ and $C_X > C_L$.

The overall utilisation is minimised by choosing values of C_H and C_X as small as possible, and a value of C_I as large as possible. Given the constraints on CAN message sizes, the minimum occurs when we choose messages H and X to have zero data bytes, so $C_H = C_X = 55\tau_{bit}$, the intermediate messages to have 8 data bytes, so $C_I = 135\tau_{bit}$, and no lower priority messages, so $C_L = 0$.

We note that this message set is somewhat pathological, as all the intermediate priority messages have arbitrarily large periods/deadlines and correspondingly large queuing jitter. However, it does illustrate that in general the existing analysis breaks down at very low utilisation levels.

Table 4 provides an upper bound on this breakdown utilisation: the existing analysis is known to breakdown at these levels of utilisation, it may breakdown at still lower levels.

Whilst it is unlikely that real-world applications will have message configurations that replicate the pathological case discussed above, such systems may include messages with large amounts of queuing jitter. Typically these are 'gatewayed' messages that have inherited a large jitter from variability in the response time of a source message sent on another network. We conclude that, for applications characterised by non-zero queuing jitter, it is prudent to assume that there could be problems with the existing analysis, irrespective of overall bus utilisation.

In fact, for real-world CAN systems, characterised by messages with non-zero queuing jitter and consequently deadlines less than periods, overall bus utilisation is a poor indicator of system schedulability.

Margin for Error

In Section 3.5 we saw how a generalised error model could be included in the revised schedulability analysis. Bit error rates on CAN are typically very low: 10^{-11} up to 10^{-6} depending on environmental conditions (Ferreira et al., 2004). However, errors do occur and it is therefore appropriate that any commercial application of CAN schedulability analysis should include at least a simple error model to account for sporadic errors on the bus. These errors are typically caused by external sources of Electromagnetic Interference (EMI) such as mobile phones, radar, radio transmitters, and lightning as well as other possible causes such as switch contacts, and shielding or wiring faults. As such errors are typically completely uncorrelated with message transmission, it is reasonable to assume that any useful error model allows for the possibility of an error occurring at any given time, and hence the error function F(t) ≥ 1 for a time interval of any length t.

Let us now consider the situation where the schedulability analysis given by Tindell and Burns (1994) and Tindell et al. (1994b, 1995) has been used along with an error model with F(t) = 1 to determine the schedulability of a system. The recurrence relation used by the existing analysis is given below:

$$w_m^{n+1} = B_m + E_m\left(w_m^n + C_m\right) + \sum_{\forall k \in hp(m)} \left\lceil \frac{w_m^n + J_k + \tau_{\text{bit}}}{T_k} \right\rceil C_k \tag{21}$$

Given that F(t) ≥ 1, then from Eq. (18), the maximum additional delay to message m due to the error recovery mechanism is always longer than the transmission time of message m, i.e. $E_m(t) > C_m$. Substituting C_m for $E_m(t)$ in Eq. (21) gives:

$$w_m^{n+1} = B_m + C_m + \sum_{\forall k \in hp(m)} \left\lceil \frac{w_m^n + J_k + \tau_{\text{bit}}}{T_k} \right\rceil C_k \tag{22}$$

We note that as $E_m(t) > C_m$, the solution to Eq. (22) cannot be larger than the solution to Eq. (21).

Recall that Eq. (16) provides a correct sufficient but not necessary schedulability test for the case where there are no errors on the CAN bus. Comparing Eqs. (22) and (16), we observe that, as $\max(B_m, C_m) \leq B_m + C_m$, the solution to Eq. (16) cannot be larger than the solution to Eq. (22) and hence cannot be larger than the solution to Eq. (21). This means that if message m is deemed to be schedulable given the queuing delay computed by Eq. (21) for the case where there are errors on the bus, then it must also be schedulable given the queuing delay computed via Eq. (16) for the case where there are no errors on the bus.

This is an important result. It means that if the existing analysis showed that every message was schedulable in the presence of any reasonable error model, with F(t) ≥ 1, then, despite the flaw in the existing analysis, every message is actually guaranteed to be schedulable when no errors are present. Put another way, the engineering margin for error provided by the error model is sufficient to account for the error in the analysis.

We observe however, that the robustness of systems analysed using the schedulability analysis given by Tindell and Burns (1994), and Tindell et al. (1994b, 1995) may not be all that was expected. Flaws in the existing analysis could lead to message configurations that will miss their deadlines in the presence of errors at a rate within the parameters of the specified error model, even though we can be sure that they will not miss their deadlines when no errors are present on the bus.

Message Omission

Many CAN applications allow for 8 data byte diagnostic messages that are not transmitted during the normal mode of operation. These messages are transmitted only when the system is in diagnostic mode and linked to service equipment. In this section we consider whether the omission of diagnostic messages provides sufficient reduction in interference/blocking to ensure that messages do not miss their deadlines during normal operation, despite being given potentially optimistic worst-case response times by the existing analysis.

To answer this question, we consider a system that is deemed to be schedulable by the existing analysis. We assume that this system includes an 8 data byte diagnostics message Y, which is only transmitted when the system is in diagnostic mode. We note that as message Y has the maximum number of data bytes, its transmission time is equivalent to the largest possible blocking factor, so $C_Y = B^{MAX}$. The blocking factor for each message m of higher priority than Y , is therefore given by $B_m = B^{MAX}$, which means that the existing analysis based on Eq. (7) computes exactly the same worst-case response time for each higher priority message m, as the correct sufficient but not necessary schedulability analysis test based on Eq. (17). The existing analysis cannot therefore result in optimistic worst-case response times for messages of higher priority than Y.

For each message of lower priority than Y, the interference due to message Y is at least B^{MAX}. Comparing Eq. (7) and (17), we observe that the solution to Eq. (7), with diagnostic message Y included in the set of higher priority messages, is at least as large as the solution to Eq. (17) when message Y is excluded. This means that if a lower priority message m is deemed to be schedulable by the existing analysis when message Y is present, then it must also be schedulable according to the correct sufficient but not necessary schedulability analysis when message Y is omitted.

We conclude that the omission of a single maximum length message of arbitrary priority provides sufficient reduction in interference/blocking to ensure that the flaw in the existing analysis cannot lead to any of the remaining messages missing their deadlines.

Message Guarantees not at Risk

In this section we consider the circumstances under which the first instance of a message in the busy period is guaranteed to have the longest response time. Under these circumstances, despite its flaws, the existing analysis gives correct results.

Assuming that message deadlines do not exceed their periods, then Eq. (16) in Section 3.4 provides an upper bound on the queuing delay for the second and subsequent instances of message m in the busy period. Comparing Eqs. (7) and (16), we observe that if $B_m \geq C_m$, then the first instance of message m is guaranteed to have a response time at least as long as subsequent ones. From the definition of B_m given in Eq. (5), we conclude the following important result: the existing analysis gives the correct response time for any message where there exists at least one lower priority message with equal or longer transmission time/message length.

PRIORITY ASSIGNMENT POLICIES

The analysis presented in Section 3 is applicable irrespective of the priority ordering of CAN messages. However, choosing an appropriate priority ordering is important in obtaining a schedulable system, and in maximising robustness to errors.

Table 5. CAN messages highlighting non-optimal priority assignment

Message	Period	Deadline	Number of bits	TX time
A	3.0 ms	3.0 ms	135	1.08 ms
B	4.0 ms	4.0 ms	135	1.08 ms
C	4.5 ms	4.5 ms	65	0.52 ms

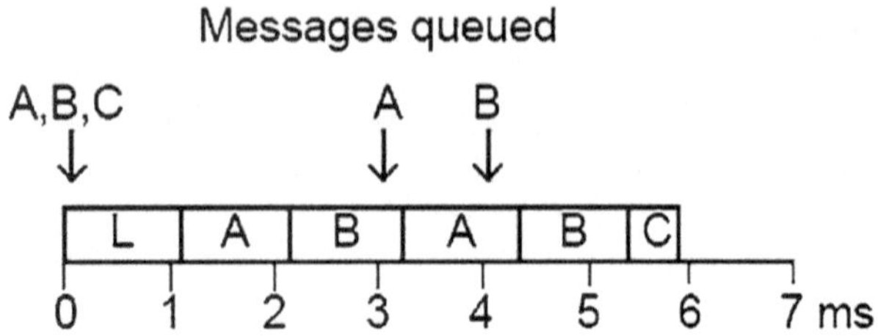

Figure 8. Message response times with "optimal" priority assignment.

Priority ordering is determined by a priority assignment policy. A priority assignment policy P is referred to as optimal if there are no systems that are schedulable using any other priority assignment policy that are not also schedulable using policy P.

Tindell and Burns (1994) and Tindell et al. (1995) claimed that deadline monotonic priority assignment (Leung and Whitehead, 1982) and "deadline minus jitter" or (D-J)-monotonic priority assignment (Zuhily, 2006) was optimal for CAN; however, whilst these policies are optimal for fixed priority pre-emptive scheduling, assuming deadlines no greater than periods, they are not optimal for fixed priority non-preemptive scheduling (George et al., 1996), and are therefore not optimal for CAN. This is illustrated by the following example using the set of messages given in Table 5.

This example assumes a 125 Kbit/s network and 11-bit identifiers. Messages A and B contain 8 data bytes and message C contains 1 data byte, giving transmission times of 1.08, 1.08 and 0.52 ms respectively, assuming worst-case bit stuffing. In addition, there are a number of lower priority messages, each containing 8 data bytes, which are also sent on the network; their transmission times are also 1.08 ms.

Setting message priorities in the order A—highest, then B, then C results in an unschedulable system. The worst-case response times of messages A and B are 2.16 ms and 3.24 ms respectively; however, in the worst case, message C does not even begin transmission before its deadline.

Figure 8 illustrates the long delays that message C is subject to before transmission. Messages A, B and C are assumed to be queued just too late to enter into arbitration at time t = 0 and hence the low priority message L is transmitted first.

The priority ordering A, B, C corresponds to both deadline monotonic and also (D-J)-monotonic priority ordering—as all the messages have zero queuing jitter. If these priority assignment policies are optimal then we should not be able to find another priority ordering which results in all of the deadlines being met; however, if we use the priority ordering A, C, B then the worst-case response times of the messages are: $R_A = 2.16$ ms, $R_C = 2.68$ ms and $R_B = 3.76$ ms, as illustrated in Fig. 9. With this priority ordering, all of the messages meet their deadlines.

The reason that the revised priority ordering results in a schedulable system is that giving the shortest message a higher priority enables all three messages to start transmission within 3 ms of being queued; hence none of them are subject to interference from a second instance of message A and subsequently a second instance of message B. This example shows that the priority assignment policies assumed by Tindell and Burns (1994) and Tindell et al. (1995) to be optimal are not.

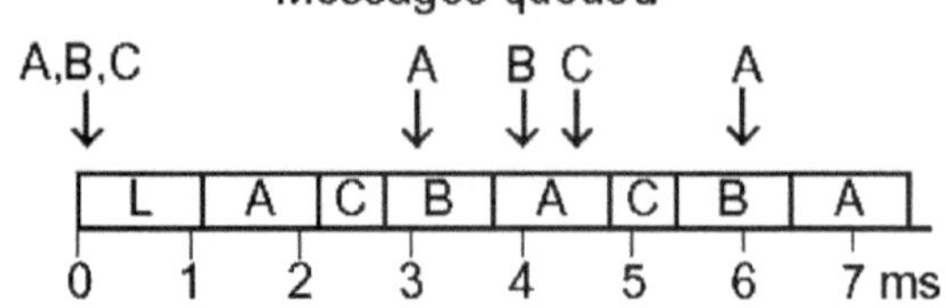

Figure 9. Message response times with an alternative priority assignment.

George et al. (1996) claimed that deadline monotonic priority assignment is optimal for non-pre-emptive systems with no jitter, provided that deadlines and execution times are in the same order i.e. $D_i < D_j$ implies $C_i \leq C_j$. The proof, given by George et al. (1996), assumes that "as $\forall i$, $D_i \leq T_i$ the worst-case response time of any task is found in its first instance"; however, this assumption is false, as we have seen with the simple example in Section 1.4, and so the proof is undermined. The theorem may or may not still be true.

George et al. (1996) also showed that the optimal priority assignment algorithm devised by Audsley (1991) is applicable to non-pre-emptive systems. In general, Audsley's algorithm is applicable provided that the worst-case response time of a message:

1. Does not depend upon the specific priority ordering of higher priority messages and,
2. Does not get longer if the message is given a higher priority.

Inspection of the various equations presented in this paper shows that both of the above conditions hold. Neither the length of the queuing delay, nor the length of the busy period depends upon the specific priority order of higher priority messages. Similarly, although the blocking term can get larger with increased priority this is always counteracted by a decrease in interference that is at least as large; hence the length of the busy period and the length of the queuing delay cannot increase with increasing message priority. The optimal priority ordering of CAN messages can therefore be determined using Audsley's priority assignment algorithm, given below.

```
Optimal Priority Assignment Algorithm

for each priority level, lowest first
{
   for each unassigned message m
   {
       if m is schedulable at this priority
       {
           assign m this priority
           break (continue outer loop)
       }
   }
    return unschedulable
}
return schedulable
```

For n messages, Audsley's algorithm performs at most $n(n - 1)/2$ schedulability tests and is guaranteed to find a schedulable priority assignment if one exists. It does not however specify an order in which messages should be tried at each priority level. This order heavily influences the priority assignment chosen if there is more than one ordering that is schedulable. In fact, a poor choice of initial ordering can result in a priority assignment that leaves the system only just schedulable. We therefore suggest that, as a useful heuristic, messages are tried at each priority level in (D-J) order, largest value of (D-J) first, with ties broken according to message length, longest first.

IMPLICATIONS AND RECOMMENDATIONS

In this section we discuss the implications of flaws in existing CAN schedulability analysis on commercial CAN schedulability analysis tools and deployed CAN applications.

CAN Schedulability Analysis Tools

CAN schedulability analysis tools need to take account of the findings presented in this paper. This will involve checking, and if necessary updating, the analysis they employ; ensuring that it cannot provide optimistic worst-case response times and false guarantees.

The sufficient but not necessary schedulability tests given in Section 3.4 provide a "quick-fix" solution with minimal changes required to the existing analysis. These tests are however pessimistic and implementing the revised analysis given in Section 3 would potentially lead to a better technical solution.

Whilst "deadline minus jitter" or (D-J)-monotonic priority ordering is still a good heuristic to use, it is not necessarily the optimal priority assignment policy for CAN. Implementing priority ordering based upon Audsley's optimal priority assignment algorithm would ensure that a schedulable priority ordering is found whenever one exists.

Commercial CAN Applications System

Designers configuring commercial CAN applications often take the engineering approach that all messages in the system should remain schedulable given the addition of any number of low priority messages that can be used for development and test purposes. Such analysis based on Tindell and Burns (1994) and Tindell et al. (1994b, 1995) would assume that every message is subject to the maximum blocking factor, as per the sufficient schedulability test given by Eq. (17). This schedulability test computes a correct upper bound on the actual response time of each message, and so provides a correct guarantee that the configured messages will meet their deadlines.

Given the flaws in the existing schedulability analysis, it would however be prudent for System Designers to check the precise details of the analysis used to compute worstcase response times for their systems. If the analysis used has the potential to compute erroneous worst-case response times, then the feasibility of all the CAN configurations designed, developed and deployed using that analysis should be checked to ensure that they are in fact schedulable, and robust to errors at the rate specified by the prescribed error model.

Faults in Deployed Systems

Many deployed CAN systems, for example those in automotive applications, will have been analysed using the pragmatic engineering approach described in the previous section. The flaws in the existing analysis cannot lead to a problem with a deployed system in this case.

Many CAN applications allow for maximum length (8 data byte) diagnostic messages that are not transmitted during normal operation. Assuming that the existing analysis deemed the deployed system to be schedulable with these diagnostic messages present, then Section 4.4 showed that the omission of a single diagnostic message provides sufficient reduction in interference/blocking to ensure that the flaws in the existing analysis cannot lead to other messages missing their deadlines during normal operation.

In Section 4.5 we saw that the existing analysis gives the correct response time for any message where there is at least one lower priority message with equal or longer transmission time/message length. Many CAN applications use exclusively 8 data byte messages as a means of addressing the high ratio of overhead to useful data on CAN. In this case, the existing analysis is guaranteed to compute correct response times for all but the lowest priority message.

Even if a message has the potential to be given an erroneous worst-case response time by the existing analysis, then, unless that message is close to being unschedulable, the computed worst-case response time is still likely to be the true value. Even if an optimistic value is computed, then the true value may still be less than the message deadline. Finally, for a deadline miss to actually happen in a deployed system requires that the worst-case message phasing occurs, and at that point a number of messages take close to their maximum transmission times. This requires worst-case or near worst-case bit stuffing to occur which is, in itself, highly unlikely (Nolte et al., 2002).

Normal practice with commercial CAN configurations is to ensure that the schedulability analysis used includes provision for a plausible error model. In this case, Section 4.3 showed that such systems are guaranteed to be schedulable when no errors are present on the CAN bus provided that they were deemed to be schedulable in the presence of errors by the existing analysis.

We conclude that deadline misses in deployed CAN systems due to flaws in the existing analysis are extremely unlikely. Any such deadline failures are more likely to occur due to errors occurring on the bus at a higher rate than that accounted for by the error model.

We note that embedded CAN-based systems are built to be resilient to some messages missing their deadlines, and to much simpler forms of error such as wiring faults. CAN is not used, in its basic form, for safety critical systems due to known issues such as the "double receive" and "babbling idiot" problems (Rufino et al., 1998; Broster and Burns, 2003; Rufino, 2002).

SUMMARY AND CONCLUSIONS

In this paper we highlighted a significant flaw in long-standing, highly cited, and widely used schedulability analysis of CAN. We showed how this flaw could lead to the computation of optimistic worst-case response times for CAN messages, broken guarantees, and deadline misses. This paper provides revised analysis that can be used to calculate correct worst-case response times for CAN.

In addition, we showed that:

1. The existing analysis can provide optimistic worst-case response times for messages from the 3rd highest priority to the lowest priority.
2. The existing analysis can lead to broken guarantees and hence deadline misses in systems with low bus utilisation.
3. Where an error model has been considered, the flaw in the existing analysis is not sufficient to lead to CAN configurations that will result in missed deadlines when no errors are present on the bus. The desired robustness to errors may not however be achieved.
4. The omission of a single maximum length diagnostic message, accounted for by the existing analysis, reduces interference/blocking enough to ensure that the deadlines of all the remaining messages are met during normal operation.
5. Despite its flaws, the existing analysis gives the correct response time for any message where there is at least one lower priority message with the same or longer transmission time/message length.

We discussed the implications of these results for commercial CAN systems developed using flawed analysis and provided two simple, sufficient schedulability tests enabling a "quick-fix" to be made to commercial CAN schedulability analysis tools.

Finally, we showed that neither deadline monotonic nor (D-J)-monotonic priority assignment is optimal for CAN. Audsley's priority assignment algorithm is however optimal for fixed priority non-pre-emptive systems and can be used to obtain a schedulable priority ordering for CAN whenever one exists.

Future Work

A considerable body of academic work has grown up from Tindell's seminal analysis of CAN. The flaws in that original work may have partly undermined some of the subsequent research built upon it. Authors that have cited the original CAN analysis in their work are therefore encouraged to check the implications. In particular the academic work most likely to be affected is that which extends the original analysis and pushes system schedulability to its limits, for example work on error models.

Postscript

Volcano Network Architect (VNA) is a commercial CAN design and analysis tool originally developed by Volcano Communications Technologies AB, and now owned by Mentor Graphics Corporation. By 2004, over 20 million cars, with an average of 20 ECUs per car, had been programmed using this technology (Oswald, 2004).

The engineering team responsible for Volcano Network Architect was given early visibility of this paper, enabling them to check the validity of the schedulability analysis used in their commercial products.

The analysis provided by Volcano Network Architect was found to be sufficient: it assumes the longest possible blocking time irrespective of message priority (Horvath, 2006) and therefore computes an upper bound on response times, similar to that given by the sufficient schedulability tests described in Section 3.4.

ACKNOWLEDGMENTS

This work was partially funded by the UK EPSRC funded DIRC project, the EU funded FRESCOR project and the IST-004527 funded ARTIST 2 network of excellence on Embedded Systems Design.

REFERENCES

1. Audsley NC (1991) Optimal priority assignment and feasibility of static priority tasks with arbitrary start times. Technical Report YCS 164, Dept. Computer Science, University of York, UK
2. Bosch (1991) CAN Specification version 2.0. Robert Bosch GmbH, Postfach 30 02 40, D-70442 Stuttgart
3. Bril RJ (2006) Existing worst-case response time analysis of real-time tasks under fixed-priority scheduling with deferred pre-emption is too optimistic. CS-Report 06-05, Technische Universiteit Eindhoven (TU/e), The Netherlands
4. Bril RJ (2006) Existing worst-case response time analysis of real-time tasks under fixed-priority scheduling with deferred pre-emption is too optimistic. CS-Report 06-05, Technische Universiteit Eindhoven (TU/e), The Netherlands
5. Bril RJ, Lukkien JJ, Davis RI, and Burns A (2006a) Message response time analysis for ideal Controller Area Network (CAN) refuted. CS-Report 06-19, Technische Universiteit Eindhoven (TU/e), The Netherlands.
6. Bril RJ, Lukkien JJ, Davis RI, Burns A (2006b) Message response time analysis for ideal Controller Area Network (CAN) refuted. In: Proceedings of the 5th International Workshop on Real-Time Networks (RTN'06)
7. Bril RJ, Lukkien JJ, Verhaegh WFJ (2006c) Worst-case response time analysis of real-time tasks under fixed priority scheduling with deferred preemption revisited. CS Report 06-34, Technische Universiteit Eindhoven (TU/e), The Netherlands
8. Broster I (2003) Flexibility in dependable communication. PhD Thesis, Department of Computer Science, University of York, UK
9. Broster I, Burns A, Rodríguez-Navas G (2002) Probabilistic analysis of CAN with Faults. In: Proceedings of the 23rd IEEE real-time systems symposium (RTSS'02), pp 269–278
10. Broster I, Burns A (2003) An analysable bus-guardian for event-triggered communication. In: Proceedings of the 24th real-time systems symposium. IEEE Computer Society Press, pp 410–419
11. Broster I, Burns A, Rodriguez-Navas G (2005) Timing analysis of real-time communication under electromagnetic interference. Real-Time Systems 30(1–2):55–81
12. Burns A (1994) Pre-emptive priority based scheduling: An appropriate engineering approach. In: S. Son (ed), Advances in Real-Time Systems, Prentice-Hall, pp 225–248
13. Casparsson L, Rajnak A, Tindell K, Malmberg P (1998/1) Volcano—a revolution in on-board communications. Volvo Technology Report
14. DeMeis R (2005) Cars sag under weighty wiring. Electronic Times, 10/24/2005
15. Ferreira J, Oliveira A, Fonseca P, Fonseca JA (2004) An experiment to assess bit error rate in CAN. In: Proceedings of 3rd international workshop of real-time networks (RTN2004). Cantania, Italy, pp 15–18
16. Frischkorn H-G (2005) Automotive architecture requirements. In: Proceedings of the summer school on architectural paradigms for dependable embedded systems. Vienna, Austria,. Vienna University of Technology, pp 45–74
17. George L, Rivierre N, Spuri M (1996) Pre-emptive and non-pre-emptive real-time uni-processor scheduling. Technical Report 2966, Institut National de Recherche et Informatique et en Automatique (INRIA), France
18. Hansson H, Nolte T, Norstrom C, Punnekkat S (2002) Integrating reliability and timing analysis of CAN-based Systems. IEEE Transactions on Industrial Electronics 49(6):1240–1250
19. Harbour MG, Klein MH, Lehoczky JP (1991) Fixed priority scheduling of periodic tasks with varying execution priority. In: Proceedings 12th IEEE real-time systems symposium. IEEE Computer Society Press, pp 116–128
20. Horvath I (2006) Private communication with the authors
21. ISO 11898-1 (1993) Road Vehicles—interchange of digital information—controller area network (CAN) for high-speed communication. ISO Standard-11898, International Standards Organisation (ISO)
22. Lehoczky J (1990) Fixed priority scheduling of periodic task sets with arbitrary deadlines. In: Proceedings 11th IEEE real-time systems symposium. IEEE Computer Society Press, pp 201–209
23. Leohold J (2004) Communication requirements for automotive systems. Keynote speech 5th IEEE international workshop on factory communication systems, Vienna, Austria, Vienna University of Technology
24. Leohold J (2005) Automotive system architecture. In: Proceedings of the summer school on architectural paradigms for dependable embedded systems. Vienna, Austria, Vienna University of Technology, pp 545–591
25. Leung JY-T, Whitehead J (1982) On the complexity of fixed-priority scheduling of periodic real-time tasks. Performance Evaluation 2(4):237–250
26. LIN Consortium (2003) LIN Protocol Specification, Revision 2.0. September. www.lin-subbus.org
27. Liu CL, Layland JW (1973) Scheduling algorithms for multiprogramming in a hard-real-time environment. Journal of the ACM 20(1):46–61
28. Motorola Inc (1998) MSCAN Block Guide V03.01 Document No. SV12MSCANV3/D. FreeScale Semiconductor Inc. (Revised July 2004)
29. Nolte T, Hansson H, Norstrom C (2002) Minimizing CAN response-time analysis jitter by message manipulation. In: Proceedings 8th IEEE real-time and embedded technology and applications symposium (RTAS'02), pp 197–206
30. Nolte T, Hansson H, Norstrom C (2003) Probabilistic worst-case response-time analysis for the Controller Area Network. In: Proceedings of the 9th IEEE real-time and embedded technology and applications symposium (RTAS'03), pp 200–207

31. Nolte T (2006) Share-driven scheduling of embedded networks. PhD Thesis, Malardalen University Press
32. Oswald M (2004) Efficient automotive electronics. Automotive Engineer
33. Punnekkat S, Hansson H, Norstrom C (2000) Response time analysis under errors for CAN. In: Proceedings 6th real-time technology and applications symposium. IEEE Computer Society Press, pp 258–265
34. Regehr J (2002) Scheduling tasks with mixed pre-emption relations for robustness to timing faults. In: Proceedings 23rd real-time systems symposium. IEEE Computer Society Press, pp 315–326
35. Rufino J, Verissimo P, Arroz G, Almeida C Rodrigues L (1998) Fault-tolerant broadcasts in CAN. In Digest of Papers, The 28th IEEE international symposium on fault-tolerant computing (FTCS'98), pp 150–159
36. Rufino J (2002) Computational system for real-time distributed control. PhD-Thesis, Technical University of Lisbon, Instituto Superior
37. Society of Automotive Engineers (1993) Class C application requirement considerations, recommended practice, SAE Technical Report J2056/1
38. Tindell KW, Burns A (1994) Guaranteeing message latencies on Controller Area Network (CAN). In: Proceedings of 1st international CAN conference, pp 1–11
39. Tindell KW, Burns A, Wellings AJ (1994a) An extendible approach for analysing fixed priority hard real-time systems. Journal of Real-Time Systems 6(2):133–152
40. Tindell KW, Hansson H, Wellings AJ (1994b) Analysing real-time communications: Controller area network (CAN). In: Proceedings 15th real-time systems symposium (RTSS'94). IEEE Computer Society Press, pp 259–263
41. Tindell KW, Burns A, Wellings AJ (1995) Calculating Controller area network (CAN) message response times. Control Engineering Practice 3(8):1163–1169
42. Wang Y, Saksena M (1999) Scheduling fixed priority tasks with pre-emption threshold. In: Proceedings of the 6th international workshop on real-time computing systems and applications (RTCSA'99), pp 328–335
43. Zuhily A (2006) Optimality of (D-J)-monotonic priority assignment. Technical Report YCS404. Dept. of Computer Science, University of York, UK

CHAPTER 8 HIGH-SPEED HIGH-SECURITY SIGNATURES

Daniel J. Bernstein, Niels Duif, Tanja Lange, Peter Schwabe, and Bo-Yin Yang

Keywords: Elliptic curves, Edwards curves, Signatures, Speed, Software side channels, Foolproof session keys

INTRODUCTION

This paper introduces software for public-key signatures with several attractive features:

- ***Fast single-signature verification.*** The software takes only 273364 cycles to verify a signature on Intel's widely deployed Nehalem/Westmere lines of CPUs. (This performance measurement is for short messages; for very long messages, verification time is dominated by hashing time.) Nehalem and Westmere include all Core i7, i5, and i3 CPUs released between 2008 and 2010, and most Xeon CPUs released in the same period.
- ***Even faster batch verification.*** The software performs a batch of 64 separate signature verifications (verifying 64 signatures of 64 messages under 64 public keys) in only 8.55 million cycles, i.e., under 134000 cycles per signature. The software fits easily into L1 cache, so contention between cores is negligible: a quad-core 2.4GHz Westmere verifies 71000 signatures per second, while keeping the maximum verification latency below 4 milliseconds.
- ***Very fast signing.*** The software takes only 87548 cycles to sign a message. A quad-core 2.4GHz Westmere signs 109000 messages per second.
- ***Fast key generation.*** Key generation is almost as fast as signing. There is a slight penalty for key generation to obtain a secure random number from the operating system; /dev/urandom under Linux costs about 6000 cycles.
- ***High security level.*** This system has a 2^{128} security target; breaking it has similar difficulty to breaking NIST P-256, RSA with $\approx$ 3000-bit keys, strong 128-bit block ciphers, etc. (The same techniques would also produce speed improvements at other security levels.) The best attacks known actually cost more than 2^{140} bit operations on average, and degrade quadratically in success probability as the number of bit operations drops.
- ***Foolproof session keys.*** Signatures in this paper are generated deterministically; key generation consumes new randomness but new signatures do not. This is not only a speed feature but also a security feature, directly relevant to the recent collapse of the Sony PlayStation 3 security system. See Section 2 for further discussion.
- ***Collision resilience.*** Hash-function collisions do not break this system. This adds a layer of defense against the possibility of weakness in the selected hash function.
- ***No secret array indices.*** The software never reads or writes data from secret addresses in RAM; the pattern of addresses is completely predictable. The software is therefore immune to cache-timing attacks, hyperthreading attacks, and other side-channel attacks that rely on leakage of addresses through the CPU cache.
- ***No secret branch conditions.*** The software never performs conditional branches based on secret data; the pattern of jumps is completely predictable. The software is therefore immune to side-channel attacks that rely on leakage of information through the branch-prediction unit.
- ***Small signatures.*** Signatures fit into 64 bytes. These signatures are actually compressed versions of longer signatures; the times for compression and

Citation: Bernstein, D.J., Duif, N., Lange, T. P. Schwabe, and B. Y. Yang (2012). High-speed high-security signatures, Journal of Cryptographic Engineering, September 2012; Volume 2, Issue 2, pp 77–89. DOI:10.1007/s13389-012-0027-1.

decompression are included in the cycle counts reported above.

- ***Small keys***. Public keys consume only 32 bytes. The times for compression and decompression are again included.

We have submitted our software to the eBATS project [15] for public benchmarking, and placed the software into the public domain to maximize reusability. The numbers 87548 and 273364 shown above are from the eBATS reports for our software on a Westmere CPU (Intel Xeon E5620, hydra2).

Our signatures are elliptic-curve signatures, carefully engineered at several levels of design and implementation to achieve very high speeds without compromising security. Section 2 specifies the signature system; Section 3 explains the techniques we use for finite-field arithmetic; Section 4 discusses fast signatures; Section 5 discusses fast verification

Comparison to Previous ECC Work

Carrying out highsecurity elliptic-curve signature verification in only 134000 cycles on a single core of a typical Intel CPU is unprecedented. The following paragraphs discuss previous work.

Readers should be aware of several difficulties in comparing ECC performance results. First, most papers on fast ECC have been limited to ECDH (variable-base-point singlescalar multiplication) and have not implemented ECC signature verification, although there are certainly some exceptions—for example, [21] reported verification 1.33× slower than ECDH, and [34] reported verification 1.36× slower than ECDH. Second, most implementations use secret array indices and secret branch conditions and therefore must be assumed to be breakable by side-channel attacks, as illustrated by the successful OpenSSL attack in [23]; this is not an issue for public-key signature verification but it is an issue for signing and for ECDH. Third, most papers report results for only a few CPUs, so anyone without access to the same CPUs must engage in error-prone extrapolation from one CPU to another; this is not an issue for systems included in the eBATS benchmarks, but we are aware of two recent ECC implementations (discussed below) that are not included in eBATS.

Intel's "Turbo Boost" and AMD's "Turbo Core" have added a further difficulty for new CPUs. Typical benchmarking frameworks measure performance on a single CPU core, and Turbo Boost fools most of these frameworks into reporting excessively low Westmere cycle counts—speeds that the CPU cannot actually achieve when a sensible workload is keeping all cores busy. The eBATS reports include explicit warnings regarding Turbo Boost. This corruption does not occur on hydra2: Turbo Boost is disabled on that computer.

Before this paper, the closest system to ours in eBATS was ecdonaldp256: ECDSA signatures using the NIST P-256 elliptic curve. On hydra2 this system takes 1690936 cycles for key generation, 1790936 cycles for signing, and 2087500 cycles for verification. Better speeds were reported for ECDH:

- Third place was curve25519, an implementation by Gaudry and Thomé [35] of Bernstein's Curve25519 [12].
- Second place was 307180 cycles for ecfp256e, an implementation by Hisil [40] of ECDH on an Edwards curve with similar security properties to Curve25519.
- First place was 278256 cycles for gls1271, an implementation by Galbraith, Lin, and Scott [34] of ECDH on an Edwards curve with an endomorphism.

The recent papers [38] and [44] point out security problems with endomorphisms in some ECC-based protocols, but as far as we can tell those security issues are not relevant to ECDH with standard hashing of the ECDH output, and are not relevant to ECC signatures.

Longa and Gebotys in [52] claimed 281000 cycles on a Core 2 Duo E6750 (C2 65nm) for ECDH on a curve similar to ecfp256e, and 229000 cycles for ECDH on a curve similar to gls1271. The software in [52] is not included in the eBATS benchmarks and apparently is not publicly available, so we are unable to benchmark it on a Westmere. More recently Käsper in [46] reported 457813 cycles for side-channel-protected ECDH on the NIST P-224 curve on a Core 2 Duo E8400 (C2 45nm); this software has been integrated into OpenSSL but not yet into eBATS.

To aid comparisons we also implemented ECDH, specifically curve25519, with the same side-channel defenses as our signature software (no secret array indices, and no secret branch conditions). We submitted our ECDH software to eBATS, which reports that the software uses 226872 cycles on hydra2 for variable-base-point single-scalar multiplication. This is a new speed record for public ECDH software, a new speed record for side-channel-protected ECDH (out of all the papers mentioned above, the only ones that report sidechannel protection are [12] and [46]), a new speed record for ECDH with single-size public keys (32 bytes at this security level instead of 64 bytes), and a new speed record for ECDH without endomorphisms.

We do not claim that curve25519 will maintain its current position on top of eBATS: we would expect ECDH with endomorphisms (especially with 64-byte keys and without side-channel protection) to be somewhat faster than ECDH without endomorphisms on many platforms. This expectation seems to be supported by the very recent paper [42] by Hu, Longa, and Xu: [42, Table 2] claims 194000 cycles for a curve with endomorphisms on an Intel Core 2 Duo E6750, and the accompanying web site [51] cited in [42, reference 18] claims 182000 cycles for a curve with endomorphisms on an Intel Core i5 540M (Westmere). The same web site also claims 215000 Westmere cycles for a curve without endomorphisms. However, like the software in [52], the software in [42] and [51] does not appear to be publicly available. The resulting lack of verifiability raises questions regarding accuracy.We are particularly skeptical of theWestmere speed claims, given the Turbo Boost issues discussed above. After we wrote this paragraph, the same web site was updated to claim 250000 cycles for the same software on another Westmere CPU.

Given our 226872-cycle ECDH speed, given the ECDHto-verification slowdowns reported in [21] and [34], and given the extra costs that we incur for decompressing keys and signatures, one would expect a verification speed close to 400000 cycles. We do better than this for several reasons, the most important reason being our use of batching. This requires careful design of the signature system, as discussed later in this paper: ECDSA, like DSA and most other signature systems, is incompatible with fast batch verification.

Comparison to Other Signature Systems

The eBATS benchmarks cover 42 different signature systems, including various sizes of RSA, DSA, ECDSA, hyperelliptic-curve signatures, and multivariate-quadratic signatures. This paper beats almost all of the signature times and verification times (and key-generation times, which are an issue for some applications) by more than a factor of 2. The only exceptions are as follows:

- Multivariate-quadratic signatures are competitive in speed. For example, sflashv2 takes 124740 cycles to sign and 165884 cycles to verify; mqqsig256 takes 4212 cycles to sign and 134900 cycles to verify; smaller mqqsig versions are even faster. However, sflashv2 was broken by Dubois, Fouque, Shamir, and Stern in [30]. We are not aware of any security evaluation of mqqsig, which was introduced last year in [36], but we disregard mqqsig256 for the simple reason that it has a 789552-byte public key.
- donald512 (512-bit DSA) takes 334508 cycles to verify. This is comparable to our single-signature verification speed but much slower than our batch verification speed. This is also at a far lower security level, breakable in about 2^{60} simple operations.
- Some RSA-type systems provide faster verification— but this advantage decreases as the security level increases, and for many applications the advantage is outweighed by much slower signatures and much larger keys. For example, rwb0fuz1024 (1024-bit Rabin–Williams) uses 12304 cycles to verify but 1751284 cycles to sign and 128 bytes for a public key; ronald1024 (1024-bit RSA) uses 60300 cycles to verify but 2171124 cycles to sign and 128 bytes for a public key; ronald3072 (3072-bit RSA) uses 231536 cycles to verify but an astonishing 31456912 cycles to sign and 384 bytes for a public key. This paper uses 134000 cycles to verify (in batches), 87548 cycles to sign, and 32 bytes for a public key.

The conventional wisdom is that RSA signatures are much better than ECC signatures in applications where each signature is verified many times, since RSA verification is much faster than ECC verification. Our ECC speed results call this conventional wisdom into question. We do not claim that our verification speeds cannot be beaten by RSA at the same security level, but we do claim that they are fast enough to make ECC an attractive option even for verification-intensive applications such as [71].

THE SIGNATURE SYSTEM

This section specifies the signature system used in this paper, and a generalized signature system EdDSA that can be used with other choices of elliptic curves.

There is an extensive literature on variants of the classic signature system introduced by ElGamal in [33]; notable variants include Schnorr's signature system [73], DSA, and ECDSA. Our generalized system is another of these variants. We do not claim novelty for any of the individual modifications that we use, but we emphasize that selecting a good combination of modifications is critical for top performance. The most obvious modification is that we use twisted Edwards curves rather than Weierstrass curves; this explains our choice of the name EdDSA (Edwards-curve Digital Signature Algorithm).

EdDSA Parameters

EdDSA has seven parameters: an integer $b \geq 10$; a cryptographic hash function H producing 2b-bit output; a prime power q congruent to 1 modulo 4; a (b − 1)-bit encoding of elements of the finite field F_q; a non-square element d of F_q; a prime ℓ between 2^{b-4} and 2^{b-3} satisfying an extra constraint described below; and an element $B \neq (0, 1)$ of the set

$$E = \left\{(x, y) \in \mathbf{F}_q \times \mathbf{F}_q : -x^2 + y^2 = 1 + dx^2y^2\right\}.$$

The condition that d is not a square implies that $d \notin \{0, -1\}$, so this set E forms a group with neutral element 0 = (0, 1) under the twisted Edwards addition law

$$(x_1, y_1) + (x_2, y_2) = \left(\frac{x_1y_2 + x_2y_1}{1 + dx_1x_2y_1y_2}, \frac{y_1y_2 + x_1x_2}{1 - dx_1x_2y_1y_2}\right)$$

introduced by Bernstein, Birkner, Joye, Lange, and Peters in [13]. Completeness of the addition law— the fact that the denominators $1 \pm dx_1x_2y_1y_2$ are nonzero— follows as explained in [13, Section 6]: −1 is a square in F_q (since q is congruent to 1 modulo 4), so this addition law on E is F_q-isomorphic to the Edwards addition law on the Edwards curve $x^2 + y^2 = 1 - dx^2 y^2$, which is complete by [14, Theorem 3.3] since −d is not a square in F_q. The latter follows from d being a non-square and −1 being a square in F_q. The extra constraint mentioned above is that $\ell B = 0$, where nB means the nth multiple of B in this group.

We use the encoding of F_q to define some field elements as being negative: specifically, x is negative if the (b−1)-bit encoding of x is lexicographically larger than the (b −1)-bit encoding of −x. If q is an odd prime and the encoding is the little-endian representation of {0, 1,..., q − 1} then the negative elements of F_q are {1, 3, 5,..., q − 2}.

An element (x, y) ∈ E is encoded as a b-bit string $\underline{(x, y)}$, namely the (b − 1)-bit encoding of y followed by a sign bit; the sign bit is 1 iff x is negative. This encoding immediately determines y, and it determines x via the equation $x = \pm\sqrt{(y^2 - 1)/(dy^2 + 1)}$

EdDSA Keys and Signatures

An EdDSA secret key is a b-bit string k. The hash $H(k) = (h_0, h_1,..., h_{2b-1})$ determines an integer

$$a = 2^{b-2} + \sum_{3 \leq i \leq b-3} 2^i h_i \in \left\{2^{b-2}, 2^{b-2}+8, \ldots, 2^{b-1}-8\right\},$$

which in turn determines the multiple A = aB. The corresponding EdDSA public key is $\underline{A}$. Bits $h_b,..., h_{2b-}1$ of the hash are used as part of signing, as discussed in a moment.

The signature of a message M under this secret key k is defined as follows. Define $r = H(h_b,..., h_{2b-1}, M) \in \{0, 1,..., 2^{2b} - 1\}$; here we interpret 2b-bit strings in littleendian form as integers in $\{0, 1,..., 2^{2b} - 1\}$. Define R = r B. Define $S = (r + H(\underline{R}, \underline{A}, M)a) \bmod \ell$. The signature of M under k is then the

2b-bit string ($\underline{R}$, $\underline{S}$), where $\underline{S}$ is the bbit little-endian encoding of S. Applications wishing to pack data into every last nook and cranny should note that the last three bits of signatures are always 0 because ℓ fits into b − 3 bits.

Verification of an alleged signature on a message M under a public key works as follows. The verifier parses the key as $\underline{A}$ for some A ∈ E, and parses the alleged signature as ($\underline{R}$, $\underline{S}$) for some R ∈ E and S ∈ {0, 1,..., ℓ − 1}. The verifier computes H($\underline{R}$, $\underline{A}$, M) and then checks the group equation 8SB = 8R + 8H($\underline{R}$, $\underline{A}$, M)A in E. The verifier rejects the alleged signature if the parsing fails or if the group equation does not hold.

To see that signatures pass verification, simply multiply B by the equation S = (r + H($\underline{R}$, $\underline{A}$, M)a) mod ℓ, and use the fact that ℓB = 0, to see that SB = r B + H($\underline{R}$, $\underline{A}$, M)aB = R + H(R, A, M)A. The verifier is permitted to check this stronger equation and to reject alleged signatures where the stronger equation does not hold. However, this is not required; checking that 8SB = 8R + 8H($\underline{R}$, $\underline{A}$, M)A is enough for security.

Weak Keys

Forgeries are trivial if A is a known multiple of B. For example, an attacker who knows that A = 37B can choose r and compute S = (r + 37H($\underline{R}$, $\underline{A}$, M)) mod ℓ. As an even more extreme example, an attacker who knows that A = 0B can choose r and compute S = r mod ℓ, independently of M. We could declare that 0B and 37B are "broken" by these two "attacks" and that users must check for, and reject, these "weak keys"; but the same confused logic would require rejecting all keys in all cryptosystems, and would have no relevance to the standard definition of signature security.

Legitimate users choose A = aB, where a is a random secret; the derivation of a from H(k) ensures adequate randomness. These users have negligible chance of generating any particular multiple of B targeted by the attacker (and no chance of generating 0B). The chance of the attacker randomly guessing a is far smaller than the chance of the attacker computing a by known discrete-logarithm algorithms; standard elliptic-curve security criteria are designed so that the latter algorithms have negligible chance of succeeding in any reasonable amount of time.

Malleability

We also see no relevance of "malleability" to the standard definition of signature security. For example, if we slightly modified the system then replacing S by −S and replacing A by −A (a slight variant of the "attack" of [76]) would convert one valid signature into another valid signature of the same message under a new public key; but it would still not accomplish the attacker's goal, namely to forge a signature on a new message under a target public key. One such modification would be to omit $\underline{A}$ from the hashing; another such modification would be to have $\underline{A}$ encode only |A|, rather than A.

Choice of Curve

Our recommended curve for EdDSA is a twisted Edwards curve birationally equivalent to the curve Curve25519 from [12]. Any efficiently computable birational equivalence preserves ECDLP difficulty, so the well-known difficulty of computing ECDLP for Curve25519 immediately implies the difficulty of computing ECDLP for our curve. We use the name Ed25519 for EdDSA with this particular choice of curve.

Specifically, Ed25519-SHA-512 is EdDSA with the following parameters: b = 256; H is SHA-512; q is the prime $2^{255} - 19$; the 255-bit encoding of $\mathbf{F}_{2^{255}-19}$ is the usual littleendian encoding of {0, 1,..., $2^{255} - 20$} ℓ; is the prime 2^{252} +27742317777372353535851937790883648493 from [12]; d = −121665/121666 ∈ F_q ; and B is the unique point (x, 4/5) ∈ E for which x is positive.

Curve25519 from [12] is the Montgomery curve $v^2 = u^3 + 486662u^2 + u$ over the same field F_q. Bernstein and Lange pointed out in [14, Section 2] that Curve25519 is birationally equivalent to an Edwards curve, specifically $x^2 + y^2 = 1 + (121665/121666)x^2y^2$; the equivalence is $x = \sqrt{486664}u/v$ and y = (u − 1)/(u + 1). As above this Edwards curve is isomorphic to $-x^2 + y^2 = 1 - (121665/121666)x^2y^2$ since −1 is a square in F_q . Our choice of base point B corresponds to the choice u = 9 made in [12].

Pseudorandom Generation of r

ECDSA, like many other signature systems, asks users to generate not merely a random long-term secret key, but also a new random secret session key r for each message to be signed. Ifr becomes public then, assuming H($\underline{R}$, $\underline{A}$, M) mod ℓ ≠ 0, the long-term secret key a can be simply computed as a = (S −r)/H($\underline{R}$, $\underline{A}$, M) mod ℓ. If the same value r is ever used for 2 different messages the secret key can be computed as well, as ElGamal noted in [33, Note 2]. It was reported in [24] that the latter failure had occurred in Sony's ECDSA implementation for codesigning for the PlayStation3, immediately revealing Sony's long-term secret key.

Furthermore, it is well known that ECDSA's session keys are much less tolerant than the long-term key of slight deviations from randomness, even if the session keys are not revealed or reused. For example, Nguyen and Shparlinski in [62] presented an algorithm using lattice methods to compute the long-term ECDSA key from the knowledge of as few as 3 bits of r for hundreds of signatures, whether this knowledge is gained from side-channel attacks or from non-uniformity of the distribution from which r is taken.

EdDSA avoids these issues by generating r = H(h_b, ..., h_{2b-1}, M), so that different messages will lead to different, hard-to-predict values of r. No per-message randomness is consumed. This idea of generating random signatures in a secretly deterministic way, in particular obtaining pseudorandomness by hashing a long-term secret key together with the input message, was proposed by Barwood in [9]; independently by Wigley in [80]; a few months later in a patent application [58] by Naccache, M'Raïhi, and Levy-dit-Vehel; later by M'Raïhi, Naccache, Pointcheval, and Vaudenay in [56]; and much later by Katz and Wang in [47]. The patent application was abandoned in 2003.

Standard PRF hypotheses imply that this pseudorandom session key r is indistinguishable from a truly random string generated independently for each M, so there is no loss of security. Well-known length-extension properties prevent secret-prefix SHA-512 from being a PRF, but also do not threaten the security of Ed25519-SHA-512, since r is not visible to the attacker. All remaining SHA-3 candidates are explicitly designed to be PRFs, and we will not hesitate to recommend Ed25519-SHA-3 after SHA-3 is standardized. It would of course also be safe to generate r with a cipher such as AES, combined with standard PRF-stretching mechanisms to support a long input; but we prefer to reuse H to save area in hardware implementations.

EdDSA samplesr from the interval $[0, 2^{2b} - 1]$, ensuring almost uniformity of the distribution modulo ℓ. The guideline [2, Section 4.1.1, Algorithm 2] specifies that the interval should be of size at least $[0, 2^{b+61} - 1]$, i.e., 64 bits more than ℓ; for Ed25519 there are 259 extra bits.

Comparison to Previous ElGamal Variants

The original ElGamal system [33, Section III] predated elliptic-curve cryptography; it instead used the multiplicative group F_q^*. ElGamal took a large non-prime ℓ, specifically $\ell = q-1$, and focused on the case of prime q. ElGamal's signatures were pairs (R, S) of integers between 0 and q − 2 inclusive satisfying $B^{H(M)} = A^R R^S$ in F_q^*. See [33, equation (3)]; see also [33, Attack 6] for the introduction of H. The signer, given M, generates a randomr coprime to and computes the signature (R, S), where $R = B^r$ and $S = r^{-1}(H(M) - Ra) \bmod \ell$.

Schnorr in [73] pointed out that one could safely work in an order-ℓ subgroup of F_q^* with a prime ℓ much smaller than q, saving most of the space for S. Schnorr also introduced several other improvements to ElGamal's system, as discussed below.

ElGamal's verification equation involves R as an element of the group F_q^* and as a scalar, the exponent for A. For more general groups one needs a function x mapping group elements to scalars. ECDSA works this way: it replaces F_q^* with an order-ℓ subgroup of an elliptic-curve group over F_q and defines x(R) as the x-coordinate of R. ECDSA also replaces A with −A, changing the signer's subtraction into an addition and obtaining the verification equation H(M)B + x(R)A = S R. ECDSA replaces this threescalar equation with the equivalent two-scalar equation $S^{-1}H(M)B + S^{-1}x(R)A = R$ at the expense of requiring S to be invertible modulo ℓ; note that both the signer and the verifier compute inverses here.

Schnorr used a cryptographic hash function for x. This has minimal expense and eliminates any concerns regarding the mathematical structure of simpler functions x. Schnorr also compressed the group element R to the scalar x(R): a Schnorr signature is (x(R), S) rather than (R, S). Given a compressed signature (x(R), S), the verifier recomputes R as $S^{-1}H(M)$ $B + S^{-1}x(R)A$ and checks that x(R) matches; at this point the verifier knows a valid uncompressed signature (R, S), so the compression cannot reduce security.

Schnorr also merged the hashing of R with the hashing of M. One way to understand this merging is to replace S with x(R)S, and to impose the extra constraint $x(R) \neq 0$, obtaining the verification equation $x(R)^{-1}H(M)B + A = S R$. There is no need for the multiplicative structure of $x(R)^{-1}H(M)$ here: one can instead use the verification equation $H(\underline{R}, M)B + A = S R$, with the signer obtaining S as $r^{-1}(H(\underline{R}, M) + a) \bmod \ell$. Schnorr actually used the equation $SB = R + H(\underline{R}, M)A$, eliminating all inversions both for the signer and for the verifier; this is an obvious advantage, saving time and reducing code size.

The presence of R as input to the hash function provides collision resilience: the attacker cannot break Schnorr's system by merely finding hash collisions. Neven, Smart, and Warinschi in [61] proposed taking advantage of collision resilience by choosing H to output only b/2 bits, reducing the size of compressed signatures by 25%; but the same proposal had actually appeared twenty years earlier in Schnorr's original paper [73, Section 2].

Practical use of Schnorr's system was hampered by a patent (which expired in 2008), but the system became well known to theoreticians, because the hashing of $\underline{R}$ allowed various security proofs. Some proofs use the "forking lemma" to show that any generic-hash attack against Schnorr's system (i.e., any attack that works for arbitrary functions H) can be converted into a DLP algorithm with a polynomially bounded, although often quite severe, loss of efficiency. There are also theorems with a different loss of efficiency for generic-group attacks (i.e., attacks that work for arbitrary groups) under mild assumptions on H, and theorems with no loss of efficiency for generic-group generic-hash attacks. See, for example, [68], [75], [11], and [61].We do not mean to exaggerate the real-world relevance of "provable security", but we find it obvious that Schnorr's system is a conservative, well-studied signature system.

Our verification equation is the same as Schnorr's verification equation with double-size hashing instead of half-size hashing, with $\underline{A}$ inserted as an extra hash input, and without Schnorr's compression of R. These modifications obviously do not compromise security. The use of double-size hashing helps alleviate concerns regarding hash-function security; the use of $\underline{A}$ is an inexpensive way to alleviate concerns that several public keys could be attacked simultaneously; and the avoidance of compression allows an important verification speedup, as discussed in Section 5. We also reuse the double-size hash to alleviate concerns regarding nonce randomness, as discussed above.

FAST ARITHMETIC MODULO $2^{255} - 19$

This section explains how our software represents elements of the field $\mathbf{F}_{2^{255}-19}$, and how our software performs efficient field arithmetic. The machine instructions used in the software are available on all 64-bit Intel and AMD CPUs, but we target Intel's Nehalem/Westmere CPUs.

Multipliers on Nehalem CPUs

Field multiplications (and squarings) are the main bottlenecks in elliptic-curve performance on most CPUs. The most important tool for fast field multiplication is a fast CPU multiplication instruction. Nehalem CPUs offer three different multiplication instructions that can be used to implement high-speed field arithmetic:

- The mulpd instruction can perform two double-precision floating-point multiplications in SIMD fashion every cycle. Newer Sandy Bridge CPUs include a vmulpd instruction that can perform up to 4 double-precision floating-point multiplications per cycle, but this instruction is not available on our target CPUs.

- Themulinstruction can multiply two 64-bit unsigned integers, producing a 128-bit result, every two cycles.
- The pmuldq/pmuludq instructions can perform two multiplications of 32-bit integers, producing 64-bit results, every cycle. The pmuldq instruction performs signed multiplication; the pmuludq instruction performs unsigned multiplication.

Multiplication and Edwards-curve arithmetic involve datalevel parallelism that we could exploit with mulpd and pmuldq, but this approach would incur a serious overhead of shuffle instructions needed to arrange data in registers as described in, e.g., [26] and [60]. This overhead is eliminated when several independent computations are run in parallel, but two 64-bit results every cycle are not fundamentally better than one 128-bit result every two cycles. We therefore decompose field multiplication into multiplications of 64-bit unsigned integers.

Radix-2^{64} Representation

The standard way to split 255-bit values into 64-bit limbs is a 4-limb, radix-2^{64} representation. Each element x of the field is represented as (x_0, x_1, x_2, x_3) with $x = \sum_{i=0}^{3} x_i 2^{64i}$. The multiplication of two elements x and y is decomposed into 16 multiplications of 64-bit unsigned integers; the 128-bit results are added up to produce the result in 8 limbs $r_0,...,r_7$. Reduction modulo $2^{255} - 19$ exploits the fact that $2^{256} \equiv 38$, so $38 \cdot r_4$ is added to r_0, $38 \cdot r_5$ to r_1 and so on.

A detail worth noting of this representation is that it uses 256 bits to represent 255-bit field elements. We use this one extra bit and do not always reduce modulo $2^{255} - 19$ but modulo $2^{256} - 38$. For a similar representation this has been shown to be useful for example in [17].

Our implementation of the signature scheme based on this representation of field elements yields high performance on many microprocessors such as AMD K10 or 65-nm Intel Core 2 processors. However, on our target platform, the Intel Nehalem/Westmere CPUs, this representation triggers a serious bottleneck. Every 128-bit result of the mul instruction is produced in two 64-bit registers. Adding two of these results requires two addition instructions. In the field multiplication most of these additions produce carries; the carry bits need to be handled by subsequent additions. The Intel Nehalem and Westmere CPUs can perform three additions per cycle, but only if these additions do not have to handle a carry bit from a previous addition (addinstruction). An add with carry (adc instruction) can only be done once every two cycles; i.e., carry bits decrease addition throughput by a factor of 6. This bottleneck is triggered not only inside field multiplication and squaring but also inside additions.

Radix-2^{51} Representation

To reduce the number of expensive adc/subc instructions, we instead represent an element x of $\mathbf{F}_{2^{255}-19}$ as $(x_0, x_1, x_2, x_3, x_4)$ with $x = \sum_{i=0}^{4} x_i 2^{51i}$.

The 5 limbs are unsigned integers. We can represent each element of the field $\mathbf{F}_{2^{255}-19}$ with each $x_i \in [0,..., 2^{51} - 1]$. In fact our implementation does not enforce these bounds except for comparisons. Multiplication accepts inputs with each limb having up to 54 bits and produces results of which each limb can be only slightly larger than 251.

Multiplication and Squaring

Schoolbook multiplication of two field elements x and y, each represented in 5 unsigned integers, takes 25 mul instructions. The results are again produced in two 64-bit integer registers, but as both inputs have only up to 54 bits, the value in the upper result register has only up to 44 bits. Adding two multiplication results now takes only one adc and one add instruction. Furthermore reduction can be carried out simultaneously to multiplication. For example, we do not compute a coefficient r_5. Whenever the result of a mul instruction belongs to r_5, for example in the multiplication of $x_2 \cdot y_3$, we multiply one of the inputs by 19 and add the result to r_0. Similarly we do not compute r_6, r_7, r_8 and r_9 but directly add into $r_1,...,r_4$. Multiplying one input by 19 yields a result with less than 64 bits so we can use the faster imul instruction for these multiplications. The 5 result coefficients require 10 64-bit registers; the AMD64 architecture has 15 such registers, so we can keep the result coefficients inside registers throughout the computation.

After the multiplication we need to reduce (carry) the 5 coefficients to obtain a result with coefficients that are at most slightly larger than 2^{51}. Denote the two registers holding coefficient r_0 as r_{00} and r_{01} with $r_0 = 2^{64} r_{01} + r_{00}$. Similarly denote the two registers holding coefficient r_1 as r_{10} and r_{11}. We first shift r_{01} left by 13, while shifting in the most significant bits of r_{00} (shld instruction) and then compute the logical and of r_{00} with $2^{51} - 1$. We do the same with r_{10} and r_{11} and add r_{01} into r_{10} after the logical and with $2^{51} - 1$. We proceed this way for coefficients $r_2,...,r_4$; register r_{41} is multiplied by 19 before adding it to r_{00}. Now all 5 coefficients fit into 64-bit registers but are still too large to be used as input to another multiplication. We therefore carry from r_0 to r_1, from r_1 to r_2, from r_2 to r_3, from r_3 to r_4, and finally from r_4 to r_0. Each of these carries is done as one copy, one right shift by 51, one logical and with $2^{51} - 1$, and one addition.

Squaring needs only 15 mul instructions. Some inputs are multiplied by 2; this is combined with multiplication by 19 where possible. The coefficient reduction after squaring is the same as for multiplication.

Multiplication and squaring are implemented as separate functions, but calls to these functions are used only for inversion (see below). Edwards-curve arithmetic uses inlined functions for point addition and doubling.

Addition, Subtraction, and Inversion

The results of additions do not have to be reduced if they are used as input to a multiplication. Long sequences of additions that let coeffi- cients grow larger than 54 bits would be a problem but we do not have such sequences of additions. Field addition is therefore nothing but 5 integer additions without carries (add instruction). Subtraction is slightly more expensive because we use unsigned coefficients. Therefore we first add a multiple of q and then perform subtraction. This costs 5 add and 5 sub instructions.

Inversion is implemented as exponentiation with exponent $q - 2$. It uses the same sequence of 255 squarings and 11 multiplications as [12].

SIGNING MESSAGES

Signature generation has three steps: (1) computing $r = H(h_b, \ldots, h_{2b-1}, M)$; (2) computing $R = r B$; (3) computing $S = (r + H(\underline{R}, \underline{A}, M)a) \bmod \ell$. Our primary concern is with short messages M, obviously the top concern for a server trying to keep up with a given volume of data; longer messages take more cycles per signature but far fewer cycles per byte. The computations of H take negligible time for short messages. The reduction modulo ℓ also takes negligible time with standard branchless techniques. For the rest of this section we focus on the main signing bottleneck, namely computing r B given r.

High-level Strategy

We begin by computing the 253-bit integer r mod ℓ. We then write r mod ℓ as $r_0 + 16r_1 + \cdots + 16^{63}r_63$ with

$r_i \in \{-8, -7, -6, -5, -4, -3, -2, -1, 0, 1, 2, 3, 4, 5, 6, 7\}$.

For each i we look up $16^i|r_i|B$ in a precomputed table, and then conditionally negate $16^i|r_i|B$ to obtain $16^i r_i B$. Finally we compute r B as $\sum_i 16^i r_i B$.

There is nothing new in our computation at this level. Computing r B as a sum of precomputed pieces is a special case of a standard scalar-multiplication algorithm published by Pippenger in [65] (subsequently reinvented in [19] and [50]); allowing negative coefficients is a standard tweak. The devil lies in the lower-level details — choosing the optimal radix 16, and computing $16^i r_i B$ and $\sum_i 16^i r_i B$ as efficiently as possible. These details are discussed below.

Low Level, Part 1: Table Lookups

Recall that, as a side-channel defense, we prohibit secret array indices. In particular, we cannot use $|r_i|$ as an array index. We instead load all table entries 0B, 16^i B, $2 \cdot 16^i$ B, $3 \cdot 16^i$ B, $4 \cdot 16^i$ B, $5 \cdot 16^i$ B, $6 \cdot 16^i$ B, $7 \cdot 16^i$ B, $8 \cdot 16^i$ B and use arithmetic operations, without branching, to combine the table entries into $16^i|r_i|B$. We similarly use arithmetic operations to compute $16^i r_i B$ from $6^i|r_i|B$ and $-6^i|r_i|B$.

We actually store table entries only for $i \in \{0, 2, 4, \ldots, 62\}$, at the expense of 4 elliptic-curve doublings. The table then contains $8 \cdot 32 = 256$ curve points (aside from 0B, which is not stored). Each point is represented as three integers (see below) modulo $2^{255}-19$. Each integer in turn is represented as five 8-byte words. Overall the table consumes 30 kilobytes of RAM.

We could instead use radix 32 or larger. Radix 32 would involve twice as many table loads (since we load all table entries), and twice as much arithmetic to combine table entries, but these costs would be outweighed by the bene- fit of fewer elliptic-curve additions. A more serious concern is that the table would be twice as large, consuming 60KB instead of 30KB. This is only a minor issue for a typical cryptographic speed test on our target CPUs (each Nehalem/Westmere core has its own fast 256KB L2 cache effi- ciently handling our sequential loads), but 30KB is clearly more attractive inside a larger application that needs to fit several different subroutines into L2 cache.

In the opposite direction, we could chop the table in half again at the expense of 8 more doublings; we could also switch to radix 8, 4, or 2. These changes would also allow reasonably fast signing on much smaller CPUs.

Low Level, Part 2: Elliptic-curve Addition

We use extended coordinates for the twisted Edwards curve $-x^2 + y^2 = 1 + dx^2 y^2$, as proposed by Hisil, Wong, Carter, and Dawson in [41]. These coordinates are (X : Y : Z : T) with XY = Z T representing x = X/Z and y = Y/Z. The addition formulas from [41, Section 3.1] are complete for our curve and use just 9 field multiplications to add a table entry (x_0, y_0) into (X : Y : Z : T). Note that these formulas rely on the −1 in $-x^2$; this is why EdDSA uses the −1 twist.

One of the field multiplications is a multiplication by d = −121665/121666. We could replace this with a small number of multiplications by 121665 and 121666, as in [13, Section 6], but our current software treats d as a generic field element to save code size. We considered switching to a new curve using a small integer d (such as 646, which has a nearprime group order; note that we do not need the twist security of Curve25519), but decided that the resulting speedup was too small to justify departing from an established curve.

A different way to save a multiplication is to use the dual addition formulas from [41, Section 3.2]. However, those formulas are not complete; they would require a detailed analysis of intermediate results in our computation to see whether any of the intermediate additions could trigger any of the exceptional cases in the formulas.

Instead we represent a precomputed point (x_0, y_0) as $(y_0 - x_0, y_0 + x_0, 2dx_0 y_0)$. These values depend only on x_0 and y_0 and are usually computed in the first part of addition in extended coordinates; providing them as part of the precomputation saves the multiplication by d, the multiplication $x_0 y_0$, and 2 field additions, at the expense of increasing the storage requirements by a factor of 1.5.

Template Attacks

We comment that for hardware implementations this type of precomputation reduces the information exposed to template attacks trying to link multiple uses of the same precomputed point.

Consider, for example, an attacker monitoring the power consumption of a device with very limited memory. Assume that the device designer has reduced the table described above to just B, 2B,..., 8B at the expense of many doublings, and has saved more memory by storing a table entry as simply (x_0, y_0). The addition formulas then begin by computing y_0-x_0, y_0+x_0, etc. If the same table entry is used again later then the same subtraction, addition, etc. will be performed again, resulting in exactly the same power trace. The attacker can therefore partition the loaded points into (at most) 16 different groups, obtaining 55 bits of information on average, as discussed in [31, Section 5.1.2].

Precomputing $y_0 - x_0$, $y_0 + x_0$, etc. guarantees (for these addition formulas) that all operations involving the precomputed point also involve the intermediate point, which varies unpredictably between different uses of the same table entry. A closer look at field arithmetic sometimes reveals lowerlevel operations that depend on only one input, such as the preliminary additions in Karatsuba's method; the results of those operations can be similarly precomputed.

Of course, there is much more to say about countermeasures to hardware side-channel attacks; we do not claim that any single countermeasure is adequate by itself. The software situation is simpler, since the side channels exposed to an attacker are much more limited.

Results

Overall we spend a bit less than 1000 cycles for each iteration of our main signing loop, i.e., for one table lookup and one elliptic-curve mixed addition. We also spend about 21000 cycles to invert Z at the end of the computation. The complete signing procedure for a short message takes 87548 cycles.

VERIFYING SIGNATURES

Fast signature verification seems considerably more difficult than fast signature generation, for two reasons. First, the verifier has to recover the elliptic-curve points A and R from the compressed points A and R. Second, checking SB = R+H(R, A, M)A seems to require not merely a fixedbase scalar multiplication SB but also a much more expensive variable-base scalar multiplication H(R, A, M)A. This section explains several techniques that we use to address these problems.

Fast Decompression

Recall that the encoding R of a point R = (x, y) contains a straightforward encoding of y but contains only a sign bit for x. One must therefore recover x via the equation $x = \pm\sqrt{(y^2-1)/(dy^2+1)}$; note that $dy^2+1 \neq 0$ since $-d$ is not a square. The division and square root here seem to involve two exponentiations, about twice as expensive as the usual Weierstrass-curve decompression.

Of course, we could use Montgomery's trick to merge the two divisions involved in decompressing two points, but two square roots and a division are still more expensive than two Weierstrass-curve decompressions. We could also skip the compression and decompression for applications willing to use 64-byte keys and 96-byte signatures; but we think that 32-byte keys and 64-byte signatures are considerably more attractive.

To save time we look more closely at the standard computation of square roots in F_q. The prime $q = 2^{255} - 19$ is congruent to 5 modulo 8, so any square $\alpha \in F_q$ satisfies $\alpha^2 = \beta^4$ where $\beta = \alpha^{(q+3)/8}$, i.e., $\pm\alpha = \beta^2$. The standard computation is a single exponentiation to compute β, followed by a quick multiplication of β by $\sqrt{-1}$ if $\beta^2 = -\alpha$.

In the decompression context we are given α as a fraction u/v, where $u = y^2-1$ and $v = dy^2+1$. Instead of computing α we merge the division with the square-root computation:

$$\beta = (u/v)^{(q+3)/8} = u^{(q+3)/8}v^{q-1-(q+3)/8}$$
$$= u^{(q+3)/8}v^{(7q-11)/8} = uv^3(uv^7)^{(q-5)/8}.$$

We check whether $\beta^2 = -\alpha$ by checking whether $v\beta^2 = -u$, and if so we multiply β by $\sqrt{-1}$. The entire computation of $\sqrt{u/v}$, starting from u and v, takes just a few multiplications more than a single exponentiation. In other words, Edwards-curve decompression is as inexpensive as Weierstrass-curve decompression.

Fast Single-signature Verification

To verify a single signature we use standard techniques for double-scalar multiplication to compute SB − H(R, A, M)A, and we then check whether the result is the same as R. (We actually check whether the encoding of the result is the same as the encoding of R, so that we can skip decompression of R.) The speed of Edwards-curve addition, especially with the −1 twist, makes these techniques particularly efficient; using the tables discussed in Section 4 does not seem to offer any advantage. This computation fits in very little space.

We have also considered the verification method suggested by Antipa, Brown, Gallant, Lambert, Struik, and Vanstone in [7], but our very efficient elliptic-curve arithmetic makes the overheads in this method— extra decompression and a Euclidean computation— much more troublesome. In the batch context discussed below, the only extra overhead of the method of [7] would be the Euclidean computation, but the benefit would also be much smaller.

Fast Batch Verification

For any system bottlenecked by signature verification, the problem is not to verify one signature at a time, but to verify many signatures as quickly as possible.

Naccache, M'Raïhi, Vaudenay, and Raphaeli in [59, Section 2.2] proposed verifying a batch of linear signature equations by verifying a random linear combination of the equations. This proposal is not directly applicable to ElGamal, DSA, Schnorr, ECDSA, etc., because all of those systems require computing linear functions (to compute R) rather than merely verifying linear functions; but if R is transmitted instead of H(···), as suggested in [59], then this problem disappears.

Unfortunately, the verification algorithm in [59] was quite slow: [59, Table 1] reported "29n" multiplications to verify n signatures from the same signer at a highly questionable 2^{20} security level. If the same technique were adapted to ECDSA and increased to a 2^{128} security level then it would require nearly 200 elliptic-curve additions for each signature from the same signer—somewhat faster than verifying each signature separately, but not much.

The followup paper [10] by Bellare, Garay, and Rabin proposed a more complicated verification technique using, e.g., 3200 multiplications to verify 100 exponentiations, or 6480 multiplications to verify 100 DSA signatures, in both cases at a substandard 2^{60} security level. See [10, Appendix A.1]. The number of multiplications per signature begins to drop as the batch size grows towards 1000— see [10, Figure 3] — but such large batches do not fit into cache on typical CPUs.

The unimpressive theoretical performance of these batchverification techniques can be traced directly to the naive exponentiation algorithms used in [59] and [10]. We do much better by using random linear combinations, as in [59], together with state-of-the-art scalar-multiplication techniques.

Specifically, we start from a batch of (M_i, A_i, R_i, S_i) where ($\underline{R}_i$, $\underline{S}_i$) is an alleged signature of M_i under key $\underline{A}_i$. We choose independent uniform random 128-bit integers z_i, compute H_i = H($\underline{R}_i$, $\underline{A}_i$, M_i), and verify the equation

$$\left(-\sum_i z_i S_i \bmod \ell\right)B+\sum_i z_i R_i+\sum_i (z_i H_i \bmod \ell)A_i=0$$

by a multi-scalar multiplication. It is important to choose new independent integers z_i uniformly at random in each batch verification. There are two reasonable choices of scalar-multiplication methods here, namely Pippenger's method in [65] and the Bos–Coster method reported in [27, Section 4]. We use the Bos–Coster method because it fits into less storage; see below for details. Note that z_i is not secret, so side-channel protection is not required.

The number of scalars here is 2n+1. Half of the scalars are 253-bit and half are 128-bit. If public keys appear repeatedly, the situation considered in [59] and [10], then we could save some time by merging the 253-bit scalars; this merging also explains why we do not use the similar signature equation SB = A + H($\underline{R}$, $\underline{A}$, M)R, which would allow only merging 128-bit scalars. Our software focuses on general-purpose verification with arbitrary keys.

If verification succeeds then we are confident that $8S_i B = 8R_i + 8H_i A_i$ for each i, i.e., that each signature is valid. The logic is simple: the differences $P_i = 8R_i + 8H_i A_i - 8S_i B$ are elements of a cyclic group of prime order ℓ, and have been verified to satisfy $\sum_i z_i P_i = 0$; but this equation cannot hold with probability more than 2^{-128} unless all $P_i = 0$. For example, if P_4 is nonzero then the choices of $z_1, z_2, z_3, z_5, z_6, \ldots$ determine exactly one choice of z_4 satisfying $\sum_i z_i P_i = 0$, and z_4 has chance at most 2^{-128} of matching that choice.

If verification fails then there must be at least one invalid signature. We then fall back to verifying each signature separately. There are several techniques to identify a small number of invalid signatures in a batch, but all known techniques become slower than separate verification as the number of invalid signatures increases; separate verification provides the best defense against denial-of-service attacks.

Fast Multi-scalar Multiplication

The Bos–Coster method mentioned above is as follows: to compute $n_1P_1+n_2P_2+\cdots$, where $n_1 \geq n2 \geq \cdots$, we recursively compute $(n_1-n_2)P_1+ n_2(P_1+ P_2)+\cdots$. For n_1 much larger than n_2, say $2^{k+1}n_2 > n_1 \geq 2^k n_2$, we could gain speed by instead recursively computing $(n_1 - 2^k n_2)P_1 + n_2(2^k P_1 + P_2) +\cdots$, but we have found this to occur so rarely that checking for it is not worthwhile.

We keep the scalars n_i in a heap so that identifying the two largest scalars is easy. The usual method to replace the root of a heap is top-down, starting at the root and swapping down for a variable number of steps. We instead use Floyd's 1964 bottom-up algorithm discussed in [48, Exercise 5.2.3–18] (often miscredited to [25] and [79]): start at the root, swap down to the bottom, and then swap up for a variable number of steps. This has the advantage of somewhat reducing the number of comparisons, and the not-so-well-known advantage of drastically reducing the number of branches, especially for balanced heaps.

RESULTS

The complete verification procedure takes under 134000 cycles per signature for batch size 64. Our batchverification software is included in, although not yet benchmarked by, the public eBATS benchmarking framework.

Doubling the batch size to 128 no longer fits into L1 cache but still improves performance on our target CPU, taking under 125000 cycles per signature. Larger batches take under 114000 cycles per signature while still fitting into L2 cache. Our software spends about 44000 cycles on decompression, so verification of uncompressed signatures (32 extra bytes) using uncompressed public keys (another 32 extra bytes) would take only about 81000 cycles for batch size 128, even faster than signing. However, in this paper we have emphasized the performance that we obtain without using so much space.

REFERENCES

1. (no editor): 17th annual symposium on foundations of computer science, IEEE Computer Society, 1976. MR 56:1766. See [65]
2. (no editor): Technical guideline TR-03111, elliptic curve cryptography (2009). URL:https://www.bsi.bund.de/SharedDocs/Downloads/DE/BSI/Publikationen/TechnischeRichtlinien/TR03111/BSI-TR-03111_pdf.pdf?_blob=publicationFile. Citations in this document: §2
3. (no editor): SPEED: software performance enhancement for encryption and decryption, 2007. URL: http://www.hyperelliptic.org/SPEED. See [35]
4. (no editor): Proceedings of the 6th ACM symposium on information, computer and communications security, Hong Kong, March 22–24, 2011, Association for Computing Machinery, 2011. ISBN 978-1-4503-0564-8. See [71]
5. Abdalla, M., Barreto, P.S.L.M. (editors): Progress in cryptology—LATINCRYPT 2010, first international conference on cryptology and information security in Latin America, Puebla, Mexico, August 8–11, 2010, proceedings, Lecture Notes in Computer Science, 6212, Springer, 2010. ISBN 978-3-642-14711-1. See [60]
6. Abe, M. (editor): Advances in cryptology—ASIACRYPT 2010, 16th international conference on the theory and application of cryptology and information security, Singapore, December 5–9, 2010, proceedings, Lecture Notes in Computer Science, 6477, Springer, 2010. ISBN 978-3-642-17372-1. See [38]
7. Antipa, A., Brown, D.R.L., Gallant, R.P., Lambert, R.J., Struik, R., Vanstone, S.A.: Accelerated verification of ECDSA signatures, in SAC 2005 [70] (2006), 307–318. MR 2007d:94044. URL:http://www.cacr.math.uwaterloo.ca/techreports/2005/tech_reports2005.html. Citations in this document: §5, §5
8. Atluri, V., Jaeger, T. (program chairs): Proceedings of the 10th ACM conference on computer and communications security, ACM Press, 2003. ISBN 1-58113-738-9. See

[47]

9. Barwood, G.: Digital signatures using elliptic curves, message 32f519ad. 19609226@news.dial.pipex.com posted to sci.crypt (1997). URL:http://groups.google.com/group/sci.crypt/msg/b28aba37180dd6c6. Citations in this document: §2

10. Bellare, M., Garay, J.A., Rabin, T.: Fast batch verification for modular exponentiation and digital signatures, in Eurocrypt '98 [63] (1998), 236–250. URL:http://cseweb.ucsd.edu/~mihir/papers/batch.html. Citations in this document: §5, §5, §5, §5, §5

11. Bellare, M., Neven, G.: Multi-signatures in the plain public-key model and a general forking lemma, in CCS 2006 [45] (2006), 390–399. URL:http://cseweb.ucsd.edu/~mihir/papers/multisignatures.html. Citations in this document: §2

12. Bernstein, D.J.: Curve25519: new Diffie-Hellman speed records, in PKC 2006 [82] (2006), 207–228. URL: http://cr.yp.to/papers.html#curve25519. Citations in this document: §1, §1, §2, §2, §2, §2, §3

13. Bernstein, D.J., Birkner, P., Joye, M., Lange, T., Peters, C.: Twisted Edwards curves, in Africacrypt 2008 [78] (2008), 389–405. URL: http://eprint.iacr.org/2008/013. Citations in this document: §2, §2, §4

14. Bernstein, D.J., Lange, T.: Faster addition and doubling on elliptic curves, in Asiacrypt 2007 [49] (2007), 29–50. URL: http://eprint.iacr.org/2007/286. Citations in this document: §2, §2

15. Bernstein, D.J., Lange, T. (editors): eBACS: ECRYPT Benchmarking of Cryptographic Systems, accessed 19 September 2011 (2011). URL: http://bench.cr.yp.to. Citations in this document: §1

16. Blakley, G.R., Chaum, D. (editors): Advances in cryptology, proceedings of CRYPTO '84, Santa Barbara, California, USA, August 19–22, 1984, proceedings, Lecture Notes in Computer Science, 196, Springer, Berlin, 1985. ISBN 3-540-15658-5. MR 86j:94003. See [32]

17. Bos, J.W.: High-performance modular multiplication on the Cell processor, in WAIFI 2010 [39] (2010), 7–24. Citations in this document: §3

18. Brassard, G. (editor): Advances in cryptology—CRYPTO '89, 9th annual international cryptology conference, Santa Barbara, California, USA, August 20–24, 1989, proceedings, Lecture Notes in Computer Science, 435, Springer, Berlin, 1990. ISBN 3-540-97317-6. MR 91b:94002. See [73]

19. Brickell, E.F., Gordon, D.M., McCurley, K.S., Wilson, D.B.: Fast exponentiation with precomputation (extended abstract), in Eurocrypt '92 [72] (1993), 200–207; see also newer version [20]. URL: http://cr.yp.to/bib/entries.html#1993/brickell-exp. Citations in this document: §4

20. Brickell, E.F., Gordon, D.M., McCurley, K.S., Wilson, D.B.: Fast exponentiation with precomputation: algorithms and lower bounds (1995); see also older version [19]. URL:http://research.microsoft.com/~dbwilson/bgmw/

21. Brown, M., Hankerson, D., López, J., Menezes, A.: Software implementation of the NIST elliptic curves over prime fields (2000); see also newer version [22]. URL:http://www.cacr.math.uwaterloo.ca/techreports/2000/corr2000-56.ps. Citations in this document: §1, §1

22. Brown, M., Hankerson, D., López, J., Menezes, A.: Software implementation of the NIST elliptic curves over prime fields, in CT-RSA 2001 [57] (2001), 250–265; see also older version [21]. MR 1907102

23. Brumley, B.B., Hakala, R.M.: Cache-timing template attacks, in Asiacrypt 2009 [54] (2009), 667–684. Citations in this document: §1

24. "Bushing", Hector Martin "marcan" Cantero, Segher Boessenkool, Sven Peter, PS3 epic fail (2010). URL:http://events.ccc.de/congress/2010/Fahrplan/attachments/1780_27c3_console_hacking_2010.pdf. Citations in this document: §2

25. Carlsson S.: Average-case results on heapsort. BIT 27, 2–17 (1987) Citations in this document: §5

26. Costigan, N., Schwabe, P.: Fast elliptic-curve cryptography on the Cell Broadband Engine, in Africacrypt 2009 [69] (2009), 368–385. URL:http://cryptojedi.org/users/peter/#celldh. Citations in this document: §3

27. de Rooij, P.: Efficient exponentiation using precomputation and vector addition chains, in Eurocrypt '94 [28] (1995), 389–399. MR 1479665. Citations in this document: §5

28. De Santis, A. (editor): Advances in cryptology—EUROCRYPT '94, workshop on the theory and application of cryptographic techniques, Perugia, Italy, May 9–12, 1994, proceedings, Lecture Notes in Computer Science, 950, Springer, Berlin, 1995. ISBN 3-540-60176-7. MR 98h:94001. See [27], [59]

29. Desmedt, Y. (editor): Advances in cryptology—CRYPTO '94, 14th annual international cryptology conference, Santa Barbara, California, USA, August 21–25, 1994, proceedings, Lecture Notes in Computer Science, 839, Springer, Berlin, 1994. ISBN 3-540-58333-5. See [50]

30. Dubois, V., Fouque, P.-A., Shamir, A., Stern, J.: Practical cryptanalysis of SFLASH, in Crypto 2007 [55] (2007), 1–12. URL: http://eprint.iacr.org/2007/141. Citations in this document: §1

31. Duif, N.: Smart card implementation of a digital signature scheme for Twisted Edwards curves, M.A. thesis, Technische Universiteit Eindhoven, 2011. URL:http://www.nielsduif.nl/2011_05_20_report_final.pdf. Citations in this document: §4

32. ElGamal, T.: A public key cryptosystem and a signature scheme based on discrete logarithms, in Crypto '84 [16] (1985), 10–18; see also newer version [33]. MR 87b:94037

33. ElGamal, T.: A public key cryptosystem and a signature scheme based on discrete logarithms, IEEE Transactions on Information Theory 31 (1985), 469–472; see also older version [32]. ISSN 0018-9448. MR 86j:94045. Citations in this document: §2, §2, §2, §2, §2

34. Galbraith, S., Lin, X., Scott, M.: Endomorphisms

for faster elliptic curve cryptography on a large class of curves, in Eurocrypt 2009 [43] (2009), 518–535. URL:http://eprint.iacr.org/2008/194. Citations in this document: §1, §1, §1

35. Gaudry, P., Thomé, E.: The mpFq library and implementing curvebased key exchanges, in SPEED [3] (2007), 49–64. URL: http://www.loria.fr/~gaudry/papers.en.html. Citations in this document: §1

36. Gligoroski, D., Odegøard, R.S., Jensen, R.E., Perret, L., Faugère, J.-C., Knapskog, S.J., Markovski, S.: The digital signature scheme MQQ-SIG (2010). URL:http://eprint.iacr.org/2010/527.pdf.Citations in this document: §1

37. Goh, E.-J., Jarecki, S., Katz, J., Wang, N.: Efficient signature schemes with tight reductions to the Diffie–Hellman problems, Journal of Cryptology 20 (2007), 493–514. URL:http://www.cs.umd.edu/~jkatz/papers.html. See [47]

38. Granger, R.: On the static Diffie–Hellman problem on elliptic curves over extension fields, in Asiacrypt 2010 [6] (2010), 283–302. URL: http://eprint.iacr.org/2010/177. Citations in this document: §1

39. Hasan, M.A., Helleseth, T. (editors): Arithmetic of finite fields, third international workshop, WAIFI 2010, Istanbul, Turkey, June 27–30, 2010, proceedings, Lecture Notes in Computer Science, 6087, Springer, 2010. ISBN 978-3-642-13796- 9. See [17]

40. Hisil, H.: Elliptic curves, group law, and efficient computation, Ph.D. thesis, Queensland University of Technology, 2010. URL: http://eprints.qut.edu.au/33233. Citations in this document: §1

41. Hisil, H., Wong, K.K.-H., Carter, G., Dawson, E.: Twisted Edwards curves revisited, in Asiacrypt 2008 [64] (2008), 326–343. URL: http://eprint.iacr.org/2008/522. Citations in this document: §4, §4, §4

42. Hu, Z., Longa, P., Xu, M.: Implementing 4-dimensional GLV method on GLS elliptic curves with j-invariant 0, 15 June 2011 version, accessed 11 July 2011 (2011). URL:http://eprint.iacr.org/2011/315. Citations in this document: §1, §1, §1, §1

43. Joux, A. (editor): Advances in cryptology—EUROCRYPT 2009, 28th annual international conference on the theory and applications of cryptographic techniques, Cologne, Germany, April 26–30, 2009, proceedings, Lecture Notes in Computer Science, 5479, Springer, 2009. ISBN 978-3-642-01000-2. See [34]

44. Joux, A., Vitse, V.: Elliptic curve discrete logarithm problem over small degree extension fields. Application to the static Diffie–Hellman problem on E(Fq5)E(Fq5) (2010). URL:http://eprint.iacr.org/2010/157. Citations in this document: §1

45. Juels, A., Wright, R.N., De Capitani di Vimercati, S. (editors): Proceedings of the 13th ACM conference on computer and communications security, CCS 2006, Alexandria, VA, USA, October 30–November 3, 2006, Association for Computing Machinery, (2006). See [11]

46. Käsper, E.: Fast elliptic curve cryptography in OpenSSL, in 2nd Workshop on Real-Life Cryptographic Protocols and Standardization (RLCPS 2011), to appear (2011). Citations in this document: §1, §1

47. Katz, J., Wang, N.: Efficiency improvements for signature schemes with tight security reductions, in CCS 2003 [8] (2003), 155–164; portions incorporated into [37]. URL:http://www.cs.umd.edu/~jkatz/papers.html. Citations in this document: §2

48. Knuth, D.E.: The art of computer programming, volume 3: sorting and searching, 2nd edition, Addison-Wesley, Reading, 1998. ISBN 0-201-89685-0. Citations in this document: §5

49. Kurosawa, K. (editor): Advances in cryptology—ASIACRYPT 2007, 13th international conference on the theory and application of cryptology and information security, Kuching, Malaysia, December 2–6, 2007, proceedings, Lecture Notes in Computer Science, 4833, Springer, 2007. ISBN 978-3-540-76899-9. See [14]

50. Lim, C.H., Lee, P.J.: More flexible exponentiation with precomputation, in [29] (1994), 95–107. Citations in this document: §4

51. Longa, P.: Speed benchmarks for elliptic curve scalar multiplication, accessed 11 July 2011 (2011). URL: http://www.patricklonga.bravehost.com/speed_ecc.html. Citations in this document: §1, §1

52. Longa, P., Gebotys, C.H.: Efficient techniques for high-speed elliptic curve cryptography, in CHES 2010 [53] (2010), 80–94. Citations in this document: §1, §1, §1

53. Mangard, S., Standaert, F.-X. (editors): Cryptographic hardware and embedded systems, CHES 2010, 12th international workshop, Santa Barbara, CA, USA, August 17–20, 2010, proceedings, Lecture Notes in Computer Science, 6225, Springer, 2010. ISBN 978-3-642-15030-2. See [52]

54. Matsui, M. (editor): Advances in cryptology—ASIACRYPT 2009, 15th international conference on the theory and application of cryptology and information security, Tokyo, Japan, December 6–10, 2009, proceedings, Lecture Notes in Computer Science, 5912, Springer, 2009. ISBN 978-3-642-10365-0. See [23]

55. Menezes, A. (editor): Advances in cryptology—CRYPTO 2007, 27th annual international cryptology conference, Santa Barbara, CA, USA, August 19–23, 2007, proceedings, Lecture Notes in Computer Science, 4622, Springer, 2007. ISBN 978-3-540-74142-8. See [30]

56. M'Raïhi, D., Naccache, D., Pointcheval, D., Vaudenay, S.: Computational alternatives to random number generators, in SAC '98 [77] (1999), 72–80. URL:http://www.di.ens.fr/~pointche/Documents/Papers/1998_sac.pdf. Citations in this document: §2

57. Naccache, D. (editor): Topics in cryptology—CT-RSA 2001: the cryptographers' track at RSA Conference 2001, San Francisco, CA, USA, April 2001, proceedings, Lecture Notes in Computer Science, 2020, Springer, 2001. ISBN 3-540-41898-9. MR 2003a:94039. See [22]

58. Naccache, D., M'Raïhi, D., Levy-dit-Vehel, F.: Patent application WO/1998/051038: pseudo-random generator based on a hash coding function for cryptographic systems requiring random drawing (1997). URL: http://www.wipo.int/pctdb/en/ia.jsp?IA=FR1998000901.

Citations in this document: §2

59. Naccache, D., M'Raïhi, D., Vaudenay, S., Raphaeli, D.: Can D.S.A. be improved? Complexity trade-offs with the digital signature standard, in Eurocrypt '94 [28] (1994). Citations in this document: §5, §5, §5, §5, §5, §5, §5

60. Naehrig, M., Niederhagen, R., Schwabe, P.: New software speed records for cryptographic pairings, in Latincrypt 2010 [5] (2010), 109–123. URL:http://cryptojedi.org/users/peter/#dclxvi. Citations in this document: §3

61. Neven, G., Smart, N.P., Warinschi, B.: Hash function requirements for Schnorr signatures, Journal of Mathematical Cryptology 3 (2009), 69–87. URL:http://www.zurich.ibm.com/~nev/papers/schnorr.html. Citations in this document: §2, §2

62. Nguyen P.Q., Shparlinski I.:: The insecurity of the elliptic curve digital signature algorithm with partially known nonces. Designs, Codes and Cryptography 30, 201–217 (2003) Citations in this document: §2

63. Nyberg, K. (editor): Advances in cryptology—EUROCRYPT '98, international conference on the theory and application of cryptographic techniques, Espoo, Finland, May 31–June 4, 1998, proceedings, Lecture Notes in Computer Science, 1403, Springer, 1998. ISBN 3-540-64518-7. See [10]

64. Pieprzyk, J. (editor): Advances in cryptology—ASIACRYPT 2008, 14th international conference on the theory and application of cryptology and information security, Melbourne, Australia, December 7–11, 2008, Lecture Notes in Computer Science, 5350, 2008. ISBN 978-3-540-89254-0. See [41]

65. Pippenger, N.: On the evaluation of powers and related problems (preliminary version), in FOCS '76 [1] (1976), 258–263; newer version split into [66] and [67]. MR 58:3682. URL:http://cr.yp.to/bib/entries.html#1976/pippenger. Citations in this document: §4, §5

66. Pippenger, N.: The minimum number of edges in graphs with prescribed paths, Mathematical Systems Theory 12 (1979), 325–346; see also older version [65]. ISSN 0025-5661. MR 81e:05079. URL: http://cr.yp.to/bib/entries.html#1976/pippenger

67. Pippenger, N.: On the evaluation of powers and monomials, SIAM Journal on Computing 9 (1980), 230–250; see also older version [65]. ISSN 0097-5397. MR 82c:10064. URL:http://cr.yp.to/bib/entries.html#1976/pippenger

68. Pointcheval, D., Stern, J.: Security arguments for digital signatures and blind signatures, Journal of Cryptology 13 (2000), 361–396. URL:ftp://ftp.di.ens.fr/pub/users/pointche/Papers/2000_joc.pdf. Citations in this document: §2

69. Preneel, B. (editor): Progress in cryptology—AFRICACRYPT 2009, second international conference on cryptology in Africa, Gammarth, Tunisia, June 21–25, 2009, proceedings, Lecture Notes in Computer Science, 5580, Springer, 2009. See [26]

70. Preneel, B., Tavares, S.E. (editors): Selected areas in cryptography, 12th international workshop, SAC 2005, Kingston, ON, Canada, August 11–12, 2005, revised selected papers, Lecture Notes in Computer Science, 3897, Springer, 2006. ISBN 3-540-33108-5. MR 2007b:94002. See [7]

71. Rangasamy, J., Stebila, D., Boyd, C., Nieto, J.G.: An integrated approach to cryptographic mitigation of denial-of-service attacks, in ASIACCS 2011 [4] (2011). URL:http://www.douglas.stebila.ca/files/research/papers/RSBG11.pdf. Citations in this document: §1

72. Rueppel, R.A. (editor): Advances in cryptology—EUROCRYPT '92, workshop on the theory and application of cryptographic techniques, Balatonfüred, Hungary, May 24–28, 1992, proceedings, Lecture Notes in Computer Science, 658, Springer, Berlin, 1993. ISBN 3-540-56413-6. MR 94e:94002. See [19]

73. Schnorr, C.P.: Efficient identification and signatures for smart cards, in Crypto '89 [18] (1990), 239–252; see also newer version [74]. Citations in this document: §2, §2, §2

74. Schnorr, C.P.: Efficient signature generation by smart cards, Journal of Cryptology 4 (1991), 161–174; see also older version [73]. URL: http://www.mi.informatik.uni-frankfurt.de/research/papers.html

75. Schnorr, C.P., Jakobsson, M.: Security of discrete log cryptosystems in the random oracle + generic model (2000). URL: http://www.mi.informatik.uni-frankfurt.de/research/papers.html. Citations in this document: §2

76. Stern, J., Pointcheval, D., Malone-Lee, J., Smart, N.P.: Flaws in applying proof methodologies to signature schemes, in Crypto 2002 [81] (2002), 93–110. Citations in this document: §2

77. Tavares, S., Meijer, H. (editors): Selected areas in cryptography, 5th annual international workshop, SAC98, Kingston, Ontario, Canada, August 17–18, 1998, proceedings, Lecture Notes in Computer Science, 1556, Springer, 1999. ISBN 3-540-65894-7. See [56]

78. Vaudenay, S. (editor): Progress in cryptology—AFRICACRYPT 2008, First international conference on cryptology in Africa, Casablanca, Morocco, June 11–14, 2008, proceedings, Lecture Notes in Computer Science, 5023, Springer, 2008. ISBN 978-3-540-68159-5. See [13]

79. Wegener, I.: Bottom-up-heapsort, a new variant of heapsort, beating, on average, quicksort (if n is not very small), Theoretical Computer Science 118 (1993), 81–98. Citations in this document: §5

80. Wigley, J.: Removing need for rng in signatures, message 5gov5dpad@wapping.ecs.soton.ac.uk posted to sci.crypt (1997). URL:http://groups.google.com/group/sci.crypt/msg/a6da45bcc8939a89. Citations in this document: §2

81. Yung, M. (editor): Advances in cryptology—CRYPTO 2002, 22nd annual international cryptology conference, Santa Barbara, California, USA, August 18–22, 2002, proceedings, Lecture Notes in Computer Science, 2442, Springer, 2002. ISBN 3-540-44050-X. See [76]

82. Yung, M., Dodis, Y., Kiayias, A., Malkin, T. (editors): Public key cryptography—9th international conference on theory and practice in public-key cryptography, New York, NY, USA, April 24–26, 2006, proceedings, Lecture Notes in Computer Science, 3958, Springer, 2006. ISBN 978-3-540-33851-2. See [12]

CHAPTER 9 POWER-AWARE SCHEDULING OF PRE-EMPTIBLE JOBS ON IDENTICAL PARALLEL PROCESSORS TO MINIMIZE MAKESPAN

R. Różycki and J. Węglarz

Keywords: Power-aware computing, Job model, Preemptable jobs, Makespan minimization

INTRODUCTION

Green computing, as a general philosophy aimed at efficient and environmentally friendly usage of computer resources, gained recently an increasing interest among OR and IT specialists. One of the most promising directions of green computing is ecologically aware management of power/energy on different levels of computer systems. We consider the problem on the level of a single computer system, where a particular attention is paid on processors (CPU) as one of the most energy consuming elements of such systems. One of the research directions towards an advanced power management on the level of a single computer system is the use of variable speed processors (VSP) (Boyer et al. 2006). Modern operating systems utilizing e.g. Intel's Speedstep or Foxton as well as AMD's PowerNow! technologies are able to control the speed of processors to prolong the battery life or to improve computing efficiency. Therefore, an operating system of a computer can decide not only which job to perform at the moment but also at what processing rate. We assume for simplicity that considered variable speed processors are able to adjust their clock period at each cycle at no cost.

Since the processing rate of a job is related to the power/energy used during its execution, several job processing models have been proposed in literature to represent this relation (Bansal et al. 2005; Irani and Pruhs 2005; Pruhs et al. 2005). The power vs. speed model is the most popular, where a single continuous, strictly convex function represents the relation.

In this paper we consider a speed vs. power model, in which jobs are characterized by (different) continuous, strictly concave functions relating their processing speeds to the amount of power allotted at a time (W˛eglarz 1981).

Of course, the assumption that speed function is continuous remains only an approximation of discrete speed increments in real processors.

Schedule length of a given set of jobs processed on processors of a single computer system is to be minimized for a given level of energy available. Such a problem belongs to the class of laptop problems (Bunde 2006), which are typical for portable electronic devices driven by energy accumulated in batteries with limited capacity. However, in contrary to the power vs. speed model in which power is not limited, we consider it as a doubly constrained resource for which both: consumption (i.e. energy), and temporary availability are constrained. The similar assumption, defined as bounded speed model, where a single processor can vary its speed between 0 and a fixed maximum value, was proposed in Bansal et al. (2008). Nevertheless, our model is more natural in the multiprocessor setting, since it enables to share power among processors as a common renewable resource.

We consider a set of precedence-related and preemptable jobs as well as parallel identical processors, since it is a natural way to model a set of computer applications that have to be scheduled on contemporary computer system equipped with an advanced multicore processor.

We represent the precedence relation among jobs by a task-on-arc directed graph. This representation, commonly used in project scheduling, has been also used for solving machine

Citation: Różycki, R. and Węglarz, J. (2014). Power-aware scheduling of pre-emptable jobs on identical parallel processors to minimize makespan, *Annals of Operations Research*, February 2014; Volume 213, Issue 1, pp 235–252. DOI:10.1007/s10479-011-0957-5.

scheduling problems by transforming them to LP problems (see Bła˙zewicz et al. 2007 as survey) for a given ordering of nodes (i.e. events) in such a digraph. If in a digraph each two nodes are connected by a path then the ordering of nodes is unique and corresponding scheduling problem is polynomially solvable. In general, the optimal solution depends on the ordering of nodes, and thus finding an ordering that leads to the best schedule is of crucial importance. Unfortunately, no polynomial algorithm for finding such an ordering is known even for the simplest case of parallel, identical machines. Thus, it is advisable to consider various heuristic or metaheuristic approaches.

Although we consider a single computer system, presented approach may be applied to the case of a cluster of identical computational nodes driven by a common power source and managed on the level of a single metascheduler.

In Sect. 2 we describe and compare two models of job processing. Scheduling problem is formulated in Sect. 3. In Sect. 4 we present a methodology for finding optimal schedules for independent jobs. This methodology is generalized in Sect. 5 for the case of dependent jobs. In this section we also propose two heuristics for finding an ordering of nodes leading to possibly short schedules. A computational experiment and its results are described in Sect. 6. Section 7 contains final remarks and future research directions.

JOB PROCESSING MODELS

Let us start with the model commonly used in the literature (e.g. Bansal et al. 2005; Irani and Pruhs 2005) concerning power management in microprocessor systems. The following form of the function (power usage function) is utilized to express the relation between processing speed s of a processor, and power p consumed during job processing:

$$p(s) = s^{\alpha}, \quad \alpha > 1 \tag{1}$$

It is easy to notice that function p(s) is strictly convex. In particular, for microprocessors based on CMOS technology α is assumed to be equal to 3 (Bunde 2006).

Moreover, it is assumed that job i is characterized by the parameter $w_i > 0$ being the number of CPU cycles needed by job i to be processed. Of course, processing time of job i depends on its size w_i and on the processing speed of a processor executing this job. A job i started at time a_i is accomplished at completion time C_i if the following equation is fulfilled:

$$\int_{a_i}^{C_i} s(t)dt = w_i \tag{2}$$

During the execution of job i the energy amount E_i given by the formula

$$E_i = \int_{a_i}^{C_i} p(s(t))dt \tag{3}$$

is consumed.

It is worth to notice that in model (1) the processing speed of a processor is treated as a decision variable and determines power p. Moreover, s is not limited, i.e. s ∈ [0,∞). As a consequence, the resulting power is theoretically unlimited too. However, power directly determines the temperature of a microprocessor (nearly all energy consumed by a processor is released as heat). Thus, no limits for power may lead to a processor overheating and, in consequence, to a serious damage of a computer system.

Let us pass to the model which determines a temporal rate of the job execution. As a consequence, instead of power usage functions we have so called processing speed functions (speed functions in short). Formally the model is expressed as follows:

$$\dot{x}_i(t) = \frac{dx_i(t)}{dt} = s_i(p_i(t)), \quad x_i(0) = 0,\ x_i(C_i) = w_i \tag{4}$$

where

- $x_i(t)$—is the state of job i at time t, i.e. the amount of data processed by t for job i,
- $s_i(\cdot)$—is a continuous, strictly concave function, $s_i(0) = 0$,
- $p_i(t)$—is an amount of power allotted to job i at time t.

Model (4) was studied in W˛eglarz (1981) in the problem of allocating continuously divisible, doubly constrained resource among project activities. In this problem power p(t) allotted to a job at time t is doubly constrained, i.e.

$$\sum_{i=1}^{n} p_i(t) \le P \quad \text{for every } t > 0 \tag{5}$$

$$\sum_{i=1}^{n} \int_0^{C\max} p_i(t)dt \le E \tag{6}$$

where P > 0, E > 0 are known.

The constraint (5) guarantees that a total usage of power at every moment does not exceed the available amount P, whereas the constraint (6) ensures that a total consumption of energy up to the end of the last job (C_{max}) in a schedule meets the known limit E.

It is easy to notice that the models (1) and (4), (5), (6) are equivalent if

$$s_i(\cdot) = s(\cdot) = p^{-1}(\cdot) = p_i^{1/\alpha}, \quad i = 1, 2, \ldots, n, \alpha > 1$$

However, the model (4), (5), (6) is much more general. First of all, time t enters directly into this model, and thus power allotted to jobs can vary over time. Secondly, functions $s_i(\cdot)$ can be different for different jobs. This is very important since power consumption percycle may differ for various instructions executed by a processor designed basing on one of modest architectures (see e.g. Boyer et al. 2006).

Moreover, functions $s_i(\cdot)$ can be also not only power functions but arbitrary continuous, strictly concave ones. Thus this model may be useful even when future technologies and microcomputer architectures will require another type of functions.

Since power is limited: $p_i(t) \in [0,P]$ in model (4), (5), (6), thus temperature of a microprocessor executing computational jobs is limited too.

PROBLEM FORMULATION

Consider a set of n precedence-related, preemptive jobs and m parallel identical processors. All jobs are ready to be processed at time 0. Each job requires for its processing a processor and an amount of power. Each job is performed by at most one processor at a time and a processor is able to process at most one job at a time. Processing rate of a job depends on the amount of power $p_i(t)$ allotted to job i at a time t and this relation is expressed by (4). Job i is characterized by the speed function $s_i(\cdot)$ and the size w_i. Precedence relations among jobs are represented by a task-on-arc digraph. Both power and energy are limited and available in amounts P and E, respectively.

This formulation allows to model practical situations, where a set of dependent modules of programs have to be executed on a multiprocessor portable device with processors driven by the common energy source—a battery of limited capacity. To prevent the computer system from overheating a power usage limit is established. We assume that processing rate function of job i in model (4) is increasing and strictly concave since this corresponds to the real relation between the temporal power usage and the processing rate in contemporary microprocessor systems.

The objective is to find a vector function

$p^*(t) = [p_1^*(t), p_2^*(t), \ldots, p_n^*(t)], p_i^*(t) \geq 0, \; i = 1, 2, \ldots, n$

which, under the constraints imposed, minimizes the schedule length $C_{max} = T$. Knowing $p^*(t)$ we are able to calculate the energy consumption for job i by integrating $p_i^*(t)$ up to the completion time C_i. We will show that finding $p^*(t)$ may be decomposed into two strongly interrelated subproblems: the subproblem of determining of the optimal sequence of jobs on processors and the subproblem of optimal power allocation to jobs already sequenced.

INDEPENDENT JOBS

Assume hypothetically that a job does not require any processor to be performed and may be processed using an amount of power only. In consequence, we deal with the subproblem of T -minimal allocation (respecting limits P and E) of power to jobs only and the following theorem is valid (Węglarz 1981).

Theorem 1

If jobs are independent and $s_i(\cdot)$, i = 1, 2,...,n are strictly concave then all jobs finish simultaneously and are processed using constant amounts of power given by the formula

$$p_i^*(t) = p_i^* = s_i^{-1}(w_i/T^*), \quad i = 1, \ldots, n, \; t \in \langle 0, T^* \rangle \tag{7}$$

where T^* is the (unique) positive root of the equation:

$$T\sum_{i=1}^{n} s_i^{-1}(w_i/T) = E \quad if \sum_{i=1}^{n} s_i^{-1}(w_i/T) \leq P \tag{8}$$

or

$$\sum_{i=1}^{n} s_i^{-1}(w_i/T) = P \tag{9}$$

otherwise. $T^*(w_1,w_2,\ldots,w_n)$ is a convex function.

Notice that (8) and (9) are direct consequence of (6) and (5). In the first case the active constraint is from the side of E, not P. It means that (7), with T^* calculated from (8), is the optimal solution of the problem since it does not exceed the power limit P. It may happen however that P is a critical constraint for the given instance of the problem, and then the minimum schedule length is found from (9).

Let us comment now on Theorem 1 from the view point of finding optimal solutions. To find the schedule of the minimum length it is better to start with (8), since it is usually of simpler form than (9). For example, for functions:

$$s_i(\cdot) = c_i p_i^{1/\alpha_i}, \quad c_i > 0, \; \alpha_i \in \{2, 3, 4\} \tag{10}$$

both (8) and (9) are analytically solvable algebraic equations of an order less than or equal to 4, however (8) is of an order less by 1, and thus is analytically solvable also for $\alpha_i = 5$. If the sum of power allocated to jobs using (8) exceeds available amount P, we have to solve (9), but the information obtained by solving (8) is valuable anyway.

Let us pass now to the problem formulated in Sect. 3 assuming that jobs are independent. It is easy to notice that if $n \leq m$, then the optimal schedule is defined by (7), (8), (9) with n = m.

However, if n>m, then for finding the optimal schedule we have to consider all melement combinations from the set of n jobs. Each combination represents a single group of jobs that may be performed in parallel on limited number of processors. Since the processors are identical, an order of jobs in such combination may be neglected. Although the jobs are preemptable, we assume that preemption of a job within a combination is forbidden. However, a job is still able to migrate among processors because in consecutive combinations it may be allotted to different processors at no additional cost. Denote combinations by $Z_k, k = 1, 2, \ldots, r = \binom{n}{m}$. Denote also by K_i, i = 1, 2,...,n, the set of indices of Z_k's containing job i and w_{ik} part of job i processed in Z_k. Let T_k and E_k, k = 1, 2,...,r, denote respectively: the length of the part of the schedule corresponding to Z_k and an amount of energy consumed for processing parts of jobs from Z_k. Notice that for a given level of P, $T_k = T_k(\{w_{ik}\}_{i \in Z_k}, E_k)$, $k = 1, 2, \ldots, r$, where w_{ik}, E_k are unknown a priori. Since for each Z_k we have n = m, we can apply Theorem 1. For the simplest situation of processing speed functions (10) where $\alpha_i = \alpha > 1$, $c_i = 1$, i = 1, 2,...,n, T_k is calculated from:

$$T_k = \max\left(\sqrt[\alpha-1]{\frac{1}{E_k}\sum_{i \in Z_k}(w_{ik})^\alpha}, \sqrt[\alpha]{\frac{1}{P}\sum_{i \in Z_k}(w_{ik})^\alpha}\right)$$

Finally, we get the following corollary for the general case of arbitrary strictly concave processing speed functions.

Corollary 2 For strictly concave $s_i(\cdot)$, i = 1, 2,...,n, the optimal schedule for independent, preemptable jobs is obtained by solving the following convex programming problem:

Minimize:

$$T=\sum_{k=1}^{r} T_k\left(\{w_{ik}\}_{i\in Z_k}, E_k\right) \tag{11}$$

Subject to:

$$\sum_{k=1}^{r} E_k \leq E \tag{12}$$

$$\sum_{k\in K_i} w_{ik}=w_i \quad i=1,2,\ldots,n \tag{13}$$

$$w_{ik}\geq 0, \qquad T_k\geq 0, \quad i=1,2,\ldots,n;\, k\in K_i \tag{14}$$

where T_k are calculated as functions of E_k and $\{w_{ik}\}$ for $i \in Z_k$ using Theorem 1. Of course, in (8), (9) it has to be assumed that n = m. Values of $p^*(t)$ (i.e. p^*_{ik}, $i \in Z_k$ and k = 1, 2,...,r) are calculated for the optimal values of w_{ik}, E_k and T_k, using (7) for each Z_k.

Corollary 2 describes the simplest way for finding an optimal solution of the problem of scheduling independent jobs for arbitrary strictly concave $s_i(\cdot)$. Of course, it may be applied to small instances only, since the number of variables in (11)–(14), grows exponentially with n.

DEPENDENT JOBS

We assume that a task-on-arc digraph represents the precedence relation in the set of jobs. Moreover, the nodes are numbered in such a way that, if arc (j_a,j_b) represents a job, then $j_a < j_b$. Nodes j_a, j_b are called respectively head and tail ones for this job. Of course, for each directed graph without cycles such an ordering is always possible, and may be found in $O(n^2)$ time. We assume, without loss of generality, that node 1 (beginning node) is the only node without predecessors, and node q (terminal node) is the only node without successors. Moreover, two nodes may be linked by at most one arc in a task-on-arc digraph. Let Q_j j = 1, 2,...,q − 1, denote the set of jobs that may be processed between the occurrences of events represented by nodes j and j + 1. Such sets are called main sets. Of course, parts of jobs processed in particular Q_j are independent. Notice that there exist only one sequence $S = [Q_1,Q_2,\ldots,Q_{q-1}]$ of main sets for a particular ordering of nodes O. An example of the task-on-arc digraph and related main sets is presented in Fig. 1.

Optimal Solution for a Given Ordering of Nodes

Let us assume that the sequence S is given representing ordering of nodes O. Since parts of jobs processed in particular Q_j's are independent, we can apply the approach described in Sect. 4 for each Q_j, j = 1, 2,...,q − 1. This means that we have to consider all m-element combinations from all Q_j's for which $|Q_j| > m$. Let us keep for these combinations the denotations Z_k, k = 1, 2,...,r. If we denote by A the set of main sets with the size greater than m, then total number r of all combinations Z_k for a given ordering of nodes O equals:

$$r=q-|A|-1+\sum_{Q_j\in A}\binom{|Q_j|}{m}$$

An example of combinations Z_k for a given sequence of main sets and m = 2 is showed in Fig. 2.

In consequence, for finding an optimal schedule for a given ordering of nodes O, one has to solve the problem (11)–(14) where Z_k, k = 1, 2,...,r are constructed as presented above.

Using Theorem 1 one can formulate two properties of an optimal schedule

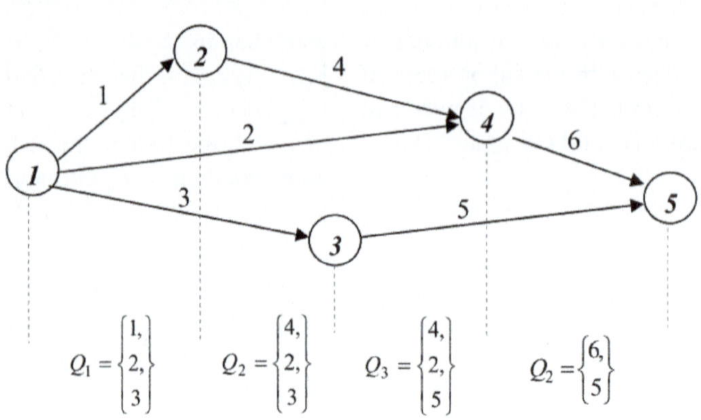

Figure 1. Example of a sequence of main sets.

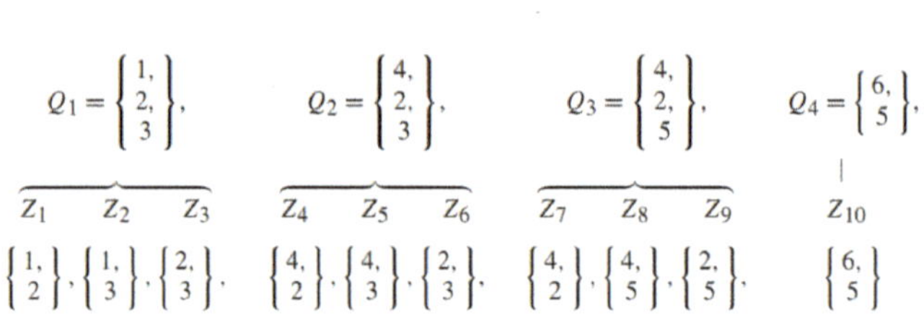

Figure 2. Example of a set of combinations Z_k.

Corollary 3 The minimum energy level which ensures the minimum schedule length for a given power level P, ordering of nodes O and corresponding combinations Z_k, is equal to:

$$E_{\min} = \sum_{k=1}^{r} T_k^* \sum_{i \in Z_k} s_i^{-1}(w_{ik}^* / T_k^*) \tag{15}$$

where $T_k^*, w_{ik}^*, k = 1, 2, \ldots, r, i \in Z_k$ are the optimal values obtained by solving problem (11), (13), (14) where T_k^* are calculated for the level of power P from the equations:

$$\sum_{i \in Z_k} s_i^{-1}(w_{ik}/T_k) = P, \quad k = 1, 2, \ldots, r$$

Corollary 4 The minimum level of power which ensures the minimum schedule length for a given energy level E, ordering of nodes O, and combinations Z_k, is equal to:

$$P_{\min} = \max_k \left\{ \sum_{i \in Z_k} s_i^{-1}(w_{ik}^* / T_k^*) \right\} \tag{16}$$

where $T_k^*, w_{ik}^*, k = 1, 2, \ldots, r, i \in Z_k$ are the optimal values obtained by solving problem (11)–(14) in which "=" should be put in (12), and T_k^* are calculated from the equations:

$$T_k \cdot \sum_{i \in Z_k} s_i^{-1}(w_{ik}/T_k) = E_k, \quad k = 1, 2, \ldots, r$$

Let us underline another interesting aspect of numerical solving of the convex programming problem (11)–(14). It is easy to observe, that quite often a situation may appear where for a given sequence of main sets S, two or more processor feasible sets contain the same combination of job indices. Such a situation may also be observed in Fig. 2 (e.g. see combinations Z_3 and Z_6). It is easy to show that an optimal schedule obtained by replacing all duplicated combinations by a single combination will never be longer than the original one. Thus it is practically justified to propose a simple procedure (Algorithm 1) of generating combinations without repetitions.

The procedure in Algorithm 1 assumes in step 4 that effective algorithm of generating of all combinations in lexicographic order is known. An example of such an algorithm may be found in Er (1985).

Heuristics for Nodes Ordering

To find an optimal schedule for the problem with dependent jobs it is necessary to find an ordering O* of graph nodes which leads to an optimal sequence of main sets S*. To this end, the full enumeration approach over the full spectrum of feasible node orderings is needed. Alternatively, one can apply a metaheuristic approach where searching over the set of all feasible node orderings is directed by algorithm-specific rules. A metaheuristic method produces results relatively fast but quality of found solutions is unpredictable and highly depends on the proper preliminary tuning of parameters and/or operators of the algorithm. Of course, both approaches are applicable for small instances of the problem only or for the instances with a dense task-on-arc precedence graph, since for each ordering a problem (11)–(14) has to be solved.

Algorithm 1

1. $j = 1$; $B = \emptyset$
2. Let set C contains all the jobs beginning in node j,
3. Index anew (starting from 1) all the jobs from C in arbitrary order, if $B \neq \emptyset$ index jobs from B consecutively (starting from $|C| + 1$) in arbitrary order.
4. if ($|B| + |C| \leq m$)
   then
      generate only one combination of all jobs from B and C,
   otherwise
      generate all combinations of m jobs over set $B \cup C$ in lexicographic order of its indices until the first job from set B appears on the first position in a new combination (the combination triggering stopping criteria is excluded from the current combination set).
5. if ($j < q$)
   then
      $j = j + 1$; let the set B contain all the jobs not completed in node j; go back to step 2;
   otherwise
      stop.

Below we propose several simple constructive heuristics which create a single ordering of nodes for a given problem instance.

Notice that the processing time of a job is unknown a priori in the considered scheduling problem. Instead, the size of a job may be taken as its base metrics used in a heuristic. Unfortunately, the relation between power allocation for a job and its effective processing time defined by (4) is ignored in this approach. Therefore, a job size may have only an auxiliary meaning when a power allocation is unknown. Thus, a preliminary allocation of power to jobs is necessary in order to respect processing times of jobs in a heuristics. We propose the approach where the preliminary allocation of power is found by solving (11)–(14) for an arbitrary feasible ordering of nodes in a graph. In consequence, we are able to calculate both: a preliminary processing time, as well as an energy consumption for each job. These two metrics may be further directly utilized in the heuristics.

We will consider three variants of the proposed heuristics, each of them taking into account one of the following parameters as a job metrics:

variant 1—size w_i,

variant 2—processing time τ_i,

variant 3—energy consumption e_i.

It is worth stressing that the metrics of jobs in variants 2 and 3 base on some preliminary ordering O. It is an open question, how far a particular ordering O determines the final ordering produced by a heuristics. We will try to estimate this relation on the bases of the computational experiment described in Sect. 6.

Level Based Heuristics

Levord is a method which has proved its efficiency for the classical problem of scheduling preemptable jobs on parallel identical processors (Józefowska et al. 2004). It exploits the intuition that jobs with greater sizes should be performed longer and thus they have to be present in a greater number of main sets than the jobs with small sizes.

We propose two methods of calculating a node level.

Let us denote by pLevord(a) the method where the level of node j is calculated as the length of the longest path from the node j to node q assuming that the weight of each arc on this path is equal to the chosen metrics a, $a \in \{w_i, \tau_i, e_i\}$ of the represented job.

Second method, denoted by sLevord(a), finds the level of node j as the sum of weights a of all arcs belonging to at least one path from the node j to node q.

Load Based Heuristic

An impact of the limited number of processors m on a constructed schedule is not respected in the level based heuristics described in Sect. 5.2.1. We will use the Nodord algorithm proposed in Słowinski (1978) as a load driven heuristic for node ordering. The algorithm was originally developed for the resource constrained project scheduling problem and further successfully applied to the aforementioned machine scheduling problem (Józefowska et al. 2004). We propose three variants of the algorithm denoted by Nodord(a) for the considered problem, each one basing on a single job metrics a, $a \in \{w_i, \tau_i, e_i\}$. Let us present the original algorithm (Słowinski 1978) with necessary modifications.

Algorithm 2 *Nodord*(a)

1. Number the nodes in the directed graph from 1 to q, so that for each arc (j_a, j_b), $j_a < j_b$.
2. Number the arcs (jobs) in the graph from 1 to n according to the increasing number of their head nodes, and if two or more arcs share the same head node, solve ties according to the increasing tail-node number.
3. Using a heuristic method for scheduling nonpreemptable jobs on m identical parallel processors and assuming that processing time of a job is equal to its value of metrics a, find job start times t_i $i = 1, 2, \ldots, n$ that satisfy precedence constraints and the nonpreemptability condition.
4. For each node except the beginning and terminal ones calculate the value of the parameter

$$d_j = \min_{i \in D_j}\{t_i\}, \quad j = 2, 3, \ldots, q-1,$$

where D_j is the set of jobs beginning in node j and d_j is the latest possible time of occurrence of node j for the task start times defined in step 3.

5. Number the nodes from 2 to $q-1$ according to increasing values of parameter d_j.

The heuristic method used in step 3 determines the quality of the Nodord(a) algorithm. It constructs a nonpreemptable schedule on m machines via a series of partial schedules (Józefowska et al. 2004). A precedence feasible job scheduled as next in a partial schedule is chosen basing on its priority. Applying the results of Józefowska et al. (2004), we calculate a priority of a job as the total length (defined by metrics a) of all its successors.

COMPUTATIONAL EXPERIMENT

To evaluate efficiency of the proposed heuristic approaches we have performed a computational experiment. The algorithms were implemented in C++ and run on SUN Fire V490 equipped with UltraSPARC IIIi processor and 32 GB RAM memory. Due to the fact that no standard library of instances is available for the considered problem, the test instances were generated randomly.

The experiment was limited to small instances since we were interested in comparing the results with the optimal ones. Thus we assumed that the number of nodes $q \in \{6, 7\}$, and the number of processors m equals 2 and 3. The density μ of the graph was set to the three values: 0.3, 0.5, 0.7. Of course, the number of jobs n is related to the number of nodes q and the density of the graph μ and was calculated as Round($q(q-1) \cdot \mu/2$). For each set of parameters m, q, μ, (instances for $q = 7$ were limited only to the case of $\mu = 0.5$) 10 instances were generated. The job sizes were generated uniformly from the interval [0,100]. Moreover, each job i was characterized by a processing speed function $s_i(\cdot)$ of the form: $s_i(\cdot) = p_i^{1/\alpha_i}$, where α_i have been set to the one of values: 2 or 3, with equal probability.

Different combinations of amounts of power P and energy E were tested. First, we set these values in the way that each one of constraints (5) or (6) was active exclusively. Moreover, we established P and E to the levels where both the constraints remained active for tested instances.

We used CFSQP 2.5 solver (Lawrence et al. 1997) for solving the non-linear programming problem (11)–(14). Algorithm 1 was applied to minimize the number of the variables in (11)–(14).

We utilized optimal solutions as the reference results in our experiment. Optimal schedules have been found by a full enumeration procedure which explored the set of all precedence feasible job orderings for a given problem instance. In Table 1 we show data representing the relative range R of schedule lengths for different problem sizes estimated basing on all precedence feasible nodes ordering.

Table 1. Relative range of schedule lengths and average computational time (E = 1000)

#	q/m	μ	R [%] ($AvgTime$ [s])		
			$P = 5$	$P = 30$	$P = 100$
1	6/2	0.3	1.57 (1.60)	1.67 (2.53)	2.11 (2.34)
2	6/2	0.5	7.20 (22.52)	7.65 (9.91)	5.0 (7.76)
3	6/2	0.7	4.31 (51.75)	4.22 (16.94)	3.16 (18.89)
4	6/3	0.3	3.7 (1.33)	3.6 (0.82)	3.8 (1.07)
5	6/3	0.5	4.6 (139.21)	5.1 (33.12)	7.0 7.0 (31.40)
6	6/3	0.7	3.1 (572.83)	4.5 (406.43)	4.6 (379.53)
7	7/2	0.5	13.74 (295.42)	8.22 (165.17)	10.88 (153.49)
8	7/3	0.5	10.09 (293.72)	16.70 (1120.31)	18.86 (1089.71)

This value was calculated as:

$$R = \frac{1}{H} \sum_{h=1}^{H} \frac{C_{h\max}^{W} - C_{h\max}^{*}}{C_{h\max}^{*}} \cdot 100\%$$

where

$C_{h\max}^{*}$—is the best schedule length for instance h

$C_{h\max}^{W}$—is the worst schedule length for instance h

H—is the number of tested instances.

In Tables 2, 3, 4 the results obtained by heuristics are presented. In consecutive rows the following metrics for each algorithm are placed:

- Average relative deviation from the optimum,
- Maximum relative deviation from the optimum,
- Number of instances for which the algorithm found an optimal solution.

Each single row in Tables 2, 3, 4 represents the values of above metrics for a specific problem size.

For each instance size we mark the best value of a particular metric by bold font.

The full enumeration approach applied for finding optimal orderings of nodes in a graph allows to recognize characteristics of a search space of all precedence feasible orderings from the schedule length point of view. Schedules most distant from the optimum were obtained for instances with an average value of graph density. The graphs with density μ set to 0.5 led to the variability of schedule lengths, measured by R, up to approx. 19%. It may be justified by the fact that the number of different feasible schedules takes the highest values in this case. Another interesting property is that the smaller amount of available power P we set in a instance, the smaller value of average computational time for finding optimum we get. Since the main part of computational time is consumed for the solution of problem (11)–(14), it may be interpreted in such a way that finding the optimal power allocation for instances with relatively small value of P by a non-linear solver is probably simpler.

A general observation resulting from the computational experiment is that efficiency of the proposed heuristics do not differ much. It is especially visible if we take into account the maximum relative deviation from the optimum. One can also conclude that in most cases all the heuristics are misled by the same instance of the problem. Notice that in the worst case (see the seventh instance size in Table 4) the schedule related to the node ordering generated by all heuristics was longer by 12.4%

than the optimal one. It is obvious, that various mechanisms implemented in our heuristics are insufficient for such malicious problem instances. However, all the heuristics found the results worse on average by approx. 2.5% only than the optimal ones. In general, the number of the optimal orderings found ranges from 10% to 100% and it depends mainly on the density of the graph and amounts of P and E.

Let us analyze sensitivity of proposed heuristics on the values of P and E. It is easy to observe that when the amount of power P is relatively small compared to an amount of energy E, then the heuristics found relatively better results than in the case of big P . It may denote that the instances with an amount of energy E as most tight limit are harder to solve then instances where power P should be respected only.

It is difficult to compare a computational time of the proposed heuristic algorithms since their computations are significantly dominated by solving the non-linear programming problem (11)–(14). The time needed by heuristics for constructing a single, potentially optimal ordering may be taken as negligibly small.

Table 2. Power related instances (P = 5, E = 1000)

#	q/m	μ	$pLevord(w_i)$	$pLevord(\tau_i)$	$pLevord(e_i)$	$sLevord(w_i)$	$sLevord(\tau_i)$	$sLevord(e_i)$	$Nodord(w_i)$	$Nodord(\tau_i)$	$Nodord(e_i)$
1	6/2	0.3	0.0016	**0.0**	0.0016	0.0016	**0.0**	0.0016	0.0016	**0.0**	0.0016
			0.0157	**0.**	**0.0157**	**0.157**	**0.0**	**0.0157**	**0.0157**	**0.0**	**0.0157**
			9	**10**	9	9	**10**	9	9	**10**	9
2		0.5	**0.001**	0.0011	**0.0010**	**0.0010**	0.0011	**0.0010**	**0.0010**	0.0011	**0.0010**
			0.021	0.021	**0.021**	**0.021**	0.021	**0.021**	**0.021**	0.021	**0.021**
			8	**8**	**8**	**8**	**8**	**8**	**8**	**8**	**8**
3		0.7	**0.003**	0.004	0.004	0.006	0.006	0.006	**0.003**	0.004	**0.003**
			0.01	**0.01**	**0.01**	0.025	0.025	0.0251	**0.01**	**0.01**	**0.01**
			8	7	7	6	6	7	**8**	6	7
4	6/3	0.3	0.0057	**0.003**	0.0053	0.0057	**0.003**	0.0053	0.0053	**0.003**	0.0053
			0.0304	**0.0304**	**0.0304**	**0.0304**	**0.0304**	**0.0304**	**0.0304**	**0.0304**	**0.0304**
			7	**9**	7	8	**9**	8	8	**9**	8
5		0.5	**0.015**	0.017	0.023	**0.015**	0.017	0.023	**0.015**	0.017	0.023
			0.045	**0.045**	0.080	**0.045**	**0.045**	0.080	**0.045**	**0.045**	0.080
			3	2	2	**3**	2	2	**3**	2	2
6		0.7	0.012	0.018	0.011	0.018	0.018	0.014	0.011	0.015	**0.010**
			0.032	0.044	**0.032**	0.044	0.044	0.033	**0.032**	0.044	**0.032**
			3	3	**4**	3	3	**4**	3	3	**4**
7	7/2	0.5	0.014	0.013	0.015	0.014	**0.012**	0.014	0.014	**0.012**	0.014
			0.02	**0.02**	**0.02**	**0.02**	**0.02**	**0.02**	**0.02**	**0.02**	**0.02**
			1	**2**	1	1	**2**	1	1	**2**	1
8	7/3	0.5	0.015	0.018	0.022	0.015	0.015	0.015	0.015	0.015	**0.013**
			0.023	0.025	0.025	**0.023**	**0.023**	**0.023**	**0.023**	**0.023**	0.024
			3	3	3	3	3	3	3	3	**4**

Table 3. Mixed, power/energy related instances (P = 30, E = 1000)

#	q/m	μ	$pLevord(w_i)$	$pLevord(\tau_i)$	$pLevord(e_i)$	$sLevord(w_i)$	$sLevord(\tau_i)$	$sLevord(e_i)$	$Nodord(w_i)$	$Nodord(\tau_i)$	$Nodord(e_i)$
1	6/2	0.3	0.004	0.007	**0.0**	0.004	0.007	**0.0**	**0.0**	0.004	0.004
			0.0375	0.0375	**0.0**	0.0375	0.0375	**0.0**	**0.0**	0.0375	0.0375
			9	8	**10**	9	8	**10**	**10**	9	9
2		0.5	**0.018**	0.019	**0.018**	**0.018**	0.019	**0.018**	**0.018**	0.019	**0.018**
			0.031	**0.031**	**0.031**	**0.031**	**0.031**	**0.031**	**0.031**	**0.031**	**0.031**
			7	6	**7**	**7**	6	**7**	**7**	6	**7**
3		0.7	0.004	**0.003**	0.004	0.014	0.014	0.015	0.004	**0.003**	0.004
			0.030	**0.030**	**0.030**	0.114	0.114	0.114	**0.030**	**0.030**	**0.030**
			7	**8**	7	8	8	7	7	**8**	7
4	6/3	0.3	0.0043	0.0053	**0.0037**	0.0043	0.0053	**0.0037**	**0.0037**	0.0053	0.0043
			0.0241	**0.0241**	**0.0241**	**0.0241**	**0.0241**	**0.0241**	**0.0241**	**0.0241**	**0.0241**
			6	5	**7**	6	5	**7**	**7**	5	6
5		0.5	**0.015**	0.016	**0.015**	**0.015**	0.018	**0.015**	**0.015**	0.016	**0.015**
			0.042	0.033	0.042	0.033	0.042	0.028	**0.022**	0.024	0.028
			4	**4**	**4**	**4**	**4**	**4**	**4**	**4**	**4**
6		0.7	0.0021	0.0024	0.0029	0.0025	0.0025	0.0033	**0.0016**	0.0021	0.0025
			0.01	**0.01**	**0.01**	**0.01**	**0.01**	**0.01**	**0.01**	**0.01**	**0.01**
			7	**7**	6	**7**	**7**	5	**7**	**7**	6

7	7/2	0.5	0.017 **0.034** **1**	0.018 **0.034** **1**	**0.016** **0.034** **1**	0.018 **0.034** **1**	0.017 **0.034** **1**	**0.016** **0.034** **1**	0.017 **0.034** **1**	**0.016** **0.034** **1**	**0.016** **0.034** **1**
8	7/3	0.5	0.014 0.04 2	0.017 0.04 2	0.017 0.04 2	0.014 0.04 2	0.014 0.04 2	0.014 0.04 2	0.014 0.04 2	0.014 0.04 **3**	**0.007** **0.025** **3**

Table 4. Energy related instances (P = 100, E = 1000)

#	q/m	μ	$pLevord(w_i)$	$pLevord(\tau_i)$	$pLevord(e_i)$	$sLevord(w_i)$	$sLevord(\tau_i)$	$sLevord(e_i)$	$Nodord(w_i)$	$Nodord(\tau_i)$	$Nodord(e_i)$
1	6/2	0.3	**0.0** **0.0** **10**	0.0051 **0.0408** 9	0.0042 **0.0408** 9	**0.0** **0.0** **10**	0.0051 **0.0408** 9	0.0042 **0.0408** 9	0.0042 **0.0408** 9	0.0051 **0.0408** 9	**0.0** **0.0** **10**
2		0.5	**0.011** **0.047** **7**	0.013 **0.047** 6	**0.011** **0.047** **7**	**0.011** **0.047** **7**	0.012 **0.047** 6	**0.011** **0.047** **7**	**0.011** **0.047** **7**	0.012 **0.047** 6	**0.011** **0.047** **7**
3		0.7	0.018 **0.099** 3	0.017 **0.099** **4**	0.018 **0.099** 3	**0.015** **0.099** **4**	**0.015** **0.099** **4**	0.016 **0.099** 3	0.018 **0.099** 3	0.017 **0.099** **4**	0.023 **0.099** 3
4	6/3	0.3	**0.004** **0.038** **8**	0.005 **0.038** 7	0.0042 **0.038** 7	**0.004** **0.038** **8**	0.005 **0.038** 7	**0.004** **0.038** 7	**0.004** **0.038** 7	0.005 **0.038** 7	**0.004** **0.038** **8**
5		0.5	0.035 **0.105** **3**	**0.034** **0.105** **3**	**0.034** **0.105** **3**	**0.034** **0.105** **3**	0.035 **0.105** **3**	**0.034** **0.105** **3**	**0.034** **0.105** **3**	0.037 **0.105** **3**	**0.034** **0.105** **3**
6		0.7	0.030 0.104 2	0.048 0.104 2	0.028 0.092 **3**	0.049 0.104 2	0.032 0.104 2	0.030 0.082 **3**	0.031 0.104 2	0.030 0.104 2	**0.012** **0.081** **3**
7	7/2	0.5	0.034 **0.124** **1**	0.033 **0.124** **1**	0.034 **0.124** **1**	0.034 **0.124** **1**	0.033 **0.124** **1**	0.034 **0.124** **1**	0.034 **0.124** **1**	0.034 **0.124** **1**	**0.026** **0.124** **1**
8	7/3	0.5	0.021 **0.124** 2	0.037 0.125 2	0.038 0.125 2	0.021 **0.124** 2	0.021 **0.124** 2	0.021 **0.124** 2	0.021 **0.124** 2	**0.015** **0.124** **3**	0.031 **0.124** 2

Even if it is difficult to indicate the best heuristic for the considered problem, we propose the following recommendation. Use Nodord algorithm in variant 1, 2 and 3, for each instance of the problem. From among three orderings obtained choose the one giving the smallest schedule length. As it is ease to notice in Tables 2, 3, 4, such a combining effect of the three heuristics guarantees the best values of algorithm metrics for each problem size and combination of parameters P/E.

SUMMARY

In this paper the problem of power-aware scheduling of preemptable jobs on parallel identical processors to minimize the schedule length was studied. Two job processing models: one based on a power vs. processor speed function, and the second based on a job processing speed vs. temporary amount of power function have been compared. In the sequel the second, more general model, has been applied to study basic properties of optimal schedules for independent and precedence-related jobs. Methods for finding optimal schedules using these properties have been presented. Although these methods, based on solving convex programming problems, can be practically solved for rather small problem sizes, they are important from the viewpoint of evaluating heuristic vs. optimal solutions. Heuristics for finding ordering of nodes in a task-on-arc digraph which led to possibly short schedules from among all possible orderings in the general case of precedence constraints among jobs have been proposed and tested experimentally.

Further research can be directed towards the construction of heuristics for finding schedules for independent jobs, as well as for dependent ones for a given ordering of nodes. Some concepts of the heuristics proposed in Józefowska et al. (2002) for the case of a nonpreemptive scheduling of independent tasks seem to be a good starting point for such interesting study. Combinations of similar heuristics with the ones presented in this paper, tested experimentally, should lead for an efficient finding of good quality power-aware schedules in the general case. Let us underline that situation where the non-linear solver is substituted by a good heuristic may also be profitable in combination with a metaheuristic approach effectively exploiting dependencies among jobs.

ACKNOWLEDGEMENTS

We are grateful to anonymous referees for their helpful comments. The research was partially supported by a grant from the Ministry of Science and Higher Education, Poland. The computational resources for the experiment were granted by Poznan Supercomputing and Networking Center.

REFERENCES

1. Bansal, N., Kimbrel, T., & Pruhs, K. (2005). Dynamic speed scaling to manage energy and temperature. In *Lecture notes in computer science* (Vol. 3404, pp. 460–471).

2. Bansal, N., Chan, H. L., Lam, T. W., & Lee, L. K. (2008). Scheduling for speed bounded processors. In *Proceedings of international colloquium on automata, languages and programming (ICALP)* (pp. 409–420).
3. Boyer, F. R., Epassa, H. G., & Savaria, Y. (2006). Embedded power-aware cycle by cycle variable speed processor. In *IEE Proceedings: Vol. 153/4. Computers and digital techniques* (pp. 283–290).
4. Błażewicz, J., Ecker, K., Pesch, E., Schmidt, G., & Węglarz, J. (2007). *Handbook on scheduling.* Berlin: Springer.
5. Bunde, D. P. (2006). Power-aware scheduling for makespan and flow. In *Proceedings of the eighteenth annual ACM symposium on parallelism in algorithms and architectures* (pp. 190–196). Cambridge, MA, USA.
6. Er, M. C. (1985). Lexicographic ordering, ranking and unranking of combinations.*International Journal of Computer Mathematics*, *3*, 277–283.
7. Irani, S., & Pruhs, K.b (2005). Algorithmic problems in power management. *SIGACT News*,*36*(2), 63–76.
8. Józefowska, J., Mika, M., Różycki, R., Waligóra, G., & Węglarz, J. (2002). A heuristic approach to allocating the continuous resource in discrete-continuous scheduling problems to minimize the makespan. *Journal of Scheduling*, *5*(6), 487–499.
9. Józefowska, J., Mika, M., Różycki, R., Waligóra, G., & Węglarz, J. (2004). An almost optimal heuristic for preemptive C_{max} scheduling of dependent tasks on identical parallel processors.*Annals of Operation Research*, *129*, 205–216.
10. Lawrence, C., Zhou, J. L., & Tits, A. L. User's guide for CFSQP version 2.5: A C code for solving (large scale) constrained nonlinear (minimax) optimization problems, generating iterates satisfying all inequality constraints (1997). Available at:http://www.isr.umd.edu/Labs/CACSE/FSQP/c_manual.ps.
11. Pruhs, K., van Stee, R., & Uthaisombut, P. (2005). Speed scaling of tasks with precedence constraints. In *Proceedings of WAOA* (pp. 307–319).
12. Słowiński, R. (1978). A node ordering heuristic for network scheduling under resource constraints. *Foundations of Control Engineering*, *3*, 19–27.
13. Węglarz, J. (1981). Project scheduling with continuously-divisible doubly constrained resources.*Management Science*, *27*(9), 1040–1053.

CHAPTER 10

ENERGY DISSIPATION AND TRANSPORT IN NANOSCALE DEVICES

Eric Pop

Keywords: Nanowire, Nanotube, Interface, Thermal transport, Rectification, Thermoelectric, Power dissipation

INTRODUCTION

Some of the greatest challenges of modern society are related to energy consumption, dissipation, and waste. Among these, present and future technologies based on nanoscale materials and devices hold great potential for improved energy conservation, conversion, or harvesting. A prominent example is that of integrated electronics, where power dissipation issues have recently become one of its greatest challenges. Power dissipation limits the performance of electronics from handheld devices (~10^{-3} W) to massive data centers (~10^{9} W), all primarily based on silicon micro/nanotechnology. Put together, the energy use of the United States information technology (IT) infrastructure is currently in excess of 20 GW, or 5%–10% of our national electricity budget, with an approximate breakdown as shown in Figs. 1(c) and 1(d) [1]. Importantly, the figures for data center energy consumption have doubled in five recent years, with waste heat requiring drastic cooling solutions (Fig. 1(c)). Such challenges are also evident at the individual microprocessor (CPU) level, where the race to increase operating frequency beyond a few GHz recently stopped when typical dissipated power reached 100 W/cm^2 (Fig. 1(b)), an order of magnitude higher than a typical hot plate [2]. Such electronic power and thermal challenges have negative impacts in areas from massive database servers to new applications like wearable devices, medical instrumentation, or portable electronics. In the latter situations, there is a basic trade-off between the available functionality and the need to carry heavy batteries to power it.

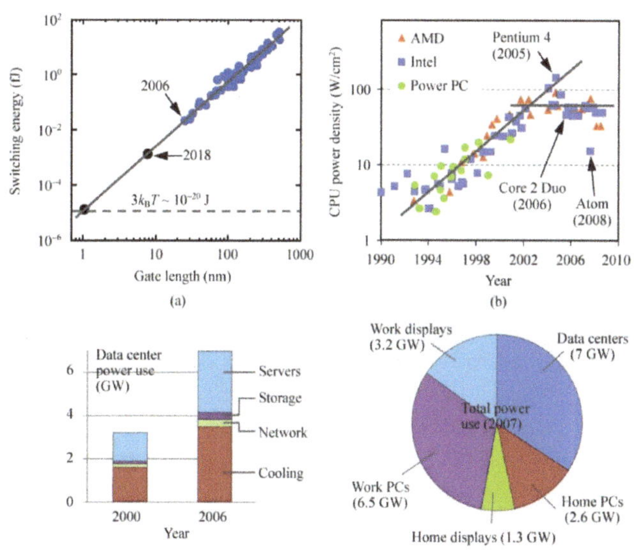

Figure 1. Energy and power dissipation from transistors, to CPUs, to data centers. (a) Switching energy of modern silicon transistors is still over 1000× higher than fundamental physical limits, but on a trend toward them [3]. (b) Near-exponential increase of CPU power density in recent decades has flattened with the introduction of multi-core CPUs (solid lines show approximate trends); by comparison, the power density on the surface of the sun is approximately 6000 W/cm^2 [2]. (c) Data center power use in the U. S. doubled in six years, with an extremely large proportion devoted to cooling. (d) Approximate breakdown of total power used by digital electronics in the U. S., where data are available [1, 6, 8].

Citation: Pop, E. Energy dissipation and transport in nanoscale devices, Nano Research, March 2010, Volume 3, Issue 3, pp 147–169. DOI:10.1007/s12274-010-1019-z.

Despite tremendous progress over the past three decades, modern silicon transistors are still over three orders of magnitude (>1000×) more energy inefficient than fundamental physical limits, as shown in Fig. 1(a). These limits have been estimated as approximately $3k_BT \approx 10^{-20}$ J at room temperature for a binary switch with a single electron and energy level separation k_BT, where k_B is the Boltzmann constant and T is the absolute temperature [3]. In the average modern microprocessor the dissipated power is due, in approximately equal parts, to both leakage (or "sleep") power and active (dynamic) switching power [4], as detailed in Section 2. Power dissipation is compounded at the system level, where each CPU Watt demands approximately 1.5× more for the supply, PC board, and case cooling [1]. Such power (mis)use is even more evident in systems built on otherwise power-efficient processors, e.g., in the case of the Intel Atom N270 (2.5 W power use) which is typically paired up with the Intel 945GSE chipset (11.8 W power use) [5]. At the other extreme, data centers require 50%–100% additional energy for cooling (Fig. 1(c)), which is now the most important factor limiting their performance, not the hardware itself.

If present growth trends are maintained, data center and overall electronics power use could reach one third of total U. S. consumption by 2025 [1]. Worldwide, the growth trends could be even steeper, given that technologically developed regions such as the U. S., Western Europe, and Japan currently account for 58% of the world's computers, but only 15% of the world population [6]. Such energy challenges for the electronics infrastructure stem not only from the power supply side which calls for new energy sources, efficient batteries, or thermoelectrics, but also from the demand side, i.e., the need for more energyefficient computing devices. To put it in financial terms, data centers consumed more than 7 GW or $4.5 billion in 2006 (Fig. 1(c)) [1]. The estimated 108 million PCs in offices across the U. S. had an annual energy cost of $4.2 billion in 2008 [6]. More troubling, the amount of equivalent CO_2 emissions generated is approximately equivalent to that of 5 million cars, comparable to the entire state of Maryland.

In fact, such estimates are likely to be conservative, missing the power consumed by routers, networks, and the internet backbone where detailed studies are not available (however, a thought-provoking study of energy consumption by e-mail SPAM has recently been made [7]). Nevertheless, the overall negative impact of these trends (to predicted 2025 levels) on energy supplies, budgets, and the environment in terms of equivalent CO_2 emissions is staggering. Thus, breakthroughs in our understanding and improvement of energy efficiency in nanoelectronics will also have a global effect, impacting the entire structure of modern society.

On a broader scale, just over half the man-made energy in the world is wasted as heat (10^{13} W), from power plants and factories to car engines and the power bricks on our laptops. Efficiently reclaiming even a small percentage of such wasted heat would itself nearly satisfy the electricity needs of our planet [8]. The fundamental issues at hand are, in fact, a two-sided problem: on one side, there is a significant need for low-energy computing devices, which is perhaps the biggest challenge in micro/nanoelectronics today. On the other side there is the challenge of waste heat dissipation, guiding, or conversion into useful electricity.

This review discusses several aspects of the above, from the nanoscale circuits to device and materials perspective, as follows: Section 2 examines energy dissipation and optimization in nanoscale circuits, with focus on leakage vs. active power in nanoscale technologies. Section 3 presents energy dissipation in nanoscale devices (e.g., silicon transistors) from the diffusive to the ballistic regime. Section 4 reviews concepts of steady-state and transient thermal transport in the context of nanoscale devices with sub-nanosecond switching times. Finally, recent directions regarding energy transport are reviewed, including electrical and thermal conductivity of carbon nanostructures and nanowires (Section 5), thermal rectification (Section 6), and the role of ubiquitous material interfaces (Section 7).

ENERGY DISSIPATION IN NANOSCALE CIRCUITS

Energy and power dissipation in nanoscale digital circuits is often described in the context of inverter activity, as shown in Fig. 2(a). Three components have been identified for digital power consumption: dynamic power used during switching for charging and discharging the inverter load, subthreshold leakage power, and short-circuit power (P_{SC}) [9]:

$$P = C_L V_{DD}^2 \alpha f + I_{leak} V_{DD} + P_{SC} \tag{1}$$

where α is the activity factor, C_L is the load capacitance, f is the clock frequency, and I_{leak} is the sub-threshold leakage current [10]. The short-circuit power is typically the smallest for well-designed circuits with equal rise and fall times [9, 11]. The leakage component, however, is a strong function of temperature T, and therefore implicitly dependent on the total power dissipated. Moreover, leakage current also scales exponentially with the voltage overdrive ($V_{GS} - V_T$) and is very sensitive to changes in threshold voltage:

$$I_{leak} = \frac{W}{L_{eff}} \mu_{eff} C_{ox} (m-1) \left(\frac{k_B T}{q} \right)^2 \exp\left(\frac{V_{GS} - V_T}{m k_B T / q} \right) \cdot \left[1 - \exp\left(-\frac{V_{DS}}{k_B T / q} \right) \right] \tag{2}$$

where μ_{eff} is the effective mobility, C_{ox} the gate oxide capacitance, W and L_{eff} the effective channel width and length, T the absolute temperature, V_{DS} the drain voltage, V_{GS} the gate voltage, V_T the threshold voltage, q the elementary charge, and m the subthreshold body factor [11]. With (down) scaling of technology dimensions and voltage, the role of sub-threshold leakage has become increasingly important, as shown by the trends in Figs. 2(b) and 2(c). This is due to a reduction in supply voltage V_{DD} without the ability to properly scale the threshold voltage, in addition to short-channel phenomena like drain induced barrier lowering (DIBL) [9].

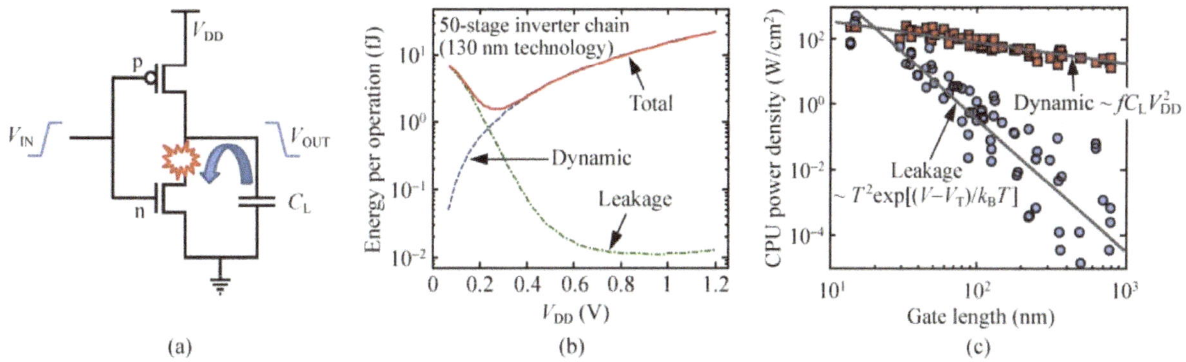

Figure 2. Power and energy dissipation in digital logic. (a) Schematic of a basic inverter. When the load capacitance C_L is discharged, its energy is dissipated in the drain of the n-FET as dynamic power $\left(\sim fC_LV_{DD}^2\right)$. (b) Energy consumed per operation (a single clock period) by a 50-stage inverter chain displays a minimum with respect to supply voltage scaling [10, 11]. (c) Scaling of dynamic and leakage power density with CMOS technology [4]. In modern CPUs total power dissipation via leakage can be equal to or greater than the dynamic (active) switching power. Lines show scaling trends.

An interesting trade-off becomes apparent if we compare the energy (rather than the power) dissipated by a simple digital circuit like an inverter chain, as shown in Fig. 2(b), here for a 130 nm technology with threshold voltage V_T = 0.4 V. Energy per operation is the more important metric for applications that are limited by battery life, such as mobile or medical devices. The dynamic energy, like the dynamic power, scales quadratically with the supply voltage V_{DD}. However, the leakage energy ($E_{leak} = I_{leak}V_{DD}t_d$) rises sharply as the supply voltage is reduced because the circuit delay (t_d) increases in sub-threshold (V_{DD} ⊠ V_T) operation. This raises an interesting point: for applications that are not concerned with maximum performance, the energy-optimal operation occurs in the sub-V_T regime for a given complementary metal oxide semiconductor (CMOS) technology, as shown in Fig. 2(b) [10, 11].

In applications that do seek optimal performance (such as desktop or server CPUs), leakage power dissipation has become significant, as illustrated in Fig. 2(c). Particularly, with respect to nanoscale devices with picosecond switching delays, Eq. (2) above merits a closer look, upon which several of its shortcomings become evident. First, the expression was derived for three-dimensional diffusive transport across an energy barrier, whereas most nanoscale devices are one- or two-dimensional (the leakage current of a carbon nanotube transistor was recently derived in Ref. [12]). Second, leakage is strongly dependent on temperature, yet this itself is highly unsteady during transient digital switching. Typical electrical transients and inverter delays are of the order of 1–10 ps, whereas device thermal time constants are ~10 ns [13, 14]. Which temperature is used in computing the leakage current is an important choice, and using an average "junction temperature" as is often done [15] can lead to significant over- or underestimates of the total leakage. Third, even if the thermal transients are properly accounted for in evaluating leakage during digital operation, the notion of temperature itself may need to be reconsidered when switching approaches the time scales of electron–phonon and phonon–phonon collisions (0.1–10 ps), leading to non-equilibrium conditions. Electrons scatter strongly with optical phonons (OP) in all materials and devices under consideration, and OP lifetimes are themselves of the order of 1–10 ps [16–20]. Optical phonons range from approximately 35 meV in Ge and GaAs, to 60 meV in silicon, and nearly 200 meV in carbon nanotubes and graphene [21], and hence the absorption of a single OP can lead to enough additional energy to immediately surpass the potential barrier in a metal oxide semicoductor field effect transistor (MOSFET)-like device (Fig. 3(c)). Recent results have also shown significant increase in bandto-band tunneling currents in carbon nanotubes due to optical phonon absorption [22]. As will be seen below, it is apparent that our present-day understanding and modeling of power dissipation in nanoscale circuits falls short of the recent advances made in nanoscale devices. This remains an area where significant progress remains to be made.

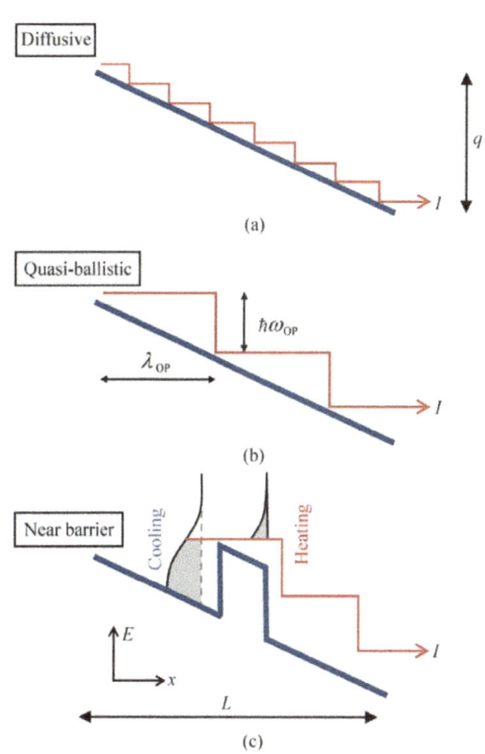

Figure 3. Schematic band diagrams and energy dissipation in (a) device under diffusive transport ($L \gg \lambda_{OP}$, $qV \gg \hbar\omega_{OP}$), (b) quasiballistic device ($L \sim \lambda_{OP}$, $qV \sim \hbar\omega_{OP}$), and (c) near an energy barrier. The barrier acts as an energy filter, resulting in lattice cooling to its left, and heating to its right (Peltier effect) [44, 45]. All cases assume a constant electric field.

ENERGY DISSIPATION IN NANOSCALE DEVICES

Diffusive, Ballistic, and Contacts

The most elementary approach for estimating the power dissipation of an electronic device is to express it as the product of the current passing through and the voltage drop across it:

$$P_D = IV = I^2R. \tag{3}$$

This is the classical expression for a device under diffusive transport (subscript D), and the voltage drop excludes the device contacts (Fig. 3(a)). Hence, this equation must be applied with care in describing the power dissipated in a structure with relatively large contact resistance (R_C), e.g., a carbon nanotube or molecular device. The intrinsic device resistance also excludes the quantum contact resistance $R_0 = h/(Mq^2) \approx 25.8/M(k\Omega)$, where q is the elementary charge, h is Planck's constant, and M is the number of modes for electrical transport (e.g., M = 4 for single-walled carbon nanotubes, accounting for band degeneracy and spin) [23–26]. In large classical devices this can be ignored, as the presence of many transport modes ensures that R_0 is extremely small.

The expression above will overestimate the total power dissipated in a quasi-ballistic device, i.e., one with dimensions comparable to or shorter than the inelastic scattering length [27, 28], as schematically shown in Fig. 3(b). In this case, carriers gain energy comparable to the applied voltage ($E \approx qV$) but do not undergo enough inelastic scattering events to equilibrate and dissipate it to the lattice by the time they exit. In order to estimate dissipation in quasi-ballistic devices (subscript QB), we introduce the inelastic mean free path (MFP) λ_{OP}, typical for carrier scattering with high-energy optical phonons ($\hbar\omega_{OP} \approx$ 200 eV in carbon nanotubes and $\approx$ 60 meV in silicon):

$$P_{QB} = \frac{\hbar\omega_{OP}}{q}\left(\frac{L}{L+\lambda_{OP}}\right)\frac{V}{R_0} \tag{4}$$

where R_0 is the quantum of electrical resistance defined above, $\lambda_{OP} = L(\hbar\omega_{OP}/qV) + \lambda_0$, and λ_0 is the spontaneous OP emission MFP for carriers with energy $E > \hbar\omega_{OP}$ [29, 30] (for example, $\lambda_0 \approx 15d$ for a carbon nanotube with diameter d [31]). Here it is assumed that the current flowing through the device is limited by the inelastic scattering length, $I = (V/R_0)\lambda_{OP}/(L + \lambda_{OP})$ and that a carrier undergoes approximately L/λ_{OP} inelastic collisions within the device, each time losing an energy $\hbar\omega_{OP}$ to the lattice. In addition, the carrier is assumed to relax to the bottom of its band immediately after the inelastic phonon emission event (Fig. 3(b)). Near the limit of zero temperature, an even sharper onset of power dissipation is expected, as soon as the applied voltage is large enough for carriers to emit a single phonon, $V > \hbar\omega_{OP}/q$, as shown for molecular junctions [32–35]. The expression above recovers the correct limit of power dissipation within a device in the ballistic limit ($P \rightarrow 0$ as $L \ll \lambda_{OP}$), although a more careful examination reveals that the scattering length itself depends on the phonon occupation and temperature at sufficiently high power levels. In this case, the electrical–thermal dissipation must be self-consistently solved, as in Refs. [30, 36]. Moreover, in quasi-ballistic transport hot carriers escape through the contact at higher voltage and a significant portion of the total power is dissipated there instead [37]. This can be interpreted as the difference between the input power and the power dissipated within the quasi-ballistic device (P_D–P_{QB} in Eq. (3) and (4) above).

A third example concerns heat dissipation during thermionic or tunneling transport across mesoscopic energy barriers, which are often encountered in transistors, diodes, heterojunctions, and device contacts (Fig. 3(c)). Such dissipative transport does not lend itself to a simple, analytic interpretation, but it has been examined with the non-equilibrium Green's functions (NEGF) approach [27, 38], by Monte Carlo simulations [39, 45], and the scattering matrix method [40]. A key feature of dissipation near barriers is the role of the energy barrier as a filter of the carrier distribution function, as schematically shown in Fig. 3(c). Carriers with higher energy have a greater transmission probability (either thermionic or tunneling) across the barrier, leading to a "cooler" carrier distribution to the left (uphill) of the barrier, and a "hotter" distribution to the right (downhill). This results in an effective cooling (negative heat dissipation) to the left, and heating of the lattice (positive heat dissipation) to the right of the barrier. This effect depends not only on the barrier shape and height, but also on the direction of the current flow. In addition, such asymmetric heating can also be observed at metal– semiconductor contacts [41, 42], at contact energy barriers between dissimilar semiconductors [43–45], and even at more subtle transitions between regions with different densities of states (DOS). The latter is a thermoelectric effect recently observed in transport between monolayer and bilayer graphene [46] and is a result of hot carriers diffusing from regions of lower to higher DOS in order to maximize the entropy (also equivalent to the classical effect of a gas that cools as it expands).

Spatial Distribution of Energy Dissipation

Returning in more detail to the case of diffusive transport (dimension L ≫ inelastic MFP), more sophisticated methods are needed to compute the spatial distribution of power dissipation, instead of the simple lumped result of Eq. (3). This is most often given by the driftdiffusion approach [47–50]:

$$P_V = \mathbf{J}\cdot\mathbf{E} + (R-G)(E_G + 3k_BT) \tag{5}$$

where J is the current density, E the electric field, (R – G) is the net non-radiative recombination rate (recombination minus generation), E_G is the semiconductor band gap, and T is the lattice temperature. The factor $3k_BT$ arises when the average energy of an electron (hole) above (below) the conduction (valence) band is $\sim 3/2k_BT$. The result above is typically implemented as a finite-element simulation on a device grid. Note the notation of P_V (power density per unit volume, i.e., W/m^3) vs. Eq. (3) above (total power in Watts). The total power dissipated can be recovered by integrating Eq. (5) over the device volume. The first term represents the Joule heating rate which is usually positive (power generation) as charge carriers move along the band structure under the influence of the electric field, giving up energy to the lattice. This equation may include higher order terms accounting for carrier drift along a temperature gradient or across a discontinuity in the band structure [43, 45, 48].

Unfortunately, this field-dependent method does not account for the microscopic nature of heat generation near a strongly peaked electric field region, such as in the drain of the transistor. Although electrons gain most of their energy at the location of the peak field, they travel several mean free paths before releasing it to the lattice, in decrements of (at most) the optical phonon energy. In silicon, the optical phonon energy is $\hbar\omega_{OP} \approx 60$ meV and in carbon nanotubes or graphene it is approximately three times greater. Typical inelastic scattering mean free paths in both silicon and carbon nanostructures are of the order $\lambda_{OP} \approx 10$–50 nm [31]. The full electron energy relaxation length is thus even longer, i.e., several inelastic mean free paths. This is illustrated near a 20 nm wide energy barrier in silicon in Fig. 4, where the drift-diffusion approach cannot capture the delocalized nature of the power dissipation region.

An improvement is provided by the hydrodynamic approach, which introduces the electron temperature (T_e) and an average electron energy relaxation time ($\tau_{e\text{-}L}$) [48]:

$$P_V = \frac{3}{2}\frac{k_B n(T_e - T_L)}{\tau_{e\text{-}L}} + (R-G)\left[E_G + \frac{3k_B}{2}(T_e + T_L)\right] \quad (6)$$

where n is the electron density and L denotes the lattice. The equation here is written for electrons as majority carriers, but the holes can be treated similarly. Unlike the drift-diffusion model, this approach is better suited for capturing transport near highly peaked electric fields. However, this suffers from the simplification of a single averaged carrier temperature and relaxation time, as scattering rates are strongly energy-dependent [51]. Neither method gives information regarding the frequencies and wave vectors of phonons emitted. Such details are important because the emitted phonons have different velocities and widely varying contributions to heat transport [52–55] and device heating [56, 57].

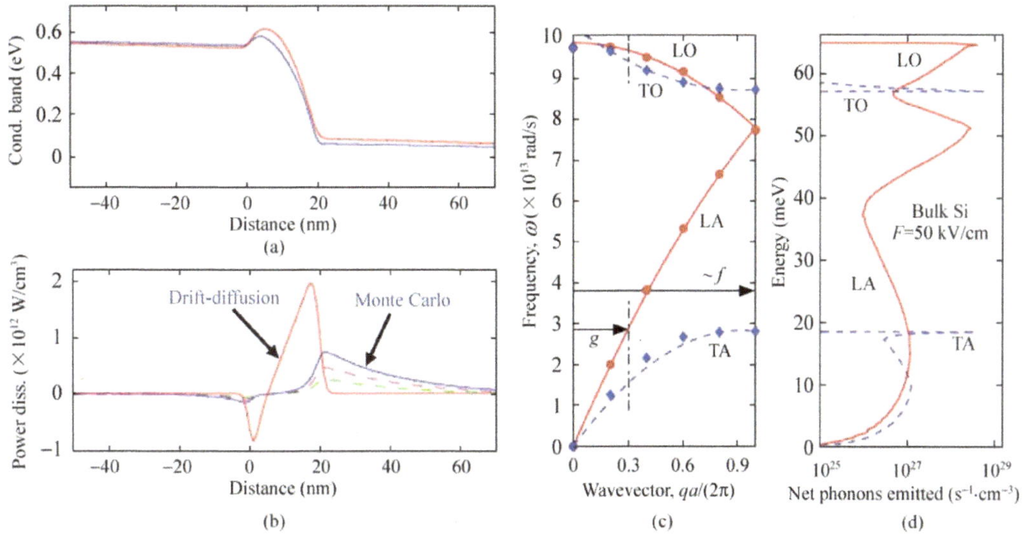

Figure 4. Computed spatial and phonon-resolved power dissipation in silicon. (a) Conduction (cond.) band diagram of a 20 nm wide energy barrier and (b) non-localized power dissipation (diss.) near it. Dashed lines represent optical (upper) and acoustic (lower) phonon emission computed with a Monte Carlo approach [51]. (c) Phonon dispersion in silicon and (d) corresponding phonon generation spectrum from Joule heating. Transverse modes are dashed lines, longitudinal modes are solid lines. Reproduced with permission from Ref. [60]. Copyright 2005 American Physical Society.

The mechanism through which lattice self-heating occurs is that of electron scattering with phonons, and therefore a model which deliberately incorporates all scattering events will also capture such energy dissipation details. Thus, the Monte Carlo method [58] originally developed for studying hot electron effects [59], is also well-suited for computing a detailed picture of energy dissipation. This was the approach adopted in Refs. [60–64], where power dissipation was computed as a sum of all phonon emission minus all phonon absorption events:

$$P_V = \frac{n}{N_{sim}\Delta t}\sum\left(\hbar\omega_{ems} - \hbar\omega_{abs}\right) \quad (7)$$

where n is the real-space carrier density, N_{sim} is the number of simulated particles (e.g., 10 000 simulated particles could be used to describe 10^{19} cm^{-3} real-space concentration), and Δt is the time. This approach has been used to investigate phonon emission as a function of frequency and mode in silicon, as well as to study heat generation near a strongly peaked electric field in a realistic device geometry. Figure 4(b) shows the comparison between a classical drift-diffusion calculation and that of the Monte Carlo result. The former tends to overestimate the peak heat generation rate and predicts a narrower heat generation region, which follows the shape of the electric field. The Monte Carlo method suggests a broader heat generation domain extending inside the device drain, and chiefly limited by the electron-phonon scattering rate there.

The Monte Carlo approach also shows that heat generation in silicon is not evenly divided among phonon modes, but that acoustic phonon modes receive approximately 1/3 and optical phonons 2/3 of the energy dissipated, as shown in Fig. 4(d). More specifically, the longitudinal optical (LO) g-type phonon is responsible for approximately 60% of the total energy dissipation [60]. Optical phonons have group velocities below 1000 m/s, and are thus much slower than the Brillouin zone center acoustic phonons typically responsible for heat transport in silicon (group velocity 5000–9000 m/s). This non-equilibrium phonon generation can lead to an energy transfer bottleneck [56]. In other words, a significant non-equilibrium phonon population may build up, particularly for the g-type LO mode. The generation rates for the other phonon modes are either smaller or their DOS is larger (the DOS is proportional to the square of the phonon wave vector, which is largest at the edge of the Brillouin zone) and non-equilibrium effects are less significant.

The self-consistent study of Rowlette and Goodson [61] built on previous work by Pop [60] and Sinha [62], and coupled the Monte Carlo method for electron transport with a Boltzmann transport equation (BTE) approach for phonon transport. Rowlette found that while extremely high heat generation densities do exist in modern silicon devices ($>10^{12}$ W/cm^3), nonequilibrium hot phonon effects are strongly dependent on the phonon relaxation time. Available data points at an optical phonon lifetime $\tau \approx 0.4$–2 ps for optical phonons near the Brillouin zone center [18], but the phonons involved in electron scattering are the g- and f-type shown in Fig. 4(c), for which no experimental lifetimes are available. The discrepancy between the LO phonon temperature computed with the self-consistent vs. the classical approach can be as high as 250% in the device drain, near the location of highest electric field [61]. However, the electron mobility is typically determined by the temperature and scattering mechanisms near the source of the channel in such quasi-ballistic devices [65], and does not appear to be strongly influenced by the additional LO population. Nevertheless, device reliability may ultimately be affected by such hot phonons, as suggested by a recent experimental study [66].

THERMAL TRANSPORT IN NANOSCALE DEVICES

Steady-state Thermal Resistance

While the previous sections focused on power dissipation in nanoscale devices and circuits, this section and the subsequent ones shift our attention to heat transport in and thermal spreading from nanoscale devices and materials. At the simplest level, heat dissipation from a lumped electronic device can be quantified by measuring its thermal resistance (R_{TH}) to the environment. This approach yields an average temperature rise of the device as:

$$\Delta T = PR_{\mathrm{TH}} \tag{8}$$

where P is the power (heat) dissipated in Watts. Note this is extremely similar to the electrical Ohm's law ($\Delta V = IR$), with temperature taking the role of voltage and power having the role of electric current flow. For instance, the thermal resistance to heat flow across a layer of thickness δ and cross-sectional area A is written analogously to its electrical resistance, as $R_{TH} = \delta/(kA)$, where k is the material thermal conductivity. As with the electrical conductance (resistance) quantum, there also exists an equivalent ballistic quantum of thermal conductance, which is directly proportional to absolute temperature, $G_0 = \pi^2 k_B^2 T\left(3h\right)$ or approximately 0.28 nW/K near 300 K [67, 68]. By comparison, the thermal conductance of an individual single-walled carbon nanotube at room temperature has been measured to be 2.4 nW/K [69].

In practice, nanoscale and semiconductor devices take more varied shapes (Fig. 5), have many modes available for thermal transport, and interfaces or three-dimensional heat spreading which make their analysis more difficult. The thermal resistance of many semiconductor devices has been measured through noise thermometry [70], gate electrode electrical resistance thermometry [71, 72], pulsed voltage measurements [13, 73], or an AC conductance method [14, 74–76]. Figure 5(a) presents a summary of such experimental data produced in the literature over the past nearly two decades, covering a wide range of device dimensions and technologies. A clear trend emerges, showing that device thermal resistance increases as a power law with reduced device dimensions, and is reaching values well above 10^5 K/W for device dimensions below 100 nm. Carbon nanotube (CNT) devices present a particularly interesting case, showing relatively high thermal resistance. This is due partly because they are usually measured on a relatively thick SiO_2 layer (50–300 nm), partly because of their much smaller "footprint" area (CNT diameter d ≪ any other typical device widths W), and partly to the thermal resistance at the CNT–SiO_2 interface [30, 77–79]. In other words, it is clear that the geometry, thermal conductivity of materials, and the thermal resistance of material interfaces play a significant role in determining the thermal impedance of a nanoscale device.

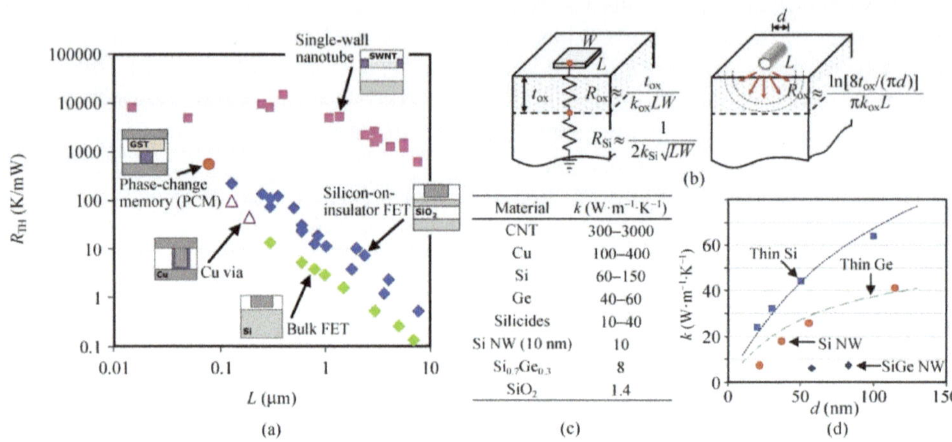

Material	k (W·m^{-1}·K^{-1})
CNT	300–3000
Cu	100–400
Si	60–150
Ge	40–60
Silicides	10–40
Si NW (10 nm)	10
$Si_{0.7}Ge_{0.3}$	8
SiO_2	1.4

(c)

Figure 5 (a) Measured thermal resistance of individual devices [13, 14, 30, 70–78, 161].. Thermal resistance generally scales inversely with device size L and substrate thermal conductivity k. (b) Simple themal models for rectangular (assuming W ≫ t_{ox}) and cylindrical devices (assuming $d \ll t_{ox}$) [30, 71, 72]. (c) Thermal conductivities at room temperature. Approximate ranges are given by grain boundary or impurity scattering, and by defects or wall number for CNT [69, 103, 162]. (d) Thermal conductivity data along thin films and nanowires (NW), showing strong decrease with nanowire diameter or film thickness due to phonon-boundary scattering [127, 132]. Dashed lines are simple models for Si and Ge [163].

To understand the scaling displayed in Fig. 5(a), consider the simplest case of a circular device with diameter D on the surface of a silicon wafer (of thickness ≫ D). Its thermal resistance will scale as $R_{TH} = 1/(2k_{Si}D)$ as that of the spreading electrical resistance from a single circular contact [80]. This can be extended to a rectangular heat source (width W, length L) by replacing $D \approx (LW)^{1/2}$ and including additional three-dimensional heat spreading shape factors [71, 81]. This case is illustrated in Fig. 5(b), along with that of a cylindrical device (e.g., nanotube, nanowire). Several online tools are also available for quick, web-based spreading resistance calculations for various shapes and substrates [82]. Many other models of varying sophistication have been published, all of which reveal various inverse length and width dependencies [71, 72, 83, 84]. Naturally, the choice of such a model in practice depends on its complexity, and on the specific geometry of the device. Care must also be taken with the limiting cases $W \gg t_{ox}$ (onedimensional heat flow through the underlying oxide) as opposed to d or $W \ll t_{ox}$ (three-dimensional spreading into the oxide), the latter being the case for all nanotube, nanowire, and even graphene nanoribbon data [85] typically available. Finally, the thermal boundary resistance (TBR) at the interfaces between the devices and its environment can also be a limiting factor, as is the case with nanotubes [30, 86], phase-change memory devices [87, 88], and to some extent graphene transistors [89, 90]. The TBR in the latter case is approximately equivalent to the thermal resistance of 20–50 nm SiO_2 at room temperature, and makes relatively little contribution for the commonly used graphene samples on 300 nm SiO_2. However, to estimate heat dissipation from graphene with thinner dielectrics, the TBR must be included in series with the additional material thermal resistances in Fig. 5(b), at the interface between the device and its adjacent dielectric. More discussion on the TBR is provided in Section 7.

Transient Heat Conduction

The thermal resistance models mentioned above are sufficient for evaluating the steady-state behavior of semiconductor devices, i.e., relevant during I–V characterization, analog operation, or to estimate the average temperature rise owed to device leakage. However, an understanding of transient heat conduction is necessary for short duration pulsed operation, such as during digital switching [9] or electrostatic discharge (ESD) events [91].

To first order, the temperature rise of a pulseheated volume can be obtained from the energy of the heating pulse (E) and the heat capacity (C) of the volume (V) being heated:

$$\Delta T\left(t_{\mathrm{p}}\right) \approx \frac{E}{CV} = \frac{Pt_{\mathrm{p}}}{CV} \tag{9}$$

where t_p is the pulse duration and P is its power. For instance, during digital operation the duration of an inverter switching event is approximately $t_p \approx$ 50–100 ps (or shorter), which is significantly less than the thermal time constant of most modern devices, $\tau \approx$ 10–100 ns [92]. This is near the "adiabatic limit", where the device can be thermally decoupled from its environment. In other words, while the device is ON there is little spread of the heated volume outside the area where the actual heating takes place (here, the channel and drain of the transistor).

The approximation above holds only for heating pulses that are short enough not to cause any significant heating outside the device volume. For a pulse of duration t_p the temperature spread extends approximately $(\pi\alpha t_p)^{1/2}$ outside the directly heated device volume [93], where $\alpha = k/C$ is the heat diffusion coefficient and k is the thermal conductivity (Fig. 5(c)). For heating pulses of t_p = 50 ps this distance is of the order ~100 nm into silicon or germanium, and ~10 nm into adjacent SiO_2 layers (like the top passivation layer, or the buried oxide below silicon-on-insulator, i.e., SOI technology). For longer heating pulses $t_p \gg$ 50 ps or device dimensions D < 100 nm, we must take into account both the energy stored as temperature rise within the device and that in the surrounding dielectric (the heated volume of which expands as $\sim t_p^{1/2}$). In this case we can estimate [15, 94]:

$$\Delta T\left(t_{\mathrm{p}}\right) \approx \frac{P}{CV}\left(2\sqrt{\tau t_{\mathrm{p}}} - \tau\right) \tag{10}$$

where $t_p > \tau$ and $\tau = D^2/(4\pi\alpha)$ is the time constant associated with the minimum device dimension D. The expression above assumes the same heat capacity for both the device and its surrounding dielectric, which is a good approximation for Si and SiO2 near room temperature ($C_{Si} \approx C_{ox} \approx 1.7\ \mathrm{J{\cdot}cm^{-3}{\cdot}K^{-1}}$). Typical estimates of the temperature rise during digital switching have shown that this dynamic value does not exceed a few degrees (e.g., 5 K) for sub-micron device technologies [14, 92]. However, the exact value for devices in the 10 nm range is unknown, and will be highly dependent on the ultimate choice of device geometry, materials, and interfaces.

For longer time scales, comparable to or larger than the device thermal time constants, several models have been proposed [81, 84, 88, 94, 95]. These bridge the time range from the adiabatic limit to the steadystate operation of a device, and are typically based on a Green's functions solution of the heat diffusion equation. This approach is faster and offers more physical insight than solutions based on finite-element (FE) solvers. The disadvantage of such methods vs. the FE approach is their applicability to only a limited range of geometries, like a heated sphere, infinite cylinder or rectangular parallelepiped, and not taking into account the full geometry and diverse materials making up a modern semiconductor device.

Finally, for time scales or power pulses longer than the thermal diffusion time through the silicon wafer backside (~0.3 ms assuming a 500 μm thick wafer) the problem returns to a simpler one of steady-state temperature rise. This was the case described in Section 4.1, and the temperature rise can be again obtained simply as a product of the (average) power input and that of the overall thermal resistance, $\Delta T = PR_{TH}$.

ENERGY TRANSPORT IN NANOTUBES, GRAPHENE, AND NANOWIRES

Carbon Nanotubes and Graphene

Carbon nanotubes (CNTs) and more recently graphene have emerged as new materials for electronics [96–98]. Graphene is a monolayer of graphite, a single sheet of hexagonally-arranged carbon atoms (Fig. 6(a)), first isolated five years ago [99]. Carbon nanotubes are cylindrical molecules with diameters ~1–3 nm, essentially "rolled-up" graphene obtained from chemical reactions [100]. Both materials have generated considerable interest because they demonstrate high mobility, thermal conductivity, and (particularly for graphene) the potential for integration with planar CMOS. Their excellent electrical and thermal properties make them especially attractive for energy efficient nanoelectronics. From this point of view, compared to traditional silicon, devices based on carbon nanomaterials benefit from:

1. Symmetric energy bands (Fig. 6(b)) and equal electron/hole mobility, both 10–100 times higher than silicon [101, 102], indicating less scattering and lower power dissipation.
2. Strong sp^2 bonds leading to thermal conductivity ~10–20 times higher than silicon, 5–10 times higher than copper, and comparable to that of isotopically pure diamond [69, 103, 104].
3. Optical phonon (OP) energies three times higher than in silicon (~180 meV vs. ~60 meV, Fig. 6(c)), suggesting they are less occupied and less likely to scatter electrons at low fields and near room temperature [30].

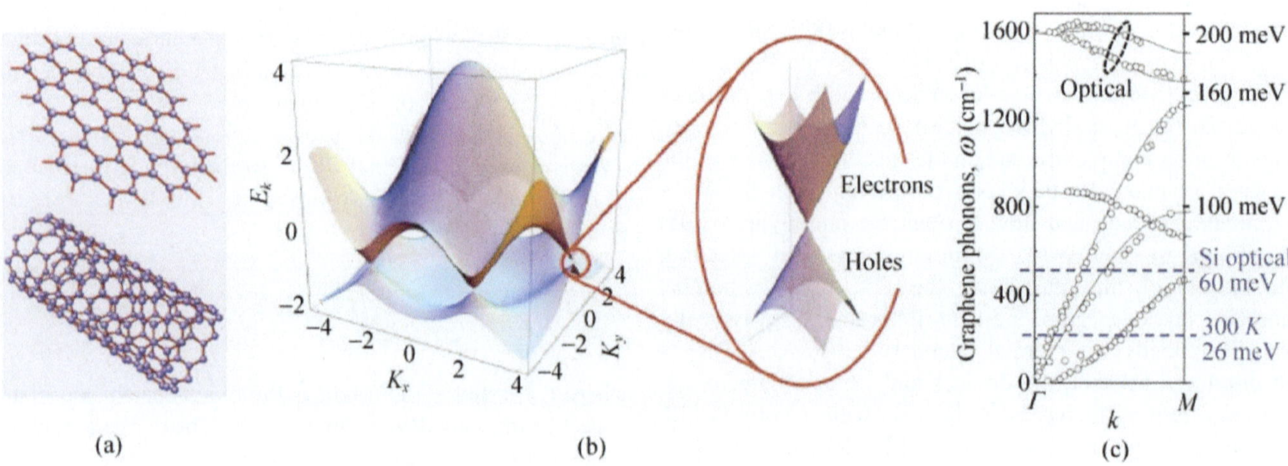

Figure 6. (a) Lattice structure of graphene (top) and carbon nanotubes (bottom) [96]. (b) Energy dispersion in graphene. Right: energy bands close to the Dirac point show linear dispersion. Reproduced with permission from Ref. [164]. Copyright 2009 American Physical Society. (c) Phonon dispersion in graphene, showing optical phonon energy much greater than that in silicon, and much above the room temperature energy (k_BT ~ 26 meV). Modified and reproduced with permission from Ref. [165]. Copyright 2005 John Wiley and Sons.

Energy dissipation in CNTs and graphene occurs when electric fields accelerate charge carriers (electrons or holes), which then scatter with lattice phonons. It has been generally assumed that the strongest highfield relaxation mechanism is that with OPs, which have high energy $\hbar\omega_{OP} \approx$ 160–200 meV [105–109]. For instance, heating and scattering with hot OPs is responsible for the negative differential conductance observed in suspended CNTs [36, 110–112], as shown in Fig. 7(d). For substrate-supported CNTs, theoretical work has suggested that energy dissipation could also occur directly with "remote" polar surfaceoptical (SO) phonon modes in the substrate, bypassing heating of the CNT [113, 114] (Fig. 7(b)). For example, the lowest SO phonons of several substrates have energies of approximately $\hbar\omega_{SO} \approx$ 48 meV for Al_2O_3, 56 meV for SiO_2, and 81 meV for AlN [55]. These are significantly lower than the CNT optical phonons, and may allow the device to directly dissipate part of its energy into the nearby insulator. A similar result has also been predicted for graphene on insulators like SiO_2 or HfO_2 [115], and is well known in silicon inversion layers with high-κ dielectrics [55, 116]. However, evidence of direct energy dissipation with substrate phonons is presently lacking, perhaps in large part due to the expected exponential dependence of electron-SO coupling on the CNT-substrate distance [114]. Thus, surface roughness, which for CNTs and graphene can be of the order of the device "thickness," or adsorbed water and impurities could significantly mask the electron-SO coupling. Nevertheless, a few studies do reveal phonon non-equilibrium in CNTs [86] and a reduction of mobility in graphene [102, 117] which have been attributed to scattering with substrate SO phonons. Interestingly, although not yet confirmed, a trade-off appears to exist between the SO scattering effect on mobility and power dissipation. Soft, low-energy SO phonons limit mobility, but at the same time enhance direct CNT cooling with the substrate.

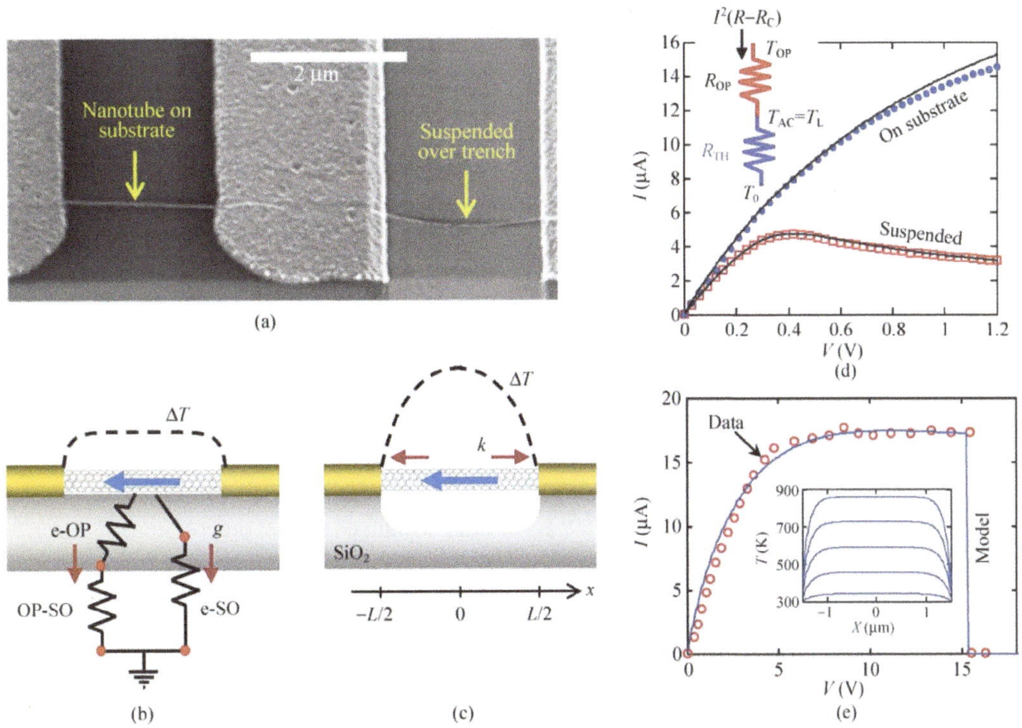

Figure 7. Energy dissipation in substrate-supported and freely suspended metallic CNTs. (a) SEM image of a single CNT that is part suspended and part supported. Reproduced with permission from Ref. [36]. Copyright 2005 American Physical Society. Schematic energy dissipation in (b) supported and (c) suspended CNTs. Red arrows show direction of heat flow. In the supported CNT, electrons scatter with device OPs, which then couple with the substrate. Electrons may also dissipate energy directly to SO phonons of the substrate [114]. (d) Measured data (symbols) and model (lines) of the CNT segments in (a). (e) Excellent agreement between data and model up to breakdown for another substrate-supported CNT (L = 3 μm). Inset shows the temperature profile along the CNT at applied voltages from 3 to 15 V, in 3 V increments. Reproduced with permission from Ref. [30]. Copyright 2007 American Institute of Physics.

More can be understood about energy dissipation and transport in one-dimensional CNT and graphene devices by measuring their temperature during Joule heating by electrical current flow. Assuming a uniform heating rate per unit length $P/L \approx I^2(R-R_C)/L$, where R is the total electrical resistance and R_C is the combined electrical resistance of the two contacts, the temperature profile T(x) of a substrate-supported metallic CNT (Fig. 7(b)) can be analytically obtained as [30, 79]:

$$T(x) = T_0 + \frac{P}{gL}\left[1 - \frac{\cosh(x/L_H)}{\cosh(L/2L_H)}\right] \text{ for } -L/2 < x < L/2 \quad (11)$$

where T_0 is the temperature of the contacts at the two ends, $L_H = (kA/g)^{1/2} \approx 0.2$ μm is the characteristic thermal length along the CNT [118], k is the thermal conductivity, and g is the net heat dissipated to the substrate per unit length. In practice, a direct measurement of this temperature profile is very challenging due to the small CNT diameter. However, a scanning thermal microscopy technique (SThM) was recently demonstrated [79], showing agreement with the "flattened" temperature profile sketched in Fig. 7(b) and obtained when the CNT length is much longer than the characteristic thermal length ($L \gg L_H$). The drawbacks of SThM on CNTs are the relatively large uncertainties in its calibration and in the temperature values obtained.

Another measurement technique that can be applied to suspended CNTs uses the temperature dependence of the Raman-active phonon G-bands [86, 119, 120]. The temperature profile of a freely suspended metallic CNT (Fig. 7(c)) is a simple inverted parabola expressed as [36]

$$T(x) = T_0 + \frac{P}{2kAL}\left[\left(\frac{L}{2}\right)^2 - x^2\right] \quad (12)$$

where again $-L/2 < x < L/2$ and T_0 is the temperature of the contacts at the two ends, and the power dissipation is assumed uniform along the CNT length. Analytic expressions for this temperature profile can also be obtained for a few cases of temperaturedependent thermal conductivity k [121]. The Ramanthermometry technique has directly measured this steeply varying temperature profile in suspended CNTs [119], and has confirmed the phonon non-equilibrium between optical and acoustic modes previously suggested on the basis of electrical characteristics alone (Fig. 7(d)) [36]. In the same study [119], the authors also provided direct measurement of uneven thermal contact resistance between CNTs and their metal electrodes, with $R_{TH,L} \approx 8 \times 10^7$ K/W (left electrode) and $R_{TH,R} \approx 10^7$ K/W (right electrode) for the particular CNT under study. These are consistent with other estimates based on SThM (1.4×10^7 –13.8×10^7 K/W) [79] and breakdown thermometry (~1.2×10^7 K/W) [78], but provide insight into the uneven quality of the two thermal contacts. The drawbacks of the Raman thermometry approach are its applicability to only those few CNTs resonant with the laser energy, and the relatively larger lateral resolution (~0.5 μm) compared to the SThM approach.

A third approach for substrate-supported CNTs can be referred to as "breakdown thermometry" and relies on measuring the CNT current at high voltage, including the Joule breakdown power. The voltage is gradually raised until the CNT breaks (Fig. 7(e)), while its I–V curve and breakdown power are compared to a transport model which includes heat dissipation. This was the approach adopted by Maune [77] and Pop [30, 78]. The breakdown temperature of CNTs is known to be approximately $T_{BD} \approx 600°C$ in air [122] and correlating it with the dissipated power yields the thermal coupling between CNT and substrate. This quantity is often given as a thermal conductance per CNT length, g in $W{\cdot}m^{-1}{\cdot}K^{-1}$ (Eq. (11) above). The three approaches have yielded somewhat different values of CNT–SiO2 thermal coupling, with SThM suggesting g ≈ 0.007–0.06 $W{\cdot}m^{-1}{\cdot}K^{-1}$ [79], Raman thermometry yielding 0.03–0.11 $W{\cdot}m^{-1}{\cdot}K^{-1}$ [86], and breakdown studies giving 0.1–0.2 $W{\cdot}m^{-1}{\cdot}K^{-1}$ [77, 78]. This discrepancy is not yet resolved, because the electronic contribution to thermal coupling, if any, has not yet been isolated (Fig. 7(b)), and the various measurements were made on CNTs with different diameters, substrate surface roughnesses, and ambient temperatures. For instance, breakdown thermometry yields the thermal coupling at the elevated breakdown temperatures, closer to 600 ° C (although the average CNT temperature is lower). Moreover, it is relevant to point out that such nanoscale thermal coupling will be strongly dependent on the CNT diameter and substrate surface roughness of the measured samples, as recently revealed by molecular dynamics (MD) simulations [123]. The dissipation will also depend on the substrate material itself (e.g., quartz vs. SiO_2) as the phonon dispersion, surface optical phonons, and interface bonding between CNTs and substrate are expected to change.

Nanowires

Unlike carbon nanotubes, nanowires are not truly onedimensional structures. Rather, they can be thought of as shrunken three-dimensional rods which exhibit one-dimensional features (e.g., energy sub-bands) when their diameters are brought to sizes comparable to the electron or phonon wavelength (~5–10 nm). In addition, nanowire surfaces are typically very good electron and phonon scatterers, such that nanowire mobility and thermal conductivity are universally lower than the bulk mobility or thermal conductivity of the same material (Figs. 5(c), 5(d), and Fig. 8). For instance, the electrical resistivity of Cu interconnects with line widths below 100 nm increases by more than a factor of two from the bulk Cu resistivity [124], and the resistivity for line widths near 40 nm was shown to increase by at least a factor of four [125, 126] at room temperature.

The thermal conductivity of nanowires is similarly influenced by their surface. The thermal conductivity of bulk crystalline silicon is nearly 150 $W{\cdot}m^{-1}{\cdot}K^{-1}$ at room temperature, which is reduced by one order of magnitude in nanowires with diameter below $D \approx 35$ nm (Fig. 5(d) and Fig. 8(a)) [127]. In simple terms this can be understood if we express the thermal conductivity as

$$k = \frac{1}{3}Cv\lambda \approx \frac{1}{3}\int_i C_\omega v_\omega \left(\frac{A_1}{D} + \frac{A_2}{\lambda_0} + A_3 N_i \right)^{-1} d\omega$$

(13)

where C_ω is the heat capacity per unit frequency mode, v_ω is the mode velocity, λ_0 is the phonon mean free path (MFP) in the bulk material, N_i is the impurity concentration, and A_1–A_3 are fitting parameters [128– 131]. The term in parenthesis represents the various contributions to the phonon MFP, including classical boundary scattering which becomes the limiting term when the diameter D falls below the range of the bulk phonon MFP, $\lambda_0 \approx$ 0.1–1 μm [54]. Variations on the simple expression above have generally been successful in reproducing experimental data on thermal conductivity of silicon nanowires [127] and thin films [132].

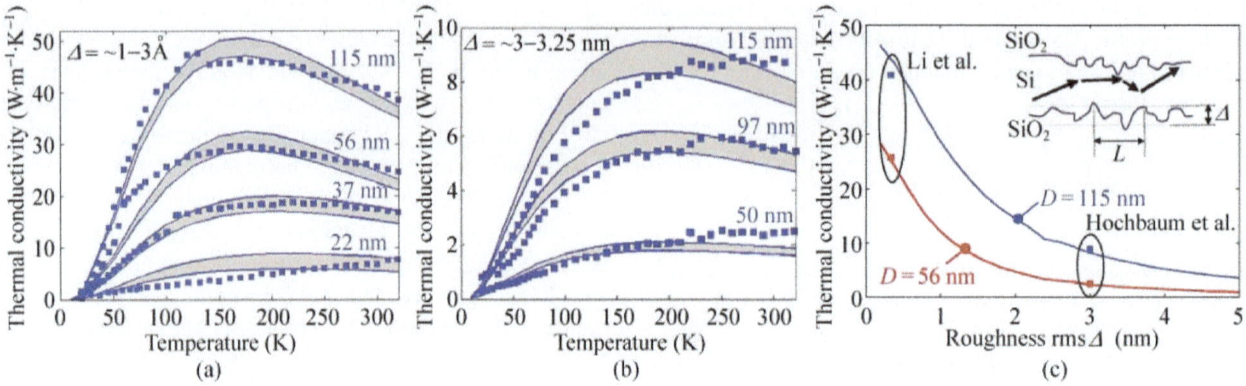

Figure 8. Measured (symbols) and modeled (lines) thermal conductivity of silicon nanowires (NWs) with various diameters D. (a) Thermal conductivity of smooth-surface NWs. Shaded areas correspond to surface roughness parameter Δ = 1–3 Å. (b) Thermal conductivity of rough-surface NWs. Shaded areas correspond to Δ = 3.00–3.25 nm. (c) Strong dependence of thermal conductivity on surface roughness Δ at room temperature. Inset shows a schematic of the nanowire cross-section. Experimental data from Ref. [127, 133], modeling described in Ref. [131]. Reproduced with permission. Copyright 2009 American Physical Society..

A new twist on our understanding of thermal transport in surface-limited nanowires was presented by the recent results of Chen, Hochbaum [133, 134], and Boukai [135], who found that silicon nanowires of diameter D < 50 nm with very rough surfaces have thermal conductivity one hundred times lower than that of bulk crystalline silicon (Fig. 8(b)). This approaches the thermal conductivity of amorphous silicon, and cannot be understood on the basis of classical boundary scattering alone (the 1/D term in Eq. (13)). To explain it, Martin et al. [131] proposed an additional scattering term which relies on a description of the surface through a roughness height Δ and autocovariance length L (Fig. 8(c)). This accounts for the fact that phonons "see" the rough nanowire as a series of constrictions along their propagation direction. Unlike classical boundary scattering, the roughness scattering rate is strongly frequency dependent ($\sim\omega^2$), with little impact on long-wavelength phonons. Using a full phonon dispersion for silicon, Martin et al. [131] found an approximately quadratic dependence of nanowire thermal conductivity on diameter and roughness as $(D/\Delta)^2$. While the exact surface roughness in the experiments is difficult to measure, it appears that the enhanced model does predict a very strong role for it, particularly for smaller diameter nanowires (Figs. 8(b) and 8(c)). If the electrical conductivity can be preserved, arrays of such nanostructures could form efficient thermoelectric devices, a domain of ongoing investigations.

THERMAL RECTIFICATION

In addition to impeding or enhancing electrical and thermal transport, nanoscale structuring could also be used to introduce thermal flow asymmetry. This is the thermal equivalent of the electrical p–n diode, a two-terminal device that achieves greater heat flux in one direction than another, e.g., $Q_{BA} > Q_{AB}$ for the same temperature difference ΔT, where A and B are its two terminals (Fig. 9) [136]. Thus, we can define the rectification ratio $\gamma = Q_{BA}/Q_{AB} > 1$ which will be referred to below. In the context of electronics and systems, thermal rectifiers could enable thermal flow to be processed independently from electron current, or suggest the possibility of phononic devices and thermal logic [137]. In the context of heat engines or buildings, such thermal rectifiers could act as energy harvesting materials, exploiting naturally occurring temperature gradients with the environment.

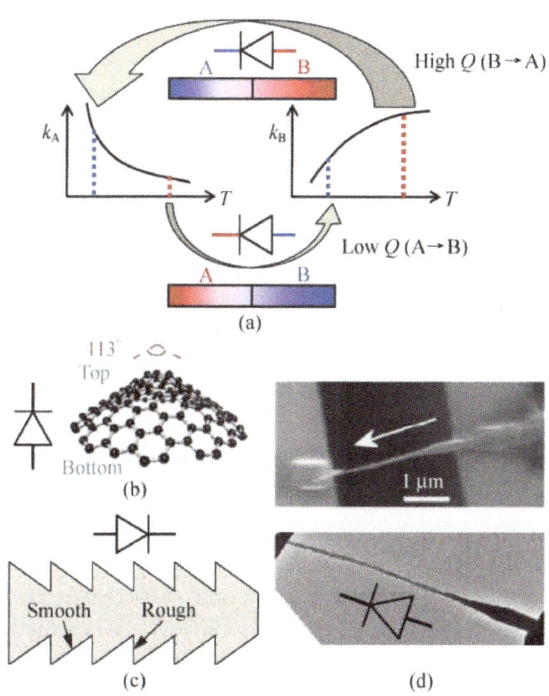

Figure 9. Thermal rectification concepts, the thermal equivalent of a p–n diode. (a) Asymmetric heat flow can be achieved between two bulk materials with different temperature dependence of thermal conductivity [138, 141]. Some proposed theoretical concepts rely on phonon filtering or bottlenecks induced by asymmetric geometry, e.g., (b) carbon nanocone. (Reproduced with permission from Ref. [142]. Copyright 2008 American Physical Society.) or (c) sawtooth nanowire [144]. (d) In practice, the only nanostructured thermal rectifier has been achieved with asymmetrically mass-loaded carbon (top) and boron nitride (bottom) nanotubes. The diode symbol illustrates the rectification direction. Reproduced with permission from Ref. [145]. Copyright 2006 AAAS..

A certain amount of thermal rectification can be achieved between two bulk materials with strongly different thermal conductivity dependence on temperature [138]. This is in fact a classical Fourier law effect, first observed in the 1970s [139, 140] and recently reexamined by Dames [138] and Kobayashi et al. [141]. As shown in Fig. 9(a), the two sides of the junction require two materials with strongly decreasing (A) and increasing (B) temperature dependence of thermal conductivity. When B is the "hot" and A is the "cold" terminal ($\Delta T = T_B - T_A$), both materials are in a regime of high thermal conductivity, and heat flows more easily from B to A (top, wider arrow in Fig. 9(a)). When the terminals are reversed ($\Delta T = T_A - T_B$), both materials are in a regime of lower thermal conductivity, and heat flow is reduced between A and B (bottom, thinner arrow). This leads to an asymmetric heat flux, $Q_{BA} > Q_{AB}$ for the same temperature difference ΔT. Such thermal rectification was recently observed by Kobayashi et al. [141], who noted a heat flux ratio $\gamma \approx 1.43$ between $LaCoO_3$ and $La_{0.7}Sr_{0.3}CoO_3$ in the temperature range 70–100 K.

Suggestions of nanostructures for thermal rectification have only appeared in the past few years. Several theoretical proposals [142, 143] have been put forward, estimating thermal rectification as high as $\gamma \approx 100$–1000 between two directions. Thermal rectification could be achieved through nanoscale structuring, e.g., inducing nano-indentations that preferentially scatter phonons [144], through geometrical heat "funneling", e.g., in the case of a carbon nanocone [142], or by asymmetric mass-loading [145], as illustrated in Figs. 9(b)–9(d). Most suggestions based on nanoscale thermal rectification have been purely theoretical, requiring engineering of coupling and layout between individual atoms, which are very difficult to achieve in practice. The only nanoscale thermal rectifier demonstrated to date at room temperature [145] has shown thermal rectification ratios $\gamma \approx 1.02$–1.07. This device relied on suspended, several microns-long multiwalled carbon or boron nitride nanotubes that were unevenly mass-loaded with a much heavier molecular species, trimethyl(methylcylopentadienyl) platinum ($C_9H_{16}Pt$), (Fig. 9(d)). In this case, heat flow preference was from the high- to the low-mass regions of the loaded nanotube. Thermal rectification at very low-temperature (0.08 K) and high magnetic field (10 T) has also been demonstrated in single quantum dots, with heat flow asymmetry $\gamma \approx 1.11$ [146]. Clearly, much work remains to be done to achieve a practical, room-temperature thermal diode with rectification up to theoretically-predicted ($\gamma \approx 100$–1000) limits.

INTERFACES

Nanoscale one- and two-dimensional devices have very little "bulk" and thus their behavior is strongly dominated by their interfaces in terms of energy transport and dissipation. This is particularly evident for thermal conductivity in nanowires, or thermal dissipation in nanotubes, as outlined above. While the experimental data and theoretical understanding of nanoscale device interfaces are just beginning to take shape, a wealth of experimental data exists for energy transport across interfaces between otherwise bulk solids. This is summarized in Fig. 10. The interface thermal conductance relates the heat flux Q crossing an interface to the temperature drop ΔT at the interface, $Q = G\Delta T$. Thermal interfaces are most often studied with pump-probe optical techniques such as time-domain thermoreflectance (TDTR), picosecond transient absorption [147–152], or the 3-omega method [153–157]. As summarized in Fig. 10, these studies have shown that the thermal conductance of bulk solid interfaces at room temperature spans a relatively limited range, but depends on composition of the interface at the level of a single molecular layer [158].

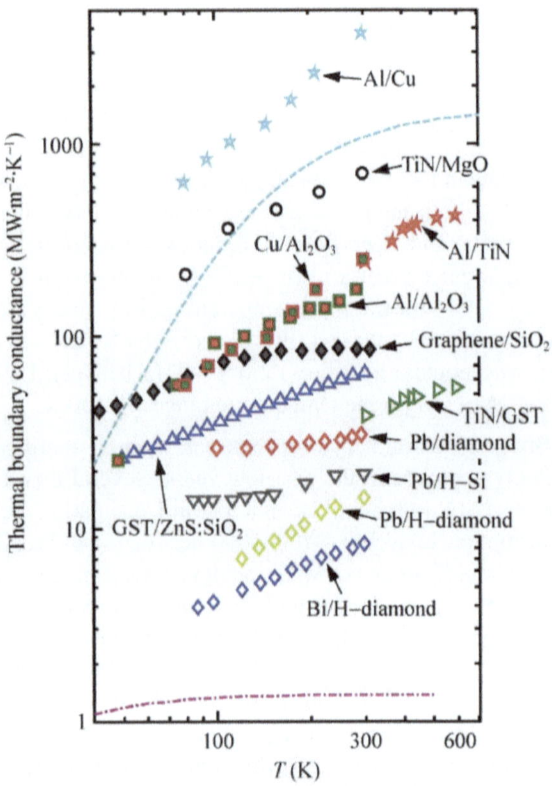

Figure 10. Thermal conductance of various material interfaces, compiled from Refs. [90, 148, 149, 157, 159, 166, 167] [GST = $Ge_2Sb_2Te_5$]. The dashed upper line is a diffuse-mismatch model calculation of the TiN/MgO interface [148], the dashed lower line is the radiation limit for the Bi/H–diamond interface [149]. The highest known conductance is at good metal interfaces (Al/Cu) where electrons dominate the thermal exchange [159]. The lowest known conductance is between materials with highly mismatched phonon modes and Debye temperatures (Bi/H–diamond) [149]. Compared to other properties of materials, the thermal conductance of solid material interfaces spans a relatively modest range (~500×), and most interfaces fall in the range 10–100 MW·m^{-2}·K^{-1} at room temperature.

The range of available thermal conductances at material interfaces appears to approach two limits. At the upper end, the best measured thermal interfaces are between metals with good intrinsic thermal properties, e.g., Al/Cu. Thermal transport across these interfaces appears to be mediated by electrons, and the phonon contribution is less than 10% [159]. The best metal-dielectric thermal interfaces occur between materials with similar Debye temperatures Θ_D (such as highly annealed, epitaxial TiN on single crystal oxides [148]) and approach the limit of a phonon diffuse mismatch model (DMM) [160]. At the low end, a large Debye temperature mismatch between the two materials is expected to lead to significant thermal impedance, such as between Pb or Bi ($\Theta_D \approx 110$ K) and diamond ($\Theta_D \approx 2200$ K) [149]. The Debye temperature is a typical figure of merit for the differences in vibrational spectrum and the phonon density of states between the two sides of the interface. With regard to Fig. 10, it is interesting to note the physical meaning of the thermal interface conductance by relating it to the equivalent thermal impedance of a dielectric layer with thickness d and thermal conductivity k, i.e., $G = k/d$. Thus, an interface with $G = 10$ MW·m^{-2}·K^{-1} is equivalent to the thermal impedance of a 140 nm film of SiO_2 ($k_{ox} = 1.4$ W·m^{-1}·K^{-1}) and $G = 100$ MW·m^{-2}·K^{-1} is equivalent to the thermal impedance of 14 nm of SiO_2. This simple comparison also highlights the importance of such material interfaces in all nanometer scale devices and structures.

Among one- or two-dimensional conductors, some of the available data on interface thermal resistance for carbon nanotubes were already presented in Section 5.1, and data for the graphene/SiO_2 thermal interface are shown in Fig. 10. In particular, given the comparable Debye temperature of CNTs and graphene with that of diamond, one expects a significant vibrational mismatch between them and other materials with typically lower Θ_D. To compare the data available for CNTs with that of other material interfaces, the range of thermal conductance per unit length $g \approx 0.007$–0.2 W·m^{-1}·K^{-1} [77–79, 86] can be converted to $G \approx 4$–100 MW·m^{-2}·K^{-1} per unit area when normalized by a 2 nm CNT diameter, well within the range of the experimental data in Fig. 10. In particular, given the large intrinsic thermal conductivity of CNTs it appears that their thermal resistance is nearly always dominated by their interface with the environment, both for solid [151] as well as liquid composites [150]. Nevertheless, little is yet known about how to engineer one-dimensional interfaces between dissimilar materials such as CNTs and their dielectric environments, and much work remains to be done in this area.

CONCLUSIONS

Understanding and controlling energy dissipation and transport in nanostructures will continue to be an area of rapid improvement and discovery, with applications ranging from low-power electronics to energy-efficient data centers, and thermal energy harvesting. For instance, reducing power dissipation in electronics will impact a wide range of applications from portable devices (10^{-3} W) to massive data centers (10^9 W). Such developments would also improve weight payloads for mobile electronics and medical instrumentation (fewer heavy batteries), and potentially exploit ubiquitous thermal gradients—from sources as varied as car engines, power plants, or even body-environment temperature differences—as supplies of electrical energy.

On a large scale, by simple estimates even a transistor twice as energy-efficient could lower nationwide power use by over 10^{10} W if implemented today, which is a significant percentage of our national power budget. Given that energy use by electronics is on a trend that could reach 30% of the national electricity consumption by 2025, such progress is crucial to maintaining progress in a post-CMOS world, and has great environmental implications as well. In addition, just over half the world energy is wasted as heat (~10^{13} W), from nuclear power plants and factories to car engines and the power bricks on our laptops. Efficiently reclaiming even a small percentage of such wasted heat would itself nearly satisfy the electricity needs of our planet.

ACKNOWLEDGEMENTS

I am indebted to Profs. D. Cahill, H. Dai, C. Dames, D. Jena, W. King, J. -P. Leburton, J. Lyding, and U. Ravaioli for many valuable discussions. I also thank Dr. M. -H. Bae for feedback on an earlier manuscript draft. This review was in part supported by the Nanoelectronics Research Initiative (NRI), the DARPA Young Faculty Award (No. HR0011-08-1-0035), the Office of Naval Research (No. N00014-09-1-0180), the National Science Foundation (No. CCF 08-29907), Intel Corp., and Northrop Grumman.

REFERENCES

1. EPA Report on Server and Data Center Energy Efficiency [Online].http://www.energystar.gov/index.cfm?c=prod_development.server_efficiency_study(accessed sep 22, 2009)
2. Pop, E.; Sinha, S.; Goodson, K. E. Heat generation and transport in nanometer-scale transistors. Proc. IEEE 2006, 94, 1587–1601.
3. Cavin, R.; Zhirnov, V.; Herr, D.; Avila, A; Hutchby, J. Research directions and challenges in nanoelectronics. J. Nanopart. Res. 2006, 8, 841–858.
4. Haensch, W.; Nowak, E. J.; Dennard, R. H.; Solomon, P. M.; Bryant, A.; Dokumaci, O. H.; Kumar, A.; Wang, X.; Johnson, J. B.; Fischetti, M. V. Silicon CMOS devices beyond scaling.IBM J. Res. Dev. 2006, 50, 339–361.
5. Intel products [Online]. http://ark.intel.com (accessed sep 22, 2009)
6. PC Energy Report 2009 [Online]. http://www.climatesavers-computing.org (accessed sep 22, 2009)
7. The Carbon Footprint of Email Spam Report [Online]. http://www.mcafee.com (accessed sep 22, 2009)
8. Energy & Environment Department at LBNL [Online]. http://eed.llnl.gov/flow/02flow.php (accessed sep 22, 2009)
9. Roy, K.; Prasad, S. Low-power CMOS VLSI Circuit Design; Wiley-VCH: Weinheim 2000.
10. Hanson, S.; Zhai, B.; Bernstein, K.; Blaauw, D.; Bryant, A.; Chang, L.; Das, K. K.; Haensch, W.; Nowak, E. J.; Sylvester, D. M. Ultralow-voltage minimum-energy CMOS. IBM J. Res. Dev. 2006, 50, 469–490.
11. Zhai, B.; Blaauw, D.; Sylvester, D.; Flautner, K. The limit of dynamic voltage scaling and insomniac dynamic voltage scaling. IEEE T. VLSI Syst. 2005, 13, 1239–1252.
12. Akinwande, D.; Liang, J.; Chong, S.; Nishi, Y.; Wong, H. S. P. Analytical ballistic theory of carbon nanotube transistors: Experimental validation, device physics, parameter extraction, and performance projection. J. Appl. Phys. 2008, 104, 124514.
13. Jenkins, K. A.; Rim, K. Measurement of the effect of self-heating in strained-silicon MOSFETs. IEEE Electr. Device L. 2002, 23, 360–362.
14. Tenbroek, B.; Lee, M. S. L.; Redman-White, W.; Bunyan, R. J. T.; Uren, M. J. Self-heating effects in SOI MOSFETs and their measurement by small signal conductance techniques. IEEE Trans. Electron Dev. 1996, 43, 2240–2248.
15. Clemente, S. Transient thermal response of power semiconductors to short power pulses. IEEE Trans. Power Electr. 1993, 8, 337–341.
16. Yan, H.; Song, D.; Mak, K. F.; Chatzakis, I.; Maultzsch, J.; Heinz, T. F. Time-resolved Raman spectroscopy of optical phonons in graphite: Phonon anharmonic coupling and anomalous stiffening. Phys. Rev. B 2009, 80, 121403.
17. Menéndez, J.; Cardona, M. Temperature dependence of the first-order Raman scattering by phonons in Si, Ge, and α-Sn: Anharmonic effects. Phys. Rev. B 1984, 29, 2051.
18. Letcher, J. J.; Kang, K.; Cahill, D. G.; Dlott, D. D. Effects of high carrier densities on phonon and carrier lifetimes in Si by time-resolved anti-Stokes Raman scattering. Appl. Phys. Lett. 2007, 90, 252104.
19. Sinha, S.; Schelling, P. K.; Phillpot, S. R.; Goodson, K. E. Scattering of g-process longitudinal optical phonons at hotspots in silicon. J. Appl. Phys. 2005, 97, 023702.
20. Ledgerwood, M. L.; van Driel, H. M. Picosecond phonon dynamics and self-energy effects in highly photoexcited germanium. Phys. Rev. B 1996, 54, 4926–4935.
21. Ioffe Institute [Online]. http://www.ioffe.rssi.ru/SVA/NSM/Semicond/index.html(accessed sep 22, 2009)
22. Koswatta, S. O.; Lundstrom, M. S.; Nikonov, D. E. Band-to-band tunneling in a carbon nanotube metal-oxide-semiconductor field-effect transistor is dominated by phonon-assisted tunneling. Nano Lett. 2007, 7, 1160–1164.
23. Buttiker, M. Role of quantum coherence in series resistors. Phys. Rev. B 1986, 33, 3020–3026.
24. Das, M. P.; Green, F. Landauer formula without Landauer's assumptions. J. Phys. -Condens. Mat. 2003, 15, 687–693.
25. Datta, S. Electronic Transport in Mesoscopic Systems; Cambridge University Press: Cambridge, 1995.
26. Datta, S. Quantum Transport: Atom to Transistor; Cambridge University Press: Cambridge, 2006.
27. Lake, R.; Datta, S. Energy balance and heat exchange in mesoscopic systems. Phys. Rev. B1992, 46, 4757–4763.
28. Ouyang, Y.; Guo, J. Heat dissipation in carbon nanotube transistors. Appl. Phys. Lett.2006, 89, 183122.
29. Park, J. Y.; Rosenblatt, S.; Yaish, Y.; Sazonova, V.; Ustunel, H.; Braig, S.; Arias, T. A.; Brouwer, P. W.; McEuen, P. L. Electron-phonon scattering in metallic single-walled carbon nanotubes. Nano Lett. 2004, 4, 517–520.
30. Pop, E.; Mann, D.; Goodson, K.; Dai, H. Electrical and thermal transport in metallic single-wall carbon nanotubes on insulating substrates. J. Appl. Phys. 2007, 101, 093710.

31. Liao, A.; Zhao, Y.; Pop, E. Avalanche-induced current enhancement in semiconducting carbon nanotubes. Phys. Rev. Lett. 2008, 101, 256804.

32. Chen, Y. -C.; Zwolak, M.; Di Ventra, M. Local heating in nanoscale conductors. Nano Lett.2003, 3, 1691–1694.

33. Segal, D.; Nitzan, A. Heating in current carrying molecular junctions. J. Chem. Phys.2002, 117, 3915–3927.

34. Galperin, M.; Saito, K.; Balatsky, A. V.; Nitzan, A. Cooling mechanisms in molecular conduction junctions. Phys. Rev. B 2009, 80, 115427.

35. Galperin, M.; Ratner, M. A.; Nitzan, A. Molecular transport junctions: Vibrational effects.J. Phys. -Condens. Mat. 2007, 103201.

36. Pop, E.; Mann, D.; Cao, J.; Wang, Q.; Goodson, K. E.; Dai, H. J. Negative differential conductance and hot phonons in suspended nanotube molecular wires. Phys. Rev. Lett.2005, 95, 155505.

37. D'Agosta, R.; Sai, N.; Di Ventra, M. Local electron heating in nanoscale conductors. Nano Lett. 2006, 6, 2935–2938.

38. Koswatta, S. O.; Lundstrom, M. S.; Nikonov, D. E. Influence of phonon scattering on the performance of p-i-n band-to-band tunneling transistors. Appl. Phys. Lett. 2008, 92, 043125.

39. Pop, E.; Rowlette, J.; Dutton, R. W.; Goodson, K. E. Joule heating under quasi-ballistic transport conditions in bulk and strained silicon devices. In Intl. Conf. on Simulation of Semic. Proc. and Dev. (SISPAD), Tokyo, Japan, 2005, pp. 307–310.

40. Stettler, M. A.; Alam, M. A.; Lundstrom, M. S. A critical examination of the assumptions underlying macroscopic transport equations for silicon devices. IEEE Trans. Electron Dev.1993, 40, 733–740.

41. Vashaee, D.; Shakouri, A. Improved thermoelectric power factor in metal-based superlattices. Phys. Rev. Lett. 2004, 92, 106103.

42. Mahan, G. D.; Woods, L. M. Multilayer thermionic refrigeration. Phys. Rev. Lett. 1998, 80, 4016–4019.

43. Pipe, K. P.; Ram, R. J.; Shakouri, A. Internal cooling in a semiconductor laser diode. IEEE Photonic Tech. L. 2002, 14, 453–455.

44. Shakouri, A.; Bowers, J. E. Heterostructure integrated thermionic coolers. Appl. Phys. Lett. 1997, 71, 1234–1236.

45. Shakouri, A.; Lee, E. Y.; Smith, D. L.; Narayanamurti, V.; Bowers, J. E. Thermoelectric effects in submicron heterostructure barriers. Microscale Therm. Eng. 1998, 2, 37–47.

46. Xu, X.; Gabor, N. M.; Alden, J. S.; van der Zande, A. M.; McEuen, P. L. Photo-thermoelectric effect at a graphene interface junction. Nano Lett. 2010, 10, 562–566

47. Wachutka, G. K. Rigorous thermodynamic treatment of heat generation and conduction in semiconductor device modeling. IEEE Trans. Comput. Aid. Des. 1990, 9, 1141–1149.

48. Lindefelt, U. Heat generation in semiconductor devices. J. Appl. Phys. 1994, 75, 942–957.

49. Sverdrup, P. G.; Ju, Y. S.; Goodson, K. E. Sub-continuum simulations of heat conduction in silicon-on-insulator transistors. J. Heat Transf. 2001, 123, 130–137.

50. Lai, J.; Majumdar, A. Concurrent thermal and electrical modeling of sub-micrometer silicon devices. J. Appl. Phys. 1996, 79, 7353–7361.

51. Pop, E.; Dutton, R. W.; Goodson, K. E. Analytic band Monte Carlo model for electron transport in Si including acoustic and optical phonon dispersion. J. Appl. Phys. 2004, 96, 4998–5005.

52. Ju, Y. S.; Goodson, K. E. Phonon scattering in silicon thin films with thickness of order 100 nm. Appl. Phys. Lett. 1999, 74, 3005–3007.

53. Mazumder, S.; Majumdar, A. Monte Carlo study of phonon transport in solid thin films including dispersion and polarization. J. Heat Trans. 2001, 123, 749–759.

54. Henry, A. S.; Chen, G. Spectral phonon transport properties of silicon based on molecular dynamics simulations and lattice dynamics. J. Comput. Theor. Nanos. 2008, 5, 141–152.

55. Fischetti, M. V.; Neumayer, D. A.; Cartier, E. A. Effective electron mobility in Si inversion layers in MOS systems with a high-κ insulator: The role of remote phonon scattering. J. Appl. Phys. 2001, 90, 4587–4608.

56. Artaki, M.; Price, P. J. Hot phonon effects in silicon field-effect transistors. J. Appl. Phys.1989, 65, 1317–1320.

57. Lugli, P.; Goodnick, S. M. Nonequilibrium longitudinal-optical phonon effects in GaAs-AlGaAs quantum wells. Phys. Rev. Lett. 1987, 59, 716–719.

58. Jacoboni, C.; Reggiani, L. The Monte Carlo method for the solution of charge transport in semiconductors with applications to covalent materials. Rev. Mod. Phys. 1983, 55, 645–705.

59. Fischetti, M. V.; Laux, S. E. Monte Carlo analysis of electron transport in small semiconductor devices including band-structure and space-charge effects. Phys. Rev. B1988, 38, 9721–9745.

60. Pop, E.; Dutton, R. W.; Goodson, K. E. Monte Carlo simulation of Joule heating in bulk and strained silicon. Appl. Phys. Lett. 2005, 86, 082101.

61. Rowlette, J. A; Goodson, K. E. Fully coupled nonequilibrium electron-phonon transport in nanometer-scale silicon FETs. IEEE Trans. Electron Dev. 2008, 55, 220–232.

62. Sinha, S.; Pop, E.; Dutton, R. W.; Goodson, K. E. Nonequilibrium phonon distributions in sub-100 nm silicon transistors. J. Heat Transf. 2006, 128, 638–647.

63. Raleva, K.; Vasileska, D.; Goodnick, S. M.; Nedjalkov, M. Modeling thermal effects in nanodevices. IEEE Trans. Electron Dev. 2008, 55, 1306–1316.

64. Vasileska, D.; Raleva, K.; Goodnick, S. M. Modeling heating effects in nanoscale devices: The present and the future. J. Comput. Electron. 2008, 7, 66–93.

65. Lundstrom, M.; Ren, Z. Essential physics of carrier transport in nanoscale MOSFETs. IEEE Trans. Electron Dev. 2002, 49, 133–141.

66. Wang, Y.; Cheung, P.; Oates, A.; Mason, P. Ballistic phonon enhanced NBTI. In IEEE International Reliability

Physics Symposium (IRPS), Phoenix, USA, 2007, pp. 258–263.

67. Rego, L. G. C.; Kirczenow, G. Quantized thermal conductance of dielectric quantum wires.Phys. Rev. Lett. 1998, 81, 232–235.

68. Schwab, K.; Henriksen, E. A.; Worlock, J. M.; Roukes, M. L. Measurement of the quantum of thermal conductance. Nature 2000, 2000, 974–977.

69. Pop, E.; Mann, D.; Wang, Q.; Goodson, K. E.; Dai, H. J. Thermal conductance of an individual single-wall carbon nanotube above room temperature. Nano Lett. 2006, 6, 96–100.

70. Bunyan, R. J. T.; Uren, M. J.; Alderman, J. C.; Eccleston, W. Use of noise thermometry to study the effects of selfheating in submicrometer SOI MOSFETs. IEEE Electr. Device L.1992, 13, 279–281.

71. Mautry, P. G.; Trager, J. Investigation of self-heating in VLSI and ULSI MOSFETs. In Intl. Conf. on Microelectronic Test Struct., San Diego, CA, USA, 1990, pp. 221–226.

72. Su, L. T.; Chung, J. E.; Antonadis, D. A.; Goodson, K. E.; Flik, M. I. Measurement and modeling of self-heating in SOI NMOSFETs. IEEE Trans. Electron Dev. 1994, 41, 69–75.

73. Jenkins, K. A.; Sun, J. Y. -C. Measurement of I-V curve of silicon-on-insulator (SOI) MOSFETs without self-heating. IEEE Electr. Device L. 1995, 16, 145–147.

74. Jin, W.; Liu, W.; Fung, S. K. H.; Chan, P. C. H.; Hu, C. SOI thermal impedance extraction methodology and its significance for circuit simulation. IEEE Trans. Electron Dev. 2001, 48, 730–736.

75. Lee, T. -Y.; Fox, R. M. Extraction of thermal resistance for fully-depleted SOI MOSFETs. InIEEE Int. SOI Conf., Nara, Japan, 1995, pp. 78–79.

76. Reyboz, M.; Daviot, R.; Rozeau, O.; Martin, P.; Paccaud, M. Compact modeling of the self heating effect in 120 nm multifinger body-contacted SOI MOSFETs for RF circuits. In IEEE Int. SOI Conf., 2004, pp. 159–161.

77. Maune, H.; Chiu, H. -Y.; Bockrath, M. Thermal resistance of the nanoscale constrictions between carbon nanotubes and solid substrates. Appl. Phys. Lett. 2006, 89, 013109.

78. Pop, E. The role of electrical and thermal contact resistance for Joule breakdown of single-wall carbon nanotubes. Nanotechnology 2008, 19, 295202.

79. Shi, L.; Zhou, J.; Kim, P.; Bachtold, A.; Majumdar, A.; McEuen, P. L. Thermal probing of energy dissipation in current-carrying carbon nanotubes. J. Appl. Phys. 2009, 105, 104306.

80. Yovanovich, M. M.; Culham, J. R.; Teertstra, P. Analytical modeling of spreading resistance in flux tubes, half spaces, and compound disks. IEEE Trans. Compon. Hybr.1998, 21, 168–176.

81. Joy, R. C.; Schlig, E. S. Thermal properties of very fast transistors. IEEE Trans. Electron Dev. 1970, 17, 586–594.

82. Micro Heat Transfer Lab (U. Waterloo) [Online].http://www.mhtl.uwaterloo.ca/RScalculators.html (accessed sep 22, 2009).

83. Darwish, A. M.; Bayba, A. J.; Hung, H. A. Accurate determination of thermal resistance of FETs. IEEE Trans. Microw. Theory 2005, 53, 306–313.

84. Rinaldi, N. On the modeling of the transient thermal behavior of semiconductor devices.IEEE Trans. Electron Dev. 2001, 48, 2796–2802.

85. Wang, X.; Ouyang, Y.; Li, X.; Wang, H.; Guo, J.; Dai, H. Room-temperature all-semiconducting sub-10-nm graphene nanoribbon field-effect transistors. Phys. Rev. Lett.2008, 100, 206803.

86. Steiner, M.; Freitag, M.; Perebeinos, V.; Tsang, J. C.; Small, J. P.; Kinoshita, M.; Yuan, D.; Liu, J.; Avouris, P. Phonon populations and electrical power dissipation in carbon nanotube transistors. Nat. Nanotechnol. 2009, 4, 320–324.

87. Reifenberg, J. P.; Kencke, D. L.; Goodson, K. E. The impact of thermal boundary resistance in phase-change memory devices. IEEE Electr. Device L. 2008, 29, 1112–1114.

88. Chen, I. R.; Pop, E. Compact thermal model for vertical nanowire phase-change memory cells. IEEE Trans. Electron Dev. 2009, 56, 1523–1528.

89. Freitag, M.; Steiner, M.; Martin, Y.; Perebeinos, V.; Chen, Z.; Tsang, J. C.; Avouris, P. Energy dissipation in graphene field-effect transistors. Nano Lett. 2009, 9, 1883–1888.

90. Chen, Z.; Jang, W.; Bao, W.; Lau, C. N.; Dames, C. Thermal contact resistance between graphene and silicon dioxide. Appl. Phys. Lett. 2009, 95, 161910.

91. Amerasekera, A.; Duvvury, C. ESD in Silicon Integrated Circuits, 2nd Edn.; Wiley-VCH: Weinheim, 2002.

92. Jenkins, K. A.; Franch, R. L. Impact of self-heating on digital SOI and strained-silicon CMOS circuits. In IEEE Int. SOI Conf., Newport Beach, CA, USA, 2003, pp. 161–163.

93. Banerjee, K.; Amerasekera, A.; Cheung, N.; Hu, C. High-current failure model for VLSI interconnects under short-pulse stress conditions. IEEE Electr. Device L. 1997, 18, 405–407.

94. Dwyer, V. M.; Franklin, A. J.; Campbell, D. S. Thermal failure in semiconductor devices.Solid-State Electron. 1990, 33, 553–560.

95. Min, Y. J.; Palisoc, A. L.; Lee, C. C. Transient thermal study of semiconductor devices.IEEE Trans. Compon. Hybr. 1990, 13, 980–988.

96. Castro Neto, A.; Guinea, F.; Miguel, N. Drawing conclusions from graphene. Phys. World2006, 19, 33–37.

97. Dresselhaus, M. S.; Dresselhaus, G.; Avouris, P. Carbon Nanotubes: Synthesis, Structure, Properties and Applications; Springer: Berlin, 2001.

98. Geim, A. K.; Novoselov, K. S. The rise of graphene. Nat. Mater. 2007, 6, 183–191.

99. Novoselov, K. S.; Geim, A. K.; Morozov, S. V.; Jiang, D.; Zhang, Y.; Dubonos, S. V.; Grigorieva, I. V.; Firsov, A. A. Electric field effect in atomically thin carbon films. Science2004, 306, 666–669.

100. Iijima, S. Helical microtubules of graphitic carbon. Nature 1991, 354, 56–58.

101. Durkop, T.; Getty, S. A.; Cobas, E.; Fuhrer, M. S. Extraordinary mobility in semiconducting carbon nanotubes. Nano Lett. 2004, 4, 35–39.

102. Morozov, S. V.; Novoselov, K. S.; Katsnelson, M. I.; Schedin, F.; Elias, D. C.; Jaszczak, J. A.; Geim, A. K. Giant intrinsic carrier mobilities in graphene and its bilayer. Phys. Rev. Lett. 2008, 100, 016602.

103. Yu, C.; Shi, L.; Yao, Z.; Li, D.; Majumdar, A. Thermal conductance and thermopower of an individual single-wall carbon nanotube. Nano Lett. 2005, 5, 1842–1846.

104. Nika, D. L.; Pokatilov, E. P.; Askerov, A. S.; Balandin, A. A. Phonon thermal conduction in graphene: Role of Umklapp and edge roughness scattering. Phys. Rev. B 2009, 79, 155413.

105. Yao, Z.; Kane, C. L.; Dekker, C. High-field electrical transport in single-wall carbon nanotubes. Phys. Rev. Lett. 2000, 84, 2941–2944.

106. Javey, A.; Guo, J.; Paulsson, M.; Wang, Q.; Mann, D.; Lundstrom, M.; Dai, H. J. High-field quasiballistic transport in short carbon nanotubes. Phys. Rev. Lett. 2004, 92, 106804.

107. Kuroda, M. A.; Cangellaris, A.; Leburton, J. -P. Nonlinear transport and heat dissipation in metallic carbon nanotubes. Phys. Rev. Lett. 2005, 95, 266803.

108. Hasan, S.; Alam, M. A.; Lundstrom, M. S. Simulation of carbon nanotube FETs including hot-phonon and self-heating effects. IEEE Trans. Electron Dev. 2007, 54, 2352–2361.

109. Perebeinos, V.; Tersoff, J.; Avouris, P. Electron-phonon interaction and transport in semiconducting carbon nanotubes. Phys. Rev. Lett. 2005, 94, 086802.

110. Wang, X.; Zhang, L.; Lu, Y.; Dai, H.; Kato, Y. K.; Pop, E. Electrically driven light emission from hot single-walled carbon nanotubes at various temperatures and ambient pressures.Appl. Phys. Lett. 2007, 91, 261102.

111. Mann, D.; Kato, Y. K.; Kinkhabwala, A.; Pop, E.; Cao, J.; Wang, X.; Zhang, L.; Wang, Q.; Guo, J.; Dai, H. Electrically driven thermal light emission from individual single-walled carbon nanotubes. Nat. Nanotechnol. 2007, 2, 33–38.

112. Mann, D.; Pop, E.; Cao, J.; Wang, Q.; Goodson, K. Thermally and molecularly stimulated relaxation of hot phonons in suspended carbon nanotubes. J. Phys. Chem. B 2006, 110, 1502–1505.

113. Petrov, A. G.; Rotkin, S. V. Energy relaxation of hot carriers in single-wall carbon nanotubes by surface optical phonons of the substrate. JETP Lett. 2006, 84, 156–160.

114. Rotkin, S. V.; Perebeinos, V.; Petrov, A. G.; Avouris, P. An essential mechanism of heat dissipation in carbon nanotube electronics. Nano Lett. 2009, 9, 1850–1855.

115. Fratini, S.; Guinea, F. Substrate-limited electron dynamics in graphene. Phys. Rev. B2008, 77, 195415.

116. Chau, R.; Datta, S.; Doczy, M.; Doyle, B.; Kavalieros, J.; Metz, M. High-κ/metal-gate stack and its MOSFET characteristics. IEEE Electr. Device L. 2004, 25, 408–410.

117. Chen, J. -H.; Jang, C.; Xiao, S.; Ishigami, M.; Fuhrer, M. S. Intrinsic and extrinsic performance limits of graphene devices on SiO_2. Nat. Nanotechnol. 2008, 3, 206–209.

118. Xiong, F.; Liao, A.; Pop, E. Inducing chalcogenide phase change with ultra-narrow carbon nanotube heaters. Appl. Phys. Lett. 2009, 95, 243103.

119. Deshpande, V. V.; Hsieh, S.; Bushmaker, A. W.; Bockrath, M.; Cronin, S. B. Spatially resolved temperature measurements of electrically heated carbon nanotubes. Phys. Rev. Lett. 2009, 102, 105501.

120. Hsu, I. K.; Kumar, R.; Bushmaker, A.; Cronin, S. B.; Pettes, M. T.; Shi, L.; Brintlinger, T.; Fuhrer, M. S.; Cumings, J. Optical measurement of thermal transport in suspended carbon nanotubes. Appl. Phys. Lett. 2008, 92, 063119.

121. Huang, X. Y.; Zhang, Z. Y.; Liu, Y.; Peng, L. M. Analytical analysis of heat conduction in a suspended one-dimensional object. Appl. Phys. Lett. 2009, 95, 143109.

122. Hata, K.; Futaba, D. N.; Mizuno, K.; Namai, T.; Yumura, M.; Iijima, S. Water-assisted highly efficient synthesis of impurity-free single-walled carbon nanotubes. Science 2004,306, 1362–1364.

123. Ong, Z. -Y.; Pop, E. Molecular dynamics simulation of thermal boundary conductance between carbon nanotubes and SiO_2. Phys. Rev. B 2009, in press,http://arxiv.org/abs/0910-2747.

124. Plombon, J. J.; Andideh, E.; Dubin, V. M.; Maiz, J. Influence of phonon, geometry, impurity, and grain size on copper line resistivity. Appl. Phys. Lett. 2006, 89, 113124.

125. Steinhögl, W.; Schindler, G.; Steinlesberger, G.; Engelhardt, M. Size-dependent resistivity of metallic wires in the mesoscopic range. Phys. Rev. B 2002, 66, 075414.

126. Steinhögl, W.; Schindler, G.; Steinlesberger, G.; Traving, M.; Engelhardt, M. Comprehensive study of the resistivity of copper wires with lateral dimensions of 100 nm and smaller. J. Appl. Phys. 2005, 97, 023706.

127. Li, D.; Wu, Y.; Kim, P.; Shi, L.; Yang, P.; Majumdar, A. Thermal conductivity of individual silicon nanowires. Appl. Phys. Lett. 2003, 83, 2934–2936.

128. McConnell, A. D.; Uma, S.; Goodson, K. E. Thermal conductivity of doped polysilicon layers. J. Microelectromech. S. 2001, 10, 360–369.

129. Srivastava, G. P. Theory of thermal conduction in nonmetals. MRS Bull. 2001, 445–450.

130. Glassbrenner, C. J.; Slack, G. A. Thermal conductivity of silicon and germanium from 3 K to the melting point. Phys. Rev. 1964, 134, A1058–A1069.

131. Martin, P.; Aksamija, Z.; Pop, E.; Ravaioli, U. Impact of phonon-surface roughness scattering on thermal conductivity of thin Si nanowires. Phys. Rev. Lett. 2009, 102, 125503.

132. Liu, W.; Asheghi, M. Thermal conduction in ultrathin pure and doped single-crystal silicon layers at high temperature. J. Appl. Phys. 2005, 98, 123523.

133. Hochbaum, A. I.; Chen, R.; Delgado, R. D.; Liang, W.; Garnett, E. C.; Najarian, M.; Majumdar, A.; Yang, P. Enhanced thermoelectric performance of rough silicon nanowires.Nature 2008, 451, 163–167.

134. Chen, R.; Hochbaum, A. I.; Murphy, P.; Moore, J.; Yang, P.; Majumdar, A. Thermal conductance of thin silicon nanowires. Phys. Rev. Lett. 2008, 101, 105501.

135. Boukai, A. I.; Bunimovich, Y.; Tahir-Kheli, J.; Yu, J. -K.; Goddard III, W. A.; Heath, J. R. Silicon nanowires as efficient thermoelectric materials. Nature 2008, 451, 168–171.

136. Li, B.; Wang, L.; Casati, G. Thermal diode: Rectification of heat flux. Phys. Rev. Lett.2004, 93, 184301.

137. Saira, O. -P.; Meschke, M.; Giazotto, F.; Savin, A. M.; Mottonen, M.; Pekola, J. P. Heat transistor: Demonstration of gate-controlled electronic refrigeration. Phys. Rev. Lett.2007, 99, 027203.

138. Dames, C. Solid-state thermal rectification with existing bulk materials. J. Heat Transf.2009, 131, 061301.

139. Jezowski, A.; Rafalowicz, J. Heat flow asymmetry on a junction of quartz with graphite.Phys. Status Solidi A 1978, 47, 229–232.

140. Marucha, C.; Mucha, J.; Rafalowicz, J. Heat flow rectification in inhomogeneous GaAs.Phys. Status Solidi A 1975, 31, 269–273.

141. Kobayashi, W.; Teraoka, Y.; Terasaki, I. An oxide thermal rectifier. Appl. Phys. Lett.2009, 95, 171905.

142. Yang, N.; Zhang, G.; Li, B. Carbon nanocone: A promising thermal rectifier. Appl. Phys. Lett. 2008, 93, 243111.

143. Casati, G. Device physics: The heat is on—and off. Nat. Nanotechnol. 2007, 2, 23–24.

144. Roberts, N. A.; Walker, D. G. Monte Carlo study of thermal transport of frequency and direction dependent reflecting boundaries in high Kn systems. In 11th Intersociety Conference on Thermal and Thermomechanical Phenomena in Electronic Systems (ITHERM), Orlando, Florida, USA, 2008, pp. 993–998.

145. Chang, C. W.; Okawa, D.; Majumdar, A.; Zettl, A. Solid-state thermal rectifier. Science2006, 314, 1121–1124.

146. Scheibner, R.; Konig, M.; Reuter, D.; Wieck, A. D.; Gould, C.; Buhmann, H.; Molenkamp, L. W. Quantum dot as thermal rectifier. New J. Phys. 2008, 083016.

147. Wang, Z.; Carter, J. A.; Lagutchev, A.; Koh, Y. K.; Seong, N. -H.; Cahill, D. G.; Dlott, D. D. Ultrafast flash thermal conductance of molecular chains. Science 2007, 317, 787–790.

148. Costescu, R. M.; Wall, M. A.; Cahill, D. G. Thermal conductance of epitaxial interfaces.Phys. Rev. B 2003, 67, 054302.

149. Lyeo, H. -K.; Cahill, D. G. Thermal conductance of interfaces between highly dissimilar materials. Phys. Rev. B 2006, 73, 144301.

150. Huxtable, S. T.; Cahill, D. G.; Shenogin, S.; Xue, L.; Ozisik, R.; Barone, P.; Usrey, M.; Strano, M. S.; Siddons, G.; Shim, M. et al. Interfacial heat flow in carbon nanotube suspensions. Nat. Mater. 2003, 2, 731–734.

151. Panzer, M. A.; Zhang, G.; Mann, D.; Hu, X.; Pop, E.; Dai, H.; Goodson, K. E. Thermal properties of metal-coated vertically aligned single-wall nanotube arrays. J. Heat Transf.2008, 130, 052401.

152. Stevens, R. J.; Smith, A. N.; Norris, P. M. Measurement of thermal boundary conductance of a series of metal-dielectric interfaces by the transient thermoreflectance technique. J. Heat Transf. 2005, 127, 315–322.

153. Tong, T.; Majumdar, A. Reexamining the 3-omega technique for thin film thermal characterization. Rev. Sci. Instrum. 2006, 77, 104902.

154. Yamane, T.; Nagai, N.; Katayama, S.; Todoki, M. Measurement of thermal conductivity of silicon dioxide thin films using a 3ω method. J. Appl. Phys. 2002, 91, 9772–9776.

155. Lee, S. -M.; Cahill, D. G. Heat transport in thin dielectric films. J. Appl. Phys. 1997, 81, 2590–2595.

156. Dames, C.; Chen, G. 1ω, 2ω, and 3ω methods for measurements of thermal properties.Rev. Sci. Instrum. 2005, 76, 124902.

157. Kim, E. -K.; Kwun, S. -I.; Lee, S. -M.; Seo, H.; Yoon, J. -G. Thermal boundary resistance at $Ge_2Sb_2Te_5$/ZnS:SiO_2 interface. Appl. Phys. Lett. 2000, 76, 3864–3866.

158. Ge, Z.; Cahill, D. G.; Braun, P.V. Thermal conductance of hydrophilic and hydrophobic interfaces. Phys. Rev. Lett. 2006, 96, 186101.

159. Gundrum, B. C.; Cahill, D. G.; Averback, R. S. Thermal conductance of metal-metal interfaces. Phys. Rev. B 2005, 72, 245426.

160. Swartz, E. T.; Pohl, R. O. Thermal boundary resistance. Rev. Mod. Phys. 1989, 61, 605–668.

161. Pop, E. Self-Heating and Scaling of Thin-Body Transistors. Ph.D. Thesis, Stanford Univ., Stanford, CA; 2005. [Online]. http://poplab.ece.illinois.edu (accessed Sep 22, 2009).

162. Kim, P.; Shi, L.; Majumdar, A.; McEuen, P. L. Mesoscopic thermal transport and energy dissipation in carbon nanotubes. Physica B Condens. Mat. 2002, 323, 67–70.

163. Pop, E; Chui, C. O.; Sinha, S.; Dutton, R.; Goodson, K. Electro-thermal comparison and performance optimization of thin-body SOI and GOI MOSFETs. In Intl. Electron Dev. Mtg. (IEDM), San Francisco CA, USA, 2004, pp. 411–414.

164. Castro Neto, A. H.; Guinea, F.; Peres, N. M. R.; Novoselov, K. S.; Geim, A. K. The electronic properties of graphene. Rev. Mod. Phys. 2009, 81, 109–154.

165. Yanagisawa, H.; Tanaka, T.; Ishida, Y.; Matsue, M.; Rokuta, E.; Otani, S.; Oshima, C. Analysis of phonons in graphene sheets by means of HREELS measurement and ab initiocalculation. Surf. Interface Anal. 2005, 37, 133–136.

166. Reifenberg, J. P.; Chang, K. -W.; Panzer, M. A.; Kim, S.; Rowlette, J. A.; Asheghi, M.; Wong, H. S. P.; Goodson, K. E. Thermal boundary resistance measurements for phase-change memory devices. IEEE Electr. Device L. 2010, 31, 56–58.

167. Stoner, R. J.; Maris, H. J. Kapitza conductance and heat flow between solids at temperatures from 50 to 300 K. Phys. Rev. B 1993, 48, 16373–16387.

CHAPTER 11 WAVEFORM DESIGN FOR COGNITIVE RADAR WITH DETERMINISTIC EXTENDED TARGETS IN THE PRESENCE OF CLUTTER

Vahid Karimi, Reza Mohseni, Yaser Norouzi and Mohammad Javad Dehghani

Keywords: Signal to Noise Ratio, Mutual Information, Cognitive Radar, Deterministic Extended Target, Clutter

INTRODUCTION

Cognitive radar is a radar system which selects its transmitted waveform to adapt to the radar environment by using feedback structure from the receiver to the transmitter. In these systems waveforms can be adaptively optimized based on preceding knowledge about the targets and environments; it leads to the improvement of the total performance of system [1]. Design of transmitting waveform has an important effect on the performance and efficiency of radar system. Adaptive waveform design for target detection and recognition has been developed during the past decade, and also recently most studies have been devoted to radar waveform optimization. In these approaches, one method is based on signal-to-noise ratio (SNR) maximizing under a particular model of the system, interference, clutter and targets [2]. Another approach is based on mutual information which is first proposed by Bell [3]. Bell shows for estimating the parameters of a target from a given ensemble, the radar waveform should be designed to maximize the mutual information between the received signal and the target ensemble [4]. SNR-based optimum matched waveform for a specific target via frequency domain approach is proposed in [5]. An optimal waveform for T-72 and M1 main battle tanks detection is proposed in [6].

The information-based approach in the presence of signal-dependent clutter is investigated in [7]. In their work, optimum waveform is investigated both from SNR maximization and information-theoretic approaches. The optimal waveform for detecting extended target in signal-dependent interference is proposed in [8]. They derive optimal waveform energy spectral density (ESD) by maximizing SNR. A method for waveform design based on mutual information is proposed in [9] in which the general water-filling method is utilized to solve the waveform design for the recognition of multiple extended targets. They have shown that their proposed method has the higher classification rate at higher SNRs and higher detection performance in lower SNRs as compared to LFM and water-filling signals. In [10], the performance of waveform design in cognitive radar is discussed and a new iterative algorithm is proposed to synthesize a constant modulus waveform to maximize SNR and MI simultaneously. Designing matched waveforms based on maximizing SNR and MI for both deterministic and stochastic targets are discussed in [11] [12]; using multiple transmission of obtaining waveform a target identification problem is investigated. A new technique is proposed in [13], which improves the classification performance of SNR-based matched waveforms. In [7] [14], a closed-loop strategy is applied to discriminate target classes rather than a finite ensemble of known targets. In [15], optimization problem based on mutual information for waveform design in signal dependent clutter is discussed and a new waveform using interior-point method to carry out the optimization task is proposed. In previous papers such as [8] [16], SNR improvement for proposed waveforms over LFM waveform is investigated and in this paper, we evaluate SNR and MI values for different waveforms and compare the results with LFM waveforms and show the best performance of obtaining waveforms.

The contributions of this paper include an analysis of applying both the information-based and SNR-based

Citation: Vahid Karimi, Reza Mohseni, Yaser Norouzi, and Mohammad Javad Dehghani (2016). Waveform Design for Cognitive Radar with Deterministic Extended Targets in the Presence of Clutter. DOI:10.4236/ijcns.2016.96023.

approach to different deterministic extended targets and clutter scenarios considering the energy constraint. In this study, energy distribution for optimum waveform in clutter and noise scenario with the energy constraint have been addressed and we focus on the point that waveform should put a considerable amount of available energy on the frequency bands in which the target spectrum frequency components is considerable while the clutter is negligible. We derive various waveforms for each scenario and finally compare our results with linear frequency modulated (LFM) waveform to verify the performance of the proposed waveform.

This paper is organized as follows: first a radar system model is described in Section 2 and then the derivation of MI-based and SNR-based waveforms in the presence of Gaussian clutter and other scenarios are derived. In Section 3, using the obtained relations in Section 2, transmit waveform spectrum for two deterministic extended targets are illustrated and the energy allocation realized by obtaining waveforms are briefly discussed. Section 4 generally discusses the performance of obtaining waveforms and other predefined waveforms and a new closed-loop structure for detecting targets in different scenarios is proposed. Finally, in Section 5, we provide a general conclusion.

PROBLEM FORMULATION

Signal, Clutter and Target Model

To analyze a radar system, all constituent blocks should be defined. We consider a signal model of a Gaussian extended target in the presence of clutter as shown in Figure 1. Now we define the essential target and clutter models to analyze the overall performance of a radar system.

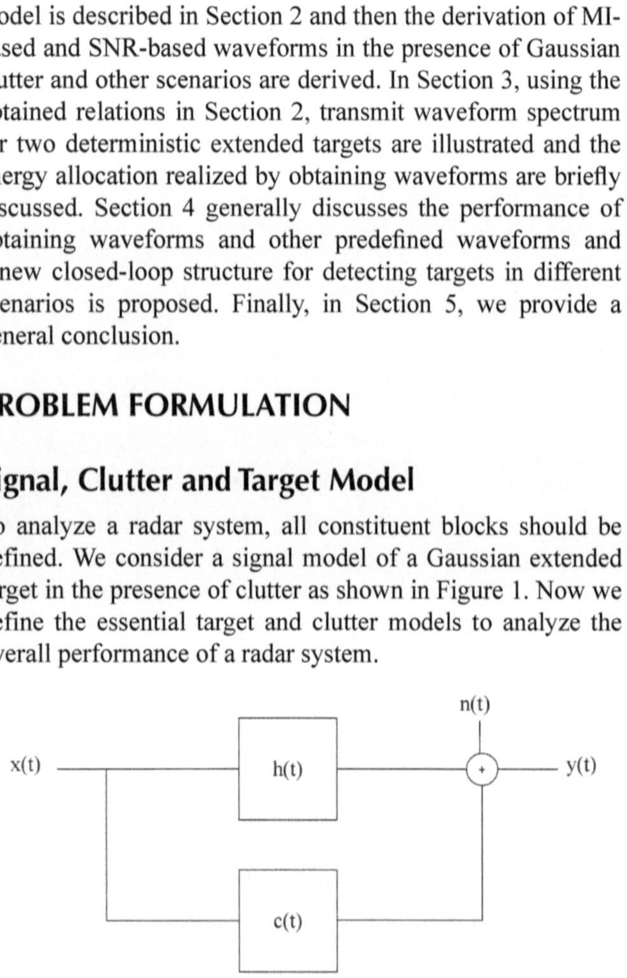

Figure 1. Known Gaussian extended target model in the presence of clutter [7].

Let x(t) be a finite-energy waveform with duration T. We assume x(t) is energy-limited and non-zero only in the time interval $|t| \le T/2$ which has a bandwidth w; The energy content of this signal is:

$$E_x = \int_{-T/2}^{T/2} |x(t)|^2 \, dt \tag{1}$$

c(t) is assumed to be sea clutter with known power spectral density (PSD) which can be approximated as follows:

$$P_c(f) = g_c \exp\left(-\frac{(f - f_c)^2}{2\sigma_c^2} \right) \tag{2}$$

Here f_c is the peak locations of the Gaussian function; g_c and σ_c^2 are parameters which are specified according to problem conditions. h(t) is the impulse response of target. Once the impulse responses are generated, it is assumed that they are exactly known to the transmitter. In this study, target is assumed to have a Gaussian mixture shape with the following power spectral density:

$$|H(f)|^2 = a_0 + \sum_{i=1}^{N} a_i \exp\left(-\frac{(f - f_i)^2}{2\sigma_T^2} \right) \tag{3}$$

where a_i and f_i are respectively determined between a specific amplitude range and an appropriate interval due to clutter and noise specifications. The variance σ_T^2 is supposed to have a constant value. Finally n(t) denotes the noise, which is a complex zero mean WSS Gaussian random process with known PSD $P_n(f)$. Now let r(t) be the received signal that is equal to:

$$r(t) = x(t) * h(t) + x(t) * c(t) + n(t) \tag{4}$$

Waveform Derivation

In this part, the Energy Spectral Density (ESD) of MI-based is computed with the assumption of known target in different scenarios and compared with SNR-based waveform discussed in [8]. Like [11], using Lagrangian multiplier, we have derived optimum waveform ESD based on MI criteria in both signal-dependent interference and signal-independent noise. The MI between a Gaussian target ensemble and the received signal without the presence of clutter is equal to:

$$MI\left(r(t); h(t) \mid x(t)\right) = \int_W \ln\left[1 + \frac{|X(f)|^2 \sigma_h^2(f)}{P_n(f)}\right] df \tag{5}$$

where $\sigma_h^2(f)$ is the continuous target signature. Now let $\varepsilon_x(f)$ be the energy spectral density (ESD) of the waveform, so we have:

$$MI = \int_W \ln\left[1 + \frac{\varepsilon_x(f)\sigma_h^2(f)}{P_n(f)}\right] df \tag{6}$$

The desired waveform ESD is found by solving the constrained optimization problem:

$$\hat{\varepsilon}_x(f) = \max_{\varepsilon_x(f)} MI \tag{7}$$

With the energy constraint:

$$\int_W \hat{\varepsilon}_x(f) df = E_x \tag{8}$$

Now using the Lagrangian multiplier method, the waveform ESD will be obtained as follows:

$$L(\varepsilon_x(f)) = \int_W \ln\left[1 + \frac{\varepsilon_x(f)\sigma_h^2(f)}{P_n(f)}\right] df + \lambda\left(E_x - \int_W \varepsilon_x(f) df\right)$$
$$= \int_W \left(\ln\left[1 + \frac{\varepsilon_x(f)\sigma_h^2(f)}{P_n(f)}\right] - \lambda\varepsilon_x(f) \right) df + \lambda E_x \tag{9}$$

We can equivalently maximize:

$$\tilde{L}(\varepsilon_x(f)) = \ln\left[1+\frac{\varepsilon_x(f)\sigma_h^2(f)}{P_n(f)}\right] - \lambda\varepsilon_x(f)$$

(10)

The first and the second derivatives of $\tilde{L}(\varepsilon_x(f))$ with respect to $\varepsilon_x(f)$ are given by:

$$\frac{d\tilde{L}(\varepsilon_x(f))}{d\varepsilon_x(f)} = \frac{\frac{|H(f)|^2}{P_n(f)}}{1+\varepsilon_x(f)\frac{|H(f)|^2}{P_n(f)}} - \lambda$$

$$\frac{d^2\tilde{L}(\varepsilon_x(f))}{d^2\varepsilon_x(f)} = -\frac{\left(\frac{|H(f)|^2}{P_n(f)}\right)^2}{\left(1+\varepsilon_x(f)\frac{|H(f)|^2}{P_n(f)}\right)^2} \tag{11}$$

Since the second derivative is negative for all $\varepsilon_x(f)$, $\tilde{L}(\varepsilon_x(f))$ is a concave function and by putting equating (10) to zero the desired $\varepsilon_x(f)$ is obtained as follows:

$$\hat{\varepsilon}_x(f) = \max\left[0, \frac{\left(\frac{|H(f)|^2}{\lambda P_n(f)} - 1\right)}{\frac{|H(f)|^2}{P_n(f)}}\right] \tag{12}$$

where the parameter λ is found by solving the following equation:

$$\int_W \max\left[0, \frac{\left(\frac{|H(f)|^2}{\lambda P_n(f)} - 1\right)}{\frac{|H(f)|^2}{P_n(f)}}\right] = E_x \tag{13}$$

In Equation (12), we see that the waveform ESD will be equal to zero for those frequencies satisfying:

$$\frac{|H(f)|^2}{\lambda P_n(f)} - 1 \le 0 \text{ or } |H(f)|^2 \le \lambda P_n(f) \quad \forall f \tag{14}$$

Thus, we can consider $\lambda P_n(f)$ as a threshold which depends on coefficient λ which means the waveform only puts energy into those frequencies where $|H(f)|^2 > \text{Threshold}$. Thus, for a radar system to be cognitive, it should customize the transmit waveform in such a way that it emphasizes frequency bands where the target spectrum is greater than the obtained threshold and deemphasizes frequency bands where the target is negligible as compared with clutter.

The mutual information between the received signal and the target ensemble with the presence of clutter is:

$$MI(r(t),h(t)|x(t)) = \int_W \ln\left[1+\frac{|X(f)|^2\sigma_h^2(f)}{P_n(f)+P_c(f)|X(f)|^2}\right]df \tag{15}$$

And like previous step for deriving non-clutter case we have:

$$\begin{aligned} L(\varepsilon_x(f)) &= \int_W \ln\left[1+\frac{\varepsilon_x(f)\sigma_h^2(f)}{P_n(f)+P_c(f)\varepsilon_x(f)}\right]df + \lambda\left(E_x - \int_W \varepsilon_x(f)df\right) \\ &= \int_W \left(\ln\left[1+\frac{\varepsilon_x(f)\sigma_h^2(f)}{P_n(f)+P_c(f)\varepsilon_x(f)}\right] - \lambda\varepsilon_x(f)\right)df + \lambda E_x \end{aligned} \tag{16}$$

We can equivalently maximize:

$$\tilde{L}(\varepsilon_x(f)) = \ln\left[1+\frac{\varepsilon_x(f)\sigma_h^2(f)}{P_n(f)+P_c(f)\varepsilon_x(f)}\right] - \lambda\varepsilon_x(f) \tag{17}$$

Now the first and second derivatives of $\tilde{L}(\varepsilon_x(f))$ with respect to $\varepsilon_x(f)$ are computed and by following a procedure like the previous step for deriving MI-based waveform without the presence of clutter, leading to the waveform described by:

$$\hat{\varepsilon}_x(f) = \max\left[0, -A(f) + \sqrt{A^2(f) + B(f)\left(\frac{1}{\lambda} - \frac{P_n(f)}{|H(f)|^2}\right)}\right] \tag{18}$$

where

$$A(f)=\frac{P_n(f)\left(2P_c(f)+|H(f)|^2\right)}{2P_c(f)\left(P_c(f)+|H(f)|^2\right)}$$

$$B(f)=\frac{P_n(f)|H(f)|^2}{P_c(f)\left(P_c(f)+|H(f)|^2\right)} \tag{19}$$

And the parameter λ is found by solving the energy constraint equation:

$$\int_W \max\left[0,-A(f)+\sqrt{A^2(f)+B(f)\left(\frac{1}{\lambda}-\frac{P_n(f)}{|H(f)|^2}\right)}\right]df = E_x \tag{20}$$

In Equation (18), we see that the waveform ESD will put energy into those frequencies satisfying:

$$-A(f)+\sqrt{A^2(f)+B(f)\left(\frac{1}{\lambda}-\frac{P_n(f)}{|H(f)|^2}\right)}\geq 0$$

$$\rightarrow A^2(f)+B(f)\left(\frac{1}{\lambda}-\frac{P_n(f)}{|H(f)|^2}\right)\geq A^2(f)$$

$$\rightarrow B(f)\left(\frac{1}{\lambda}-\frac{P_n(f)}{|H(f)|^2}\right)\geq 0 \xrightarrow{B(f)\geq 0\ \forall f} |H(f)|^2\geq \lambda P_n(f)\quad \forall f \tag{21}$$

Here again the threshold $\lambda P_n(f)$ is obtained by solving the energy constraint Equation (20) showing that the parameter λ is totally different for each energy constraint presented in this paper. A noteworthy point is that although the thresholds are absolutely similar, their values are different due to parameter λ. Consequently, the waveform energy allocation is different from the non-clutter case. It has been shown in simulation results that this waveform distributes most of its energy into frequency bands which the target PSD is high and the clutter PSD is low.

The waveform ESD for SNR-based waveform for a known target with the presence of clutter is described in [8]. SINR for a deterministic target in the presence of clutter is computed as:

$$SINR=\int_W \frac{|X(f)|^2|H(f)|^2}{|X(f)|^2 P_c(f)+P_n(f)}df \tag{22}$$

And the waveform ESD using the Lagrangian multiplier technique is [8]:

$$\hat{\varepsilon}_x(f)=\max\left[0,\frac{|H(f)|\sqrt{\frac{P_n(f)}{\lambda}}-P_n(f)}{P_c(f)}\right] \tag{23}$$

where parameter λ is found by solving:

$$\int_W \max\left[0,\frac{|H(f)|\sqrt{\frac{P_n(f)}{\lambda}}-P_n(f)}{P_c(f)}\right]df=E_x \tag{24}$$

As mentioned in [8], the obtained waveform only puts energy into those frequencies where $|H(f)|^2>\lambda P_n(f)$ as we derived for MI-based waveforms and $\lambda P_n(f)$ is considered as a threshold with different values for all waveforms. The last waveform discussed here is the waveform ESD for SNR-based waveform without the presence of clutter. In this case by putting $P_c(f)=0$, SNR expression in (22) is reduced to:

$$SINR=\int_W \frac{|X(f)|^2|H(f)|^2}{P_n(f)}df \tag{25}$$

Using the Lagrange multiplier technique, we form the objective function $\tilde{L}(\varepsilon_x(f))$ like previous steps, and we see that the first derivative of this function is independent of $\varepsilon_x(f)$ and simply implies that no solution exists using this technique. However, by the technique discussed in [2] and [11], the transmit waveform x(t) with time duration T that optimizes (25) is defined by:

$$\lambda_{max}\hat{x}(t)=\int_{-T/2}^{T/2}\hat{x}(\tau)M(t-\tau)d\tau \tag{26}$$

In summary, the transmit waveform x(t) that maximizes SNR is the Eigenfunction corresponding to the maximum eigenvalue of the kernel M(t) which is described by:

$$M(t)=\int_{-\infty}^{\infty}\frac{|X(f)|^2}{P_n(f)}e^{j2\pi ft}df \tag{27}$$

And finally the SNR has been just the product of this eigenvalue and the energy in the transmit waveform as follows:

$$SNR=\lambda_{\max}E_x \tag{28}$$

Different methods are proposed to derive the optimum waveform for this scenario. One of them is the approach proposed in [16] whereby updating phase modulated waveform in such a way that the SNR at the receiver filter output is maximized and the performance of extended target detection is enhanced significantly. In this paper, we proposed a new waveform due to energy constraint and also the distribution of the target on specified bandwidth. Indeed, we are interested in waveforms which can follow the target spectrum and detect all peaks of the target in frequency domain. Regarding this assumption we defined the ESD of this waveform in such a way that is independent of noise and clutter power spectrum and only depends on the target spectrum. We name this waveform "Target Spectrum Follower (TSF)" [17]. Therefore, ESD of the proposed waveform is as follows:

$$\varepsilon_{TSF}(f)=\lambda|H(f)|^2,\quad |f|\le W/2 \tag{29}$$

where the parameter λ is found by solving:

$$\int_W \varepsilon_{TSF}(f)df=E_x \tag{30}$$

The ESD obtained using this method produces a waveform that totally follows the target spectrum. The energy of transmitted waveform is allocated to the target peaks in proportion to the amount of every peak; in other words, the proposed waveform allocates its energy in all of target spectrum frequency bands. This allocation is illustrated in Figure 3(d) and Figure 4(d). Unlike other waveforms, this waveform can detect lower target peaks which has some advantages and disadvantages that will be discussed later. Our main purpose to design such a waveform is to show the important role of energy allocation in radar transmitted waveform and the effect of this consideration on target detection scenarios.

The waveforms discussed in this paper are: SNR-based waveform in clutter (CSNR-based) [8], MI-based waveform without clutter (MI-based), MI-based waveform in clutter (CMI-based), target spectrum follower (TSF) and linear frequency modulation waveform (LFM).

NUMERICAL EXAMPLES

Numerical Waveform Examples with Specified Gausiian Clutter

We cannot find a closed-form solution for the waveform derivations. In this section, we provide two numerical examples which can be obtained using the proposed waveforms. As it was mentioned in the previous section, target and clutter are supposed to have a Gaussian shape with the power spectrum $|H(f)|^2$ and $P_c(f)$, respectively. Two targets for numerical examples are considered here. In example one, suppose that we have a Gaussian target and clutter shown in Figure 2. The ESD for CSNR-based, CMI-based, MI-based and TSF waveforms for target 1 are evaluated and shown in Figure 2. In the case of waveform design in signal-dependent interference, CSNR-based waveform tends to concentrate most of its energy on one dominant narrow frequency band in which the target PSD is comparable to clutter PSD putting the rest of its energy into frequency bands with great amount of target PSD. CMI-based waveform follows such a procedure with different amounts of energy allocation to each frequency band. Therefore, both waveforms place most of their energy where clutter power is low. We will briefly discuss the performance of CSNR-based waveform over CMI-based waveform later.

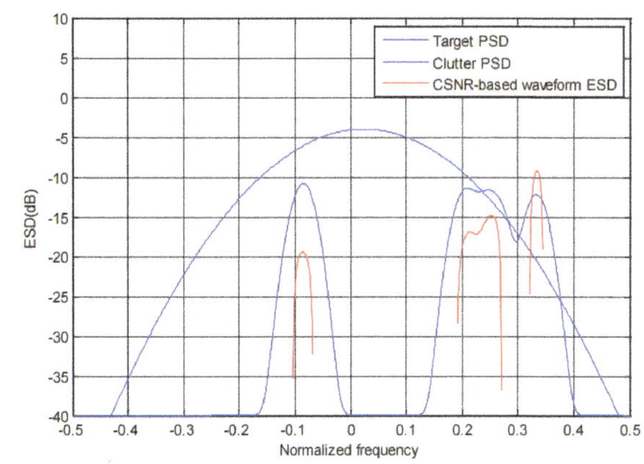

(a)

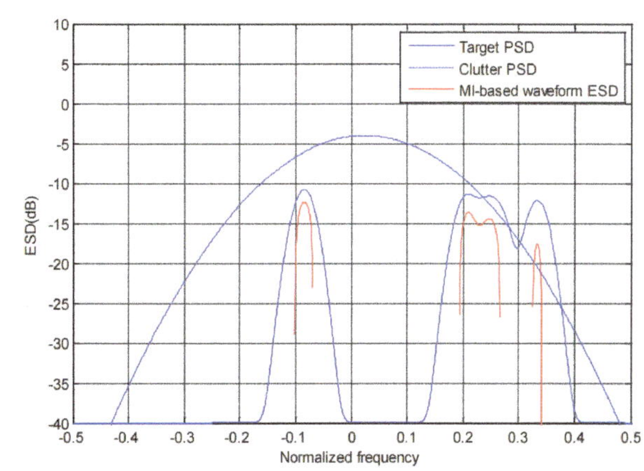

(b)

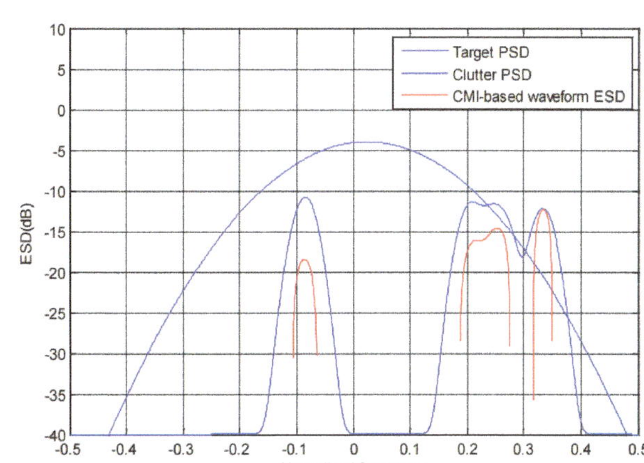

(c)

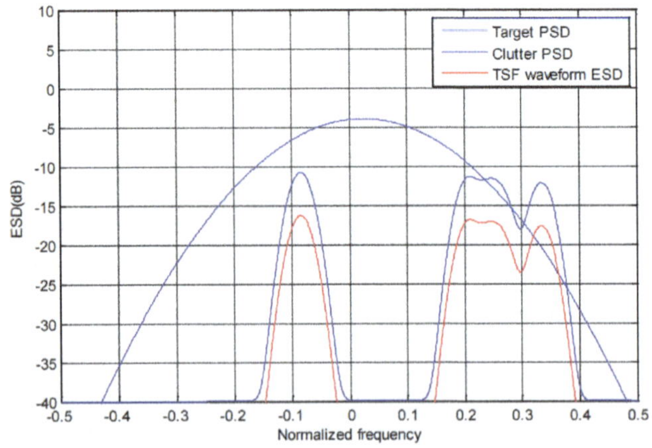

(d)

Figure 2. Corresponding ESD to Different Scenarios with Target 1. (a) ESD for CSNR-based waveform; (b) ESD for MI-based waveform; (c) ESD for CMI-based Waveform; (d) ESD for TSF Waveform.

As shown in Figure 2(b) MI-based waveform without considering clutter shape puts its energy relative to the target peak values and finally TSF waveform unlike other waveforms follows all target peak energy allocation is done in each band where the target has either small or large peaks. As shown in this figure, CSNR-based waveform compared with other waveforms distributed less energy into the frequency band [−0.2, 0], where the target is comparable to the clutter and instead puts more energy into the frequency band [0.3, 0.4], where the clutter has smaller amounts of PSD compared with the target.

The target-to-noise ratio (TNR) is defined as the ratio of the area under target PSD to the area under noise PSD [17], and also the ratio of the area under the clutter PSD to noise PSD is called clutter-to-noise ratio (CNR). The last ratio is TCR which is the area under the target PSD to clutter PSD. In this example, TNR, CNR and TCR are 1.66 dB, 10.94 dB and −9.29 dB, respectively. As we expected, with specified energy constraint, these waveforms which are designed in such a way to focus their energy into frequency bands; the target has the largest PSD value and the clutter has the smallest PSD value. As can be seen in Figure 2 for target 1, the CSNR- based and CMI-based allocated most of their energy in the frequency sub-band [0.3, 0.4], where the target is comparable in terms of PSD value with clutter while this point for MI-based and TSF waveforms are not considered. We will show later that this is the main reason for the improvement of CSNR-based and CMI-based on scenarios with low values of TCR.

We repeat the experiment with target 2 and with the same clutter and noise PSD. For this target, we have TNR = 0.51 dB, CNR = 0.95 dB and TCR = −0.44 dB and corresponding waveforms for this target scenario are computed and illustrated in Figure 3. As seen in this figure, TSF waveform follows all target peaks and puts its energy due to each target peak not consider clutter and noise for this energy allocation. CMI-based and MI-based waveforms have approximately similar response to this target and just follow two great peaks of this target putting their energy equally into the corresponding frequency bands while CSNR-based puts its main energy into frequency band which has the smallest value of clutter PSD value; this could be a reason for better performance of this waveform compared with others.

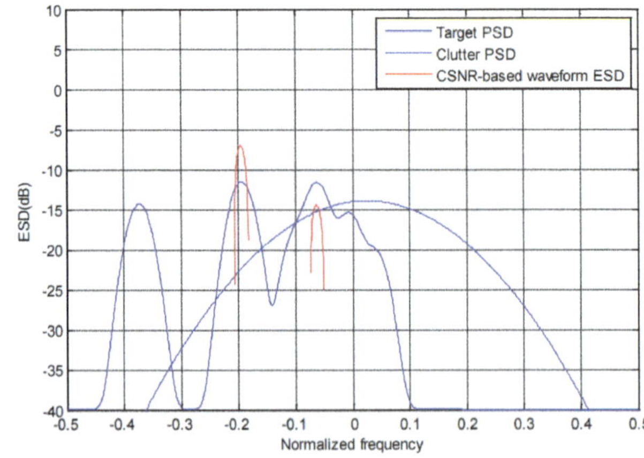

(a)

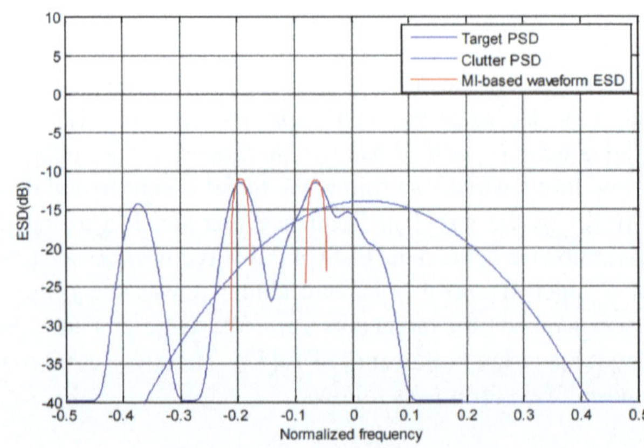

(b)

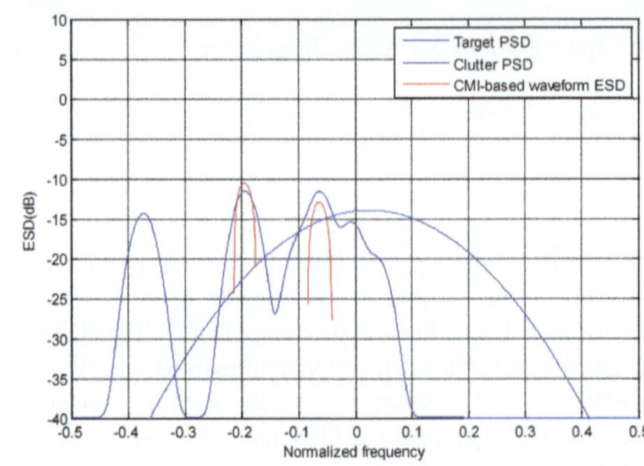

(c)

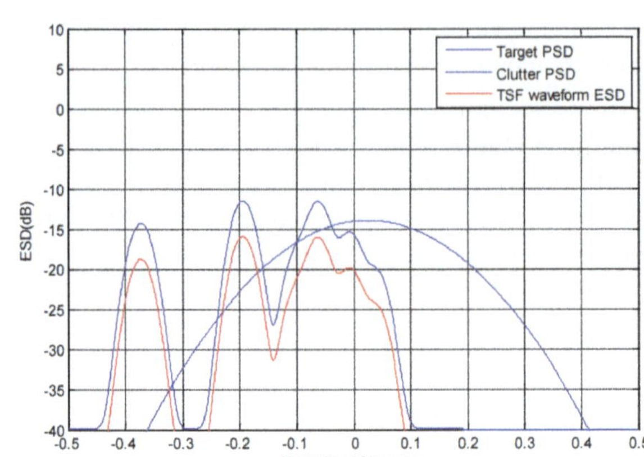

(d)

Figure 3. Corresponding ESD for different scenarios with target 2. (a) ESD for CSNR-based waveform; (b) ESD for MI-based waveform; (c) ESD for CMI-based waveform. (d) ESD for TSF waveform.

SNR and MI Comparison between Different Waveforms

As shown before, solving optimization problem requires knowledge of the target and clutter spectrum. It was shown in [8] that the probability of detection is related to the probability of false alarm by:

$$P_D = 1 - \psi_{h_1}\left(\frac{g_0}{g_1}\psi_{h_0}^{-1}\left(1 - P_{FA}\right)\right) \tag{31}$$

where $\Psi_{h_1}(\eta)$ is the cumulative density function of a chi-square random variable with h_1 degrees of freedom. It has been shown that:

$$\frac{g_0}{g_1} = \frac{1}{1 + D(x)} \tag{32}$$

With a good approximation $D(x)$ is equal to:

$$D(x) = 2\int_W \frac{|X(f)|^2 |H(f)|^2}{|X(f)|^2 P_c(f) + P_n(f)} df = 2SINR \tag{33}$$

And finally we have:

$$P_D = 1 - \psi_{h_1}\left(\frac{1}{1 + 2SINR}\psi_{h_0}^{-1}\left(1 - P_{fa}\right)\right) \tag{34}$$

Using the above equation, since maximizing P_D for arbitrary values of P_{fa} depends on the values of SINR for each waveform and by maximizing SINR, we can achieve higher values of P_D. Previous research has shown that SNR-based waveform and LFM waveform are compared through this method and shown that the proposed method had better performance in terms of maximizing the probability of detection. Here, we obtained different waveforms with both SNR and MI values and now we want to select the desired waveform based on achieving maximizing probability of detection for each target and clutter scenarios regarding the point that the desired waveform should have a sufficient MI value in comparison with other waveforms. MI and SINR which are defined in (14), (21) respectively, are considered as objective functions. We computed their values for three different targets.

For each target the corresponding waveform according to target, clutter and noise spectrum is generated and utilized for computing SNR and MI. Now we suppose a specific scenario with known target, clutter and noise PSDs. We are interested in investigating the effects of target, clutter and noise separately. So, we consider the scenario 1 which includes a specific Gaussian extended target, Gaussian clutter and AWGN noise with known PSDs, then compute CNR, TCR and TNR values for this scenario. Afterwards, by changing clutter (target and noise are fixed), scenario 2 is obtained and similarly scenario 3 and scenario 4 are obtained by changing the target and noise, respectively. Target, Clutter and Noise PSD for each scenario are illustrated in Figure 4 and also their CNR, TCR and TNR values are shown in Table 1.

After defining scenarios, ESD of each waveform in accordance with each scenario is computed due to the obtained relations. Eventually for each scenario corresponding SNR and MI values for investigating the applied changes in target, clutter and noise are evaluated by Equation (15), (22) as summarized in Table 2.

As indicated in Table 2, CSNR-based waveform compared with other waveforms has greater SNR amount for each scenario, and according to (34) this waveform should have the best performance in terms of maximizing the probability of detection. The point that should be considered is its MI value which is not maximum even in one scenario and this is the CMI-based waveform that has greater MI amounts compared with other waveforms. As mentioned in [18], LFM waveform is a good candidate for distinguishing point targets, and for an extended target case may not have good performance which we will discuss in the following paragraph.

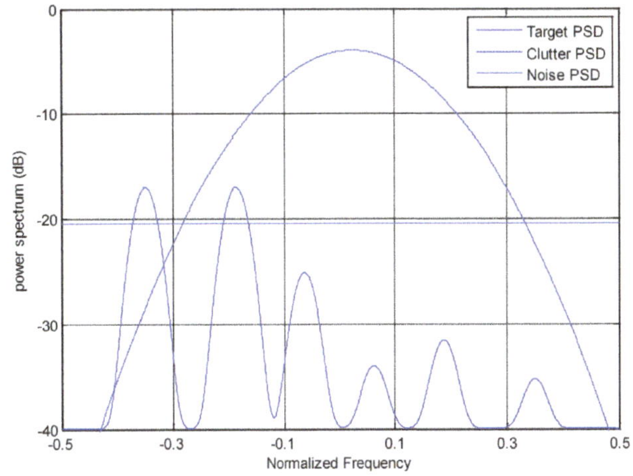

(a)

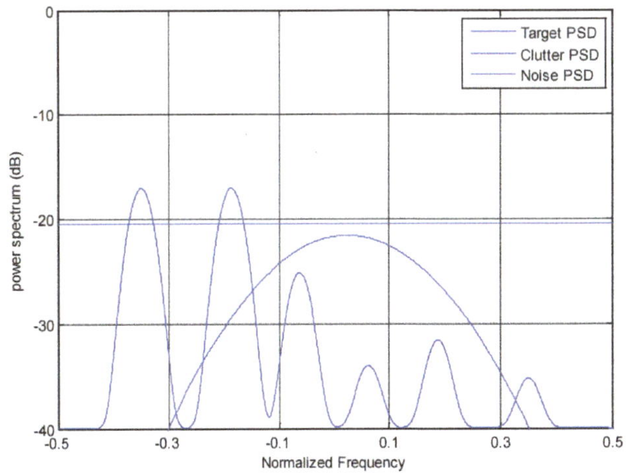

(b)

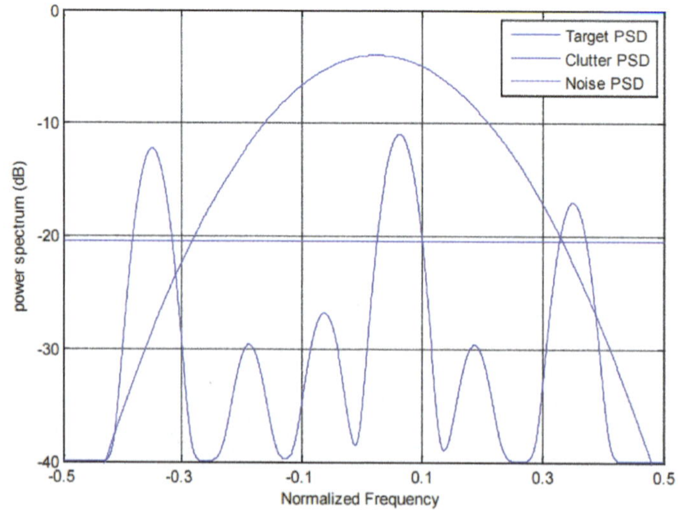

(c)

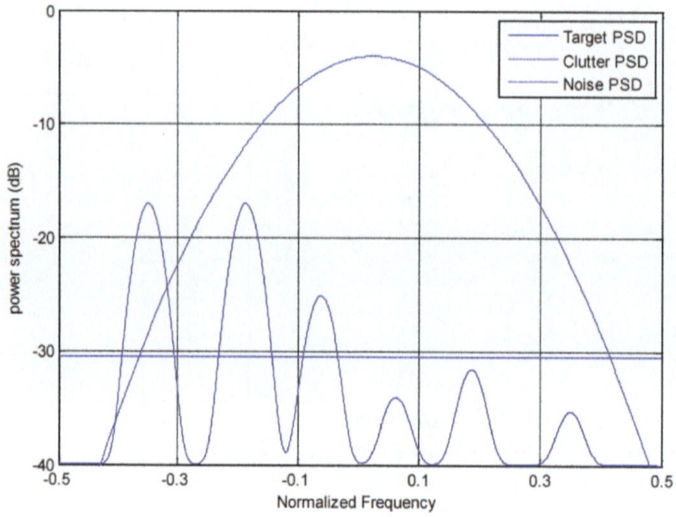

(d)

Figure 4. Target, clutter and noise PSD for each scenario. (a) Scenario 1; (b) Scenario 2; (c) Scenario 3; (d) Scenario 4.

Table 1. CNR, TCR and TNR values for different scenarios in dB.

Value(dB) **Scenario**	**CNR**	**TCR**	**TNR**
Scenario 1	10.95	−17.37	−6.42
Scenario 2	−6.62	0.2	−6.42
Scenario 3	10.95	−11.82	−0.87
Scenario 4	20.95	−17.37	3.58

Table 2. Corresponding SNR and MI values in dB for different waveforms in each scenario.

Waveform **Scenario**	**MI-based Waveform**	**CMI-based Waveform**	**CSNR-based Waveform**	**TSF Waveform**	**LFM Waveform**
Scenario 1	SNR = 17.16 MI = 16.82	SNR = 17.94 MI = 17.45	SNR = 18.09 MI = 17.1	SNR = 15.66 MI = 15.55	SNR = 8.25 MI = 8.24
Scenario 2	SNR = 18.16 MI = 17.75	SNR = 18.16 MI = 17.76	SNR = 18.26 MI = 15.53	SNR = 16.06 MI = 15.94	SNR = 8.34 MI = 8.34

Scenario 3	SNR = 19.93 MI = 19.47	SNR = 22.5 MI = 21.51	SNR = 22.87 MI = 20.67	SNR = 19.76 MI = 19.52	SNR = 13.59 MI = 13.56
Scenario 4	SNR = 25.26 MI = 24.03	SNR = 26.75 MI = 25.03	SNR = 27.18 MI = 24.47	SNR = 24.15 MI = 23.39	SNR = 17.63 MI = 17.53

In Table 2 for scenario1, we have 9.84 dB SNR improvement for CSNR-based waveform over LFM waveform and 9.69 dB SNR improvement for CMI-based waveform over LFM waveform. This improvement for the waveform is proposed in [8] over LFM waveform was 3 dB and by the method proposed in [16] SNR of the proposed waveform was about 4 dB higher than the square pulse or the LFM pulse. For TSF waveform, we have 7.41 dB improvement which can be a good choice in some scenarios. For example, for high values of TCR, this waveform has a better performance in terms of MI and approximately the same performance in terms of SNR compared with CSNR-based waveform.

It is obvious from Table 1 and Table 2 that by decreasing the clutter PSD, SNR and MI value will increase from one scenario to another. This idea is true for the case of increasing the amounts of target spectrum and decreasing noise PSD.

To evaluate the performance, we used receiver operating characteristic (ROC) curves corresponding to the obtained waveforms and LFM. Figure 5 shows these ROC curves for all scenarios. As it is clear, the improvement of obtaining waveforms over LFM is shown in this Figure for all scenarios; we see that CSNR-based and CMI-based waveforms for all scenarios have a better performance compared with other waveforms.

PROVIDING GENERAL RESULT AND DEFINING AN APPROPRIATE CLOSED-LOOP STRUCTURE

Numerical Results with Different Deterministic Extended Target, Clutter and Noise

Now to better demonstrate the importance of waveform design in cognitive radar, we simulate different targets, clutters and noise realizations. To reach an aggregation, we compute SNR and MI values for each waveform by changing TCR, TNR and CNR values in a specific range. For example in Figure 6(a) and Figure 6(b), SNR and MI changes in dB versus TCR for each waveform are illustrated. It can be concluded from this figure that all waveforms have a similar SNR performance with increasing TCR and as previously discussed have a better performance over LFM waveform.

From MI consideration, as we expected, CMI-based and MI-based have better performance with a little difference compared with CSNR-based and TSF waveforms. Finally, we can conclude that indifferent values of TCR, all waveforms have 9 dB and 7 dB improvement in SNR and MI, respectively over LFM waveform.

Figure 6(c) and Figure 6(d) show the SNR and MI value changes in dB for different TNR values. As indicated in this Figure, CSNR-based and CMI-based waveforms are completely similar in low TNRs but in high TNR values in terms of MI, CMI-based waveform is slightly better. TSF waveform also shows a moderate performance and better than MI-based and LFM waveforms especially in moderate TNRs.

The last case shows the SNR and MI value changes in dB for different CNRs. As indicated in Figure 6(e) and Figure 6(f) for different values of CNRs, similar to those we have for TNR and TCR values, CSNR-based and CMI-based waveforms are completely similar. A noteworthy point is that in low CNR values, MI-based waveform has a good performance but with increasing CNR, its performance declines and TSF waveform performance outperform this waveform in high CNR values. Finally, LFM waveform is fairly constant with CNR and performs the poorly as compared with other waveforms. So, according to the mentioned features for each waveform and due to each application, we should propose a good structure which considers all aspects of the problem conditions. Due to each waveform feature, it should utilize the best waveform to transmit into the environment.

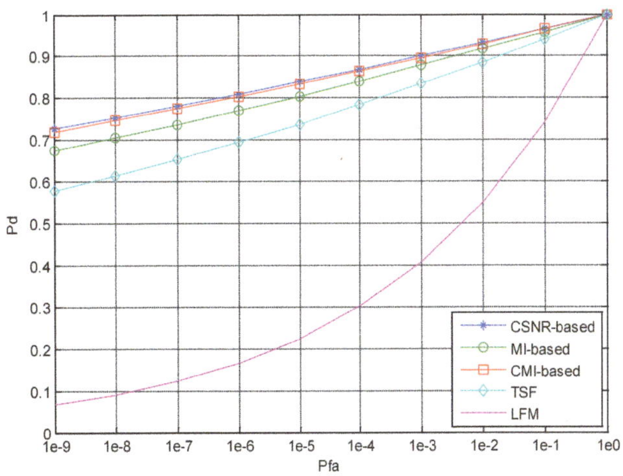

(a)

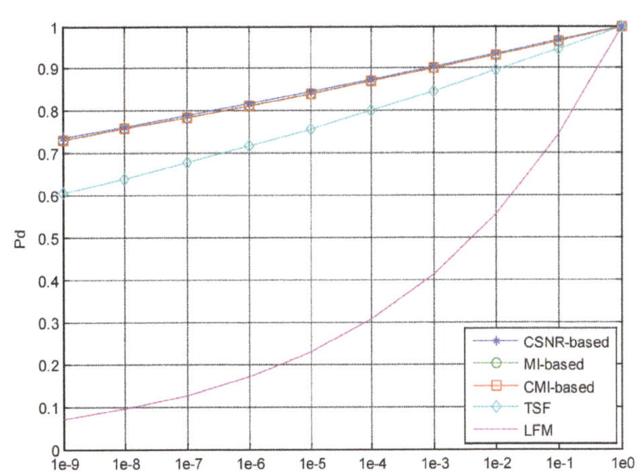

(b)

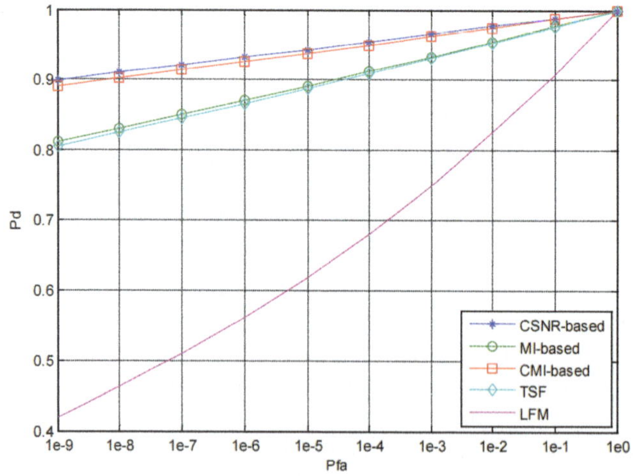

(c)

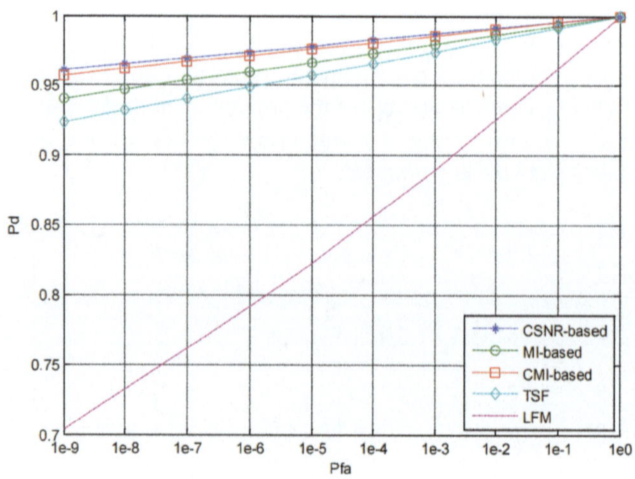

(d)

Figure 5. ROC curves corresponding to each waveform and each scenario. (a) Scenario 1. (b) Scenario 2. (c) Scenario 3. (d) Scenario 4.

Therefore, a cognitive radar should be designed in such a way that firstly evaluates the TCR, TNR and CNR values and considering to the application and also waveform features which are mainly discussed in this paper, chooses the best waveform to transmit into the environment and achieve the maximum probable values of P_D.

Determine a New Closed-Loop Structure for Detecting Deterministic Extended Targets

In Figure 7, new closed-loop structure for detecting M targets indifferent scenarios is proposed. In this structure, SNR and MI values are considered as the cost function. This means that in the transmitter, the desired waveform are selected based on its MI and SNR values.

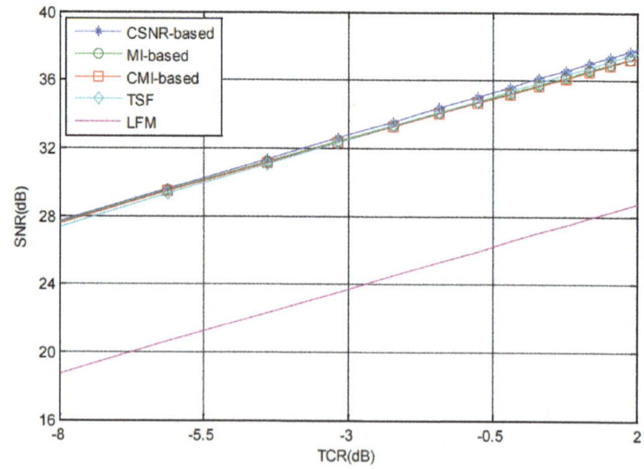

(a)

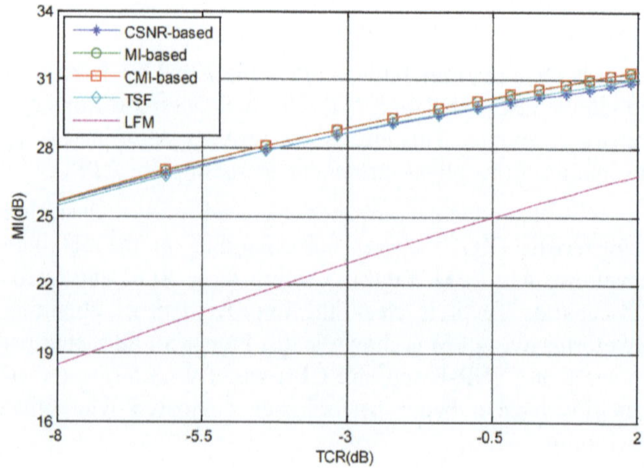

(b)

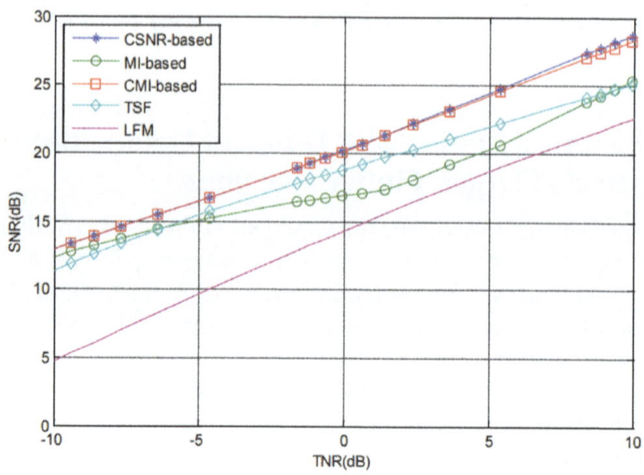

(c)

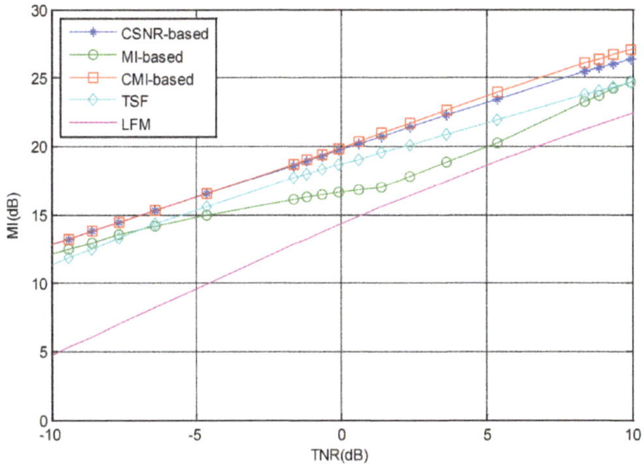

(d)

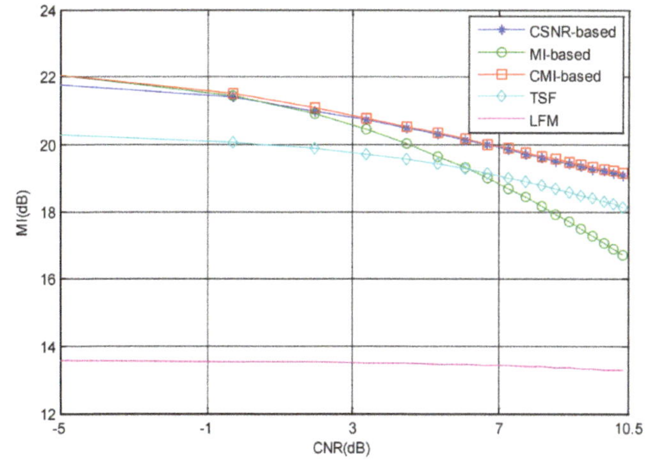

(f)

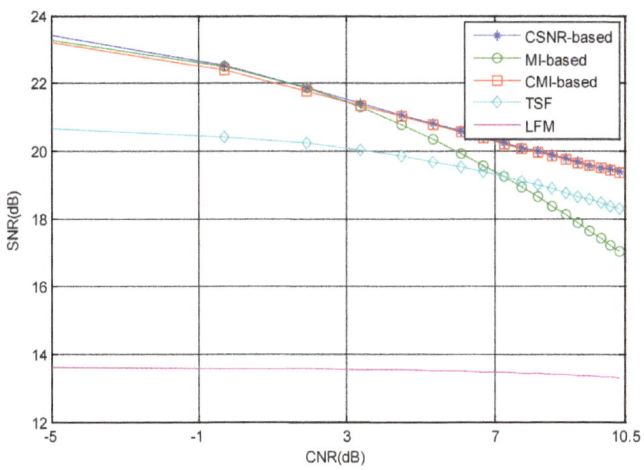

(e)

Figure 6. Comparison of SNR and MI versus TCR, TNR and CNR with deterministic extended target. (a) SNR versus TCR; (b) MI versus TCR; (c) SNR versus TNR; (d) MI versus TNR; (e) SNR versus CNR; (f) MI versus CNR.

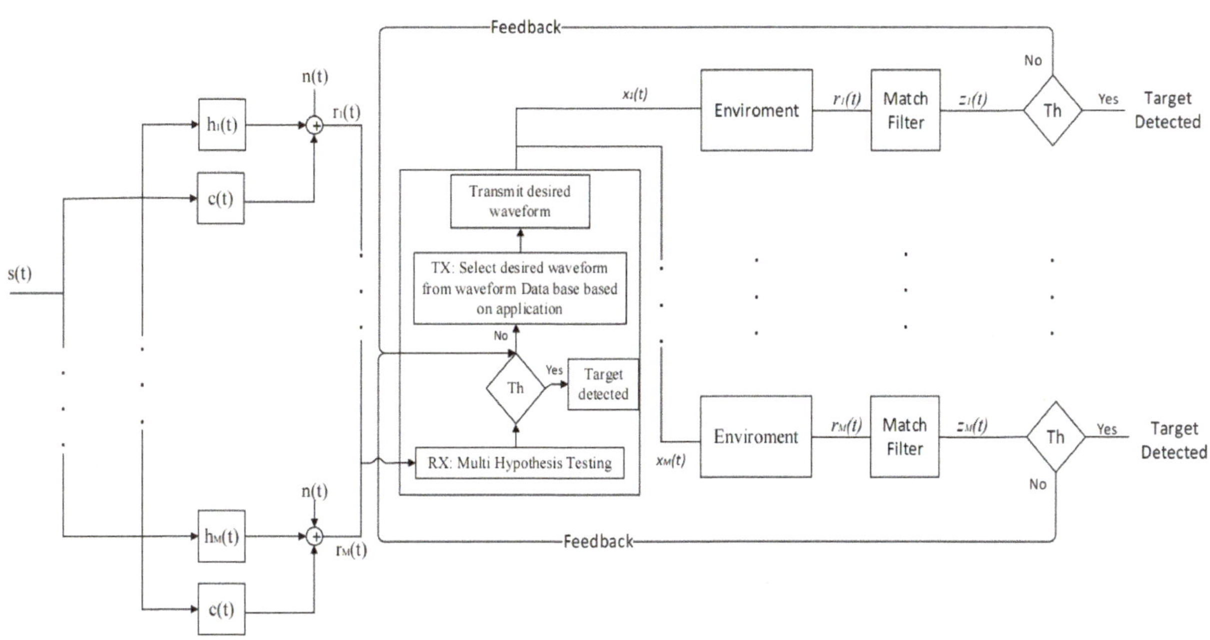

Figure 7. Modelling a new structure of cognitive radar for detecting targets in different noise and clutter scenarios.

As indicated in this figure, the proposed cognitive radar structure includes an adaptive transmitter based on the feedback from the receiver and the interactions with a defined database. The radar transmitted its initial waveform into the environment. If the target is detected in the receiver, we have reached our goal; otherwise, cognitive radar loop is established using a feedback from the receiver to the transmitter and this procedure is done as long as the target of interest is identified. Surely, we can set the number of transmitting signals to a predetermined value to prevent the cognitive radar from being in an infinitive loop.

SUMMARYAND CONCLUSION

In this paper, we considered a deterministic Gaussian extended target with Gaussian clutter case for our experiments. The transmitted waveform optimization is done by maximizing the signal to noise ratio and mutual information between the target frequency responses of different targets and the received signal. This helps achieve a better performance in target recognition. We derived different MI-based, SNR-based and TSF waveforms and used an approximate analytical expression for the probability of detection for different waveforms and compared the performance of these waveforms based on this idea. The remarkable features for designing these waveforms for deterministic extended targets, are energy allocation in which waveforms should allocate their energy into frequency bands where target has the greatest PSD value and clutter has the smallest PSD value. This idea will cause the product of target and transmitted waveform got the maximum amount and then the performance of target detection in the presence of clutter will increase.

The derivations of MI-based waveforms for deterministic Gaussian extended targets and in the case of Gaussian clutter are new results. Moreover, TSF waveform in signal-dependent interference is a new result derived to show that depending on the application such as deterministic extended target, energy allocation is not done well. In fact for the scenarios outlined in the text, it is better to allocate most of transmitting waveform energy into frequency bands in which target has considerable peak amount and clutter amount is negligible instead of distributing energy into all target spectrum peaks. In this paper, different values of SNR and MI are obtained in different TCR, CNR and TNR amounts and it is proven that the obtained waveforms have various performances in different TCR, CNR and TNR and in cognitive radar. We propose a new closed-loop structure considering all features discussed in this paper and for each scenario of target, clutter and noise intelligently find the optimum solution. This work just uses deterministic Gaussian targets case in Gaussian clutter. Future work will attempt to investigate methods of waveforms designing for stochastic extended targets in Gaussian and non-Gaussian environments and suggest a general method for each clutter and target scenarios.

REFERENCES

1. Haykin, S. (2006) Cognitive Radar: A Way of the Future. IEEE Signal Processing Magazine, 23, 30-40. http://dx.doi.org/10.1109/MSP.2006.1593335 [Paper reference:1]
2. Pillai, S.U., Oh, H.S., Youla, D.C. and Guerci, J.R. (2000) Optimum Transmit Receiver Design in the Presence of Signal-Dependent Interference and Channel Noise. IEEE Transactions on Information Theory, 46, 577-584. http://dx.doi.org/10.1109/18.825822 [Paper reference:2]
3. Bell, M.R. (1993) Information Theory and Radar Waveform Design. IEEE Transactions on Information Theory, 39, 1578-1597. http://dx.doi.org/10.1109/18.259642 [Paper reference:1]
4. Bell, M.R. (1988) Information Theory and Radar: Mutual Information and the Design and Analysis of Radar Waveforms and Systems. Ph.D. Dissertation, California Institute of Technology, Pasadena. [Paper reference:1]
5. Kay, S. (2007) Optimal Signal Design for Detection of Gaussian Point Targets in Stationary Gaussian Clutter/ Rever- beration. IEEE Journal of Selected Topics in Signal Processing, 1, 31-41. [Paper reference:1]
6. Garren, D.A., Osborn, M.K., Odom, A.C., Goldstein, J.S., Pillai, S.U. and Guerci, J.R. (2001) Enhanced Target Detection and Identification via Optimized Radar Transmission Pulse Shape. IEE Proceedings—Radar, Sonar and Navigation, 148, 130-138. http://dx.doi.org/10.1049/ip-rsn:20010324 [Paper reference:1]
7. Romero, R. and Goodman, N. (2008) Information Theoretic Matched Waveform in Signal Dependent Interference. 2008 IEEE Radar Conference, Rome, 26-30 May 2008, 1-6. http://dx.doi.org/10.1109/RADAR.2008.4720730 [Paper reference:3]
8. Deng, X., Qiu, C., Cao, Z., Morelande, M. and Moran, B. (2012) Waveform Design for Enhanced Detection of Extended Target in Signal-Dependent Interference. IET Radar, Sonar & Navigation, 6, 30-38. http://dx.doi.org/10.1049/iet-rsn.2010.0275 [Paper reference:9]
9. Jiu, B., Liu, H.W., Li, L.Y. and Wu, S.J. (2009) A Method of Waveform Design Based on Mutual Information. Frontiers of Electrical and Electronic Engineering in China, 4, 134-140. http://dx.doi.org/10.1007/s11460-009-0035-9 [Paper reference:1]
10. Zhang, J.D., Zhu, D.Y. and Zhang, G. (2011) 5Multi-Objective Waveform Design for Cognitive Radar. Proceedings of 2011 IEEE CIE International Conference on Radar, Chengdu, 24-27 October 2011, 580-583. http://dx.doi.org/10.1109/cie-radar.2011.6159607 [Paper reference:1]
11. Romero, R., Bae, J. and Goodman, N. (2011) Theory and Application of SNR and MI-Based Matched Illumination Waveforms. IEEE Transactions on Aerospace and Electronic Systems, 47, 912-927. http://dx.doi.org/10.1109/TAES.2011.5751234 [Paper reference:3]
12. Romero, R.A. and Goodman, N.A. (2009) Waveform Design in Signal-Dependent Interference and Application to Target Recognition with Multiple Transmissions. IET Radar, Sonar & Navigation, 3, 328-340. [Paper reference:1]
13. Romero, R. and Goodman, N.A. (2009) Improved Waveform Design for Target Recognition with Multiple Transmissions. 2009 International Waveform Diversity and Design Conference, Kissimmee, 8-13 February 2009, 26-30. [Paper reference:1]

14. Bae, J. and Goodman, N. (2007) Adaptive Waveforms for Target Class Discrimination. 2007 International Waveform Diversity and Design Conference, Pisa, 4-8 June 2007, 395-399. http://dx.doi.org/10.1109/WDDC.2007.4339450 [Paper reference:1]

15. Wang, B., Wang, J.K. and Song, X. (2012) A Novel Waveform Design Method Based on Mutual Information. 2012 International Conference on Environmental Engineering and Technology Advances in Biomedical Engineering, 8, 373. [Paper reference:1]

16. Wei, Y.M., Meng, H.D., Liu, Y.M. and Wang, X.Q. (2011) Radar Phase-Modulated Waveform Design for Extended Target Detection. Tsinghua Science & Technology, 16, 364-370. http://dx.doi.org/10.1016/S1007-0214(11)70053-0 [Paper reference:3]

17. Karimi, V. and Norouzi, Y. (2013) Target Detection Enhancement Based on Waveform Design in Cognitive Radar. Electronics New Zealand Conference (ENZCon), Auckland, 5-6 December 2013, 40-45. [Paper reference:2]

18. Chen, C.-Y. and Vaidyanathan, P.P. (2009) MIMO Radar Waveform Optimization with Prior Information of the Extended Target and Clutter. IEEE Transactions on Signal Processing, 57, 3533-3544. http://dx.doi.org/10.1109/TSP.2009.2021632 [Paper reference:1]

CHAPTER 12

DESIGN AND REALIZATION OF A DUAL WIDE BAND PRINTED MONOPOLE ANTENNA FOR WIFI AND WIMAX SYSTEMS

Gerard Rushingabigwi, Liguo Sun, Qi Zhu, Yulong Xia, and Yongxia Yu

Keywords: HFSS-IE, Wideband Antenna, WiFi, WiMax

INTRODUCTION

Printed circuit board (PCB) antennas, notably most of patch antennas suffer from narrow bandwidth and lowpower capacity [1].

Disadvantages encountered with Printed microstrip antennas [2]-[8] can be overcome with Printed monopole antenna, notably the narrow bandwidth which limits their uses in modern wideband wireless applications.

Considering the best of HFSS-IE simulator over normal HFSS [9], HFSS-IE simulator has been the selected design tool. The appreciable simulation results motivated us to manufacture this antenna which finally presents coherence while simulation results are compared with measurements.

As applications, this antenna would be utilized for Wireless Local Area Networks (WLAN) systems based on IEEE802.11 as well as Wireless Metropolitan Local Area Network (WMAN) systems based on IEEE802.16a standards.

According to [11] [12], WiMAX systems based on IEEE 802.16a standards are compatible with bands ranging from 2 GHz to 11 GHz, expecting the bit rate from 70 to 100 Mb/s. The literature clarifies that wideband (WB) and ultra wideband (UWB) communication systems have received great attention in the wireless world due to their merits such as high data rate, low cost for short range access and remote sensing applications [13] [14].

In this paper, the reader is noticed that WiFi is interchangeable with WLAN while WiMAX is interchangeable with WMAN.

THE PROPOSED ANTENNA DESIGN AND RESULTS

Antenna is one of the most essential elements that characterize wireless systems.

A transmitted signal is considered UWB if the return loss' absolute bandwidth at −10 dB, exceeds 500 MHz [15]. Printed monopole antennas have been characterized with many possibilities for both wideband and UWB performance [13]-[16].

A couple of monopole antennas were surveyed such as inverted F [17] [18], inverted L [19] [20] and snake-like [21].

Design Methodology and the Proposed Antenna

HFSS-IE Simulator is available with HFSS version 14 and above. HFSS-IE is based on 3D full wave method of moments (MoM) electromagnetic Integral Equation to evaluate the surface currents of the object in question; then it calculates radiation and the scattering fields using the derived current [9] [10].

In our design, Figure 1, the shorter L-shaped radiating element is meant for the frequency band with resonance at 5.6 GHz while the longer L-shaped radiating element corresponds to the frequency band whose resonance is at 2.5 GHz. To tune the antenna for WB around 2.5 GHz and for UWB around 5.6 GHz, an inverted L radiator is printed on less radiation upped area of the antenna.

Citation: Rushingabigwi, G., Sun, L., Zhu, Q., Xia, Y. and Yu, Y. (2016). Design and Realization of a Dual Wide Band Printed Monopole Antenna for WiFi and WiMAX Systems. International Journal of Communications, Network and System Sciences, 9, 184-197. DOI: 10.4236/ijcns.2016.95018.

Simulated Results and Impedance (Z) Parameters

The return loss (RL) is such an important antenna characteristic that, throughout the design process, the RL is analysed to decide on the necessary bandwidth performance requirements. According to [22], the RL is defined as a measure of how much of the available power is not delivered to the load; a matched load has a zero reflection coefficient ($\Gamma = 0$) and thus has an infinity RL.

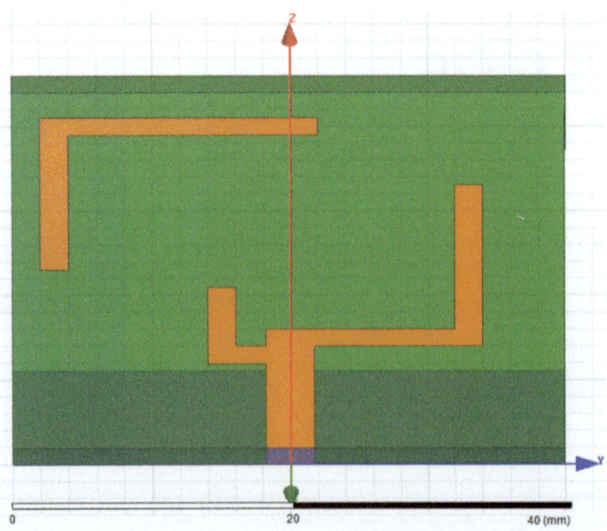

Figure 1. Three dimensional (3D) view of proposed antenna.

$$RL = -20\log|\Gamma| dB \tag{1}$$

After all the parametric analysis and optimization, our design model's RL is presented in Figure 2 and each band's impedance parameters are measured according to Figure 3.

The Smith Chart, Figure 4, shows the perfect matching of antennas' impedance with the 50 Ω feed-line, for both f_1 = 2.5GHz, and f_2 = 5.6 GHz.

Figure 5 and Figure 6 show the radiation patterns while Figure 7 and Figure 8 show the radiation fields overlay and surface currents distribution for both frequency bands.

Total Efficiency and Voltage Standing Wave Ration (VSWR)

The antenna total efficiency is defined as "the ratio of radiated power to the incident power, which is approximated to e_T [23], such that

$$e_T = e_r e_c e_d \tag{2}$$

where:

T_e is the total efficiency;

r_e is the mismatch efficiency, such that

$$e_r = 1 - |\Gamma|^2 \tag{3}$$

e_c is the conduction efficiency;

d_e is the dielectric efficiency;

$e_{cd} = e_c e_d$ is the antenna radiation efficiency.

Γ is the voltage reflection at the input antenna terminals,

$$\Gamma = \frac{Z_{in} - Z_0}{Z_{in} + Z_0} \tag{4}$$

Z_{in} is the antenna input impedance;

Z_0 is the transmission feed line's characteristic impedance;

The voltage standing wave ratio (VSWR) is generally referred to as the measure of antenna impedance matching with the feed line's impedance. Mismatches result in standing waves (SW) along the feed line. VSWR is mathematically defined.

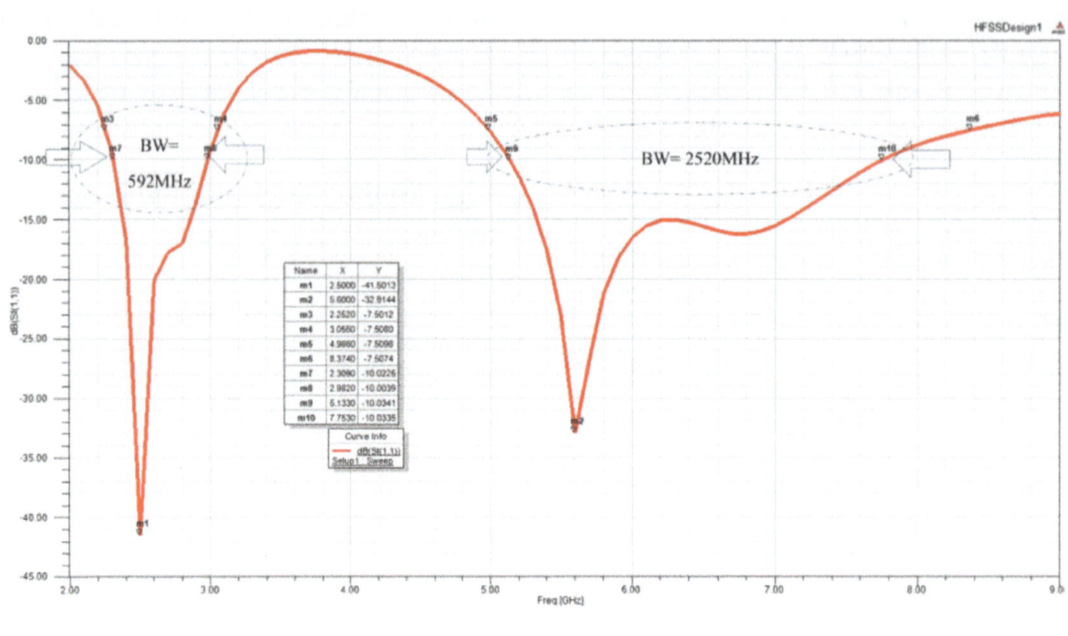

Figure 2. The simulated RL.

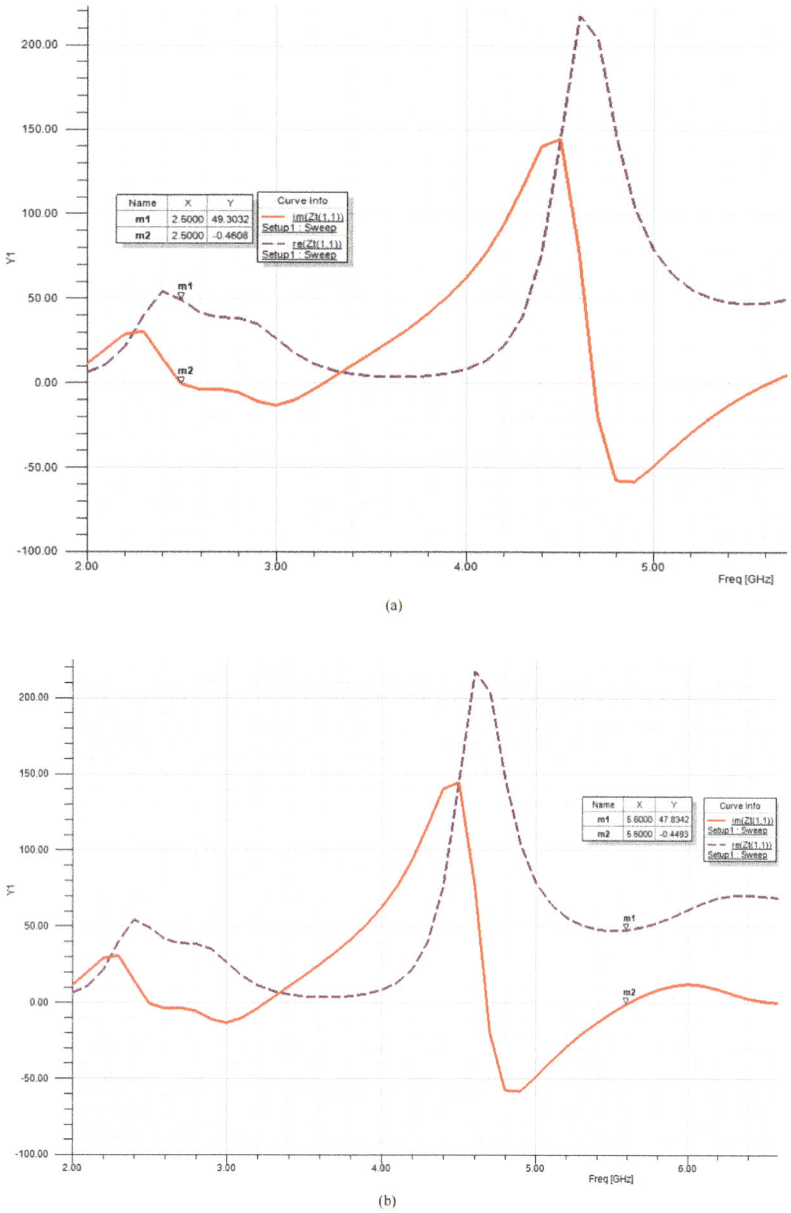

Figure 3. Impedance (Z) parameters: (a) for 2.5 GHz band; (b) for 5.6 GHz band.

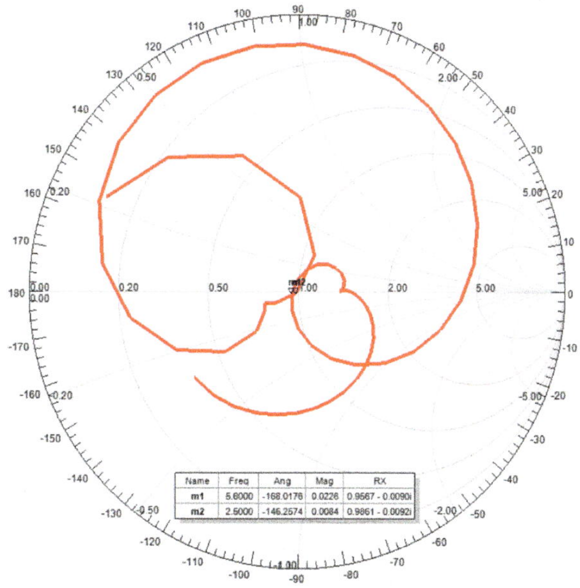

Figure 4. The Smith Chart's impedance measurement.

$$\text{VSWR} = \frac{1+|\Gamma|}{1-|\Gamma|} \quad (5)$$

In case of our design, the antenna radiation efficiency (e_{cd}) is assumed unity since the antenna is simulated under perfect electric conduction (PEC) boundary and measurements after implementation were done under well isolated environment; which means the total antenna efficiency evaluated here is equal to the mismatch efficiency.

Thus, referring to the measured antenna impedances in Figure 3; keeping in mind that the standard impedance for the microstrip feed-line is 50 Ω, the total antenna efficiency is now calculated for both 2.5 GHz and 5.6 GHz respectively, according to Equations (3) and (4); the VSWR is computed according to (5).

When the antenna is operated at 2.5 GHz,

$$\Gamma = \frac{49.3 - j0.46 - 50}{49.3 - j0.46 + 50} = -(0.007 + j0.0046)$$

$e_{T(2.5GHz)} = 1 - |\Gamma|^2 = 99.16\%$

VSWR = 1.017.

When the antenna is operated at 5.6 GHz,

$$\Gamma = \frac{47.83 - j0.45 - 50}{47.83 - j0.45 + 50} = -(0.0221 + j0.0045)$$

$e_{T(5.6GHz)} = 1 - |\Gamma|^2 = 97.74\%$ [2]

VSWR = 1.046.

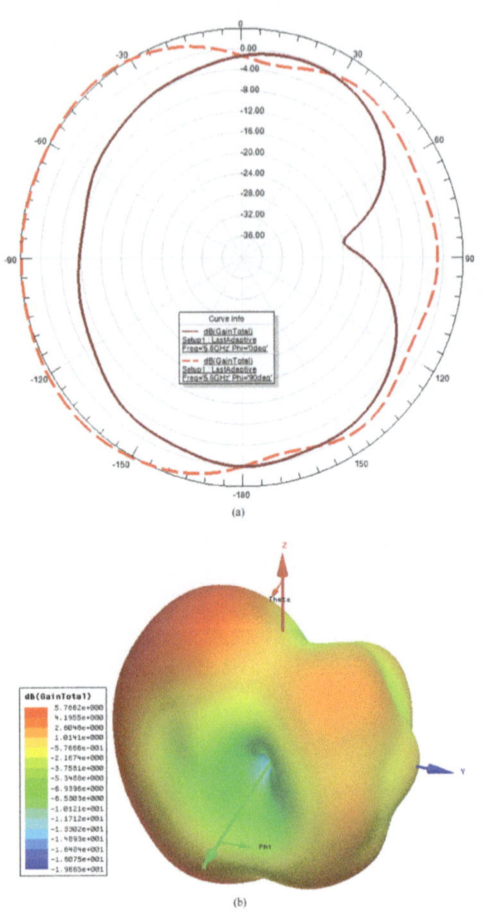

Figure 5. Radiation patterns at 5.6 GHz (a) E-H Radiation Pattern; (b) 3D Polar plot.

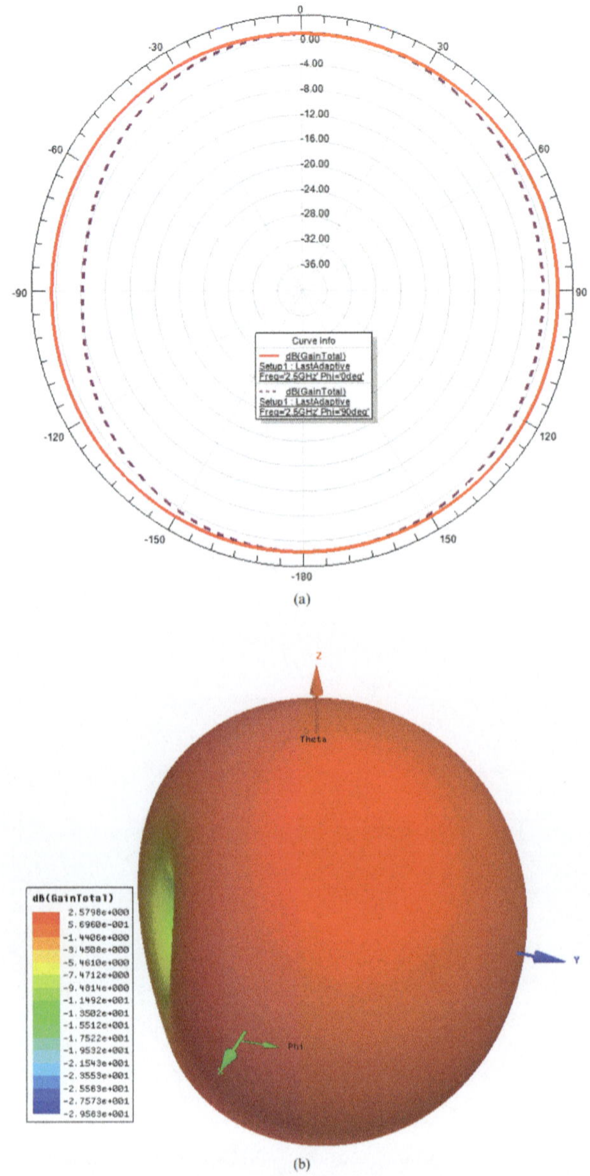

Figure 6. Radiation patterns at 2.5 GHz (a) E-H Radiation Pattern; (b) 3D Polar plot.

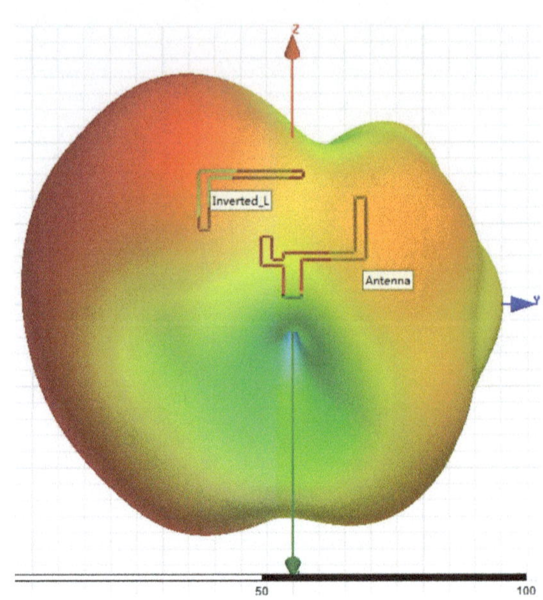

(a)

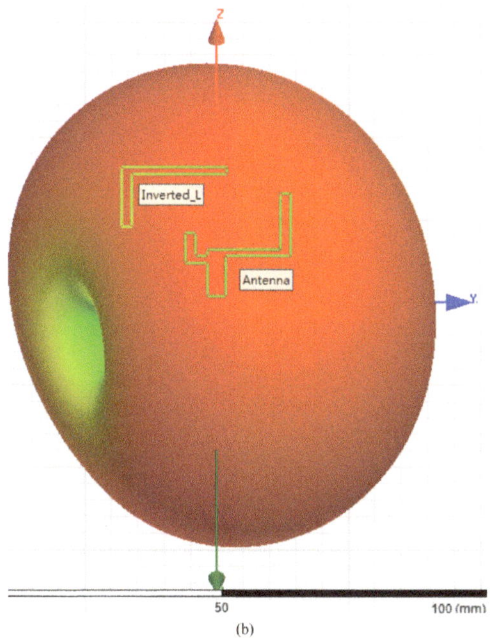

Figure 7. 40% Radiation fields overlays: (a) at 5.6 GHz; (b) at 2.5 GHz.

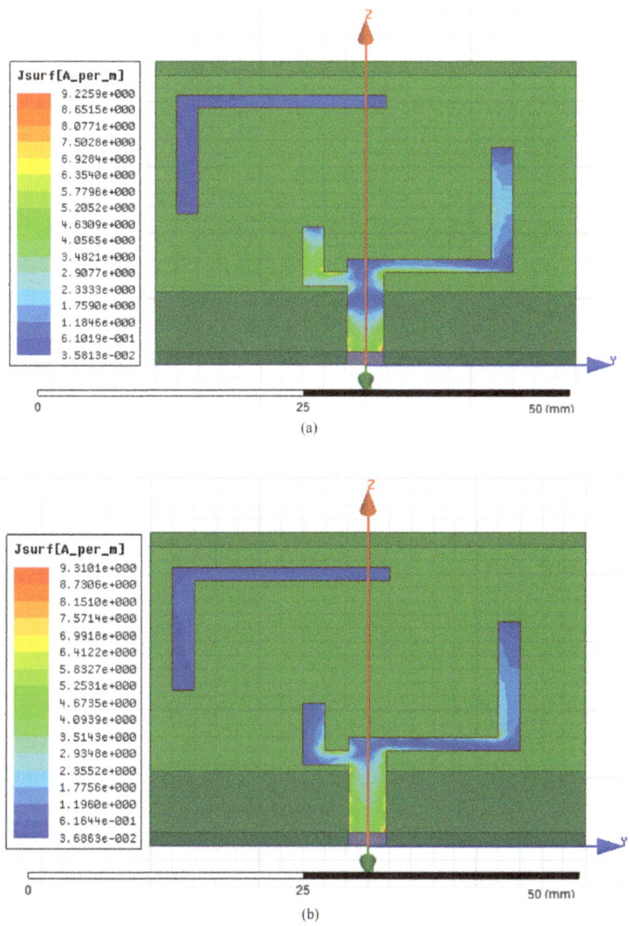

Figure 8. Surface current distribution: (a) at 5.6 GHz; (b) at 2.5 GHz.

Manufacturing Results

With the satisfactory simulation results in hands, the antenna design model was manufactured as per pictures in Figure 9. The two dimensional (2D) radiation test results for one sample product presented in Figures 10-12 are coherent with the simulated results.

DISCUSSIONS

Analyzing the simulation RL, the −10 dB bandwidth (BW) approximates to 592 MHz, or 2390 MHz - 2982 MHz in the first frequency band as well as 252 MHz, say 5133 MHz - 7753 MHz in the second band. With these bandwidths, the design qualifies for dual wideband antenna [15].

The antenna mismatch total efficiency is very good in both frequency bands.

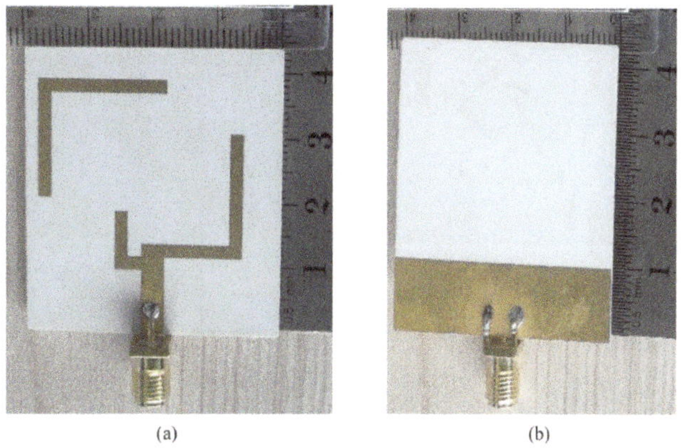

Figure 9. Picture of the manufactured antenna: (a) top view; (b) bottom view.

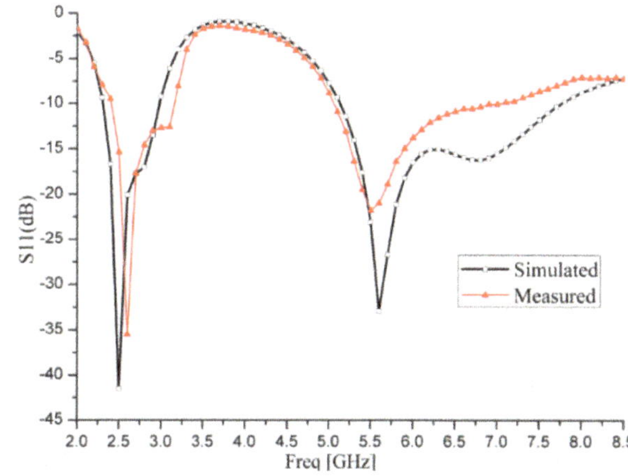

Figure 10. Simulated versus measured RL.

Looking at the measurements results, however, this antenna suffers from losses and signal degradation due to interconnect effects [24], just upon the port welding.

It has been noticed that the port may not be soldered by directly pressing it against the substrate's edge; rather, a small gap would be left or otherwise the radiated power distorts. On the other hand, when the gap in between the port and the substrate edge is slightly increased to about 1mm, the environmental conditions interfere to af-fect the integrity of the signal transmitted to the antenna.

This antenna is a prototype sample which was not packaged in any commercial product. The encountered signal integrity problems due to port soldering would be carefully solved whenever preparing this antenna for the real applications of a miniature antenna for WiFi and WiMAX systems.

CONCLUSION

All the pre-set goals have been achieved. The antenna's overall performances were proven by both simulation and manufacturing results. The antenna was manufactured by a competent company while the related measurements were conducted in the University. For the tested three samples, results are all coherent; however, only one sample's 2D measured radiation patterns are presented in this article. The designed, manufactured and tested/measured antenna system would be subjected to final product manufacturing, especially when there is any industrial request.

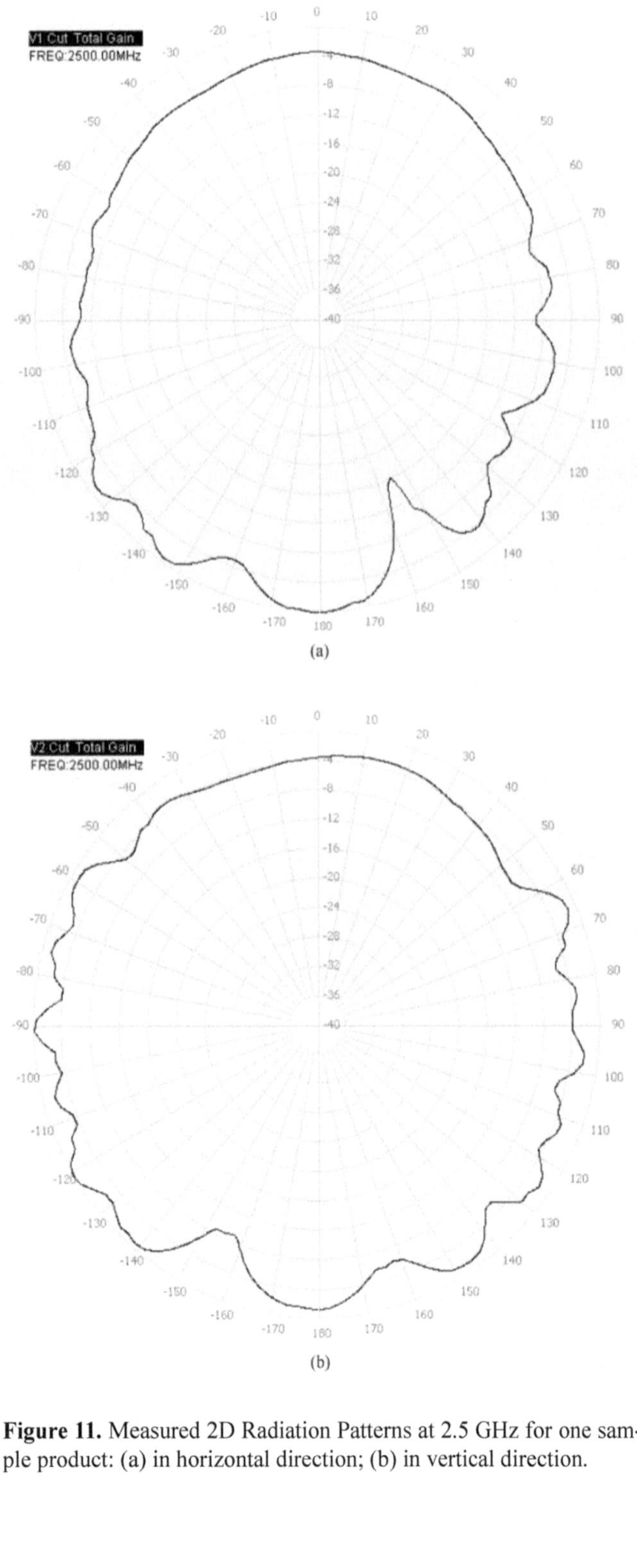

Figure 11. Measured 2D Radiation Patterns at 2.5 GHz for one sample product: (a) in horizontal direction; (b) in vertical direction.

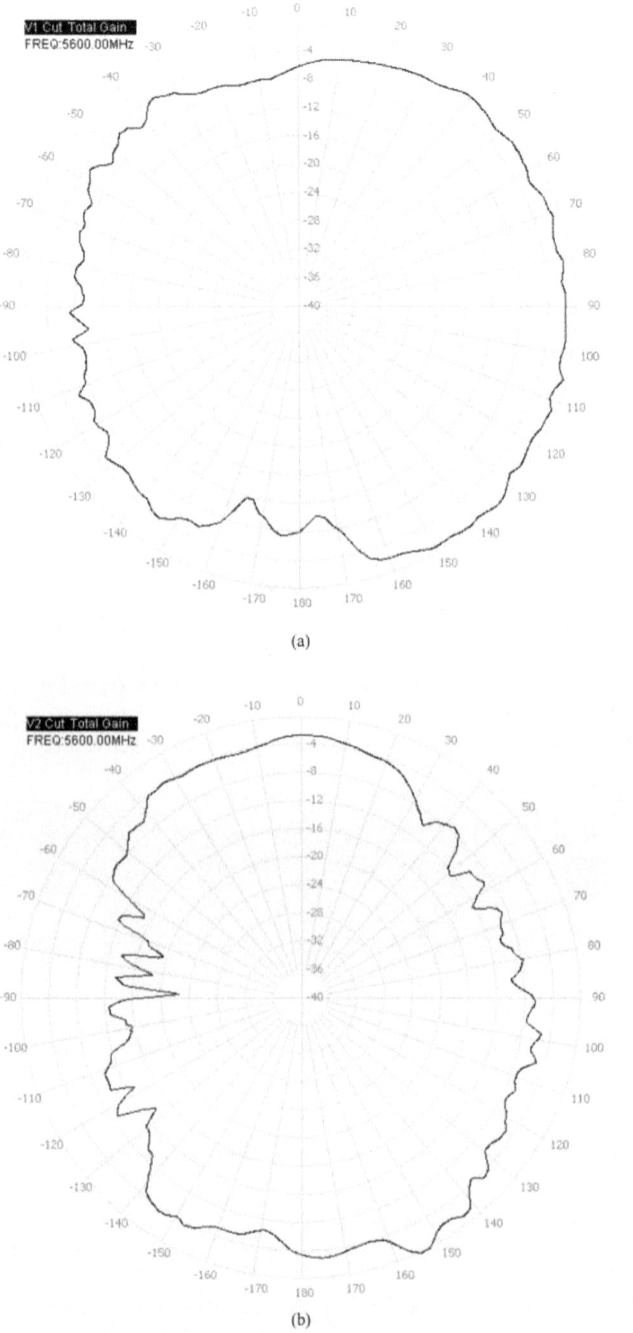

Figure 12. Measured 2D Radiation Patterns at 5.6 GHz for one sample product: (a) in horizontal direction; (b) in vertical direction.

ACKNOWLEDGEMENTS

A lot of gratitude is addressed to the Government of People's Republic of China, to have supported and strengthened engineering research activities in the University of Science and Technology of China (USTC). Many thanks also go to the University of Rwanda, college of Science and Technology (UR, CST) for a couple of valuable supports.

REFERENCES

1. Pozar, D.M. (1983) Considerations for Millimeter-Wave Printed Antennas. *IEEE Transactions on Antennas and Propagation*, 31, 740-747.

2. Balanis, C.A. (2005) Microstrip Antennas. In: *Antenna Theory Analysis and Design*, 3rd Edition, A John Wiley & Sons, Inc., Publication, Ch. 14, 811-879.
3. Lelaratne, R. and Langley, R.J. (2000) Dual-Band Patch Antenna for Mobile Satellite Systems. *IEE Proceedings - Microwaves, Antennas and Propagation*, 147, 427-430. http://dx.doi.org/10.1049/ip-map:20000864
4. Huq, K.R.M., Siraj, A.S., Khan, M.I. and Shama, N. (2014) Design of a Triple Band Microstrip Patch Antenna for Cellular and Wi-Fi Applications. 2014 *International Conference on Informatics, Electronics & Vision*, Dhaka, 1-6. http://dx.doi.org/10.1109/ICIEV.2014.6850839
5. Patel, S.K. and Kosta, Y.P. (2011) E-shape Microstrip Patch Antenna Design for GPS Application. 2011 *Nirma Uni-versity International Conference on Engineering* (*NUiCONE*), Ahmedabad, 1-4. http://dx.doi.org/10.1109/nuicone.2011.6153261
6. Ibrahim, R.A.R., Yagoub, M.C.E. and Habash, R.W.Y. (2009) Microstrip Patch Antenna for RFID Applications. *CCECE* 09 *Canadian Conference on Electrical and Computer Engineering*, St. John's, 940-943. http://dx.doi.org/10.1109/ccece.2009.5090266
7. Mokhtar, M.H., Rahim, M.K.A., Murad, N.A. and Majid, H.A. (2013) A Compact Slotted Microstrip Patch Antenna for RFID Applications. 2013 *IEEE International Conference on RFID-Technologies and Applications* (*RFID-TA*), Jo-hor Bahru, 1-4. http://dx.doi.org/10.1109/RFID-TA.2013.6694536
8. Dwivedi, S., Rawat, A. and Yadav, R.N. (2013) Design of U-Shape Microstrip Patch Antenna for WiMAX Applica-tions at 2.5 GHz. 2013 *Tenth International Conference on Wireless and Optical Communications Networks* (*WOCN*), Bhopal, 1-5. http://dx.doi.org/10.1109/WOCN.2013.6616214
9. Mingyang, L. and Liu, M., Ed. (2013) Projects in HFSS-IE Solver. In: *HFSS from Beginning to Proficiency*, Posts and Telecom Press, Beijing, Ch. 11, 338-347.
10. Mingyang, L. and Liu, M., Ed. (2014) Monopole Antenna and Dipole Antenna Design. In: *HFSS Antenna Design*, 2nd Edition, Publishing House of Electronics Industry, Beijing, Ch. 3, 22-43.
11. Zhang, Y. and Ansari, N. (2010) Wireless Telemedicine Services over Integrated. IEEE 802.11/WLAN and IEEE 802.16/WIMAX Networks. *IEEE Wireless Communications*, 17, 30-36. http://dx.doi.org/10.1109/MWC.2010.5416347
12. Li, B., Qin, Y., Low, C.P. and Gwee, C.L. (2007) A Survey on Mobile WiMAX Wireless Broadband Access. *IEEE Communications Magazine*, 45, 70-75. http://dx.doi.org/10.1109/MCOM.2007.4395368
13. Cui, Y.-H., Li, R.L. and Wang, P.A. (2013) Novel Broadband Planar Antenna for 2G/3G/LTE Base Stations. *IEEE IEEE Trans. IEEE Transactions on Antennas and Propagation*, 61, 2767-2774. http://dx.doi.org/10.1109/TAP.2013.2244837
14. Lak, H.J., Ghobadi, C. and Nourinia, J. (2011) A Novel Ultra-Wideband Monopole Antenna with Band-Stop Characte-ristic. *Wireless Engineering and Technology*, 2, 235-239. http://dx.doi.org/10.4236/wet.2011.24032
15. Tatsis, G., Raptis, V. and Kostarakis, P. (2010) Design and Measurements of Ultra-Wideband Antenna. *International Journal of Communications, Network and System Sciences*, 3, 116-118. http://dx.doi.org/10.4236/ijcns.2010.32017
16. Bakariya, P.S. and Dwari, S. (2012) A Compact Super Ultra-Wideband (UWB) Printed Monopole Antenna. *IEEE 5th International Conference on Computer and Devices for Communication* (*CODEC*), Kolkata, 1-3. http://dx.doi.org/10.1109/codec.2012.6509206
17. Tan, Q. and Erricolo, D. (2007) Comparison between Printed Folded Monopole and Inverted F Antennas for Wireless Portable Devices. *IET Microwaves, Antennas & Propagation*, Honolulu, 4701-4704.
18. Soras, C., Karaboikis, M., Tsachtsiris, G. and Makios, V. (2002) Analysis and Design of an Inverted-F Antenna Printed on a PCMCIA Card for the 2.4GHz ISM Band. *IEEE Antennas and Propagation Magazine*, 44, 37-44.
19. Ni, W. and Nakajima, N. (2010) Small Printed Inverted-L Monopole Antenna for Worldwide Interoperability for Mi-crowave access Wideband Operation. *IET Microwaves, Antennas & Propagation*, 4, 1714-1719. http://dx.doi.org/10.1049/iet-map.2009.0469
20. Chen, H.-D., Chen, J.-S. and Cheng, Y.-T. (2003) Modified Inverted-L Monopole Antenna for 2.4-5GHz Dual-Band Operations. *IET Electronics Letters*, 39, 1567-1568. http://dx.doi.org/10.1049/el:20031037
21. Rushingabigwi, G. and Sun, L.G. (2015) Design of an 868 MHz Printed S-Shape Monopole Antenna. *Journal of Computer and Communications*, 3, 49-55. http://dx.doi.org/10.4236/jcc.2015.33009
22. Steer, M. (2010) Transmission Lines. In: *Microwave and RF Design. A System Approach.* SciTech Publishing, Raleigh, Ch. 4, 196-197.
23. Balanis, C.A. (2005) Fundamental Parameters of Antennas (Antenna Efficiency). In: *Antenna Theory Analysis and De-sign*, 3rd Edition. A John Wiley & Sons, Inc., Publication, Ch. 2, 64-66.
24. Stephen, H.H., Garrett, W.H. and James, A.M.-C. (2000) Non-ideal Interconnect Issues. In: *High-Speed Digital System Design—A Handbook of Interconnect Theory and Design Practices*, John Wiley & Sons, Inc., Ch. 4, 70-92.

CHAPTER 13

INTEGRATING A TRIMBLE RECON X 400 MHZ INTEL PXA255 XSCALE CPU® MOBILE FIELD DATA COLLECTION SYSTEM USING DIFFERENTIALLY CORRECTED GLOBAL POSITIONING SYSTEM TECHNOLOGY AND A REAL-TIME BIDIRECTIONAL ACTIONABLE PLATFORM WITHIN AN ArcGIS CYBERENVIRONMENT FOR IMPLEMENTING MOSQUITO CONTROL

Benjamin G. Jacob and Robert J. Novak

Keywords: Trimble Recon X, Vector Control Management System (VCMS), Anopheles arabiensis, QuickBird Imagery

INTRODUCTION

Currently, very few mosquito control programs have collected valid outcome metrics that enable public health officials and funding agencies to quantitatively determine the degree and magnitude of their impact on time series, field or remote-sampled aquatic, larval, habitat, endemic, transmission-oriented, explanatory, parameter estimators. A principle reason for this deficiency is the absence of sufficient management systems to capture, manage and analyze large amount of diverse data being generated by various mosquito control activities. Existing management information is neither accurate nor comprehensive, a vital necessity for efficient, effective and economic mosquito management [1]. Geographic location, combined with knowledge of ecological features of multiple seasonal-sampled georeferenced vector mosquito aquatic larval habitat sites may be a potential key element for implementing efficient and cost effective larval control measures. A remote evaluation of mosquito treatment data offers a systematic way of comparing the costs and consequences of interventions in order to improve the allocation of resources by prioritizing georeferenced vector aquatic mosquito habitats based on seasonal larval/pupal productivity [2]. Treatments or habitat perturbations should be based on surveillance of larvae in the most productive areas of an ecosystem [3]. Additionally, a remote surveillance system can be used more widely to improve the uptake of existing effective control measures and ensure the maximum impact from the introduction of new technologies.

Personal Digital Assistant (PDA) technologies offer a unique way to collect and manage multiple, ecological, seasonal-sampled, vector, mosquito-related, endemic, transmission-oriented data variables. Developing and implementing streamlined data collection, aggregation and reporting methodologies using PDA handheld computers

Citation: Jacob, B. and Novak, R. (2014). Integrating a Trimble Recon X 400 MHz Intel PXA255 Xscale CPU® Mobile Field Data Collection System Using Differentially Corrected Global Positioning System Technology and a Real-Time Bidirectional Actionable Platform within an ArcGIS Cyber-environment for Implementing Mosquito Control. Advances in Remote Sensing, 3, 141-196. DOI: 10.4236/ars.2014.33012.

equipped with global positioning systems (GPS) can provide geographically detailed, real-time information when conducting surveys in remote villages [4]. While proper identification of mosquito species and knowledge of their bionomics focuses on control efforts, remote inspection of georeferenced larval habitats and weekly trapping for adult mosquitoes ensures knowledge of the mosquito population in a given municipality [3]. Accurate, georeferenced, collection data is crucial for understanding mosquito biogeography, ecology, and the impact of environmental changes, as well as for species distribution modeling, planning mosquito surveys, and for determining disease risk [2]. Field teams require a simple visual mechanism to help designate priority treatment of seasonal-sampled, georeferenced, vector, mosquito, aquatic, larval habitats within a project area, in order to pro- mote on-going tracking in remote communities, which can maximize treatment efficiency. A PDA mobile collection system combined with newer GIS software could help prioritize health interventions based on field or remote specified, vector, mosquito-related, georeferenced, explanatory, endemic, transmission-oriented, covariate coefficients in a more standardized fashion.

The evolution of PDA technology, geographic information systems (GIS), GPS, and remote sensing (RS) data may enable spatiotemporal-sampled field collections and data analysis of ecological georeferenced vector mosquito aquatic larval habitat data in ways that were not possible before. Unfortunately, the rapid development and integration of remote technologies has created many new tools which has also widened the "digital divide", leaving many vector biologists, epidemiologists and other data analysts with little understanding of the technology and potential applications of current geospatial techniques for robust, seasonal, vector, mosquito, aquatic, larval habitat predictive, endemic, transmission-oriented, epidemiological, risk modeling. For example, sub- resolution RS (e.g., panchromatic QuickBird at 0.61 m pixel spatial resolution) data can be used to provide unique thematic information regarding a variety of time series georeferenced explanatory georeferencable vector mosquito aquatic larval habitat potential explanatory endemic transmission-oriented geo-biophysical characteristics, including surface temperature, imperviousness, water habitat clarity, evapotranspiration, vegetation pigments, biomass, canopy structure, leaf area index (LAI), and soil moisture [2]. Leaf Area Index (LAI) is a dimensionless quantity that characterizes plant canopies which is defined as the one-sided green leaf area per unit ground surface area (LAI = leaf area/ground area, m^2/m^2) in broad leaf canopies [5].

There are only limited contributions in literature of robust, biophysical, geo-spatiotemporal, endemic, transmission-oriented, vector, mosquito-related, predictive, epidemiological, risk models using data collected employing PDAs, however in these contributions biophysical technologies have provided detailed, local terrestrial geographic data [e.g., vegetation land use land cover (LULC), meandering riverine pathways)]. For example, Thorp et al. (2012) employed remote sensing technology to provide spatial information on crop growth status of durum wheat (Triticum durum) study by inverting a radiative transfer model and then generating multiple time series, eco-physiological, sub-models for estimating the crop's biophysical properties. An Atmospheric radiative transfer model, code or simulator calculates radiative transfer of electromagnetic radiation through a planetary atmosphere [1]. The outcome of the model inversion procedure was influenced by the timing and availability of remote sensing data, the spectral resolution of the data, the types of models implemented, and the choice of parameters that were adjusted in the submodel residually forecasted derivatives. The objective was to investigate seasonal phenological issues by inverting a linked radioactive transfer and then quantitating the eco-physiological model derivatives to estimate LAI, canopy weight, plant nitrogen content, and yield for the durum wheat (Triticum durum). The study was conducted in central Arizona over the winter of 2010-2011. Observations of crop canopy spectral reflectance between 268 and 1095 nm were obtained weekly using a GER 1500 spectroradiometer. Other field measurements were regularly collected to describe plant growth characteristics and plant nitrogen content. Linkages were developed between the DSSAT Cropping System Model (CSM) and the PROSAIL radiative transfer model (CSM-PROSAIL) and between the DSSAT-CSM and an empirical model relating vegetation LULC parameter (CSM-Choudhury). A parameter estimation algorithm (i.e., PEST) was implemented to adjust the leaf area growth parameters of the CSM by minimizing error between measured and simulated seasonal canopy spectral reflectance. A genetic algorithm was implemented to identify the optimum combination of the remote sensing observations and to optimize simulations of multiple vegetation indices through model inversion. For example, the relative root mean squared error (RRMSE) between measured and simulated LAI was 24.1% for the CSM-PROSAIL model, whereas the stand-alone PROSAIL and CSM models simulated LAI with RRMSEs of 40.7% and 27.8%, respectively. Wheat yield was simulated with RRMSEs of 12.8% and 10.0% for the lone CSM model and the CSM-PROSAIL model, respectively. Optimized leaf area growth parameters for CSM-PROSAIL were different among cultivars ($p < 0.05$), while those for CSM- Choudhury were not. Only two observations, one at mid-vegetative growth and one at maximum vegetative growth, were required to optimize LAI simulations for CSM-PROSAIL, whereas CSM-Choudhury required four observations. Inverting CSM-PROSAIL using hyperspectral data offered several advantages as compared to the CSM-Choudhury inversion using a simple vegetation index, including better estimates of crop biophysical properties, different leaf area growth parameter estimates among cultivars ($p < 0.05$), and fewer required remote sensing observations for optimum LAI simulation.

Methods for constructing a seasonal explanatory geobiophysical geographic predictive real-time vector-re- lated remotely sensed mosquito models include: Principal Component Analysis, Unsupervised Classification, Map Algebra, Band Ratios, and Regression of reflection values [2]. Imaging georeferenced, seasonal-sampled, vector, mosquito-related, spectrometry data, in conjunction with suitable analysis techniques in a cyber-envi- ronment, may provide a basis for quantitatively measuring phenological change in seasonal, terrain-related, georeferenced, vector, mosquito, aquatic, larval habitats which results from changes in primary productivity and canopy vegetation vigor. The phenological changes of seasonal, vector, mosquito, aquatic, larval habitats may be in response to regional and/or global-scale environmental or climatic changes.

Two assumptions must be satisfied if PDA technology, GIS, GPS, and imaging spectrometry are to be useful in remote biophysical analysis of seasonal, vector, mosquito, aquatic, larval habitats. First, there must be a strong correlation between georeferencable, explanatory, operationizable, canopy-shaded, larval, habitat characteristics and the rates at which processes important to the biosphere occur. Secondly, these canopy-shaded characteristics must be successfully measured using high spectral resolution remote-sensing data. These include the measurement of visible and near infra-red (NIR) spectral position for constructing normalized difference vegetation index (NDVI), moisture stress index and shortwave infrared reflectance parameters related to canopy chemistry parameters and terrestrial imaging spectrometry.

The NDVI is a simple graphical indicator that can be used to analyze remote measurements, typically but not necessarily from a space platform, and assess whether the target being observed contains live green vegetation or not [1]. Leaf cells have also evolved to scatter solar radiation in the NIR spectral region (which carries approximately half of the total incoming solar energy), because the energy level per photon in that domain (wavelengths longer than about 700 nanometers) is not sufficient to be useful to synthesize organic molecules. Although electromagnetic waves have wavelengths that range from 1 nanometer (nm) (10^{-9} meter) for x-rays to 1 kilometer (10^3 meters) for radio waves wavelengths of visible light are in the range of 400 to 700 nanometers. Jacob et al. [2] employed the equation E = (hc)/L in a MapTP Mobile, web-based mapping applications where E was the energy of photon in a seasonal, vector, malarial, mosquito-related (e.g., Anopheles arabiensis s.s.), aquatic, larval, habitat, spectral, endmember, forecasting, predictive, epidemiological, risk model, h was the Planck's constant, 6.626 × 10^−34 J·s, c was the speed of light, 2.998 × 10^8 m/s and L was wavelength for converting nanometers to units of joules (J), whereby, 715 nm = 7.15 × 10^−7 m and then E = (6.626 × 10^−34 J·s)*(2.998 × 10^8 m/s)/(7.15 × 10^−7 m) = 2.778 × 10^−19 J. The authors noted that live green plants associated with the seasonal-sampled Anopheles arabiensis s.s. aquatic larval habitats appeared relatively dark in the photosynthetically active radiation (PAR) and relatively bright in sub-meter resolution (panchromatic QuickBird 0.61 m resolution) NIR region of the electromagnetic spectrum. Photosynthetically active radiation designates the spectral range (wave band) of solar radiation from 400 to 700 nanometers that photosynthetic organisms are able to use in the process of photosynthesis [1]. PAR explanatory covariate coefficients were employed for constructing multiple, biophysical, seasonal, An. arabiensis s.s.-related, endemic, transmission-oriented, predictive, explanatory, eco-epidemiological risk models where the forecasted derivatives were quantified as μmol photons $m^{-2} \cdot s^{-1}$, a measure of the photosynthetic photon flux (area) density, or PPFD. The derivatives were also expressed as Einstein units (i.e., $\mu E \cdot m^{-2} \cdot s^{-1}$). PAR action spectrum, revealed absorption spectra for chlorophyll-A, chlorophyll-B, and carotenoids related to the sampled aquatic larval habitats. Differential reflection in the red and infrared (IR) bands enabled remotely monitoring density and intensity of green vegetation growth using the spectral reflectivity of solar radiation. By contrast, clouds tended to be rather bright in the red (as well as other visible wavelengths) for some of the imaged, georeferenced, (e.g., Anopheles arabiensis s.s.), aquatic, larval, habitat objects and quite dark in the NIR.

Optimally, a seasonal, vector, mosquito-related, aquatic, larval habitat image should have less than 20% cloud coverage for constructing forecasting, epidemiological, endemic, transmission-oriented, risk-related, predictive models [2]. The pigment in plant leaves, chlorophyll, strongly absorbs visible light (from 0.4 to 0.7 μm) for use in photosynthesis [1] which has been known to be associated with seasonal-sampled malaria mosquito aquatic larval habitats. For example, Mwangangi et al. (2007) conducted investigations using laboratory and field data on experimental riceland plots in the Mwea, Rice Scheme, Kenya. In these trials, chlorophyll derivatives were added to the infested plots which were subsequently ingested by the An. arabiensis s.s. larvae and the accumulated photoactive compound (photosensitizer) inside the larvae body induced sunlight exposure which created an oxidation stress that resulted in organism death.

The cell structure of the leaves, on the other hand, strongly reflects near-infrared light (from 0.7 to 1.1 μm). The more leaves a plant associated to a georeferenced vector mosquito-aquatic larval habitat has, the more these wavelengths of light are affected, respectively. Since early instruments of Earth Observation, such as NASA's ERTS and NOAA's AVHRR, acquired data in visible and NIR, it was natural to exploit the strong differences in plant reflectance to determine their spatial distribution in these satellite images. Dube et al. (2011) studied repellent properties of volatiles emitted from ethnomedicinal plant leaves against malaria and yellow fever vectors in Ethiopia. In their research plant-based mosquito repellents, volatile emanations were investigated from five plant species, Corymbia citriodora, Ocimum suave, Ocimum lamiifolium, Olea europaea and Ostostegia integrifolia, traditionally used in Ethiopia as protection against mosquitoes. The behaviour of two mosquitoes, the malaria vector An. arabiensis and the arbovirus vector Aedes aegypti, was assessed towards volatiles collected from the headspace of fresh and dried leaves, and the smoke from burning the dried leaves in a two-choice landing bioassay and in the background of human odour. Volatile extracts from the smoke of burning dried leaves were found to be more repellent than those from fresh leaves, which in turn were more repellent to mosquitoes than volatiles from dried leaves. Of all smoke and fresh volatile extracts, those from Co. citriodora (52% - 76%) and Oc. suave (58% - 68%) were found to be the most repellent, Os. integrifolia (29% - 56%) to be intermediate while Ol. europaea (23% - 40%) and Os. integrifolia (19% - 37%) were the least repellent. One volatile present in each of the fresh leaf extracts of Co. citriodora, Oc. suave and Os. integrifolia was β-ocimene. The levels of β-ocimene reflected the mosquito repellent activity of these three fresh leaf extracts. Female host- seeking mosquitoes responded dose-dependently to β-ocimene, both physiologically and behaviourally, with a maximal behavioral repulsion at 14% β-ocimene. β-ocimene (14%) repelled mosquitoes in 6-minute landing assays comparable to the synthetic insect repellent N, N-diethyl-m-toluamide (10% DEET). Volatiles in the smoke of burning as well as fresh leaves of Co. citriodora and Oc. suave have significant repellent properties against host seeking An. arabiensis and Ae. aegypti mosquitoes.

The dependence of seasonal vector mosquito aquatic larval habitat canopy vegetation reflectance on sun and sensor geometry can potentially provide information on canopy properties of georeferenced seasonal-sampled vector mosquito aquatic larval habitats, but also may be a source of unmodeled systematic error in single-angle remote sensing measurements. Asner et al. (2002) investigated view angle effects on canopy reflectance and spectral mixture analysis (SMA) of coniferous forests using AVIRIS and the angular variability of reflectance measurements from the NASA Airborne Visible/Infrared Imaging Spectrometer (AVIRIS), and the resulting impact on spectral mixture analysis (SMA) using both full-range (400 - 2500 nm) and shortwave-infrared wavelengths (2080 - 2280 nm; AutoSWIR). The study was conducted in coniferous forests in Central Oregon using five AVIRIS overpasses to generate multiple view angle measurements. Canopy reflectance was highly anisotropic, with the strength of the angular signal controlled by species type, canopy cover and soil reflectance. Canopy cover estimates from full-range SMA averaged only slight decreases (~6% relative) toward the retro-solar direction for 16 field plots in the study region. Auto SWIR was even less influenced by view angle, exhibiting changes only for large differences in view angle. In addition, AutoSWIR's ability to accommodate spatiotemporal endmember variability led to stronger agreement with field cover values than full-range SMA. The results suggested that while view angle can significantly affect reflectance measurements from AVIRIS, the consequent variability in vegetation cover estimates from various LULC objects (e.g., georeferenced An. arabiensis aquatic larval habitats) SMA and AutoSWIR is low.

In addition, although many seasonal, vector, mosquito-related, endemic, transmission-oriented studies have investigated the view-angle dependence of vegetation metrics like the NDVI, the effect of viewing geometry on other remote sensing signatures and methods has not been adequately addressed. One of the most recent applications in vegetation, LULC-related, vector, mosquito-related, remote sensing is unmixing algorithms which may be used to estimate the fractional cover of major scene components, or endmembers (e.g., green vegetation, soil and shade), in an extracted spectrally decomposed image pixel. Fractional extent of vegetation is important from biophysical and biogeochemical perspectives, as well as for investigating secondary sources of spectral variation [1]. Ideally, the inclusion of photometric spectrally quantitated shade as an endmember compensates for view angle effects by modeling the portion of each pixel that is blocked from direct sunlight. However, wavelength dependent multiple scattering in vegetation canopies related to georeferenced, seasonal, vector, mosquito-related, explanatory biophysical, time series, simulation models results in shade that is not efficiently spectrally quantifiable. Therefore, there is some variability to be expected in decomposing, seasonal vector, mosquito-related products due to viewing and solar geometry.

Seasonal, canopy-shaded effects may be spectrally quantified by employing multiple unmixing algorithms and a time series, stochastic/deterministic, interpolator for generating robust, seasonal, endemic, transmission- oriented, endmember, eco-epidemiological, risk maps. For example, Jacob et al. (2011) spectrally decomposed a sub-meter spatial resolution (i.e., QuickBird) riceland, An. arabiensis s.s., aquatic, larval, habitat pixel for forecasting productive habitats based on spatiotemporal, field-sampled, count data in Karima, riceland, agro-village, complex in the Mwea Rice Scheme, Kenya using a stochastically interpolated reference signature. Initially, the authors constructed a regression model which revealed that georeferenced, tillering An. arabiensis s.s. aquatic larval habitats were the most productive. Individual decomposed, sub-pixel, endmember, spectral, reflectance estimates from a QuickBird visible and NIR at 0.61 m pixel, spatial, resolution, of a paddy-preparation, aquatic, larval habitat were then extracted by using a Li-Strahler geometric-optical model. The model used three scene components: sunlit canopy (C), sunlit background (G) and shadow (T) generated from the riceland image. The G, C, T components' classes were estimated using ENVI, an object-based classification algorithm. In the object-based technology the Digital Number (DN) of the pixel in every QuickBird band was viewed using the z-profile from a spectral library. After making an atmospheric correction from the image for the epidemiological study site, the DN was converted into ground reflectance. A geo-spatiotemporal, convex, geometrical, explanatory, predictive model was also employed for endmember validation of the spectrally, decomposed, tillered habitat. An Ordinary kriged-based interpolation was performed in GIS using the reference signature generated from the unmixing models. The basic idea of kriging is to predict the value of a function at a given point (e.g., georeferenced, vector, mosquito, aquatic, larval habitat), by computing a weighted average of the known values of the function in the neighborhood of the point. Ordinary kriging assumes stationarity of the first moment (i.e., mean) of all random variables:

$$E\{Z(x_i)\} = E\{Z(x_0)\} = m$$

where m is unknown [1] Linear, unbiased, explanatory, predictors and variance, estimates were derived for all seasonally, productive, georeferenced, immature, An. arabiensis s.s. habitats in the epidemiological study site based on the extracted, unmixed, endmember, interpolated, spectral reflectance estimates.

Spectral unmixing tools and other algorithms [e.g., linear, non-linear, LULC] may be employed in PDA-GIS-differentially corrected GPS (DGPS)-RS technologies, to spatially/spectrally quantitate and interpolate visible and NIR pixel reflectance data (target signature) of a productive, seasonal-sampled, empirical, georeferenced, vector, mosquito, aquatic, larval habitat. Thereafter, the data can be employed as a dependent variable in a robust stochastic/deterministic interpolator within a remote cyber-environment for targeting other seasonal high density foci habitat sites along with their spatially/spectrally associated continuous explanatory variables, which can help implement and target larval control strategies. By so doing, the applications of PDA-GIS-DGPS-RS technologies would also enable vector biologists, epidemiologists and other research collaborators to gain a further understanding of state-of-the-art technologies currently available for implementing control operations [e.g., Integrated Vector Management (IVM)]. DGPS uses a network of fixed, ground-based reference stations to broadcast the difference between the positions indicated by the satellite systems and the known fixed positions for broadcasting the difference between measured satellite pseudoranges and actual (i.e., internally computed) pseudo-ranges, and receiver stations (www.amsa.gov.au/navigation).

Integrated Vector Management is a decision-making process for the management of vector populations, so as to reduce or interrupt transmission of vector-borne diseases (http://www.ivmproject.net/); the strategy is based on the premise that effective control is not the sole preserve of the health sector but of various public and private agencies, including communities. Salient attributes of IVM include methods based on knowledge of factors influencing local vector biology; disease transmission and morbidity; use of a range of interventions, often in combination and synergistically; collaboration within the health sector and with other public and private sectors that impact on vectors; engagement of local communities and other stakeholders; and, a public health regulatory and legislative framework (www.who.int/mediacentre/factsheets/). Driving forces behind a growing interest in IVM for malaria eradication would include the need to overcome challenges experienced with conventional single-intervention approaches to targeted vector control as well as recent opportunities for promoting multi-sec- torial approaches to human health. IVM would encourage then effective coordination of the control activities of all sectors that would have an impact on seasonal disease transmission including health, housing agriculture and others. Commensurate benefits for non-health-sector partners make it more likely that IVM approaches will be effective (http://www.ivmproject.net/). Alternate wet/dry intermittent irrigation, combined with other vector control methods, has been effective in controlling the vectors of Japanese encephalitis in China, India, Indonesia and Sri Lanka.

Implementation of IVM using PDA-GIS-DGPS-RS cyber-analytical data, however, requires explicit understanding of geo-spatiotemporal, georeferenced, patterns of vector, mosquito-borne, immature habitats in specific ecosystems (e.g., a riceland agro-village complex), for enabling vector biologists epidemiologists, and other data analysts to characterize various seasonal environmental conditions (e.g., flooding, droughts) and link endemic, transmission-oriented, covariate coefficients with other important data (e.g., demographic, meteorological) in both space and time. Spatial autocorrelation is the correlation among values of a single variable strictly attributable to their relatively close locational positions on a two-dimensional surface, introducing a deviation from the independent observations assumption of classical statistics. Positive geo-spatial autocorrelation in a PDA-GIS- DGPS-RS derived time-series, cyber-predictive, risk map means that the empirical-sampled nearby covariate coefficients values (e.g., aggregation of georeferenced vector mosquito aquatic larval habitats with high seasonal larval counts) tend to be similar on the risk map. Habitat-based, endemic, transmission-oriented, eco-epidemiological, risk modeling of mosquito larval interventions employing spatiotemporally quantitated entomological inoculation rates has determined that not all georeferenced aquatic habitats have a significant impact on incidence and prevalence [3].

Real-time bidirectional PDA-GIS-DGPS-RS technologies may measure geo-spatial autocorrelation and other bio-ecological trends in a regressed, time series, empirical datasets of quantitated Euclidean habitat distance measurements, while optionally creating line graphs of those distances and their corresponding z-scores. A z-score is a statistical measurement of a score's relationship to the mean in a group of scores. Commonly, in geo-spatiotemporal, vector, mosquito-related, endemic, transmission-oriented, statistical, risk-based data analyses, a z-score is tabulated from an empirical dataset of georeferenced, explanatory, field or remote-specified, regressors which may be positive or negative, indicating whether a residually forecasted derivative is above or below the mean, and by how many standard deviations [2]. Thereafter, an incremental, geospatial, autocorrelation tool (e.g., weighted matrix, eigenfunction decomposition, algorithm) in a real-time, bidirectional, PDA-GIS- DGPS-RS cyber-environment may be employed for robustly quantitating a series of georeferencable, seasonal-sampled, larval habitats using a weighted matrix.

In the PDA-GIS-DGPS-RS cyber-environment, a vector, mosquito-related, seasonal, Hadamard matrix, would be a square matrix whose entries are either +1 or −1 and whose rows are mutually orthogonal. In geometric terms, this means that every two different rows in the Hadamard matrix would represent two perpendicular vectors, while in combinatorial terms, it means that every two different rows would have matching entries in exactly half of their columns and mismatched entries in the remaining columns. It is a consequence of this definition that the corresponding properties hold for columns as well as rows in the matrix. The n-dimensional parallelotope spanned by the rows of an $n \times n$ Hadamard matrix in the cyber-environment would then have the maximum possible n-dimensional volume among parallelotopes spanned by vectors whose entries are bounded in absolute value by 1 in the risk model. Equivalently, a Hadamard matrix would have a maximal determinant among matrices with entries of absolute value less than or equal to 1 in the risk model residually forecasted derivatives and so, the extremal solution of a Hadamard's maximal determinant matrix The analogous question for matrices with elements equal to 0 or 1 would then be equivalent since the maximal determinant of a $\{1, -1\}$ matrix of size n would be 2^{n-1} times the maximal determinant of a $\{0, 1\}$ vector, mosquito-related, seasonal, georeferenced, matrix of size $n-1$. The intensity of the geo-spatiotemporal autoregressive clustering would then be determined by the z-score returned in a real-time, bidirectional, PDA-GIS-DGPS-RS, cyber-endemic, transmission-oriented, risk map.

Typically, as the seasonal-sampled, georeferenced, time-series, aquatic, vector, mosquito-related, larval habitat, Euclidean, distance-based explanators increases, so would the z-score, indicating intensification of clustering in a specified geographic location in an epidemiological interventional study site. At some particular georeferenced habitat distance measurement, however, (e.g., Euclidean distance from an epidemiological capture point to a seasonal, prolific, georeferenced,malaria-related, aquatic, larval habitat), the z-score would peak in a real- time bi-directional PDA-GIS-DGPS-RS cyber-environment. Furthermore, a vector biologist, epidemiologist and/ or other research collaborator may encounter multiple peaks in an ecological dataset of georeferenced, residually, forecasted, derivatives rendered from a seasonal, predictive, georeferenced, explanatory, vector, mosquito, aquatic, larval, habitat, risk model constructed in real-time, bidirectional, PDA-GIS-DGPS-RS architecture. When regressively, quantitating an empirical-sampled, time series, explanatory, dataset of non-linear georeferenced vector mosquito aquatic, larval, habitat, field and/or remote specified, endemic, transmission-oriented, georeferenced, data, feature attributes, commonly latent autocovariate errors occur. One strategy may be then to identify an appropriate scale of the predictive, geo-spatial, autocorrelation analysis employing bidirectional, PDA-GIS-DGPS-RS, cyber-technologies so as to select the optimal, seasonal, Euclidean, habitat, distance measurements as defined from an and eco-epidemiological capture point. By so doing, seasonal-sampled, georeferenced, vector, mosquito, aquatic, larval habitats that are associated with the statistically significant peaks may be identified and quantified. The temporal scale of the satellite image in a georeferenced cluster may then be determined in a robust, real-time, bidirectional, PDA-GIS-DGPS-RS cyber-environment.

Further, geo-spatiotemporally quantizing z-scores in real-time, bidirectional, PDA-GIS-DGPS-RS technologies can render seasonal, cartographic, interpolations of explanatory, georeferencable, vector, mosquito-related, risk, model, residual forecasts from a regressed, empirical, sampled, dataset of endemic, transmission-oriented, explanatory coefficients. Interpolation is a method of constructing new, explanatory, time series, georeferenced data points within the range of a discrete set of known data points [1]. These seasonal, vector, mosquito-related, interpolated, explanatory coefficients may reflect the intensity of georeferencable time series, clustering, in specific field or remote specified, ecological, empirical datasets.

In empirical, georeferencable, explanatory datasets of seasonal, vector, mosquito-related, endemic, transmission-oriented, risk-based, data points obtained by georeferenced targeted sampling, aquatic larval habitats can represent the values of a function for a limited number of independent variables values. It is often required to interpolate the value of that function for an intermediate value of the independent variable when time series regressive modeling georeferenced immature seasonal vector mosquito-related data feature attributes [2]. This may be achieved by curve fitting or regression analysis. A different problem which is closely related to cartographic interpolation of seasonal-sampled vector mosquito-related georeferenced data is the approximation of a complicated function by a simple function. Suppose the formula for some given function for a predictive, time series, geo-spatiotemporal, malarial-related, epidemiological, risk model is known, but too complex to evaluate efficiently. A few known seasonal-sampled, larval, habitat, georeferenced,data points from the original function in a real-time, bidirectional, PDA-GIS-DGPS-RS, cyber-environment may be used to create a time series, cartographic interpolation based on a simpler function. Of course, when a simple function is employed to estimate productive, georeferenced, seasonal-sampled, larval, habitat, data points from the original, interpolation errors are usually present; however, depending on the problem domain and the time series, explanatory, interpolation method employed (e.g., block kriging in a PDA-GIS-DGPS-RS cyber-environment using sub-meter resolution satellite data), the gain in simplicity may be of greater value than the resultant loss in accuracy.

Gaussian processes in a predictive, and eco-epidemiological risk model can be powerful, explanatory, non- linear, interpolation tool in a robust, real-time, PDA-GIS-DGPS-RS cyber-environment. Although Gaussian processes have a long history in the field of seasonal, vector, mosquito-related, regression statistics, they seem to have been employed extensively only in niche areas. With the advent of kernel machines in the machine learning community, seasonal, vector, mosquito-related, forecasting, risk models based on Gaussian processes have become commonplace for problems of time series regression (e.g., kriging) and classification as well as a host of more specialized applications [e.g., spectral endmember decomposition algorithm employed for interpolating canopied-shaded larval habitat objects]. Many interpolation tools in real-time, bidirectional, PDA-GIS-DGPS- RS cyber-environments can be constructed equivalent to particular Gaussian processes. For example, Gaussian, seasonal, quantification of georeferenced, vector, mosquito, larval, habitat, remote processes can be used not only for fitting an interpolant that passes exactly through given empirical, seasonal-sampled, georeferenced, data points but also for regression analyses (e.g., for fitting a curve through noisy data).

In seasonal-sampled, empirical, ecological, georeferencable datasets of explanatory, vector, mosquito data analyses, kriging or Gaussian process regression is a method of interpolation for which the values (e.g., total, larval, habitat, seasonal, density counts) are modeled by a Gaussian process governed by prior covariances, as opposed to a spline chosen to optimize smoothness of the fitted values. Quantitating time series, explanatory, covariance estimates in a robust, real-time, PDA-GIS-DGPS-RS cyber-environment can provide a measure of the strength of the correlation between two or more sets of random variates in an empirical dataset of seasonal-sampled, vector, mosquito-related, biophysical, time series explanators. In mathematics, a spline is numeric function that is piece-wise-defined by polynomial functions, and which possesses a sufficiently high degree of smoothness at the places where the polynomial pieces connect (i.e., knots) [1]. A polynomial may have an expression consisting of explanatory georeferencable seasonal, vector, mosquito-related determinates which can involve operations of addition, subtraction multiplication for deriving explanatory, non-negative, integer exponents in a robust, real-time, bidirectional, PDA-GIS-DGPS-RS cyber-environment. For example, a polynomial of a single, seasonal, malaria, mosquito, (e.g., Anopheles gambiae s.l.)-related, indeterminate or variable tabulation could include x, is $x^2 - 4x + 7$, where x is a single-, sampled, explanatory, covariate, coefficient value (e.g., amount of daily precipitation in an eco-epidemiological, interventional, georeferenced, urbanized, study site), which then could be expressed mathematically as a quadratic polynomial in a robust, real-time, bidirectional, PDA-GIS-DGPS-RS cyber-environment.

A quadratic function in a robust, real-time, bidirectional, PDA-GIS-DGPS-RS cyber-environment, vector, mosquito-related, predictive, and eco-epidemiological, risk model would be a polynomial function of the form $f(x) = ax^2 + bx + c, a \neq 0.$ By so doing, the expression $ax^2 + bx + c$ may define a quadratic function for constructing a robust, seasonal, vector, mosquito-related, predictive, endemic, transmission-oriented, risk model in a robust cyber-environment. The model residual forecasted derivatives would then be a polynomial of degree 2, or a 2nd degree polynomial, (i.e., a quadratic polynomial). If the quadratic function in a seasonal, explanatory, georeferencable, vector-related, mosquito, larval, aquatic, habitat, seasonal, predictive, endemic, transmission- oriented, risk model in a robust real-time bidirectional PDA-GIS-DGPS-RS cyber-environment, is set equal to zero, then the result is a quadratic equation. In elementary algebra, a quadratic equation is any equation having the form $ax^2 + bx + c = 0$ where x represents an unknown, and a, b, and c are constants not equal to 0. If a = 0 in a robust, quadratic, seasonal, vector, mosquito-related, predictive, endemic, transmission-oriented, seasonal, eco- epidemiological. risk model constructed in a real-time bidirectional PDA-GIS-DGPS-RS cyber-environment, a quadratic equation is a second-order polynomial equation in a single variable x where $ax^2 + bx + c = 0$, with $a \neq 0$ and because it is a second-order polynomial equation, the fundamental theorem of algebra (i.e., every polynomial equation having complex coefficients and degree ≥ 1 has at least one complex root) guarantees that it has two solutions [3]. The parameters a, b, and c are called, respectively, the quadratic coefficient, the linear coefficient and the constant or free term [3]. Because the quadratic equation involves only one unknown, it is called "univariate" [2]. A seasonal, vector, mosquito-related, real-time, bidirectional, PDA-GIS-DGPS-RS cyber-environment-derived explanatory, predictive, quadratic equation would contain powers of x that are non-negative integers, and therefore would be a polynomial equation. Quadratic, seasonal vector, malaria, mosquito-related, predictive equations can then be solved by a process known as factoring by completing the square, by using the quadratic formula, or by graphing the residually forecasted derivatives in the PDA-GIS-DGPS-RS cyber-environment.

Under suitable assumptions on the priors, kriging vector seasonal georeferenced mosquito-related data in a robust real-time bidirectional PDA-GIS-DGPS-RS cyber-environment would render the best linearized explanatory unbiased prediction of endemic transmission oriented intermediate values (e.g., larval habitat counts). In Bayesian statistical inference, a prior probability distribution, often called simply the prior, of an uncertain p is the probability distribution that would express one's uncertainty about p before some evidence is taken into account [3]. For example, p in a real-time, bidirectional, PDA-GIS-DGPS-RS cyber-environment, could be the proportion of residually forecasted number of larvae in a georeferenced habitat for an upcoming seasonal sampling frame. The prior in a geo-spatiotemporal, vector, mosquito-related, endemic, transmission-oriented, explanatory, georeferenced, risk model is meant to attribute uncertainty rather than randomness to an uncertain quantity [2]. The unknown quantity in the real-time, bidirectional, cyber-environment may then be expressed as a sampled parameter or late explanatory, predictor, variable employing a Bayesian, probabilistic, estimation, probablistic matrix.

Bayes rule can be derived from more basic axioms of probability, specifically conditional probability [3]. Let (Ω, F, P) be a measure space with P(Ω) = 1, then (Ω, F, P) is a probability space, (e.g., a mathematical construct that models prolific, georeferenced, vector, mosquito-related habitats) with sample space Ω, event space F and probability measure P. When applied, the probabilities involved in Bayes' theorem may have any of a number of probability interpretations for seasonal, georeferenced, vector, mosquito-related, forecasting, endemic, transmission-oriented, explanatory, risk models. In one of these interpretations, the theorem may be used directly as part of a particular approach to statistical inference of the residually forecasted derivatives. In particular, the Bayesian interpretation of probability, the theorem can express how a subjective degree of belief should rationally change to account for evidence: this is Bayesian inference, which is fundamental to Bayesian statistics. Bayes' theorem has applications in a wide range of calculations involving probabilities for seasonal, vector, mosquito-related, georeferencable, predictive, endemic, transmission-oriented, explanatory, risk model derived in a cyber-environment, not just in Bayesian inference. For instance, since individual, sampled, georeferenced, aquatic, larval, habitat, vector, mosquito-related, risk outcomes might be of little practical use, more complex events (e.g., aggregation of flooded larval habitats) may be employed to characterize groups of endemic, transmission-oriented, explanatory, time series outcomes. One applies Bayes' theorem, multiplying the prior by the likelihood function and then normalizing, to get the posterior probability distribution, which is the conditional distribution of the uncertain quantity given the data [3]. A prior for a seasonal, vector, mosquito-related, predictive, eco-epidemiological, risk model is often the purely subjective assessment of an experienced expert.

Theoretically, a conjugate prior may be employed in a real-time, bidirectional, PDA-GIS-DGPS-RS, cyber-environment to make calculations of the posterior distribution in seasonal-sampled, empirical, regressed, mosquito-related, explanatory dataset easier. In Bayesian probability theory, if the posterior distributions p(θ|x) are in the same family as the prior probability distribution p(θ) in the prior and posterior are then called conjugate distributions, and the prior is called a conjugate prior for the likelihood function [3]. In seasonal, vector, mosquito-related statistics, a likelihood function (often simply the likelihood) is a function of the sampled estimators [2]. The likelihood of a set of seasonal-sampled, georeferencable, vector, mosquito-related, parameter values, θ given outcomes x, can be equal to the probability of those observed outcomes given those parameter estimators value in a PDA-GIS-DGPS-RS cyber-environment. A Gaussian prior over the mean in a vector, mosquito-related, endemic, transmission-oriented, seasonal, predictive, epidemiological, risk model may ensure that the posterior distribution is also Gaussian. This means that the Gaussian distribution would be a conjugate prior for the likelihood which would also be Gaussian in a robust, seasonal, vector, mosquito-related model employing real-time bidirectional PDA-GIS-DGPS-RS cyber-tools.

Parameters of prior distributions in a seasonal, vector, mosquito-related, endemic, transmission-oriented, predictive, epidemiological, risk model are called hyperparameters, to distinguish them from parameters of the model of the underlying-sampled data [2]. In Bayesian statistics, a hyperparameter is a parameter of a prior distribution; the term is used to distinguish them from parameters of the model for the underlying system [3]. For instance, if a vector biologist, epidemiologist or research collaborator is employing a beta distribution to model the distribution of a seasonal-sampled, vector, mosquito-related parameter p of a Bernoulli distribution in a real-time, bidirectional, PDA-GIS-DGPS-RS cyber-environment, then p would be a parameter of the underlying system (i.e., Bernoulli distribution), and α and β a would be the sampled parameters of the prior distribution (i.e., beta distribution), hence hyperparameters. In probability theory and statistics, the Bernoulli distribution is the probability distribution of a random variable which takes value 1 with success probability p and value 0 with failure probabity This is vital in cyber-environments since traditionally seasonal-sampled, georeferenced, vector, arthropod-related aquatic larval habitat values based on non-cyber-environmental derived data has not yielded the most likely intermediate, total, larval, habitat, regressable, georeferencable, explanatory, covariate, coefficient, seasonal, density, count, values.

A real-time, bidirectional, PDA-GIS-DGPS-RS, cyber-environment may utilize a geospatial analytical domain for interpolating seasonal-sampled, empirical, vector, mosquito-related, explanatory dataset. The technique is also known as Kolmogorov Wiener prediction. Linear filters are ubiquitous in applied vector mosquito research ranging from simple differencing operations, mechanical detrending devices and seasonal adjustment to autoregressive models. Many of these filters have very long and often infinite impulse response sequences, such that some approximation procedure is necessary for finite samples. There are also very different kinds of seasonal, explanatory, interpolations for statistically targeting georeferenced, vector, mosquito, aquatic, larval, habitats in PDA-GIS-DGPS-RS cyber-environments. For instance, in the domain of seasonal, vector, mosquito-related, digital, signal processing, interpolations could be performed for converting a geographically, sampled, digital signal to a higher sampling rate (e.g., upsampling) employing various digital filtering techniques (e.g., convolution with a frequency-limited impulse signal). In this application there would be a specific requirement in the real-time, bidirectional, cyber-environment that the harmonic content of the original signal be preserved without creating aliased harmonic content of the original signal (i.e., the original Nyquist limit of the signal above fs/2 of the original signal sample rate). In signal processing, the Nyquist rate is twice the bandwidth of a band limited function or a band limited channel [2].

In a robust, real-time, bidirectional, PDA-GIS-DGPS-RS cyber-environment, seasonal-sampled, georeferenced, vector, mosquito, aquatic, larval, habitat, explanatory, covariate coefficients rigorously regressed based on a response variable representing total, quantized, larval/pupal, density counts can provide additional autoregressive insights for geospatially, targeting, endemic, transmission-oriented zones (e.g., hyperendemic foci). For example, statistically significant peak z-scores computed employing real-time, bidirectional, PDA-GIS-DGPS- RS, technologies can indicate whether a georeferenced, dataset, of explanatory, aquatic, larval, habitat, Euclidean distances has endemic, transmission-oriented, geo-spatiotemporal processes promoting positive/negative, autocovariate, time series clustering. In a real-time, bidirectional, PDA-GIS-DGPS-RS cyber-environment, the seasonal specified autocovariate coefficients derived from regressing an empirical sampled dataset of explanatory, field and/or remote specified, georeferencable, covariate coefficients may be employed in autonormal, autopoisson or autologistic probabilistic, endemic, transmission-oriented, epidemiological risk model frameworks.

There are different ways to formulate the autonormal, autopoisson or autologistic model in a real-time bidirectional PDA-GIS-DGPS-RS cyber-environment depending on the biological context of the empirical sampled georeferenced vector mosquito-related field and/or remote specified explanatory endemic transmission parameter estimator datasets. For example, in a autologistic regression model framework employed in the analysis of vector mosquito species' spatial distributions, constructed in a robust real-time bidirectional PDA-GIS-DGPS-RS cyber-environment, an additional explanatory variable, (i.e., the autocovariate), may be used to correct the latent effects of the geo-spatial autocorrelation coefficients. In statistics, given a real stochastic process X(t), the autocovariance is the covariance of the variable against a time-shifted version of itself. If the process has the mean $E[X_t]=\mu_t$, then the autocovariance is given by

$$C_{XX}(t,s)=E\left[(X_t-\mu_t)(X_s-\mu_s)\right]=E\left[X_tX_s\right]-\mu_t\mu_s$$

where E is the expectation operator [6]. The values of the time series, georeferencable, explanatory, vector mosquito-related, geo-spatiotemporal, quantitated autocovariate coefficient would then depend on the sampled field or remote-specified, parameter estimator values of the response variable in the neighborhood as defined by the PDA-GIS-DGPS-RS cyber-tools. This approach has not been employed for biogeographical, risk-based, explanatory, seasonal, real-time, vector, mosquito-related, epidemiological data analyses. Nor has it been assessed for its validity and performance against artificial, simulation, endemic, transmission-oriented, explanatory, georeferenced, field and remote-specified, data variable, with known properties. A geo-spatiotemporal, robust, explanatory, PDA-GIS-DGPS-RS cyber-assessment may quantitate varying range and strength of seasonal, latent, autocorrelation coefficients in sampled, vector, mosquito data as well as the prevalence of the focal species. Autologistic regression models consistently underestimate the effect of the environmental variables in forecasting models while rendering biased, explanatory, estimates compared to a non-spatial, explanatory, logistic, regression method.

Historically, a drawback of autopoisson, or autologistic, geo-spatiotemporal, eco-epidemiological, forecasting, vector, mosquito-related, risk models constructed employing empirical-sampled, ecological, datasets of georeferenced, explanatory, endemic, transmission-oriented, field and/or remote-specified, covariate coefficients is that for positive autocorrelation the likelihood of the models is not available in closed form. A vector biologist, epidemiologist or a data analyst may investigate how this restriction can be avoided by right truncating a geo-spa- tiotemporal, explanatory, vector, mosquito-related, time series distribution datasets in a real-time, bidirectional PDA-GIS-DGPS-RS cyber-environment. Truncation is the term for limiting the number of digits right of the decimal point, by discarding the least significant ones [4] - [6]. In a robust, time series, explanatory, PDA-GIS- DGPS-RS cyber-environment, different parameter estimation techniques may be then reviewed which may be deemed applicable to seasonal, vector, mosquito-related, explanatory, field and/or remote-specified, georeferencable, simulation, explanatory, auto-models, for parameter, estimator significance testing. The residually, forecasted, time series derivatives may then be compared with other environmental, simulation, model outputs (e.g., generalized, Bayesian, hierarchical, probabilistic, estimation paridyms). The method may be easily implemented in a robust real-time, bi-directional, geo-spatiotemporal, PDA-GIS-DGPS-RS cyber-environment, digitized, orthogonal, grid matrix via standard statistical software packages and a maximum pseudo-likelihood estimate which may then subsequently render unbiased, explanatory, interpolated,point estimates in an ecological, explanatory, dataset of residually, forecasted, vector, mosquito-related, time series derivatives.

An alternative method, Monte Carlo maximum likelihood in a PDA-GIS-DGPS-RS cyber-environment may also mathematically contribute to robust, risk-related, cartographic delineations of explanatory, georeferenced, seasonal-sampled, predictive, vector, mosquito-related, real-time, bidirectional operationalizable, eco-epidemi- ological, endemic, transmission-oriented, risk model residually forecasted derivatives rendered from a time series regression equation. The Monte Carlo maximum likelihood method can render unbiased, explanatory, continuous, estimators of the likelihood function for robustly constructing a family of diffusion-related, seasonal, vector, mosquito-related, forecasting, risk models within a computationally efficient real-time, bidirectional, PDA-GIS- DGPS-RS cyber-infrastructural, regression-based framework. Furthermore, the bidirectional cyber-environment could utilize an exact simulation of diffusions for constructing the predictive, vector, mosquito-related, real-time, seasonal, endemic, transmission-oriented, operationizable, explanatory, predictive, risk model with limited discretization error. A vector biologist, research epidemiologist or data analyst can then show that under regularity conditions, the Monte Carlo maximum likelihood estimation converges to the true likelihood estimation in a forecasting probabilistic, real-time, bidirectional, endemic, transmission-oriented, regression-based vector, mosquito-related, eco-epidemiological, risk model output. For datasize n → ∞, a vector biologist, epidemiologist and/or a data analyst can then employ that the number of Monte Carlo iterations in the PDA-GIS-DGPS-RS cyber-environment which may be tuned as O(n1/2). Monte Carlo iterations can demonstrate the consistency properties of the Monte Carlo likelihood as an observational predictor of the true parameter estimator value in a geo- spatiotemporal, endemic, transmission-oriented, vector, mosquito-risk-related, explanatory, predictive, eco- epidemiological, risk model using O(n1/2) (2).

In mathematics, big O notation describes the limiting behavior of a function when the argument tends towards a particular value or infinity, usually in terms of simpler functions. It is a member of a larger family of notations that is called Landau notation, Bachmann-Landau notation or asymptotic notation. In computer science, big O notation is employed to classify algorithms by how they respond (e.g., in their processing time or working space requirements in a real-time, bidirectional, PDA-GIS-DGPS-RS cyber-environment to changes in input size. In analytic number theory, it is used to estimate the "error committed" while replacing the asymptotic size, or asymptotic mean size, of an arithmetical function, by the value, or mean value, it takes at a large finite argument. A famous example is the problem of estimating the remainder term in the prime number theorem. Big O notation can characterize seasonal, vector, mosquito-related functions according to their immature growth rates employing different functions in a real-time, bidirectional, PDA-GIS-DGPS-RS cyber-environment. The letter O is used because the growth rate of a function is also referred to as order of the function [3]. A description of a function in terms of big O notation usually only provides an upper bound on the growth rate of the function. Associated with big notation are several related notations, using the symbols o, Ω, ω, and Θ which may describe other kinds of bounds on asymptotic, vector, mosquito-related, environmental, explanatory, y growth rates [e.g., seasonally quantized land use land cover (LULC) change levels] in robust, real-time, bidirectional, PDA-GIS-DGPS-RS cyber-environments.

Therefore, if a vector biologist, epidemiologist or a data analyst lets df and g be two functions in a robust explanatory, vector, mosquito, larval, habitat, forecasting, georeferencable, risk model in a real-time, bidirectional, PDA-GIS-DGPS-RS cyber-environment, the residually, forecasted, estimates targeting seasonal, endemic, transmission-oriented areas of interest (e.g., hyperendemic transmission foci) may be defined employing a subset of the sampled, endemic, transmission-oriented explanatory values (e.g., total larval habitat density count values). A vector ecologist, or a data analyst may then use $f(x)=O(g(x))$ as $x\to\infty$ if and only if there is a positive constant M such that for all sufficiently large values of x, f(x) is at most M multiplied by the absolute value of g(x) [2]. That is, f(x) = O(g(x)) in a robust, georeferencable, autoregressive, vector, mosquito-related, predictive, PDA- GIS-DGPS-RS cyber-environment derived real-time, bidirectional, endemic, transmission-oriented, risk model if and only if there exists a positive sampled real number M and a real number x_0 such that $|f(x)|\le M|g(x)|$ for all $x\ge x_0$.

In many seasonal, explanatory, predictive, vector, mosquito-related, model contexts, the assumption that is of interest is the growth rate (e.g., seasonal larval counts) as the variable x goes to infinity is left unstated, thus f(x) = O(g(x)) may be more viable when regressing seasonal-sampled, endemic transmission-oriented, explanatory, covariate coefficients in a real-time, bidirectional, PDA-GIS-DGPS-RS cyber-environment. The notation can also be used to describe the behavior of f near some seasonal-sampled, georeferenced, operationizable, vector, mosquito-related, habitat, immature, count values in a robust, PDA-GIS-DGPS-RS, cyber-environment using a (e.g., a = 0 during the dry season). As such, $f(x)=O(g(x))$ as $x\to a$ in a bidirectional, PDA-GIS-DGPS-RS, cyber-environment if and only if there exists positive sampled, vector, mosquito-related, density, count numbers δ and M where $|f(x)|\le M|g(x)|$ for $|x-a|<\delta.$ If g(x) is non-zero for the sampled, vector, mosquito-related, seasonal, endemic, transmission-oriented, georeferencable, coefficient values of x which are sufficiently close to a, both of these definitions can be unified in a real-time bidirectional PDA-GIS-DGPS-RS cyber-environment using the limit superior: $f(x)=O(g(x))$ as $x\to a$ if and only if $\limsup_{x\to a}\left|\frac{f(x)}{g(x)}\right|<\infty.$ The formal definition of O notation is not used directly; rather, the O notation for a function f is derived by the following simplification rules. If f(x) is a sum of several terms in a robust, real-time, bidirectional, PDA-GIS-DGPS-RS cyber-environment, the one with the largest growth rate may be considered an explanatory georeferenced representative sample, and all others omitted. If f(x) is a product of several factors, any constants (terms in the product that do not depend on x in the cyber-environment may be also omitted.

Conversely, if $f(x) = 6x^4 - 2x^3 + 5$, in an explanatory, time series, quadratic, real-time, bidirectional, malaria mosquito-related, forecasting, epidemiological, risk model, for instance, and suppose a vector biologist, epidemiologist and/or another data analyst wishes to simplify this function, O notation may be employed to forecast, seasonal larval/ pupal growth rate as x approaches infinity to parsimoniously temporally quantitate the statistically significant time series parameter estimators at a certain threshold. This function then could be timely quantitated in a robust, real-time, bidirectional PDA-GIS-DGPS-RS cyber-environment employing the sum of three terms: $6x^4$, $-2x^3$, and 5. Of these three terms, the one with the highest growth rate would be the one with the largest exponent as a function of x, namely $6x^4$ in the endemic, transmission-oriented, explanatory, risk model, residually, forecasted derivatives. Thereafter, $6x^4$ may be defined as a product of 6 and x^4 in which the first factor does not depend on x) in the cyber-environment. Omitting this factor will then result in the simplified form x^4 for the risk model derivatives. Thus, f(x) may be of "big-oh" of (x^4) in the bidirectional, real-time, PDA-GIS-DGPS-RS cyber-environment.

Mathematically, $f(x) = O(x^4)$ would be the model output variable in a vector mosquito-related predictive e endemic transmission-oriented eco-epidemiological explanatory risk model. A vector biologist, an epidemiologist or a research collaborator may then confirm this calculation using the formal definition: let $f(x) = 6x^4 - 2x^3 + 5$ and $g(x) = x^4$ in the cyber-environment. Applying, the statement that $f(x) = O(x^4)$ would be then equivalent to the expansion of $|f(x)| \le M\,|g(x)|$ for some suitable choice of x_0 and M and for all $x > x_0$. This equation may be proven by letting $x_0 = 1$ and $M = 13$ in a robust vector mosquito-related seasonal risk model constructed in a real-time bidirectional PDA-GIS-DGPS-RS cyber-environment. Then, for all $x > x_0$ the output would be tabulated as: $|6x^4 - 2x^3 + 5| \le 6x^4 + |2x^3| + 5$ so $|6x^4 - 2x^3 + 5| \le 13|x^4|$.

Hence, a vector biologist, epidemiologist or a data analyst can introduce and summarize the notion of time series auto-models in real-time bidirectional PDA-GIS-DGPS-RS cyber-environments for deriving robust, geo- spatiotemporal specifications for risk modeling empirically, georeferencable, explanatory, operationizable sets of time series field or remote-specified, vector, mosquito, endemic, transmission-oriented, data that contains a response variable. This is the model specification that was recently addressed by Jacob et al. [7] for empirically regressing an eco-epidemiological dataset of randomized explanatory Gaussian geo-spatiotemporal field and remote specified sub-meter resolution immature seasonal-sampled An. arabiensis-related endemic transmission oriented georeferenced explanatory covariate coefficients employing an auto-normal model. The authors revealed that the operationalizable residually forecasted derivatives rendered from the model using the georeferenced explanatory generalized auto-model specifications, including the auto-Poisson and auto-negative binomial probability specifications, robustly risk mapped the time series malarial empirical-sampled An. arabiensis s.s.-related ecological datasets while simultaneously accommodating positive spatial autocorrelation uncertainty coefficients.

The central role of the Poisson distribution with respect to the analysis of the georeferencable, seasonal, vector, mosquito-related, immature counts is analogous to the position of the normal distribution in the context of models for continuous data [7]. Accordingly, when seasonally regressed datasets of vector, mosquito, larval/ pupal data comprises of counts, especially for rare events, the optimal probability estimation eco-epidemiologi- cal, endemic, transmission-oriented, risk model in a PDA-GIS-DGPS-RS cyber-environment would be one based upon an auto-Poisson specification, which may be written in the form of approximations in the cyber-environ- ment. The auto-log-Gaussian approximation attained from the regression of the seasonal-sampled, georeferenced, vector, mosquito data would circumvent the auto-Poisson's intractable normalizing factor, and both it and the auto-logistic approximation would circumvent the auto-Poisson's restriction to only situations involving negative spatial autocorrelation, a restriction at odds with the real world vector mosquito-related data analyses, since most georeferenced data exhibit positive spatial autocorrelation. The intractable normalizing constant can be resolved in a real-time bidirectional PDA-GIS-DGPS-RS cyber-environment using Markov Chain Monte Carlo (MCMC) procedures.

The negative autocorrelation restriction in a seasonal, explanatory, georeferencable, vector, mosquito-related, endemic, transmission-oriented, eco-epidemiological, risk model can also be resolved through Windsorizing the sampled georeferenced data feature attributes in a real-time, PDA-GIS-DGPS-RS cyber-environment. Winsorizing is the transformation of statistics by limiting extreme values in the statistical data to reduce the effect of possibly spurious outliers [1]. Negative geo-spatial autocorrelation in many instances may be inconspicuous in empirical datasets of georeferencable, seasonal, vector, mosquito-related, predictive, autoregressive, eco-epi- demiological, endemic, transmission-oriented, risk model residually forecasted derivatives. On occasion geo-spatiotemporal, latent, autocovariate traits may be quantitated parsimoniously in a geographically, weighted, regression-based, endemic, transmission-oriented matrix. By quantitating residual, explanatory, autocorrelation, uncertainty coefficients in an ecological, empirical dataset of georeferenced, regressed, vector, mosquito-related, explanators, and robust confidence intervals may be plotted reflecting seasonal distributions of productive, aquatic, larval habitats based on geo-spatiotemporal field-sampled count data in an epidemiological interventional study site (e.g., total weekly collected larval/pupal malarial mosquito data in a riceland agro-ecosystem).

Furthermore, due to the intractability of the normalizing constants, non-normalized explanatory randomized geo-spatiotemporal parameter estimates in an empirical-sampled georeferencable dataset of seasonal vector mosquito-related observational predictors in a real-time bidirectional PDA-GIS-DGPS-RS cyber-environment could be parsimoniously regressed. MCMC technique would be required although attention had been restricted to the natural exponential family of statistical distributions (e.g., auto-normal, auto-logistic, auto-binomial, auto-Pois- son, and auto-negative binomial). In statistics, MCMC methods are a class of algorithms for sampling from probability distributions based on constructing a Markov chain that has the desired distribution as its equilibrium distribution [2].

Quasi-maximum likelihood estimate (QMLE, also known as a "pseudo-likelihood estimate" or a "composite likelihood estimate") is an estimate of a parameter θ in a seasonal real-time bidirectional vector mosquito-related endemic transmission oriented epidemiological predictive risk model that may be formed by maximizing a function that is related to the logarithm of the likelihood function in a robust PDA-GIS-DGPS-RS cyber-environment. In contrast, the maximum likelihood estimate would maximize the actual log-likelihood function for a seasonal- sampled vector mosquito-related georeferencable geo-spatiotemporal endemic transmission-oriented epidemiological risk model. The function that is maximized to form a QMLE in seasonal real-time bidirectional PDA-GIS- DGPS-RS cyber-environment can be a simplified form of the actual log likelihood function. A common way to form such a simplified function may be to employ the log-likelihood function of a misspecified mosquito-related model that treats certain sampled georeferenced vector data feature attribute values in a PDA-GIS-DGPS-RS cyber-environment as being independent, even when in actuality they may not be. This would remove any time series parameter estimators from the vector mosquito-related seasonal real-time bidirectional risk model that are used to characterize the dependencies. However, performing this maximum likelihood estimation only makes sense for quantitating time series parameter estimator statistical significance in a PDA-GIS-DGPS-RS cyber- environment, if the dependency structure is a nuisance parameter with respect to the goals of the analysis. As long as the quasi-likelihood function that is maximized in the vector mosquito-related endemic transmission-oriented epidemiological risk model is not overly simplified, the QMLE or composite likelihood estimate generated would be consistent and asymptotically normal. This likelihood estimate would be less efficient than the maximum likelihood estimate in a robust seasonal real-time bidirectional PDA-GIS-DGPS-RS cyber-environment, but may only be slightly less efficient if the quasi-likelihood is constructed so as to minimize the loss of information relative to the actual likelihood. Standard approaches to statistical inference that are used with explanatory vector seasonal mosquito-related maximum likelihood estimates, such as the formation of confidence intervals, and statistics for a seasonal robust model can then be generalized to the quasi-maximum likelihood setting.

On the other hand, an exact maximum likelihood method developed for the parameter estimator statistical significance testing in a seasonal real-time bidirectional PDA-GIS-DGPS-RS cyber-environment employing a non-Gaussian vector mosquito-related nonlinear log-density function would depend on a latent Gaussian dynamic process with long-memory properties. This method would rely on the method of importance sampling based on time series explanatory linearized Gaussian approximating models constructed within a robust seasonal real-time bidirectional PDA-GIS-DGPS-RS cyber-environment from which latent processes can be simulated. Given the presence of a latent long-memory process, a vector biologist, epidemiologist or data analyst could require a modification of the sampling technique in these models. In particular, the long-memory in a seasonal real-time bidirectional PDA-GIS-DGPS-RS cyber-environment would need to be approximated by a finite dynamic linearized process. By so doing, approximations (e.g., multi-seasonal parameter estimators of a prolific georeferenced malaria-related mosquito aquatic larval habitat) may be compared within the cyber-environment derived residualized model outputs. For instance, it may be shown that auto-regression coefficients obtained from minimizing mean squared forecasting errors in a robust seasonal vector mosquito-related endemic transmission oriented seasonal real-time bidirectional risk model can lead to an effective and feasible method for rendering unbiased residual forecasts targeting prolific vector aquatic larval habitats based on geo-spatiotemporal field and/ or remote regressed georeferenced data feature attributes. Thus, in an empirical study quantitating explanatory time series dependent vector mosquito field and/or remote specified endemic transmission dynamics, a vector biologist, epidemiologist or other data analyst may quantitate log-return time series data by univariate and multivariate long-memory stochastic volatile and autoregressive models in a robust seasonal real-time bidirectional PDA-GIS- DGPS-RS cyber-environment employing distance-weighting schemes and Monte Carlo maximum likelihood estimation for constructing robust, generalized, long-memory, time series, explanatory, vector, mosquito-related, endemic, transmission-oriented, epidemiological, predictive risk models.

The Global Moran's I tool in Geospatial Analyst in seasonal real-time bidirectional PDA-GIS-GPS-RS cyber-environments may measure seasonal geo-spatial autocorrelation based on empirical datasets of regressed, georeferenced, seasonal, vector, mosquito-related, aquatic, larval habitat feature locations and feature values. Indicators of spatial association are statistics that evaluate the existence of clusters in the spatial arrangement of a given variable [6]. For instance, suppose a vector biologist, epidemiologist or other data analyst is studying seasonal tabulated malarial-related explanatory regressed monthly prevalence rates among census tracts in a given city constructs a real-time seasonal explanatory predictive epidemiological risk model in a real-time bidirectional PDA-GIS-DGPS-RS cyber-environment, employing the rates mean could efficiently reveal where areas have higher or lower rates of endemic transmission foci occur, than is to be expected by chance alone (i.e., geolocalized georeferenced cluster); that is, the values occurring above or below those of a random distribution in space.

Global spatial autocorrelation is a measure of the overall clustering of the data [2]. One of the statistics used to evaluate global spatial autocorrelation is Moran's I, defined by:

$$I = \frac{\frac{N}{S_0}\sum_i \sum_j W_{ij} Z_i Z_j}{\sum_i Z_i^2}$$

where Z_i is the deviation of the variable of interest with respect to the mean; W_{ij} is the matrix of weights that in some cases is equivalent to a binary matrix with ones in position i, j whenever observation i is a neighbor of observation j, and zero otherwise; and $S_0 = \sum_i \sum_j W_{ij}$. The matrix W is required because in order to address explanatory geo-spatial autocorrelation and also model spatial interaction, a structure to constrain the number of neighbors to be considered needs to be imposed. This is related to Tobler's first law of geography which states that Everything depends on everything else, but closer things more so-in other words, the law implies a spatial distance decay function, such that even though all sampled vector mosquito-related observational predictors have an influence on all other sampled observations, after some distance threshold (e.g., Euclidean distance from a georeferenced epidemiological capture point to a prolific sampled aquatic larval habitat) that influence can be neglected.

Distance decay in a geo-spatiotemporal explanatory seasonal vector mosquito-related field and/or remote georeferenced endemic transmission oriented predictive risk model may be graphically represented by a curving line that swoops concavely downward as distance along the x-axis increases in a seasonal residual scatter plot in a real-time bidirectional PDA-GIS-DGPS-RS cyber-environment. Distance decay can be then mathematically re- presented by the expression $I = 1/d^2$, where I is interaction and d is distance, among other forms. Global spatial analysis or global explanatory geo-spatial autocorrelation analysis would yield one statistic to summarize the whole study area in the seasonal real-time bidirectional cyber-environment. In other words, the global analysis in the PDA-GIS-DGPS-RS cyber-environment would assume homogeneity in the empirical-sampled georeferenced mosquito data. If that assumption does not hold in a vector mosquito-related endemic transmission-oriented eco- epidemiological predictive risk model, then having only one statistic to represent the geo-spatiotemporal autocorrelation coefficients in the real-time bidirectional PDA-GIS-DGPS-RS cyber-environment would not make sense, as the statistic would differ over geographical space. But if there is no global autocorrelation or no clustering in a regressed explanatory dataset of georeferenced field and/or remote seasonal-sampled vector mosquito- related endemic transmission oriented aquatic larval habitat risk model parameter estimators, a vector biologist, epidemiologist or data analyst could still find georeferencable clusters at a local level using geo-localized geo- spatial autocorrelation coefficients.

Moran's I is a summation of individual cross products which may be exploited by the "Local indicators of spatial association" (LISA) to evaluate the clustering in those individual units (e.g., prolific georeferenced vector mosquito aquatic larval habitats) by calculating Local Moran's I for each spatial unit and evaluating the statistical significance for each I_i. From the previous equation a geo-spatiotemporal real-time bidirectional vector mosquito- related endemic transmission-oriented autoregressive risk model:

$$I_i = \frac{Z_i}{m_2} \sum_j W_{ij} Z_j$$

may be robustly constructed employing PDA-GIS-DGPS-RS cyber-tools where: $m_2 = \frac{\sum_i Z_i^2}{N}$. By so doing, $I = \sum_i \frac{I_i}{N}$ I would be the Moran's I measure of global autocorrelation in the cyber-environment where, I_i is local,

and N would be the number of analysis units in the endemic transmission-oriented georeferenced time series explanatory predictive risk map. Thereafter, a vector biologist, epidemiologist and/or other data analyst may employ a default conceptualization of the geospatial explanatory seasonal vector mosquito related endemic transmission- oriented field and/ or remote specified relationships in a real-time bidirectional PDA-GIS-DGPS-RS cyber-environment for performing a robust Hot Spot Analysis tool to geographically specify all the geographically and non-geographically sampled time series larval/pupal habitat explanatory Euclidean distance values.

A Hot Spot Analysis tool embedded in a seasonal real-time bidirectional PDA-GIS-DGPS-RS cyber-envi- ronment could calculate the Getis-Ord Gi* statistic for each time series sampled explanatory field or remote specified vector mosquito aquatic larval habitat endemic transmission-oriented georeferencable data feature attribute in an empirical dataset. By so doing, the resultant z-scores and p-values could then determine where the sampled georeferenced vector mosquito larval habitat data features attributes with either high or low values (e.g., total seasonal larval habitat density counts) cluster geospatially. This seasonal real-time bidirectional PDA-GIS- DGPS-RS cyber-infrastructure tool would work by invasively examining each sampled georeferenced time series explanatory aquatic larval habitat feature attribute within the context of neighboring sampled aquatic larval habitat features. A georeferenced aquatic larval habitat explanatory data feature attribute with a high larval count value may reveal the importance of a sampled field and/or remote specified parameter estimator, but the geographic location of the sampled covariate may not be a statistically significant hot spot. To be a statistically significant hot spot, a georeferenced explanatory data feature attribute in a seasonal real-time bidirectional PDA-GIS-DGPS-RS cyber-environment would have to have a high value (e.g. total seasonal larval density count) and be surrounded by other geographically sampled features with high values as well (e.g. positive autocorrelation). The local sum for a georeferenced field and/or remote time series specified explanatory endemic transmission-oriented data feature attribute and its neighbors could then be compared proportionally to the sum of all the georeferenced vector mosquito data feature attributes in the cyber-environment.

When the local sum is very different from the expected local sum, and that difference is too large to be the result of random chance in an explanatory geo-spatiotemporal vector mosquito-related field and/or remote sampled endemic transmission-oriented eco-epidemiological predictive risk model, a statistically significant z-score would result. For instance, a time series Gi* statistic quantized in a robust seasonal explanatory real-time bidirectional PDA-GIS-DGPS-RS cyber-environment could return a z-score for each empirical sampled vector mosquito-re- lated field and/or remote specified georeferenced endemic transmission oriented data feature attribute in an empirical regressed dataset of explanatory covariate coefficients. For statistically significant positive z-scores, the larger the z-score is, the more intense the clustering of high values (i.e., hot spot), while for statistically significant negative z-scores, the smaller the z-score is, the more intense the clustering of low values (i.e., cold spot) [2]. This PDA-GIS-DGPS-RS cyber-tool could then create a new output feature class with a z-score and p-value for each sampled georeferenced aquatic larval habitat explanatory data feature attribute in an input feature class.

If there is a selection set applied to the input feature class in a seasonal real-time bidirectional explanatory PDA-GIS-DGPS-RS cyber-environment, then only selected sampled georeferenced aquatic larval habitat endemic transmission oriented data feature attributes would be analyzed, and only selected attributes would appear in the output feature class. This cyber-tool would also optimally return the z-score and p-value field names as derived output values for potential use in other custom time series, explanatory, vector, mosquito-related, field and/or remote specified, risk models and scripts derived in the cyber-environment. When this cyber-tool runs, the output georeferenced habitat feature class would automatically be added to the table of contents with default rendering applied to the z-score field. The hot to cold spot rendering applied to the geo-spatiotemporal vector mosquito-related endemic transmission-oriented risk model would then be defined by a layer file in the cyber-environment. Thereafter, a vector biologist, epidemiologist and/or other research collaborator could reapply the default rendering, if needed, by importing a template layer of symbology. The recommended and default conceptualization of spatial relationships for the Hot Spot Analysis (e.g., Getis-Ord Gi*) tool would then be expressed as a fixed distance band in the seasonal real-time bidirectional PDA-GIS-GPS-RS cyber-environment.

Fixed distance band method works well for point data as it is often a good option for polygonizing time series georeferencable data when there is a large variation in polygon size (e.g., very large seasonal epidemiological vector mosquito-related field and/or remote specified time series polygons at the edge of the epidemiological interventional study site and very small polygons at the center of the study site) (www.esri.com). Peak distances may then be appropriate for parsimoniously quantitating explanatory georeferenced time series larval habitat sampled values to use for tools with a Distance Band or Distance Radius parameters in seasonal real-time bidirectional PDA-GIS-GPS-RS cyber-environment when conducting geo-spatiotemporal endemic transmission oriented vector mosquito risk-based data analyses. Relationships between the geo-spatiotemporally quantitated georeferenced data feature attributes may then be defined using both space and time, employing several new explanatory field and/or remote specified parameter estimators. Thereafter, by performing a robust space/time analyses, a vector biologist, epidemiologist and/or research collaborator may select an option in the cyber-envi- ronment for determining the conceptualization of the geo-spatiotemporal relationships in an ecological empirical dataset of seasonal-sampled parameter estimators selected for regression in the cyber-environment. By so doing, a vector biologist, epidemiologist and/or other research collaborator could define space in a seasonal explanatory georeferenced vector mosquito-related field and/or remote specified endemic transmission oriented predictive epidemiological risk model by specifying a threshold distance value; [e.g., specifying a Date/Time Field and both a Date/Time Type (such as HOURS or DAYS) and a Date/Time Interval Value], all of which may be independent interpolatable explanatory parameter estimators in the risk model. Commonly in geo-spatiotemporal vector arthropod-related seasonal endemic transmission-oriented explanatory autoregressive covariate coefficients, the Date/Time Interval Value is an Integer [2]. For instance, if 1000 feet in entered in a robust PDA-GIS- GPS-RS cyber-environment, for an elevational-related explanatory covariate associated to a particular georeferenced vector mosquito-related larval/pupal aquatic larval habitat, selecting HOURS can provide Date/Time Interval Value features within 1000 feet and those occurring within three hours of each sampled georeferenced aquatic larval habitat other would be considered neighbors.

For performing a time series Hot Spot Analyses in a seasonal real-time bidirectional PDA-GIS-GPS-RS cyber-environment, seasonally-derived vector mosquito-related epidemiological forecasting risk model, an output source of identification may be added for use in for joining an output georeferencable explanatory dataset to the original dataset. Given a set of weighted vector geo-spatiotemporal-sampled georeferenced mosquito data attri- bute features, a PDA-GIS-GPS-RS cyber-environment can statistically identify significant hot spots and cold spots using the Getis-Ord Gi* statistic. This tool can identify statistically georeferenced seasonal vector mosquito geospatial clusters of high values (i.e., hot spots) and low values (i.e., cold spots). It can create a new Output Feature Class with a z-score and p-value for each feature in the Input Feature Class in the cyber-environment. The statistic can also return the z-score and p-value field names as robust derived output values for potential use in customized vector mosquito-related real-time predictive risk models and scripts. The z-scores and p-values are measures of statistical significance which tell you whether or not to reject the null hypothesis, feature by feature [2]. In effect, a seasonal real-time bidirectional PDA-GIS-GPS-RS cyber-environment can indicate whether the observed spatial clustering of high or low explanatory, georeferencable, operationizable, vector mosquito data values is more pronounced than one would expect in a random distribution of those same values. A high z-score and small p-value for a feature indicates a spatial clustering of high values. A low negative z-score and small p- value indicates a spatial clustering of low values [3]. The higher (or lower) the z-score of the seasonal georeferenced vector mosquito-related explanatory endemic transmission-oriented covariate coefficients the more intense the clustering. A z-score near zero of a regressed dataset of vector related would then indicate no apparent spatial clustering. The z-score is based on the randomization null hypothesis computation [3].

Seasonal, explanatory, georeferencable, real-time, bidirectional, vector, mosquito-related data calculations based on either Euclidean or Manhattan distance require projected data to accurately measure habitat distances. For line and polygon features, feature centroids may be employed in distance computations in a robust real-time bidirectional PDA-GIS-DGPS-RS cyber-environment. For multipoints, polylines, or polygons with multiple parts, the centroid is computed using the weighted mean center of all feature parts [2]. The weighting for georeferenced point-related seasonal larval habitat attribute features in the cyber-environment would thus be 1 whereby, line features would be the length of the habitat and the polygon features would be the sample area. The Input Field in the cyber-environment would contain a variety of seasonal-sampled values. The math for this statistic would require some variation in the sampled variable being analyzed; it cannot solve if all input seasonal real-time bidirectional georeferenced vector mosquito-related endemic transmission-oriented explanatory covariate coefficient values are 1. If a vector biologist, epidemiologist or data analyst wants to use this tool to delineate the spatial trend pattern of incident data in a robust PDA-GIS-GPS-RS cyber-environment. This would involve aggregating the incident data into empirical sampled geographic datasets. The conceptualization of the georeferenced explanatory vector mosquito seasonal-sampled data feature attribute spatial relationship parameter estimators would then reflect inherent relationships among the sampled habitat features analyzed.

Within the seasonal real-time bidirectional PDA-GIS-GPS-RS cyber-environment the FIXED_DISTANCE_ BAND configuration would be able to quantitate the default Distance Band or Threshold Distance which could then subsequently ensure each sampled georeferenced habitat feature attribute has at least one neighbor. But often, this default will not be the most appropriate distance to employ for robustly spatially quantitating geospatial geo- referenced explanatory vector mosquito-related covariate coefficients. Additional strategies for selecting an appropriate scale (i.e., distance band) for the analysis may be outlined in selecting a fixed distance band value in a robust seasonal real-time bidirectional PDA-GIS-GPS-RS cyber-environment.

The INVERSE_DISTANCE or INVERSE_DISTANCE_ SQUARED in a seasonal real-time bidirectional PDA-GIS-GPS-RS cyber-environment may be employed for seasonal vector mosquito-related explanatory parameter estimator statistical significance estimation. When zero is entered for the Distance Band or Threshold Distance parameter in the cyber-environment, all the sampled features may be considered neighbors of all other georeferenced data feature attributes; when this parameter is left blank, the default distance will be applied automatically in the cyber-environment. Weights for distances less than 1 become unstable when they are inverted [2]. Consequently, the weighting for seasonal-sampled vector mosquito-related explanatory georeferenced habitat feature data attributes separated by less than 1 unit of distance (common with Geographic Coordinate System Vector mosquito-related projections) may be given a weight of 1 in a real-time bidirectional PDA-GIS-GPS-RS cyber-environment.

Analysis on features with a Geographic Coordinate Systemvector mosquito-related projections is not recommended when the inverse distance-based spatial conceptualization methods (INVERSE_DISTANCE, INVERSE_DISTANCE_ SQUARED, or ZONE_OF_INDIFFERENCE) is selected in a seasonal real-time bidirectional PDA-GIS-GPS-RS cyber-environment. For the inverse distance options (INVERSE_ DISTANCE, INVERSE_DISTANCE_SQUARED, or ZONE_ OF_INDIFFERENCE), any two seasonal sampled vector mosquito-related habitat points that are neigbours will be given a weight of one to avoid zero division. This would assure the sampled georeferenced larval habitat features are not excluded from the analysis. Additional options for the conceptualization of the seasonal geo-spatiotemporal relationships in the regressed vector mosquito-related parameter estimators, including space-time relationships, may be also quantitated in a robust seasonal real-time bidirectional PDA-GIS-GPS-RS cyber-environment using the Generate Spatial Weights Matrix (SWM) or Generate Network Spatial Weights tools. To take advantage of these additional options, a vector biologist, epidemiologist or a data analyst may employ one of these tools to construct the time series spatially autoregressive vector mosquito-related weighted matrix file prior to analysis; by selecting GET_SPATIAL_WEIGHTS_FROM_FILE for parsimoniously quantitating the conceptualization of spatial relationships parameter; and for the Weights Matrix File parameter in the seasonal real-time bidirectional real-time cyber-environment while, specifying the path to the spatial weights file created.

Seasonal vector mosquito-related map layers can be used to define the Input Feature Class in the regression- related predictive risk model constructed in a seasonal real-time bidirectional PDA-GIS-GPS-RS cyber-envi- ronment. When employing a layer with a selection in the cyber-environment, however, only the selected features may be included in the analysis. If a vector biologist, epidemiologist or data analyst provides a Weights Matrix File with an spatial autoregressive extension in the cyber-environment, this tool would then be expected to conduct robust parameter estimator significance level quantification in a robust geo-spatiotemporal weights matrix file created using either the Generate Spatial Weights Matrix or Generate Network Spatial Weights tools; otherwise, this tool would be expected as an ASCII formatted spatial weights matrix file. In some seasonal real-time bidirectional PDA-GIS-GPS-RS cyber-environment derived vector mosquito-related predictive seasonal risk models, the expected behavior of the residually forecasted derivatives may be different depending on which type of spatial weights matrix file is employed for the regressive quantization process. ASCII-formatted spatial weights matrix files apply specific time series weights when constructing seasonal explanatory geo-predictive predictive vector mosquito-related endemic transmission-oriented epidemiological risk models. Missing georeferenced data feature-to-feature attribute relationships in seasonal vector mosquito-risk models may then be treated as zeros. The default weight for self potential is zero, unless you specify a Self Potential Field value, or include self potential weights explicitly [2]. If the seasonal real-time bidirectional weights are row standardized, results will likely be incorrect for analyses on selection sets. If a vector biologist, epidemiologist or a data analyst runs an analysis on a selection set, a robust parameter estimation may be conducted by converting the ASC11spatial weights file to a SWM file by reading the empirical sampled vector mosquito-related log-transformed ASCII data into a table and then converting the table using the CONVERT TABLE option within a robust Spatial Weights Matrix in the cyber-environment. If the weights are row standardized, they will be restandardized for selection sets; otherwise, weights would be the default weight for self-potential.

Running a seasonal vector mosquito-residual risk based explanatory data analyses within an ASC11formatted SWM in a seasonal real-time bidirectional PDA-GIS-GPS-RS cyber-environment would be memory intensive. For example, if there exists an empirical sampled dataset of seasonal georeferencable vector data analyses on more than 5000 attribute features, a vector biologist, epidemiologist or data analyst may consider converting the ASCII- formatted SWM file into an SWM formatted file in the cyber-environment. This can be performed by putting ASCII weights into a formatted table (using Excel, for example) in the real-time bidirectional PDA-GIS-GPS-RS cyber-environment. Next, the Generate Spatial Weights Matrix tool using CONVERT_TABLE for the Conceptualization of Spatial Relationships parameter would be run. The output would be an SWM-formatted spatial weights matrix file. When this tool runs in the cyber-environment, an Output Feature Class would be automatically added to the table of contents with default rendering applied to the z-score field. The hot-to-cold rendering applied would then be defined by a layer file in the cyber-environment. The default rendering may be re-applied, if needed, by importing the template layer symbology. The Output Feature Class in the seasonal PDA-GIS-GPS-RS cyber-environment would then include a SOURCE_ID field which would allow joining the Input Feature Classes.

When using shapefiles, however, a seasonal real-time bidirectional cyber-environment may not be able to store null values. Tools or other procedures that create shapefiles from nonshapefile inputs in the cyber-environment may store or interpret seasonally regressed vector mosquito-related null values as zero. In some cases, nulls are stored as very large negative values in shapefiles (www.esri.com). However, this can lead to unexpected seasonal residually forecasted misspecified derivatives. Conversely, a new parameter may be added to any output report file that allows the vector biologist, epidemiologist and/or other research collaborators to specify the location of the optional to PDF report. The seasonal real-time bidirectional PDA-GIS-GPS-RS cyber-environment could then include new seasonal vector mosquito field and/or remote specified data to the report and diagnostics from the many regression matrices [e.g., Ordinary Least Square (OLS)] analysis, as well as several graphics to help interpretation.

A exploratory regression tool in a robust seasonal real-time bidirectional PDA-GIS-GPS-RS cyber-environ- ment could evaluate all possible combinations of input candidate explanatory time series georeferenced vector mosquito-related field and/or specified endemic transmission oriented predictor variables for constructing OLS models that best explain a dependent variable (e.g., total seasonal-sampled aquatic larval habitat density count data) within the context of user-specified criteria. In statistics, OLS or linear least squares is a method for estimating the unknown parameters in a linear regression model [6]. This method would minimize the sum of squared vertical distances between the observed responses in an empirical-sampled dataset of georeferenced vector mosquito related field or remote specified endemic transmission oriented parameter estimators and the responses predicted by the linear approximation. The resulting estimator may then be expressed by a simple formula in the seasonal real-time bidirectional cyber-environment, especially in the case of a single repressor on the right-hand side. For instance, suppose the resulting regressed field or remote specified vector mosquito-related endemic transmission oriented explanatory risk model consists of no bservations $\{y_i,x_i\}n_i = 1$ in a robust PDA-GIS-GPS-RS cyber-environment. Each seasonal real-time bidirectional observation could then include a scalar response y_i and a vector of p endemic transmission oriented repressors x_i. In a linear vector mosquito related regression-based risk model constructed in the cyber-environment, the response variable would be a linear function of the regressors: $y_i = x_i'\beta + \varepsilon_i$ then where β is a p × 1 vector of unknown field and/or remote parameters; $\varepsilon_i's$ are unobserved scalar random variables (i.e., errors). The residually forecasted regression derivatives in the cyber-envi- ronment could then account for the discrepancy between the actually observed endemic transmission oriented responses y_i and the "predicted outcomes" $x_i'\beta$; while simultaneously denoting a matrix transpose, so that $x'\beta$ is the dot product between the vectors x and β.

In mathematics, the dot product, or scalar product (or sometimes inner product in the context of Euclidean space), is an algebraic operation that takes two equal-length sequences of numbers (usually coordinate vectors) and returns a single number [6]. This operation in a robust, time series, explanatory, vector, mosquito-related, field and/or remote specified, endemic, transmission oriented, epidemiological risk model may be defined either algebraically or geometrically in a seasonal real-time bidirectional PDA-GIS-GPS-RS cyber-environment. This model can also be written in matrix notation as $y = X\beta + \varepsilon$, where y and ε are n × 1 vectors, and X is an n × p matrix of regressors, which is also sometimes called the design matrix [6]. As a rule, in a robust explanatory vector arthropod-related field and/or remote specific endemic transmission oriented risk model the constant term is always included in the set of regressors X, by taking $x_{i1} = 1$ for all $i = 1, \ldots, n$ [2]. The coefficient β_1 corresponding to this regressor would then be the intercept. There may be some relationship between the time series dependent vector arthropod-related field and/or remote specified explanatory endemic transmission oriented regressors. For instance, the third regressor may be the square of the second regressor in the seasonal real-time bidirectional PDA-GIS-GPS-RS cyber-environment derived epidemiological, time series, predictive, risk model. In this case, assuming that the first regressor is constant, a vector biologist, epidemiologist and/or other research collaborators may generate a quadratic model specification in the second regressor. The model which would still be considered linear as it would be linear in the β seasonal real-time bidirectional PDA-GIS-GPS-RS cyber-en- vironment. In mathematics, the term quadratic describes something that pertains to squares, to the operation of squaring, to terms of the second degree, or equations or formulas that involve such terms [6].

The OLS estimator generated from a robust seasonal real-time bidirectional PDA-GIS-GPS-RS cyber-envi- ronment would be consistent when the explanatory georeferenced field or remote specified endemic transmission oriented time series regressors are exogenous and there is no perfect multicollinearity, and are optimal in the class of linear unbiased estimators when the errors are homoscedastic and serially uncorrelated. Multicollinearity refers to a situation in which two or more explanatory variables in a multiple regression model are highly linearly related [6]. In perfect multicollinearity in a seasonal vector arthropod-related regression-based empirical dataset of residually forecasted explanatory georeferencable field and/or remote specified derivatives, the correlation between two independent variables is equal to 1 or −1 [2]. In practice, vector biologists, epidemiologists and/or other research collaborators rarely face perfect multicollinearity in empirically regressed datasets of explanatory time series georeferenced field and remote specified vector mosquito-related endemic transmission oriented explanatory covariate coefficients [2]. More commonly, the issue of multicollinearity would arise in a seasonal real-time bidirectional PDA-GIS-GPS-RS cyber-environment derived vector mosquito-related endemic transmission oriented risk models when there is an approximate linear relationship among two or more independent variables in the model.

Mathematically, an empirical dataset of time series, explanatory, vector, mosquito-related, field or remote, explanatory, endemic, transmission oriented, predictive variables would be perfectly multicollinear in a robust PDA-GIS-GPS-RS cyber-environment if there exists one or more exact linear relationships among some of the sampled variables. For instance, if there exists $\lambda_o + \lambda_1 X_{1i} + \lambda_2 X_{2i} + \text{L} + \lambda_k X_{ki} = 0$ holding for all sampled malaria mosquito-related time series field and remote specified endemic transmission oriented observations i, in an empirical sampled dataset where λ_j are constants and X_{ij} is the ith observation on the jth explanatory variable. Then multicollinearity may be examined in a real-time bidirectional PDA-GIS-DGPS-RS cyber-environment by attempting to obtain estimates for the parameters of the multiple regression equation $Y_i = \beta_0 + \beta_1 X_{1i} + \text{L} + \beta_k X_{ki} + \varepsilon_i$ [1.1]. The OLS estimates would then involve inverting the geo-spatiotemporal vector malarial mosquito-related matrix $X^T X$ where

$$X = \begin{bmatrix} 1 & X_{11} & \text{L} & X_{k1} \\ \text{M} & \text{M} & \text{O} & \text{M} \\ 1 & X_{1N} & \text{L} & X_{kN} \end{bmatrix}$$

in the PDA-GIS-GPS-RS cyber-environment. If there is an exact linear relationship (e.g., perfect multicollinearity) among the independent variables in the cyber-environ- ment, the rank of X (and therefore of $X^T X$) would then be less than k + 1, and the matrix $X^T X$ would not be invertible. For instance, suppose that instead of the above equation 1.1, a vector biologist, epidemiologist and/or other research collaborators construct a vector time series malaria-mosquito (e.g., Anopheles gambiae s.l.) related endemic transmission oriented forecasting equation in a seasonal real-time bidirectional PDA-GIS-GPS-RS cyber-environment employing a modified form with an error term v_i : $\lambda_o + \lambda_1 X_{1i} + \lambda_2 X_{2i} + \text{L} + \lambda_k X_{ki} + v_i = 0$. In this case, there would be no exact linear relationship among the seasonal – sampled field and/or remote specified time series explanatory endemic transmission oriented variables, but the X_j variables would be nearly perfectly multicollinear if the variance of v_i is small for some set of the sampled covariate coefficient values for theλ's. In this case, the matrix $X^T X$ would have an inverse residually forecasted derivative, but it would be ill-conditioned so that a given computer algorithm in a robust PDA-GIS-GPS-RS cyber-environment, may or may not be able to compute an approximate inverse, and if it does the resulting computed inverse may be highly sensitive to slight variations in the sampled data (e.g., magnified effects of rounding error); thus, rendering misspecifications in the output.

In statistics, a sequence or a vector of random variables is homoscedastic if all random variables in the sequence or vector have the same finite variance [2]. The variance is a parameter that describes, in part, either the actual probability distribution of an observed population of numbers, or the theoretical probability distribution of a sample (a not-fully-observed population) of numbers [6]. In the latter case, an empirical regressed sampled dataset of georeferenced explanatory field and/or remote specified explanatory endemic transmission oriented covariate coefficient distributions tabulated in a seasonal real-time bidirectional PDA-GIS-GPS-RS cyber-environment may be used to construct an estimate of its variance: in the simplest cases this estimate can be the sample variance (e.g., homogeneity of variance). The complementary notion is called heteroscedasticity. The assumption of homoscedasticity simplifies mathematical and computational treatment in a time series, linearized, vector, mosquito-related, field and/or remote, specified, explanatory, endemic, transmission-oriented, epidemiological risk model. Serious violations in homoscedasticity (assuming a distribution of vector mosquito-related sampled data is homoscedastic when in actuality it is heteroscedastic) may result in overestimating the goodness of fit in an empirical dataset of sampled field and remote explanatory specified covariate coefficients as measured by the Pearson coefficient in a robust seasonal real-time bidirectional PDA-GIS-GPS-RS cyber-environment. In statistics, the Pearson product-moment correlation coefficient (sometimes referred to as the PPMCC or PCC or Pearson's r) is a measure of the linear correlation (i.e., dependence) between two variables X and Y, giving a value between +1 and −1 inclusive, where 1 is total positive correlation, 0 is no correlation, and −1 is total negative correlation [6].

Under some conditions, the method of OLS provides time series minimum-variance mean-unbiased estimation when the errors have finite variances. For example, under the additional assumption that the errors be normally distributed in a robust real-time bidirectional PDA-GIS-GPS-RS cyber-environment, OLS would be the maximum likelihood estimator for a seasonal filarsis mosquito-related (e.g., Culex quinquefasciatus) predictive endemic transmission-oriented explanatory epidemiological risk model. The OLS tool in the cyber-environment would produce an output feature class and optional tables with coefficient information and diagnostics for rendering robust inferences. All of these would be accessible in the robust real-time bidirectional cyber-environment Results window. The output feature class would be automatically added to the table of contents, with a hot/cold rendering scheme applied to the endemic transmission-oriented explanatory risk model forecasted derivatives. A full explanation of each output would be provided when interpreting the OLS results in the land cover data, for example, in the seasonal PDA-GIS-GPS-RS cyber-environment, to update information about the environment (e.g., natural or urban land cover) in seasonal mosquito larval control operational areas at a resolution that may not be obtained on the ground. Such data can be processed to provide potentially relevant information about seasonal environmental conditions critical for IVM implementation.

Further, to ensure a consistent scale of analysis Space-Time Window, Zone of Indifference, Contiguity, K Nearest Neighbor, and Delaunay Triangulation may also work well within a robust seasonal real-time bidirectional PDA-GIS-GPS-RS cyber-environment for parsimoniously quantitating topographic trends in an empirical ecological dataset of geo-spatiotemporal sampled vector mosquito-related explanatory specified field and/or remote endemic transmission-oriented georeferencable feature data attributes. In pattern recognition, the k-Nearest Neighbors algorithm (or k-NN for short) is a non-parametric method used for classification and regression [1]. In both cases, the input would consist of the k closest training examples in the feature space in the constructed time series vector mosquito-related predictive endemic transmission-oriented explanatory risk model. The output in the bidirectional PDA-GIS-GPS-RS cyber-environment would depend on whether k-NN is used for classification or regression. In k-NN classification, the output is a class membership [1]. An object is classified by a majority vote of its neighbors, with the object being assigned to the class most common among its k nearest neighbors (k is a positive integer, typically small). If k = 1, then the object is simply assigned to the class of that single nearest neighbor [3]. In k-NN regression, the output is the property value for the object. This value in a time series vector mosquito-related epidemiological predictive risk model parameter estimator quantitation process in the cyber-environment would be the tabulated average of the sampled field and/or remote explanatory coefficient values of its k nearest neighbors. k-NN is a type of instance-based learning, or lazy learning, where the function is only approximated locally and all computation is deferred until classification [1]. The k-NN algorithm is among the simplest of all machine learning algorithms In mathematics and computational geometry, a Delaunay triangulation for a set P of points in a plane is a triangulation DT(P) such that no point in P is inside the circumcircle of any triangle in DT(P) [3]. Delaunay triangulations maximize the minimum angle of all the angles of the triangles in the triangulation; they tend to avoid skinny triangles. For a set of seasonal-sampled vector mosquito-related aquatic larval habitat points on the same line there is no Delaunay triangulation (the notion of triangulation is degenerate for this case). For four or more points on the same circle (e.g., the vertices of a rectangle) in a real-time bidirectional PDA-GIS-GPS-RS cyber-environment, the Delaunay triangulation is not unique: each of the two possible triangulations that split the quadrangle into two triangles may not satisfy the "Delaunay condition", (i.e., the requirement that the circumcircles of all triangles have empty interiors). By considering circumscribed spheres, the notion of Delaunay triangulation in a robust PDA-GIS-GPS-RS cyber-environment would extend to three and higher dimensions in a vector mosquito-related epidemiological predictive risk model residually forecasted derivatives. Generalizations are possible to metrics other than Euclidean in a vector mosquito-related predictive risk model [2]. However in these cases a Delaunay triangulation is not guaranteed to exist or be unique.

Attention to time series spatial patterns in an empirical sampled dataset of georeferenced explanatory specified endemic transmission oriented covariate coefficients in a robust real-time bidirectional PDA-GIS-GPS-RS cyber-environment can lead to insights that may have been otherwise overlooked, while ignoring space may lead to false conclusions about ecological relationships in the sampled data. Jacob et al. [7] employed multiple time series Gaussian geo-spatiotemporally autoregressive georeferencable explanatory operationalizable field and remote specified endemic transmission-oriented predictive epidemiological risk models, fit within ArcGIS Spatial Analyst, to examine relationships in geographic space of multiple georeferenced sub-meter resolution imaged aquatic larval habitats of An. arabienisis. The time series field and remote explanatory endemic transmission oriented covariate coefficients were sampled in the Karima agro-village complex, Mwea Rice Scheme, Kenya. Interestingly, in preliminary models that ignored space, the abundance of the sampled time series vector malarial mosquito larval habitat species was correlated with both local and landscape-scale seasonal-sampled variables. These models were then modified to account for quantitated broad scale surface spatial trends employing a time series analysis and fine-scale autocorrelation in an autoregressive covariance uncertainty matrix. A covariance matrix for a seasonal, An. arabienisis, aquatic, larval habitat, field and/or remote specified, endemic, transmission oriented risk model was constructed as a square matrix whose diagonal entries were the variances of, and whose off diagonal entries which were the covariance's between, the row/column labeling variables.

Residuals from OLS regression models then revealed positive autocorrelation (i.e., georeferenced habitats of similar seasonal An. arabienisis aquatic larval habitat density counts aggregated in geographic space), indicating that the assumption of independent errors was violated in the forecasted model derivatives. In contrast, the time series autoregressive residually forecasted derivatives from the explanatory non-linear models showed little spatial pattern, suggesting that these models were appropriate. The magnitude of the sampled explanatory quantitated aquatic larval habitat seasonal effects tended to decrease, and the relative importance of different explanatory georeferenced biophysical explanatory field and remote specified georeferenced endemic transmission-oriented observational predictors shifted when the authors incorporated broad scale predictor variables and then fine-scale space into the analysis. The degree to which the random sampled An. arabiensis aquatic larval habitat effects changed when space was added to the models was roughly correlated with the amount of spatial structure in the empirical sampled aquatic larval habitat explanatory predictor variables.

Within a robust seasonal real-time bidirectional PDA-GIS-GPS-RS cyber-environment, latent time series spatial patterns in empirical explanatory datasets of regressed vector mosquito-related endemic transmission oriented residually forecasted derivatives rendered from OLS models may be parsimoniously quantitated. Failure to include or adequately measure autocorrelation in an empirical dataset of regressed georeferenced field or remote specified explanatory endemic transmission oriented geo-spatiotemporal sampled An. arabienisis aquatic larval habitat variables may lead to stochastic/deterministic interpolation misspecifications. Time series seasonal real-time bidirectional An. arabienisis species abundance and distribution data may not be explained by a simple logistic regression-related endemic transmission oriented field and/or remote specified aquatic larval habitat risk model.

As such, given a set of time series sampled explanatory seasonal real-time bidirectional vector mosquito-re- lated feature data attribute (e.g., total aquatic larval habitat density count), a robust PDA-GIS-GPS-RS cyber- environment could determine whether a seasonally regressed spatial pattern expressed in an empirical dataset of endemic transmission oriented georeferencable explanatory risk model residually forecasted derivatives (e.g., locational covariates of an aggregation of productive georeferenced aquatic larval habitats in an urban ecosystem) has latent autocorrelation traits. For example, a geo-spatial autocorrelation tool in a seasonal real-time bidirectional PDA-GIS-GPS-RS cyber-environment could test the null hypothesis that in an endemic transmission oriented a time series malaria mosquito (urban Anopheles gambiae s.l.) explanatory endemic transmission- oriented epidemiological risk model there is no spatial clustering of the georeferenced habitats based on total seasonal-sampled larval density count values for an interventional study site. When the p-value is small and the absolute value of the z score in the cyber-environment is large enough that it falls outside of the desired confidence level, the null would be rejected [3]. If the index value tabulated is greater than 0, a set of sampled georeferenced explanatory aquatic larval habitat endemic transmission-oriented field or remote specified data feature attributes would then exhibit a clustered pattern. If the value is less than 0, the set of mosquito-related data feature attributes would exhibit a dispersed pattern. A PDA-GIS-GPS-RS cyber-tool would then calculate the Moran's I value employing both a z-score and p-value to evaluate the significance of the clustering, dispersion, or randomization in the empirical dataset of georeferenced time series regressed explanatory endemic transmission oriented covariate coefficients. p-values are generally numerical approximations of the area under the curve for a known distribution, limited by the test statistic [2]. The Global Moran's I tool calculates a z-score and p-value to indicate whether or not you can reject the null hypothesis [5]. In the case of a time series robust explanatory georeferenced vector mosquito-related endemic transmission oriented risk model constructed in a PDA-GIS- GPS-RS cyber-environment, the null hypothesis would state that the sampled larval habitat sampled data feature attribute values, for instance are randomly distributed across an interventional study site area. The z-score would thus be based on the randomization null hypothesis computation in the cyber-environment. The 90 percent confidence interval for treatment effect consists of all the hypothesized values of an effect that are not rejected [1].

For each test, in a robust seasonal explanatory vector mosquito-related explanatory spatiotemporal field and/or remote specified endemic transmission-oriented risk model, a vector biologist, epidemiologist and/or other data analyst could reject the null hypothesis, if the one-tailed p-value in the most extreme direction is less than 0.05 in a seasonal real-time bidirectional PDA-GIS-GPS-RS cyber-environment. One-tailed tests are used for asymmetric distributions that have a single tail, such as the chi-squared distribution, which are common in measuring goodness-of-fit in seasonal vector arthropod-related linearized risk models, or for one side of a distribution that has two tails, such as the normal distribution, which is common in estimating georeferenced larval habitat geolocation [2]; this corresponds to specifying a direction in the cyber-environment. Measures of goodness of fit may then typically summarize the discrepancy between georeferenced observational values and the values expected in a PDA-GIS-GPS-RS cyber-environment under the seasonal predictive vector mosquito-related endemic-trans- mission oriented epidemiological risk model in question which can be used in statistical hypothesis testing, (e.g. to test for normality of forecasts in a regressed dataset of field or remote sampled endemic transmission oriented covariate coefficients), to determine whether two seasonal real-time bidirectional samples are drawn from identical distributions (e.g., Kolmogorov-Smirnov test), or whether outcome frequencies follow a specified distribution (Pearson's chi-squared test).

In georeferencable seasonal real-time bidirectional vector mosquito-related endemic transmission oriented risk-based predictive time series analyses, the chi-squared distribution (also chi-square or χ^2-distribution) with k degrees of freedom is the distribution of a sum of the squares of k independent standard normal random variables. A special case of the gamma distribution may be employed for quantitating probability distributions for aggregating inferential statistics and hypothesis testing seasonal georeferenced vector mosquito-related endemic transmission oriented epidemiological risk model confidence intervals) [2]. In a robust seasonal real-time bidirectional PDA-GIS-GPS-RS cyber-environment, the gamma distribution would be quantitated as a two-para- meter family of continuous probability distributions which may be expressed commonly employing different parameterizations in a vector mosquito-related regression-based risk model constructed in a PDA-GIS-GPS-RS cyber-environment. For example, with a shape parameter k and a scale parameter θ in a robust PDA-GIS-GPS-RS cyber-environment and a shape parameter $\alpha = k$ and an inverse scale parameter $\beta = 1/\theta$, a rate parameter for a seasonal dengue-related mosquito (Aedes egypti)-related risk model may be efficiently quantitated. With a shape parameter k and a mean parameter $\mu = k$ quantified a rejection would occur for any observed explanatory field and/or remote sampled endemic transmission oriented value employing a specified threshold estimate region within a seasonal real-time bidirectional PDA-GIS-GPS-RS cyber-environment. For example, a vector biologist, epidemiologist and/or other research collaborator would have 0.10 probability of rejecting any hypothesized effect value in a time series predictive dengue mosquito-related eco-epidemiological risk model if it was true. The seasonal real-time bidirectional model would thus render a $1 - 0.10 = 0.90$ confidence level for a robust seasonal dengue mosquito-related epidemiological field and remote specified explanatory predictive risk model constructed in a real-time bidirectional PDA-GIS-GPS-RS cyber-environment.

Optimally, a seasonal real-time bidirectional PDA-GIS-GPS-RS cyber-infrastructure for archiving seasonal vector mosquito-related regressed data collections would be a combination of data resources, network protocols, computing platforms and computational services. Most time series explanatory georeferenced vector mosquito- related ecological domains adopt intrinsic geospatial principles such as quantitated seasonal constraints in phenomena evolution (e.g., larval development) for determining robust randomized sample frames. Generally these environments require large amounts of time series data processing for performing robust optimal empirical geospatial explanatory georeferencable risk-based endemic transmission oriented risk-based analyses parsimoniously. These processes commonly include operationalized geospatial analysis of seasonal-sampled explanatory larval habitat data feature attribute relationship calculations, meteorological time series parameter estimator quantitation, land cover observational predictor estimator geovisualization, and geomorphological ecohydrological terrain-related risk-related catchment modeling. The algorithms for these data processing mechanisms may be integrated within a real-time PDA-GIS-GPS-RS cyber-environment seamlessly regardless of the size of the empirical sampled explanatory mosquito-related endemic transmission oriented field and remote specified covariate coefficient geospatial explanatory ecological datasets.

The studies of transmission dynamics based on regressed empirical-sampled explanatory georeferenced field and/ or remote specified ecohydrological and vegetation-related LULC covariate coefficients for timely seasonal vector mosquito-related risk-based eco-epidemiological data analyses have been historically more focused on the understanding of mechanisms and controls at the point/plot spatial scale rather than at the scale of topographically complex epidemiological landscapes. For example, although in riceland agro-village ecosystems where topographic vegetation and climatic gradients play a crucial role in controlling time series explanatory ecohydrological irrigated and biogeochemical explanatory land cover covariate coefficient behaviors, the effects of spatially varying georeferenced seasonal real-time bidirectional vector mosquito-related aquatic larval habitat structures based on environmental specifications such as, the role of irrigated water redistribution and terrain- related ecogeomorphological processes have not been qualitatively nor quantitatively robustly regressed at the watershed scale.

Land degradation in riceland agro-village environments can alter seasonal vector mosquito-related larval habitat productivity through time series quantitated irrigated water erosion variables driven by ecogeomorphological processes which may also alter seasonal-sampled georeferenced aquatic habitat immature density counts along transfer paths at hill slopes [8]. Seasonal soil-hydraulic conditions of the upper soil layers and the vegetation structure of the hill slope can also influence immature vector mosquitoes. These LULC processes may be geo- spatiotemporally interlinked in seasonal real-time bidirectional PDA-GIS-GPS-RS cyber-environments with each other by augmenting feedback mechanisms in such a way that a small change in land use (e.g. temporary overgrazing, cattle trails) may result in real-time robust quantitation of seasonal immature density counts. A canopied forestland-shrub land LULC to rice agriculture LULC transition zone in an eco-epidemiological interventional georeferenced agro-village complex may then be rigorously analyzed in a PDA-GIS-GPS-RS cyber- environment for generating optimal time series regressively quantitated soil-vegetation-transfer feedback mechanisms that influence seasonal aquatic larval habitat productivity. By so doing, an explanatory eco-geo-morphological time series process-based georeferenced endemic transmission oriented field or remote specified endemic transmission-oriented eco-epidemiological risk model may be developed which simulates the redistribution of sediments and nutrients during high-intensity rainstorm in 1-sec time steps in specific urban environments such as riceland agroecosystems. The soil moisture and transpiration dynamics in daily time steps, and the vegetation dynamics (i.e., establishment, immature growth, mortality) in 14-day time steps for a high-resolution grid of 1×1 m^2, for instance, may then be quantitatively/qualitatively geo-spatiotemporally regressed in a robust explanatory seasonal real-time bidirectional PDA-GIS-GPS-RS cyber-environment. Furthermore, through long-term seasonal riceland malaria mosquito (An. arabiensis s.s.)-related predictive epidemiological risk modeling and the modeling of extreme environmental conditions (e.g., prolonged droughts), numerical approaches may be employed in a in a PDA-GIS-GPS-RS cyber-environment for parsimoniously analyzing which type of feedbacks may occur and may trigger persistent seasonal vegetation change and land degradation of hill slopes associated to productive georeferencable aquatic larval habitats based on geo-spatiotemporal field and/ or remote immature sampled count data. Employing such models in a seasonal real-time bidirectional PDA-GIS-GPS-RS cyber-environment would then make it possible to couple the occurrence of geospatial time series explanatory spatial patterns of soil moisture of a hill slope georeferenced vector mosquito-related aquatic larval habitat during high-intensity rice irrigated flooding cycles. By so doing, the seasonal vector malarial mosquito field or remote specified endemic transmission oriented risk model could either investigate only individual extreme events or, model the long-term dynamics of remote classified LULC geographic sub-zones in a urbanized environment in a PDA-GIS-GPS-RS cyber-en- vironment for spatially targeting prolific riceland An. arabiensis aquatic larval habitats in an epidemiological study site without including the detailed erosion processes.

At the plot scale, numerous theoretical field experimental predictive seasonal explanatory geo-spatiotemporal autoregressive epidemiological risk-related modeling approaches have been presented in GIS literature. For instance, recently, there has been an emergence of interest in extending time series explanatory eco-hydrological real-time predictive endemic transmission-oriented field and/or remote specified risk models for monitoring and evaluating specific ecosystems by robustly spatially quantitating larger scale georeferencable seasonal explanatory biophysical empirical vector mosquito-related endemic transmission oriented data feature attributes in ArcGIS Spatial Analyst. Unfortunately, the two-way coupling between seasonal vegetation dynamics and seasonal irrigated rice water cycles was not studied in these models. By not seasonally quantitating vital seasonal relationships during specific temporally dictated sample frames (e.g., tillering in the rice cycle) in a time series vector mosquito-related endemic transmission oriented field and/or remote specified epidemiological risk model, a robust seasonal real-time bidirectional PDA-GIS-GPS-RS cyber-environment, geospatially and temporally heterogeneous conditions of watersheds in catchment sites may not be revealed in the log-transformed residually forecasted linear/non-linear regression-based derivatives thus, creating misspecifications in stochastically/de- terministically seasonal real-time bidirectional interpolated model. A distributed representation of explanatory georeferencable time series geo-spatiotemporal ecohydrological dynamics and a subsequent aggregation to the watershed scale are thus warranted for investigating time series sampled vector immature mosquito-related parameter estimators in riceland and other urban epidemiological study sites. Extending time series explanatory eco-geohydrological analysis of riceland malaria larval habitat georeferenced data feature attributes then to larger spatial scales within a robust PDA-GIS-GPS-RS cyber-environment may help resolve quantitating important uncertainty related parameter estimators (e.g., heteroscedastic residual, latent auto-correlated coefficients) explanatory endemic transmission oriented covariate coefficients.

Seasonal real-time bidirectional PDA-GIS-GPS-RS cyber-environments have domains that can collect, archive, share, analyze, visualize and simulate vector mosquito-related field and remote sampled data, information, and knowledge. Many of these domains can generate valuable predictive seasonal risk-related endemic transmission oriented epidemiological data and information for a malaria-sampled mosquito aquatic larval habitat geographic reference point. Georeferenced or geospatial data feature attributes in a real-time bidirectional robust PDA-GIS-GPS-RS cyber-environment can have spatial/spectral inter-connections that can follow geospatial time series principles/constraints, such as those required for efficient robust, explanatory, time series, endemic, transmission-oriented, vector, mosquito-related, predictive, seasonal risk modeling. A cross-cutting infrastructure that can support geospatial seasonal vector mosquito-related data processing within and across scientific domains is desirable [5]. A robust explanatory seasonal real-time bidirectional PDA-GIS-GPS-RS cyber envi- ronment infrastructure can support the collection, management, and utilization of geospatial georeferenced seasonal vector mosquito-related data, information, and knowledge for generating multiple relational seasonal geodatabases and domains.

Historically, the realization of the importance of time series cyber-infrastructures can be traced back conceptually to 1884 when the national program for topographic mapping was started, and formally to 1994 when the US Federal Geographic Data Committee (FGDC) was established to build a cross-agency National Spatial Data Infrastructure (NSDI). Since then, much progress has been made in defining standards by the Open Geospatial Consortium (OGC) and the International Organization for Standardization (ISO), implementing prototypes through various test beds, popularizing industry products through seed funding, and building applications for this infrastructure through several initiatives. For instance, in 2007, the Infrastructure for Spatial Information in the European Community (INSPIRE) directive entered into force and laid down a general framework for a Spatial Data Infrastructure (SDI) to support European Community environmental policies and activities. These initiatives presently support each other with their own unique emphases: for instance, the NSDI focuses on spatial data collection, sharing, and service, and their geodata.gov provides optimal geospatial data services. Data.gov provides all publicly available US government data for instance, with their geospatial aspects supplemented by geodata.gov.

Digital Earth is a vision popularized for advancing remote sensing technology to store, integrate, and utilize georeferenced vector seasonal sampled mosquito related endemic transmission-oriented data feature attributes to build a virtual world for multiple epidemiological risk modeling applications employing predictive seasonal real-time bidirectional PDA-GIS-GPS-RS technologies. Digital Earth is seen as a global strategic contributor to scientific and technological developments, and is considered a catalyst in finding solutions to international scientific and societal issues (www.digitalearth-isde.org/). Digital Earth can play a strategic and sustainable role in addressing such challenges to human society as natural resource depletion, food and water insecurity, energy shortages, environmental degradation, natural disasters response and infectious disease vector arthropod-related risk modeling (http://earth.google.com/outreach/showcase.html). Thus, Digital Earth may be an integral part of other advanced technologies in a robust seasonal real-time bidirectional PDA-GIS-GPS-RS cyber-environment for constructing datasets of robust, georeferencable seasonal, vector, mosquito-related, endemic, transmission- oriented, field and/or remote specified explanatory eco-epidemiological risk modeling seasonal-sampled covariate coefficients including: earth observation, geo-information systems, communication networks, sensor webs, electromagnetic identifiers, virtual reality, grid computation and others. Grid computing may be focused on distributed computers in a PDA-GIS-GPS-RS cyber-environment to optimize distributed computing while Cloud computing is focused on data, platforms, infrastructure, and software as services for end-users [5]. The Global Earth Observation System of Systems (GEOSS) is an initiative to build a system of systems for global Earth observations focused initially on nine societal benefit areas. These data systems and others then could be part of a robust PDA-GIS-GPS-RS architecture for capturing, managing, scoring and quantitating seasonal-sampled field and/or remote specified explanatory vector mosquito-related endemic transmission oriented risk variables.

Furthermore, over time, the amount and availability of geographic information has grown exponentially, and thus seasonal real-time bidirectional PDA-GIS-GPS-RS cyber-infrastructures is needed to process and integrate seasonal-sampled vector mosquito larval habitat explanatory field or remote specified endemic transmission oriented geospatial information to supply real time risk-based eco-epidemiological analyses and to support scientific and application problems solving across multiple geographic regions. Robust PDA-GIS-GPS-RS cyber-environments would utilize geospatial principles and geospatial information to transform how vector mosquito research, development, and education are conducted within and across other disciplines such as Earth sciences and linear algebra for implementing IVM.

Optimally, seasonal mosquito vector biology-related cyber-environments would be based on recent advancements in GIS, information technology, PDA technology computer networks, sensor networks, Web computing, e-research/e-science and other solutions for optimizing seasonal explanatory observational georeferencable predictive vector mosquito-related data feature attribute geovisualizations for end-users (e.g., malarialogists) to utilize and contribute to the time series cyber-infrastructures. As such, robust real-time bidirectional PDA-GIS- GPS-RS cyber-environment functions for seasonal vector mosquito-related eco-spatiotemporal data analyses, management and storage could then include the following: (a) a middleware layer to bridge time series geospatial functions, resource management, monitoring, scheduling, and other system-level functions, (b) a geospatial information integration layer to parsimoniously quantitate explanatory geospatial explanatory seasonal mosquito sampled data, information, and knowledge flow as supported by multiple field and/or remotes ampled observations, geospatial processing, and knowledge mining; and, (c) geospatial functions to provide various analytical functions for end-users. Vector seasonal mosquito-related time series explanatory predictive geospatial information has a geographic connection that spans multiple domains [9]; thus, the framework of qualitatively/quantitatively regressing empirical geographically-sampled mosquito-related time series explanatory field and/or remote specified endemic transmission oriented datasets can be leveraged to facilitate the sharing of data, information, and knowledge within and across domains [2]. The evolution of cyberspace has resulted in an increasing number of applications to support seasonal vector mosquito-related research, development, and decision making for generalized seasonal real-time bidirectional PDA-GIS-GPS-RS cyber-infrastructure transforms. The introduction of mobile devices will therefore broaden access to time series explanatory georeferenced endemic transmission-oriented field and/or remote specified cyber-infrastrures thereby improving the rate of sharing of geospatial information for robust mosquito-related location intelligence gathering.

Location Intelligence is the capacity to organize and understand complex phenomena through the use of graphical relationships inherent in all information [5]. The goal of location intelligence in a seasonal real-time bidirectional PDA-GIS-GPS-RS cyber-environment for seasonal malarial risk-based endemic transmission oriented field and/or remote specified exploratory data analyses would be to select the best "geographies of opportunity" (e.g., aggregation of georeferenced prolific larval habitats based on geo-spatiotemporal field and/or remote sampled data) by providing a fit-to-purpose geospatial time series analytic solution. Deploying location intelligence by analyzing seasonal data using a PDA-GIS-GPS-RS time series cyberinfrastructural explanatory endemic transmission oriented predictive georeferencable predictive risk model would then be a critical for success in seasonal mosquito control operations. Geolocational or seasonal real-time bidirectional PDA-GIS-GPS- RS cyber-infrastructure tools could then enable vector biologists epidemiologists or other research collaborators to collect, store, analyze and visualize explanatory georeferenced field and/or remote specified endemic transmission oriented data feature attributes (e.g., terrestrial vegetation LULC attributes, aquatic chemistry of larval mosquito abundance in catch basins) accurately. Location intelligence experts are defined by their advanced education in spatial technology and applied use of spatial methodologies [4]. Location intelligence experts can use a variety of seasonal analytical tools to in a PDA-GIS-GPS-RS cyber-environment to measure productive vector mosquito aquatic larval habitat geographical locations for treatment. An important determinant of mosquito-borne pathogen transmission is the spatial distribution of vectors [2]. Optimally, these seasonal prolific larval habitat epidemiological risk mapping applications would transform large amounts of regressed time series dependent explanatory vector mosquito-related endemic transmission-oriented georeferencable field and/or remote specified data feature attributes in a seasonal real-time bidirectional PDA-GIS-GPS-RS cyber-environment into digital and cartographic color-coded visual representations. By so doing, seasonal trends and meaningful intelligence [e.g., levels of yearly change in forested canopied LULC to urban agro-village complex LULC in an epidemiological interventional study site] may be parsimoniously derived.

Land use land cover seasonal vector mosquito-related explanatory field and/or remote specified forecasting endemic transmission oriented georeferencable risk models constructed in a robust seasonal real-time bidirectional PDA-GIS-GPS-RS cyber-environment would ensure fitness for use; would draw on research into the cognitive issues that surround increasingly personalized and flexible possibilities for risk map use with an expanded range of map forms; would respond to the state of the art in the realm of interfaces; and would drive and respond to developments in the field of vector databases and geocomputation. The creation of time series explanatory vector mosquito-related georeferenced aquatic larval habitat geographic and non-geographic location intelligence in a PDA-GIS-GPS-RS cyber-environment would be directed by domain knowledge, formal frameworks with a focus on decision support using remote sensing technologies.

It is possible to regard explanatory, predictive, seasonal, vector, time-series, mosquito-related, endemic, trans- mission-oriented, epidemiological risk modeling itself as a form of representation, although it may be assumed to be the seasonal real-time bidirectional PDA-GIS-GPS-RS cyber-environment data-handling step prior to subsequent geographic representation. Modeling of seasonal geospatialized explanatory vector mosquito-related endemic transmission oriented data is undertaken to ensure that data are captured, held, managed and manipulated in a suitable way for the application in hand [2]. The application may involve representation of all explanatory georeferenced operationalizable field and/ or remote seasonal-sampled mosquito aquatic larval habitat types in a PDA-GIS-GPS-RS cyber-environment for archival, communication and/or analytical purposes. Regardless, the visual representation of geospatial seasonal-sampled eco-geomorphological bidirectional mosquito- related field and/ or remote specified endemic transmission oriented real-time data and, in particular, the interaction of contemporary aspects of visualization and geospatial endemic transmission oriented visual time series representation can be addressed in a robust PDA-GIS-GPS-RS cyber-environment. All references for explanatory, georeferencable, time series, field and/or remote specified, endemic, transmission-oriented, malarial mosquito (e.g., immature An. gambiae s.l.) aquatic habitat representations should be broadly interpreted as referring to visual representation unless otherwise qualified [2]. Furthermore, a wide variety of methods for employing multi-source geo-ecomorphological data in digital explanatory vector mosquito-related aquatic larval habitat image classification have been developed which may be implemented in robust seasonal real-time bidirectional PDA-GIS-GPS-RS cyber-environments. These data sources may be viewed as attempts to exploit the value of amalgamating multiple sources of geo-spatiotemporal explanatory field and remote sampled data (e.g., ancillary data and/or multiple types of imagery) for extraction of vital information from seasonal georeferenced vector mosquito-related larval habitat images through image interpretation.

Hallmarks of multi-parameter explanatory regressed arthropod-related larval habitat image interpretation logic include (1) use of a systematic strategy that proceeds from "knowns" to "unknowns" (e.g., stochastic and/or deterministic time series interpolated larval habitats) and use of inference and "convergence of evidence" exploiting observed relationships between multiple data types (i.e., image and ancillary data) [2]. For instance, Jacob et al. [7] developed a spectral endmember time series predictive autoregressive spectral risk model, using decomposed 0.61 m^2sub-pixel reflectance estimates in an explanatory multi-temporal QuickBird visible and NIR image which were extracted from a georeferenced *Simulium damnosum sensu lato* species riverine larval habitat using a Li-Strahler geometric-optical model. In Africa, the parasite primarily transmitted by black flies of the *S. damnosum s.l.* species complex, a vector of onchocerciasis ("River Blindness") which develop as larvae in fast running rivers and streams. Onchocerciasis, or river blindness, has historically been one of the most important causes of blindness worldwide (http://www.who.int/topics/onchocerciasis). This procedure allowed for the creation of a spectral signature of a unit of habitat.

The model employed three scene components: sunlit canopy (C), sunlit background (G) and shadow (T) generated from the QuickBird image, to determine the sub-pixel endmember spectra associated with the known habitats. The C, G and T component classes were estimated using the ENVI software package (Exelis Visual Information Solutions, Boulder, CO) which employs an object-based classification algorithm. In the geometric-optical approach, the bidirectional reflectance emitted from the georeferenced S. damnosum s.l. riverine larval habitat was modeled as a purely geometric phenomenon that resulted when scenes of discrete, three-dimensional objects were illuminated and viewed from different positions in the hemisphere. The shapes of the within canopied shaded objects (e.g., Precambrian rock) and their immature count densities were the driving variables of the model along with the mixture of sunlit and shaded objects and background that was observed from a particular viewing direction, given a direction of illumination which in turn, controlled the brightness in the black fly riverine larval habitat image.

In Jacob et al. [7] non-parametric spatiotemporal estimators from the decomposed endmember spectra and the geometric-optical model were then used to construct a Boolean model that generated a spectral reference target signature in an ArcGIS database specific for the verified immature S. damnosum s.l. habitats. In probability theory, the Boolean-Poisson model or simply Boolean model for a random subset of the plane (or higher dimensions, analogously) is one of the simplest and most tractable models in stochastic geometry.In mathematics and telecommunications, stochastic geometry models of wireless networks refer to mathematical models based on stochastic geometry that are designed to represent aspects of wireless networks [1]. The related research consisted of analyzing these models with the aim of better understanding wireless communication networks in order to predict and control various network performance metrics. The immature S. damnosum s.l. habitats endemic transmission-oriented eco-epidemiological risk model required using techniques from stochastic geometry and related fields including point processes, spatial statistics, geometric probability, as well as methods from more general mathematical disciplines such as geometry, probability theory, stochastic processes, queuing theory, information theory, and Fourier analysis. Commonly, in seasonal vector arthropod-related field and/or remote endemic transmission-oriented eco-epidemiological risk modeling, stochastic geometry is employed to study of random spatial patterns in georeferenced larval habitat whereby the dependent variable is the immature sampled total density count data [9].

The explanatory linear and non-linear geo-spatiotemporal endmember identifiers of the S. damnosum s.l. riverine larval habitat spectral signatures were then used to predict georeferenced aquatic larval habitats along unsurveyed rivers in both Togo and Uganda. To assess the time series predictive S. damnosum s.l. habitats spectral endmember eco-epidemiological risk model's ability to forecast georeferencable prolific riverine larval habitat sites that may have become temporarily active under varying flow or flooding conditions, a second geographical model was developed. The strategic approach taken was to overlay a Digital Elevation Model (DEM) with signals characteristic of Precambrian rock plus white water, or Precambrian rock alone. The Voltaian system consists of Precambrian to Paleozoic sandstones along with joints, fractures, and quartz veins which had formed earlier in the rocks which has ecological associations with onchocerciasis-related Simulium vectors [2]. Digital elevation model is a digital model or 3D representation of a terrain's surface—commonly for a planet (including Earth), moon, or asteroid—created from terrain elevation data. In order to accomplish this, the authors employed PCI Geomatics software (PCI Geomatics, Toronto, Canada), which supported an automatic overlay of an interpolated wet and dry Precambrian rock signature along the river course. This analysis revealed the geographic locations of both active habitats (i.e. those with water flowing over Precambrian rock) as well as georeferencable sampled sites that might become active under increased flow or flooding conditions. The 3-dimendional (d), DEM is a simple tool to locate differences in elevation that would show areas where such fast flowing water could occur during different river flow conditions [5].

In Jacob et al. [7], of the 30 sites along the Sarakawa River in Northern Togo predicted to be larval habitats by the time series S. damnosum s.l. habitats spectral epidemiological endmember risk model, all (100%) were found to contain larvae. In contrast, none of the 52 sites not predicted by the model, but deemed to be potential habitat by the entomologist accompanying the verification team contained S. damnosum s.l. larvae. Together, these data suggested that the model exhibited a sensitivity and specificity approaching 100% for the prediction of S. damnosum s.l. riverine larval sites in Togo. A second model was then developed to predict such seasonally active sites. There was a complete correspondence between the sites forecasted by the first and the second model, suggesting that the model had identified all active and potentially active S. damnosum breeding sites in the savanna epidemiological riverine ecosystem epidemiological study site. To test the generality of the model, it was applied to predict S. damnosum s.l. riverine sites in northern Uganda. A total of 25 potential S. damnosum s.l. larval habitat sites were predicted. Of the 25 sites predicted to be suitable S. damnosum s.l. larval habitats by the model, 23 [92%; 95% (CI) 81% - 100%] were found to contain S. damnosum s.l. larvae. In contrast, just 2/10 (20%; 95% CI 0% - 45%) sites examined which were not predicted to represent S. damnosum s.l. aquatic habitats by the model were found to contain larvae. The spectral endmember model thus exhibited a sensitivity of 80% and a specificity of 92% when applied in Uganda, a performance that was statistically significant ($p < 0.0001$; Fisher's Exact test). The two sites that were not predicted by the model which nonetheless were found to contain larvae consisted of low hanging streamside vegetation immersed in fast flowing water. *Simulium* sibling species identified cytotaxonomically from onchocerciasis focuses consistently harbor much higher densities of larvae than the other, with three annual peaks of abundance: during the dry season and at the beginning and end of the rainy season with larvae most abundant on submerged rocks and fallen leaves, in small shallow areas characterized by high water current, high conductivity and sparse terrestrial vegetation cover [2]. The mean number of larvae found at the sites predicted by the endemic transmission oriented spectral spatiotemporal endmember risk model (21.91) was significantly greater that the mean number of larvae at the sites consisting of immersed overhanging vegetation (4.0; $p < 0.001$, Mann Whitney U test). Statistical methods may be used to develop predictive time series explanatory georeferencable epidemiological endmember equations in a seasonal real-time bidirectional PDA-GIS-GPS-RS cyber-environment, and these equations may then applied to the unsampled areas (e.g., georeferenced vector arthropod-related larval habitats) to generate a robust endmember predictive map.

Regression is a multivariate analysis that is available in ArcGrid Workstation which may be embedded in a seasonal real-time bidirectional PDA-GIS-GPS-RS cyber-environment for parsimoniously quantitating seasonal- ly vector mosquito-related georeferenced spectral explanatory endemic transmission-oriented data attributes (e.g., QuickBird visible and NIR-based 0.61m resolution reference signature). The central role of the Poisson distribution, for instance, with respect to the analysis of count-related seasonal vector mosquito endemic transmission oriented georeferencable explanatory data is analogous to the position of the normal distribution in the context of models for continuous data. Accordingly, when time series field and/or remote sampled geospatial explanatory endemic transmission-oriented data comprise immature counts, especially for rare events, the probability seasonal real-time bidirectional vector mosquito-related aquatic larval habitat georeferencable explanatory risk model may be based upon an auto-Poisson specification, which may be written in the form of approximations in a PDA-GIS-GPS-RS cyber-environment. The auto-log-Gaussian approximation in a robust seasonal real-time bidirectional PDA-GIS-GPS-RS cyber-environment could then circumvent the auto-Poisson's intractable normalizing factor, while the auto-logistic approximation would circumvent the auto-Poisson's restriction to only situations involving negative spatial autocorrelation in the sampled field and/or remote sampled endemic transmission oriented mosquito vector data. Commonly georeferencable vector mosquito data exhibit positive spatial autocorrelation [2]. The intractable normalizing constant can then be resolved using Markov Chain Monte Carlo (MCMC) procedures seasonal real-time bidirectional PDA-GIS-GPS-RS cyber-environment.

In statistics, Markov chain Monte Carlo (MCMC) methods (which include random walk Monte Carlo methods) are a class of algorithms for sampling from probability distributions based on constructing a Markov chain that has the desired distribution as its equilibrium distribution [1]. Markov chain (discrete-time Markov chain or DTMC) is a mathematical system that undergoes transitions from one state to another on a state space whereby a random process is characterized as memoryless: the next state depends only on the current state and not on the sequence of events that preceded [2]. The negative autocorrelation restriction can then be resolved through Windsorizing sampled georeferenced vector mosquito related field and remote endemic transmission oriented data in a seasonal real-time bidirectional PDA-GIS-GPS-RS cyber-environment. Winsorized mean is a measure of central tendency, much like the mean and median, and even more similar to the truncated mean which involves the calculation of the mean after replacing given parts of a probability distribution or sample at the high and low end with the most extreme remaining values [3]. The Winsorized mean in a seasonal vector mosquito-related endemic transmission-oriented eco-epidemiological risk model uses more information from the distribution or sample than the median [2]. However, unless the underlying distribution is symmetric, the Winsorized mean of a sample in a robust PDA-GIS-GPS-RS cyber-environment is unlikely to produce an unbiased sampled vector mosquito-related parameter estimator for either the mean or the median.

Furthermore, since the process of interpolated seasonal arthropod-related image data interpretations often involves use of heuristics and/or "rules" based on expert entomological knowledge and observation (e.g., rules about biogeographic relationships between vegetation zonation, elevation, slope and other spatiotemporal ecogeomorphological explanatory terrain-related georeferenced vector mosquito-related aquatic larval habitat covariates) regression residuals rendered from probabilistic interpolators [e.g., Inverse Distance Weighted (IDW) Matrices] in seasonal real-time bidirectional PDA-GIS-GPS-RS cyber-environments, may render optimal unbiased estimators.

In the mathematical field of numerical analysis, stochastic/deterministic interpolation is a method of constructing new data points within the range of a discrete set of known data points [2]. Within Geostatistical Analyst in a seasonal real-time bidirectional PDA-GIS-GPS-RS cyber-environment, a vector biologist, epidemiologist or other research collaborator could easily create a continuous surface risk-related endemic transmission- oriented explanatory map from measurements stored in a point feature layer or raster layer or by using georeferenced polygon centroids [5]. The sample larval habitat points in the endemic transmission oriented explanatory georeferenced field and/or remote specified risk model may then render seasonal measurements such as elevation, depth to the water table, or levels of pollution which also may be kriged. The Interpolation tools are generally divided into deterministic and geostatistical methods. IDW, Spline, and Trend are deterministic, while Kriging is a geostatistical method (www.esri.com). Topo to Raster and Topo to Raster by File in a robust explanatory seasonal real-time bidirectional PDA-GIS-GPS-RS cyber-environment would then employ an interpolation algorithm designed for creating continuous georeferencable larval habitat surfaces from contour lines [2]. These methods may contain properties favorable for creating surfaces for conducting robust eco-geohydrologic analysis of seasonal-sampled vector mosquito-related endemic transmission oriented field and/or remote specified explanatory covariate coefficients.

When used in conjunction with ArcMap and Geostatistical Analyst, a comprehensive set of PDA-GIS-GPS-RS cyber-tools may provide explanatory, georeferencable, interpolatable, larval habitat surfaces which may be employed to visualize, analyze, and understand spatial phenomena (e.g., prolific larval habitat development). This would include generating watershed and streams (e.g., ecohydrologic analysis), generating slope coefficients, delineating drainage networks, creating contours, working with low-pass and high-pass filters and analyzing the result associated within a geophysical urbanized environment. Further, drainage patterns (e.g., a dendritic pattern) associated with physical environments (landscape and geology map) be cartographically displayed. These models may be able to describe regions underlain by homogeneous material (e.g., flooded soils) that may help target georeferenced vector aquatic larval habitats. That is, the subsurface geology associated with georeferencable explanatory vector mosquito aquatic larval habitats may have a similar resistance to weathering so there no apparent control over the direction the tributaries take.

Surface interpolation functions create a continuous (or prediction) surface from sampled point values [5]. The continuous seasonal georeferenced vector mosquito larval habitat surface representation of an explanatory raster dataset can represent height, concentration, or magnitude—for instance, elevation of a sampled productive habitat [2]. In mathematical predictive time series risk modeling vector mosquito-related endemic transmission oriented explanatory field and/or remote specified data, bicubic interpolation would extend cubic interpolation for interpolating data points (sampled aquatic larval habitats) on a two dimensional regular grid [2]. The interpolated surface is smoother than corresponding surfaces obtained by bilinear interpolation or nearest-neighbor interpolation [1].

Bicubic interpolation of georeferencable operationzable explanatory time series mosquito data can be accomplished using either Lagrange polynomials, cubic splines, or cubic convolution algorithm in a real-time bidirectional PDA-GIS-GPS-RS cyber-environment. Given a set of k + 1 vector mosquito-related data points where no two x_j are the same, the interpolation polynomial in the Lagrange form is a linear combination $L(x)=\sum_{j=0}^{k} y_j \ell_j(x)$ of Lagrange basis polynomial

$$\ell_j(x) := \prod_{\substack{0\le m\le k \\ m\ne j}} \frac{x-x_m}{x_j-x_m} = \frac{(x-x_0)}{(x_j-x_0)} \mathrm{L} \frac{(x-x_{j-1})}{(x_j-x_{j-1})} \frac{(x-x_{j+1})}{(x_j-x_{j+1})} \mathrm{L} \frac{(x-x_k)}{(x_j-x_k)},$$

where $0 \le j \le k$. In image processing, bicubic interpolation is often chosen over bilinear interpolation or nearest neighbor in image resampling, when speed is not an issue. In contrast to bilinear interpolation, which only takes 4 pixels (2 × 2) into account, bicubic interpolation considers 16 pixels (4 × 4). Images resampled with bicubic interpolation are smoother and have fewer interpolation artifacts. Surface interpolation functions make geographic time series predictions from sample measurements for all geographic locations in a raster dataset whether or not a measurement has been taken at the location (www.esri.com). Each model would produce predictions using different calculations. Regardless, robust seasonal real-time bidirectional PDA-GIS-GPS-RS cyber-environ- ments may help improve interpolated interpretations from seasonal vector mosquito-related regressed data and georeferenced larval habitat digital images. These include image stratification, classification modification, post- classification sorting, and advanced methods for multisource data analysis.

In this study, QuickBird imagery from Digital Globe [7] was employed to display an empirical dataset of explanatory seasonal real-time bidirectional time series ecological sampled georeferenced endemic transmission oriented field and remote specified data feature attributes associated to multiple georeferenced aquatic larval aquatic habitats of An. arabiensis. The research was conducted in a riceland agro-village complex in the Mwea Rice Scheme, Kenya. The malarial vectorial system in Africa is complex, comprising typically of *Anopheles gambiae*, *An. arabiensis*, and *An. funestus* as the primary vectors and a number of complementally vectors including *An. pharoensis*, *An. coustani* and *An. rivuro.* An. arabiensis is considered a species of dry, savannah environments and sparse woodland yet it is known to occur in deforested areas (e.g., irrigated rice fields), but only where there is a history of recent land disturbance or clearance [10].

An. arabiensis aquatic larval aquatic habitats are generally small, temporary, sunlit, clear and shallow fresh water pools, although the vector mosquito species is able to utilize a variety of habitats including slow flowing, partially shaded streams and a variety of large and small natural and man-made habitats [8]. Immature An. arabiensis have been found in turbid waters and, on occasion, in brackish habitats. It readily makes use of irrigated rice fields where larval densities are related to the height of the rice, peaking when the plants are still relatively short and then dropping off substantially as the rice plants mature [2]. Such density fluctuations are also reflected in the adult population, which also peak when rice stalks are small and decline as the plants mature [8]. These patterns may be due to a preference for sunlit areas of water with relatively limited emergent vegetation, with densities decreasing as shade from the growing plants increases. There is evidence that An. arabiensis may be attracted by the application of fertilizers or by the amount of dissolved oxygen within the paddy water. However, fertilizer application occurs at the start of plant cultivation, and dissolved oxygen content is related to sunlight exposure (e.g. via increasing photosynthesis), so which factor is the primary oviposition attractant in rice fields is uncertain for anopheline immatures. An. arabiensis is described as a zoophilic, exophagic and exophilic species [10]. Importantly, An. arabiensis have a wide range of feeding and resting patterns, depending on geographical location of the aquatic larval habitats.

Georeferenced entomological surveys in riceland agroecosystems are often biased by locations with known high vector densities using sub-meter resolution technologies [2]. For remote identification of georeferenced vector mosquito aquatic larval habitats, the first step is often to construct a discrete tessellation of the region [5]. Results of previous research have demonstrated time series digitized grid cell data to be superior to orthogonal grid cell data for identification of riceland*Anopheles* aquatic habitats [9]. Building a time series explanatory GIS digitized non-orthogonal gridded geodatabase can identify relationships between riceland mosquito aquatic habitats, rice plant stage of development and agro-village-complex within the framework of georeferenced relational database technology. For example, Jacob et al. [2] constructed multiple gridded seasonal QuickBird visible and NIR classified land cover maps for optimizing grid area surveys in an agro-village complex. The classified land cover maps were generated from mosaicked 2012 and 2013 multispectral QuickBird imagery that was employed for robustly quantitating remotely sensed change detection LULC data using the back propagation artificial neural network (ANN) in IDRISI.

In computer science and related fields, ANN are computations that are capable of machine learning and pattern recognition which are usually presented as systems of interconnected "neurons" that can compute values from inputs by feeding information through the network [5]. ANN consists of artificial neurons inspired from biological neurons. Artificial neurons are typically organized in layers, so that each ANN includes an input layer. This layer can have as many neurons as the ANN's input variants in robust seasonal real-time bidirectional PDA-GIS-GPS-RS cyber-environments. Input layer neurons are then connected to the neurons of the hidden layers or to neurons in the next layer. Hidden layers may also exist in ANN in a PDA-GIS-GPS-RS cyber-en- vironments. Each hidden layer can have neurons linked in different ways to the other hidden layers or to the output layer. Hidden layer neurons can get their input through the input layer or some other hidden layer or, in some cases, even the output layer. Finally, an output layer may also be generated through which the output vector passes PDA-GIS-GPS-RS cyber-environments. This layer has as many neurons as the ANN's output variants. Output layer neurons can get their input through the input layer or the hidden layers [5].

IDRISI is the premier system for integrated neural network and machine learning solutions as it employs an advanced Multi-Layer Perceptron (MLP) neural network classifier that offers a variety of innovations (www.esri.com). Multilayer perceptron (MLP) can be feed-forward ANN models that may risk map empirical datasets of input seasonal-sampled An. arabiensis related field and remote specified data geo-spatiotemporally sampled in riceland environments onto a set of appropriate outputs in robust PDA-GIS-GPS-RS cyber-environ- ments. Key features of IDRISI seasonal real-time bidirectional PDA-GIS-GPS-RS cyber-environments for risk modeling explanatory An. arabiensis related georeferenced endemic transmission oriented field and remote specified data feature attributes spatiotemporally sampled in riceland environments include: 1) a complete GIS analysis package for basic and advanced spatial analysis, including tools for surface and statistical analysis, 2) decision support, 3) land change and prediction, and image time series analysis, 4) a complete Image Processing system with extensive hard and soft classifiers, including machine learning classifiers such as neural networks and classification tree analysis, as well as image segmentation for classification, 5) integrated modeling environments including the Earth Trends Modeler for image time series quantitation of environmental trends and Land Change Modeler for LULC analysis and prediction [5]. IDRISI is an integrated GIS and remote sensing software which relies on a PC grid-based system that offers tools for researchers and scientists engaged in analyzing earth system dynamics for effective and responsible decision making for environmental management, sustainable resource development and equitable resource allocation (www.esri.com).

ANN algorithms have been employed for vector mosquito-related endemic transmission-oriented risk modeling. In Jacob et al. [2] training sites for "water", "grass/bare soil", "built", and "paved" classes were developed using polygons digitized from visual interpretation of the 0.61 m QuickBird panchromatic band. Thereafter a robust ANN algorithm produced a geo-spatiotemporal explanatory LULC classification with an overall accuracy of 81% and Kappa of 0.74, which was more accurate than those produced by other classification algorithms evaluated, such as maximum likelihood. The maximum likelihood classification tool considered both the variances and covariances of the class signatures for assigning each cell to one of the classes represented in a signature file with the assumption that the distribution of a class sample was normal. The "built" class has 24% errors of omission and 20% errors of commission, while the "bare soil" class had 7% errors of omission and 10% errors of commission. The kappa measure of agreement is the ratio K = P(A) □ P(E)/P(E) where P(A) is the proportion of times the k raters agree, and P(E) is the proportion of times the k raters are expected to agree by chance alone [5]. Complete agreement in time series calculated kappa coefficients from the seasonally regressed time series explanatory malaria mosquito-related field and remote specified endemic transmission oriented risk model parameter estimators corresponded to K = 1, and lack of agreement (e.g., purely random coincidences of prevalence rates) corresponded to K = 0. Negative values of kappa in a seasonal explanatory seasonal real-time bidirectional predictive malaria-related mosquito risk model would mean negative agreement—that is, the propensity of raters to avoid assignments made by other raters [2].

In this research we tested the Trimble Recon® X-Series handheld computer as part of a seasonal real-time bidirectional PDA-GIS-GPS-RS cyber-environment for robustly and parsimoniously capturing analyzing and archiving multiple georeferenced An. arabiensis larval aquatic habitats and their geo-spatiotemporally sampled explanatory covariates in Karima rice-village agro-complex in the Mwea Rice Scheme in Kenya. Our Trimble Recon 400× outdoor rugged computer had a 400 MHZ processor 64 RAM/256 flash memory 4000 MAH battery bluetooth 802.11 yellow sunlight readable VGA display PD. Recon runs the Windows Mobile 6 operating system, ensuring broad support for software offered by Solution Providers world-wide (http://www.trimble.com/). Based on state-of-the-art hardware and software (200 and 400 MHz XScale processors, reflective color display, Windows Mobile), the Recon uses an innovative modular design with a Power Boot module that integrates the battery and ports into a replaceable component (http://ruggedpcreview.com/3_handhelds_tds_recon.html). The difference between the 200 and 400 models is in processor speed and amount of non-volatile NAND storage (http://www.trimble.com/). An important goal of NAND flash development has been to reduce the cost per bit and increase maximum chip capacity so that flash memory can compete with magnetic storage devices like hard disks. NAND flash has found a market in devices to which large files are frequently uploaded and replaced. MP3 players, digital cameras and USB drives use NAND flash (http://whatis.techtarget.com/definition/NAND-flash-memory).

The Recon meets rigorous MIL-STD-810F military standards for impact, vibration, humidity, altitude and extreme temperature. The X-series also comes with an IP67 rating. Timely sampling and archiving of seasonal explanatory predictive field and/or remote sampled endemic transmission oriented malarial data is vital for robust predictive risk modeling [2]. Further, the new X-Series gives users (e.g., vector biologists, epidemiologists) the option of integrated Bluetooth and 802.11 g wireless capabilities. The system presently employs Windows Mobile 5.0. Other new features included a built-in microphone, a redesigned keypad and an improved color touch screen. Recently, TDS announced a slightly more powerful battery (4000 vs. 3800 mAH) and the switch to a fully sunlight viewable LCD display. Power sources are vital for proper maintenance of PDA-GIS-GPS-RS cyber-environments [5].

QuickBird visible and near-infra red (NIR) imagery combined and displayed within a PDA-GIS-GPS-RS cyber-environment employing a Trimble Recon X 400MHz Intel PXA255 Xscale CPU® may help visually determine the geographic location of field technicians in relation to the seasonal vector mosquito-related aquatic larval habitats being treated. Incorporating an effective results-based digital management system, can become a key component for ongoing seasonal malaria treatment and monitoring programs [4]. Riceland data acquisitions, within a GPS-equipped PDA employing wireless communications may optimize a geodatabase with a related Web application for optimal An. arabiensis mosquito data analyses for developing an IVM. As such, in this research, a Trimble Recon® X PDA (400 MHz Intel PXA255 Xscale CPU), incorporating digitized rice paddy polygons and QuickBird satellite within a PDA-GIS-GPS-RS cyber-environment was tested for use in identifying treated and untreated seasonal An. arabiensis aquatic habitats in Karima rice-village complex, Mwea Rice Scheme, Kenya. Although the focus in this research is on An. arabiensis in riceland agroecosystems, the PDA- GIS-GPS-RS cyber-environment architecture may be applicable to aggregate, store and analyze geo-spati- otemporal sampled field and/or remote specified endemic transmission-oriented explanatory covariate coefficients associated to vector and nuisance arthropos-related (e.g., immature S. damnosum s.l. black fly) aquatic larval habitats.

MATERIAL AND METHODOLOGY

Study Site

The study was conducted 112 km northeast of Nairobi, Kenya in Karima rice-village complex within the Mwea Rice Scheme. Mwea occupies the lower altitude zone of the Kirinyaga District, in an expansive low-lying, formally wet savannah ecosystem. The scheme is situated on the foot hills of Mount Kenya at 37°20'E and 0°, 41'S. The Mwea Rice Scheme is located in the west central region of Mwea Division and covers an area of approximately 13,640 hectares. More than 50% of the scheme area is used for rice cultivation. The remaining area is used for subsistence farming, grazing, and community activities. The mean annual precipitation is 950 mm, with maximum rainfall occurring in April-May and October-November. The average temperatures range from 16°C to 26.5°C. Relative humidity varies from 52% to 67%. According to the 2010 Kenyan national census, the Mwea Rice Scheme has a population of 170,000, occupying 26,000 households. The Karima study site is located at the central-west region of the scheme and has 168 homesteads, with approximately 1250 residents.

In Mwea, the beginning of each cropping or growing cycle is scheduled according to the water availability through the irrigation water distribution scheme. The schedule of individual farmers' rice planting also differs within this time when water is available. Most fields are cultivated once a year, although some farmers cultivate a second crop. The typical cultivation cycle includes a sowing-transplanting period (June-August), a growing period (August-November) and a post-harvest period (November-December). The second crop is cultivated prior to the short rainy period between January and May. The duration of the rice cycle varies between 120 and 150 days, depending on the rice variety planted. After harvesting, active mosquito habitats may persist in shallow puddles left after harvest [8].

The start of the rice growing cycle begins with paddy preparation by pre-flooding the paddy with about 12 cm of water. Farmers sow rice seeds into a small section (nursery) of their paddy to obtain single stemmed plants for transplanting throughout the paddy. For transplanting, the number of seedling plants is 3 - 4 seedlings per hill with a planting density is 20 - 30 hills per m^2. Fertilizers in the form of Ammonium sulphate and triple super phosphate are applied at a rate of about 50 kg/ha and 125 kg/ha, respectively, about five days before transplanting. Studies by Muturi et al. [8] showed that dose amount of ammonium sulfate accounted for up to 40% mortality rate and one week delay in development time to An. arabiensis larvae. The actual amount of water used by farmers for land preparation, and during the crop growth period, is much higher than the actual field requirement. Paddy farmers in Mwea often store water in their fields as a back-up safety measure due to the unreliability in supply of water for irrigation [2]. This leads to a high amount of surface runoff, seepage and percolation, accounting for about 50 - 80 percent of the total water input to the field creating active mosquito aquatic larval habitats.

Aquatic Habitat Sampling

Base maps were generated in ArcGIS of the study site (Figure 1). Field sampling was conducted from July 2005 to July 2007. One hundred and fifty two temporary, permanent and semi-permanent An. arabiensis aquatic larval habitats in the Karima epidemiological study site were mapped and classified using a CSI-DGPS Max differentially corrected global positioning systems (DGPS) receiver and OmniSTAR L-Band satellite signal. Within the DGPS MAX, CSI Wireless has integrated the CSI Wireless SLX receiver, a tri-purpose GPS/WAAS/L-band receiver, and the CSI Wireless SBX, a high performance DGPS beacon receiver for optimal positional accuracy (http://www.omnistar.com/Portals/). We employed three internal differential correction services. Binary messages may be output from the DGPS MAX receiver, the Mini MAX receiver, the PowerMAX receiver, the Vector, the VectorPRO, the Vector Sensor, the Vector Sensor PRO and the Vector OEM along with NMEA 0183 data (http://www.gpsdgps.com/download/m_program.pdf). In this research, the Binary messages of the sampled An. arabiensis aquatic larval habitats were supported by the receivers which were in an Intel Little Endian format for direct read in the PDA-GIS-GPS-RS cyber-environment architecture. Each binary message begins with an 8-byte header and ended with a carriage-return line-feed pair (0x0D, 0x0A). The first four characters of the header was the ASCII sequence BIN.

Figure 1. QuickBird visible and near-infra-red data with ecologically sampled An. arabiensis aquatic larval habitats at the Karima study site.

The beacon receiver successfully obtained DGPS beacon signals for the Karima agro-village complex epidemiologist study site. The Wide Area Augmentation System (WAAS) demodulator decoded the corrected georeferenced An. arabiensis aquatic larval habitat data from the WAAS, as the L-band satellite differential receiver obtained the corrections from the Omni STAR Worldwide DGPS service. The WAAS is an air navigation aid developed by the Federal Aviation Administration (prime contractor Raytheon Company) to augment the Global Positioning System (GPS), with the goal of improving its accuracy, integrity, and availability (http://www8.garmin.com/aboutGPS/waas.html). The WAAS is an air navigation aid developed by the Federal Aviation Administration (prime contractor Raytheon Company) to augment the Global Positioning System (GPS), with the goal of improving its accuracy, integrity, and availability. Essentially, WAAS is intended to enable aircraft to rely on GPS for all phases of flight, including precision approaches to any airport within its coverage area. WAAS uses a network of ground-based reference stations, in North America and Hawaii, to measure small variations in the GPS satellites' signals in the western hemisphere. Measurements from the reference stations are routed to master stations, which queue the received Deviation Correction (DC) and send the correction messages to geostationary WAAS satellites in a timely manner (every 5 seconds or better).

In addition to real-time DGPS, the DGPS MAX also supports post-processing of malaria-related data [2]. A vector biologist, epidemiologist and/or research collaborators may configure the DGPS MAX for output of binary measurement vector mosquito geo-spatiotemporal sampled field and remote specified endemic transmission oriented data for logging with the use of an external device. By so doing, a conversion utility would be available from CSI Wireless for translation from the proprietary binary format into the Receiver Independent Exchange format (RINEX). Receiver Independent Exchange Format is a data interchange format for raw satellite navigation system data which allows the user to post-process the received data to produce a more accurate result—usually with other data unknown to the original receiver, such as better models of the atmospheric conditions at time of measurement [5]. RINEX version enables storage of our sampled seasonal field and remote specified measurements from pseudorange, carrier-phase and Doppler systems for to GPS (including modernization signals (e.g. L5 and L2C), GLONASS, Galileo, Beidou, along with data from EGNOS and WAASsatellite based augmentation systems (SBAS), QZSS, simultaneously (ftp://igscb.jpl.nasa.gov/pub/data/format/rinex210.txt). The final output of a navigation receiver is usually its position, speed or other related physical quantities. The base station then transmits the range error corrections to remote receivers in real-time. The remote receiver corrects its satellite range measurements using these differential corrections, yielding a much more accurate position (http://www.omnistar.com/Portals/). This was the predominant DGPS strategy used for our real-time An. arabiensis aquatic larval habitat-related field sampled applications. The OmniSTAR L-Band signal is a wide-area DGPS service that uses data from a worldwide network of reference stations, which are combined with precise orbit and clock information for every satellite in the NAVSTAR GPS satellite constellation. The OmniSTAR services, OmniSTAR-VBS (Virtual Base Station), OmniSTAR-HP (High Performance), OmniSTAR-XP (Extended Performance) and OmniSTAR-G2 (GPS and GLONASS), were specifically developed to satisfy the need for high accuracy positioning systems and services for land-based applications (http://www.omnistar.com/Portals/). The addition of GLONASS to the solution significantly increases the number of satellites available which is useful when faced with conditions that limit satellite visibility, such as terrain, vegetation or buildings.**OmniSTAR G2** service provides short-term accuracy of 1 - 2 inches and long term repeatability of better than 10 centimerers, 95% CEP. OmniSTAR-HP/XP/G2 is a dual frequency premium augmentation solution in the OmniSTAR family of High Performance services. OmniSTAR-HP/XP/G2 delivers centimeter accuracy derived by a number of independent positioning technologies: OmniSTAR's primary network of dual frequency reference stations provide for measurements that eliminate errors caused by signal delays in the ionosphere. The positional accuracy of the CSI DGPS Max receiver, using an OmniSTAR L-Band satellite signal, was 0.179 (+/−0.392) [9]. Jacob et al. [2] described the purpose of DGPS for mosquito aquatic larval habitat mapping as providing global absolute positioning capability with respect to a consistent terrestrial reference frame,

The Karima rice-village complex epidemiological study site was sampled for immature Anopheles, along various cross-sectional transects of potential habitats, using standard collecting techniques [11]. Areas with high and low larval productivity were categorized by habitat type; paddy, irrigation or drainage canal, marshland, riparian, ephemeral pool, seep, or other natural or man-made aquatic sites. Specific physical properties of these habitats were recorded. Temporary aquatic habitats were defined as habitats with water for 2 weeks, semi-per- manent were habitats with water for 3 months and permanent habitats had water for 6 months or longer.

Water bodies were inspected for mosquito larvae using standard dipping techniques with a 350-ml dipper to collect the mosquito larvae [11]. The number of dips per habitat was 20. All data from the habitat characterization of each aquatic larval habitat was recorded on a field sampling form. Larvae and a sample of water from each larval habitat were placed in whirl-pack bags and transported to the Mwea Research Station for further pro- cessing. All 3rd and 4th in star larvae were immediately preserved in 95% ethanol and later identified morpholo- gically to species using taxonomic keys [12]. The 1st and 2nd instars were reared in plastic pans under semi-field conditions and those that survived to 3rd in star were also preserved and identified morphologically to species. The pupae were kept in mosquito emergent cages [13] and the resultant emergent mosquitoes were identified morphologically. Anopheline larvae were separated from Culicine larvae for microscopic identification. Culicine larvae were not identified to genus and species levels as the focus of this study was on anophelines.

Aquatic Habitat Characterization

Meteorological data was acquired from a Davis Instruments 6153-IP Ethernet Wireless Weather Station placed in a secured homestead within the study site. Internet Protocol or IP Network Weather Stations update weather servers real-time, providing up to the second data without the use of a PC [5]. The VantagePro2 console and wireless receiver provided forecasting, on-screen graphing, and much more (http://www.ambientweather.com/dain61ipvafa.html). Quick view icons in the station revealed daily forecast (e.g., sunny, partly sunny, cloudy or, rain) at the Karima epidemiological rice agro-village complex while a moving ticker-tape displayed gave more specified meteorological variables (e.g., solar radiation). The integrated sensor suite combined a rain collector, temperature and humidity sensors, and anemometer into one package—making the setup for improving performance and reliability. For optimizing the weather monitoring at the study site, we added Weather Link to our Vantage Pro. The Weather Link data logger fit neatly into the Vantage Pro console storing our weather data even when the computer was off. The Weather Link software created graphs and generated summaries of meteorological variables at the Karima epidemiological rice agro-village complex. The Davis VantagePro2 employed frequency hopping spread spectrum radio to transmit and receive data up to 1000' (300 m) line of sight. The fan aspirated radiation shield improved accuracy of the seasonal meteorological georeferenced explanatory endemic transmission- oriented covariate coefficients at the Karima epidemiological study site by circulating fresh air into the temperature and humidity shield.

Environmental variables recorded for each habitat were number of aquatic animals, depth, and distance to the nearest house, canopy coverage, shade, and turbidity. Distance to the nearest house was measured with a tape when it was shorter than 100 m. When the distance exceeded 100 m, it was estimated visually. The distance to the nearest house was categorized into 7 classes (e.g., 1: 50 - 100 m, 2: 101 - 200 m, and so on, and 7 for distances greater than 600 m). Canopy cover was defined as the amount of terrestrial vegetation and other objects in the habitat. The non-mosquito aquatic invertebrates collected in the dipper samples during the sampling occasion were preserved in 100% ethanol and later identified to family using taxonomic keys of [14]. The number of individuals of each family identified were counted and recorded. Shade coverage of a habitat was measured in percentage of water surface covered by placing a square frame (1 m^2) with grids (100 cm^2) above the habitat. Turbidity was measured by placing water samples in glass test tubes and holding against a white background, and classified into 2 levels: clear and turbid.

Remote Sensing Data

Raster image data from the Digital Globe QuickBird satellite service were acquired for the periods of: 15 July 2012 February 20, and 15 July 2013 within the agro-village complex epidemiological study site area. The QuickBird image data were delivered as pan-sharpened composite products in infra-red (IR) colors. The QuickBird raster image products provided four discrete non-overlapping spectral bands, in the 0.45 μm to 0.72 μm range, with an 11-bit collected information depth. The clearest, cloud-free images available of the contiguous sub-areas of the epidemiological study site were used to identify land cover and other spatial features associated with An. arabiensis aquatic larval habitats. The Order Polygon contained 5 vertices consisting of longitude/latitude (decimal degrees) geographic coordinates using a WGS84 ellipsoid. The satellite data contained 64 km^2 of the land cover in the study site.

The QuickBird imagery was classified using the Iterative Self-Organizing Data Analysis Technique (ISODATA) unsupervised routine in ERDAS Imagine v.8.7™ [15]. Unsupervised classifications are commonly used for the identification of land covers and mosquito habitats associated with intermediate hosts and disease vectors [16] - [18]. A total number of 15 Ground Control Points (GCPs) within the Karima study site were measured. A coordinate accuracy of 0.2 - 0.4 cm in the horizontal, and 0.6 - 0.8 cm in the vertical, was obtained. Twelve of the collected GCPs were applied as part of the image rectification process; while the remaining three were used as reference coordinates for the evaluation of QuickBird data.

Non-parametric approaches were used to build a correlation matrix between QuickBird pixels in the Karima agro-village study site image and its ground locations, based on two cubic polynomials algorithms. A univariate cubic polynomial has the form $f(x) = a_3x^3 + a_2x^2 + a_1x + a_3$ [2]. An equation involving a cubic polynomial is called a cubic equation [5]. A closed-form solution known as the cubic formula exists for the malaria-related solutions of an arbitrary cubic equation [0]. A cubic polynomial is a polynomial of degree 3 [5]. In this research we derived a univariate cubic polynomial from the empirical sampled dataset of remotely sensed spatiotemporal An arabiensis related endemic transmission oriented explanatory covariate coefficients employing the form $f(x) = a_3x^3 + a_2x^2 + a_1x + a_3$. An equation involving a cubic polynomial is called a cubic equation [5]. A closed- form solution known as the cubic formula also exists for the solutions of arbitrary cubic equations for quantitating uncertainty terms in latent forecasted derivatives from remotely sensed forecasting models. Non-parametric methods were then applied for the rectification of the acquired Standard type imagery using 1) Simple linear polynomial correction with 3-Dimensional (D) ground control point GCPs, 2) rational polynomial coefficients (RPC) correction without refinement; and, 3) RPC correction with linear refinement using 3D GCPs. These non- parametric methods were then applied for the rectification of the satellite imagery using a rational polynomial coefficient correction procedure with linear refinement utilizing the GCPs.

Thereafter, the rational function model (RFM) was used as an alternative to physical sensor models for 3D ground point determination with the high-resolution QuickBird imagery. However, owing to the sensor orientation bias inherent in the vendor-provided RPCs, the geopositioning accuracy obtained from these RPCs was limited. In this paper, the performances of two schemes for orientation bias correction (i.e., RPCs modification and RPCs regeneration) were performed based on one separate-orbit QuickBird stereo image pair in the Karima rice-village complex, and four cases for bias correction, including shift bias correction, shift and drift bias correction, affine model bias correction and second-order polynomial bias correction. A 2-step least squares adjustment method was then adopted for correction parameter estimation with a comparison with the RPC bundle adjustment method. The experiment results demonstrated that in general the accuracy of the 2-step least squares adjustment method was almost identical to that of the RPC bundle adjustment method. With the shift bias correction method and minimal one GCP, the modified RPCs improved the accuracy from the original 23 m to 3m in planimetry and 17 m to 4 m in height in the Karima agro-village complex epidemiological study site. With the shift and drift bias correction method, the regenerated RPCs achieved a further improved positioning accuracy of 0.61 m in planimetry and 1 m in height with minimal 2 well-distributed GCPs. The affine model bias correction yielded a geopositioning accuracy of better than 0.50 m in planimetry and 1 m in height with 3 well-posi- tioned GCPs. Further, tests with the second-order polynomial bias correction model indicated the existence of potential high-order error signals in the vendor-provided RPCs. Thereafter, based on the condition that an adequate redundancy in GCP number was available, an accuracy of 0.42 m in planimetry and 0.80 m in height of the data was attainable.

Since biases or errors may still have existed after applying the original RPC parameters generated from the QuickBird image further refinement was performed using linear equations an sampled 3-dimensional (D) GCPs. Root Mean Square Error was then calculated for the linear polynomial rectification using the 3-D GCP's sampled in the study site (Table 1). The root-mean-square deviation (RMSD) or root-mean-square error (RMSE) is a frequently used measure of the differences between values predicted by a model or an estimator and the values actually observed. Basically, the RMSD represented the sample standard deviation of the differences between predicted values and observed values [5]. In this research, these individual differences were represented as the spatiotemporal vector mosquito larval habitat field and remote sampled endemic transmission oriented regression-based residuals when the calculations are performed over the data sample that was used for estimation, and are called prediction errors when computed out-of-sample. The use of additional 3-D GCPs for the refinement of the model allowed accurate dereferencing of the ecological sampled An. arabiensis aquatic habitat data. These types of process ensured that accurate geographic extents of the study site were able to encompass all georeferenced data.

In this research, the RMSD of a seasonal An. arabiensis larval habitat parameter estimator $\hat{\theta}$ with respect to an estimated parameter θ was defined as the square root of the mean square error: $\mathrm{RMSD}(\hat{\theta}) = \sqrt{\mathrm{MSE}(\hat{\theta})} = \sqrt{E\left((\hat{\theta}-\theta)^2\right)}$. For an unbiased estimator, the RMSD was then the square root of the variance (i.e., standard error in the QuickBird An. arabiensis aquatic larval habitat regression-based risk model). The RMSD of the predicted time series field and/or remote endemic transmission oriented explanatory larval habitat values $\hat{y}_{th}$ for times t of a regression's dependent variabley was computed for n different predictions as the square root of the mean of the squares of the deviations employing $\mathrm{RMSD} = \sqrt{\frac{\sum_{t=1}^{n}(\hat{y}_t - y_t)^2}{n}}$. In some disciplines, the RMSD is used to compare differences between two things that may vary, neither of which is accepted as the "standard" [5]. For instance, when measuring the average difference between two time series $x_{1,t}$ and $x_{2,t}$, the formula becomes $\mathrm{RMSD} = \sqrt{\frac{\sum_{t=1}^{n}(x_{1,t} - x_{2,t})^2}{n}}$.

Table 1. Calculated Root Mean Square Error (RMSE) for a linear polynomial rectification using 3-Dimensional GCP's sampled in the Karima study site.

GCP	XRi	YRi	Ri	Rx	Ry	T
1	0.451	0.322	0.592	0.512	0.162	0.892
2	0.637	−1.271	1.263			
3	0.333	−0.626	0.721			

i: number of Ground Control Points (GCP), XRi and YRi: the X and Y residual for GCPi, Ri: the RMS error for GCPi, Rx and RY: the X and Y RMS error, T: the total RMS error.

Habitat Mapping

Each An. arabiensis aquatic habitat in the study site, with their corresponding data attributes, was entered into the Vector Control Management System (VCMS®) relational database software product [19]. The VCMS provides public health professionals involved in mosquito/vector control and pest management with a comprehensive database tracking and reporting system (http://store.elecdata.com/field_data_collection/vcms.aspx). Detailed information on larval habitat site locations and data collection and trapping activities comprised the core of our data system. VCMS is an international model for public health agencies that standardizes data collection/reporting activities and can aggregate data on various types of vector borne diseases.

In this research the VCMS® database supported a mobile field data acquisition component module, called Mobile VCMS that synchronized the seasonal sampled An. arabiensis aquatic larval habitats field data from industry standard Microsoft Windows Mobile™ devices which also supported our add-on DGPS data collection. Mobile VCMS and its corresponding FieldBridge® middleware software component were then used to support both wired and wireless synchronizing of the field data for conducting the larval aquatic habitat monitoring. Field data collected with Mobile VCMS was synchronized directly into a centralized VCMS relational database repository. Additional geocoding and spatial display of the ecological data was handled using the embedded VCMS GIS Interface Kit™ that was developed utilizing ESRI's MapObjects™ 2 technology. The VCMS database supported the export of all field data, using any combination of An. arabiensis aquatic habitats sampled at the study site, in order to further process and spatially display specific data attributes in a standalone desktop GIS software package. VCMS® offers connectivity with handheld computers and field data collection devices including GPS receivers. VCMS® can interface with MapInfo GIS files via a MapX programming interface for display and manipulation of spatial data or imagery (www.esri.com). The MapInfo TAB format is a popular geospatial vector data format for geographic information systems software which is developed and regulated by MapInfo Corporation as a proprietary format (MITAB-MapInfo.TAB Read/Write Library).

ESRI ArcPad® software was installed on the Recon X® and we created an ArcPad folder on the C drive of our field computer (laptop PC) to connect and synchronize data with ArcPad on the PDA. The ArcPad® Data Manager for ArcGIS Desktop, the Datum Configuration Tool and the ArcPad®Deployment Manager were copied into this folder, along with other ArcPad® modules. ArcPad has traditional GIS functionality such as map navigation, layering, querying, hyperlinks, and so forth [5]. The key features of ArcPad include support for industry-standard data formats; display and query functionality; editing and data capture; support for optional GPS receivers; and ArcPad tools for ArcGIS Desktop, which are used to prepare data for use in ArcPad (www.esri.com). A key feature of ArcPad in seasonal mosquito data analyses is the ability to use data directly from an individual›s field-level desktop or an organization's enterprise GIS system without the need to convert it to unique portable formats [2]. ArcPad uses vector data in industry-standard shape file format (as used by ArcInfo®, ArcEditor™, ArcView®, ArcIMS®, and other ESRI software programs. In this research ArcPad directly supported the following raster image formats (world file required): JPEG, PNG, CADRG, MrSID (compressed images), and Windows® bit map (BMP) generated from the multivariate time series empirical datasets of An. arabiensis related field and remote specified endemic transmission oriented covariate coefficients. This software was used to display and manipulate the QuickBird imagery on the Recon PDA field computer.

Grid-Based Algorithm

A polygon layer outlining each An. arabiensis aquatic habitat was created by digitizing the georeferenced QuickBird imagery in GeoGrid®. Our matrix was based on GEO grid framework based on grid technology, remote sensing data, and geographic information that was initially developed at National Applied Research Laboratories (NARL). The approach was initiated by synergy of NARLpsilas core competence on environment monitoring and disaster reduction techniques which included high-resolution satellite image processing, virtual reality visualization, grid computing, and disaster mitigation technology along with the advanced cyber-infra- structure environment established within NARL. The framework was constituted by three layers, i.e. application module, service interface, and computing/data/sensor grids. A prototype platform entitled 3D GIS was constituted by using 2 m resolution FORMOSAT-2 data and 5 m resolution digital terrain model that was modeled and then displayed in 3D stereo visualization and in Web pages. The presented approach emphasized the synergy of multiple discipline with cross-field cooperation for geoscience application which then becomes a benchmark in echo to implementation of GEOSS.

Determining spatial heterogeneity in larval habitat distribution using a digitized grid cell database can have important operational significance because vector control operations can be designed to target zones where high larval densities occur [2]. Treatments or habitat perturbations should be based on surveillance of larvae in the most productive areas of the agro-ecosystem and adjacent village [25]. The impact of larval control using new formulations of insecticides should be rigorously tested using a modified longitudinally designed study employing information from digitized grid cells [7]. Laboratory studies should test Bacillus thuringiensis ssp. israelensis (Bti), Bacillus sphaericus (Neide) (Bsph) and Bti:Bsph ratios may be determined employing lethal concentration parameters on all highly productive riceland An. arabiensis aquatic habitats as identified in a stratified digitized grid matrix [9]. Controlling a small proportion of productive riceland An. arabiensis aquatic habitats may yield significant reductions in a rice environment. For larval control, we assume that treatments applied to individual habitats are 100% effective in eliminating all immatures, i.e. treated habitats produce zero contribution to the total productivity.

In this research, a cell within the grid matrix contained an attribute value, as well as a georeferenced aquatic larval habitat location coordinate, and was joined relationally to other databases (Figure 2). The spatial location of each cell was implicitly contained within the ordering of the matrix. The habitats were then characterized in relation to the ecological attributes sampled of an aquatic habitat. Each An. arabiensis aquatic larval habitat/po- lygon was assigned a unique identifier. Field attribute tables were then linked to the polygons. The polygons were used to define the sampling frame, which extended to include a 2 km buffer from the external boundary of the Karima rice-village study site [20]. A mark-release-recapture study in an area near Bamako, Mali, showed that An. arabiensis generally does not disperse further than 1 km [21]. In this research, digitized grid cells were stratified based on LULC transition throughout the rice cycle and defined as: ploughed, flooded, post-trans- planted, tillering, and fallow/ post-harvest.

1. Ploughing: field preparation prior to transplanting of rice seedlings.
2. Flooding: comprised of areas of intensive use with much of the land covered by water.
3. Post-transplanting: a period following rice seedlings transplanting.
4. Tillering: extends from the appearance of the first tiller until the maximum tiller (5 - 9) number is reached. Stem elongation occurs and the tillers continue to increase in number and height, with increasing ground cover and canopy formation.
5. Fallow/post-harvest: after harvesting, the land is left bare waiting the next crop cycle.

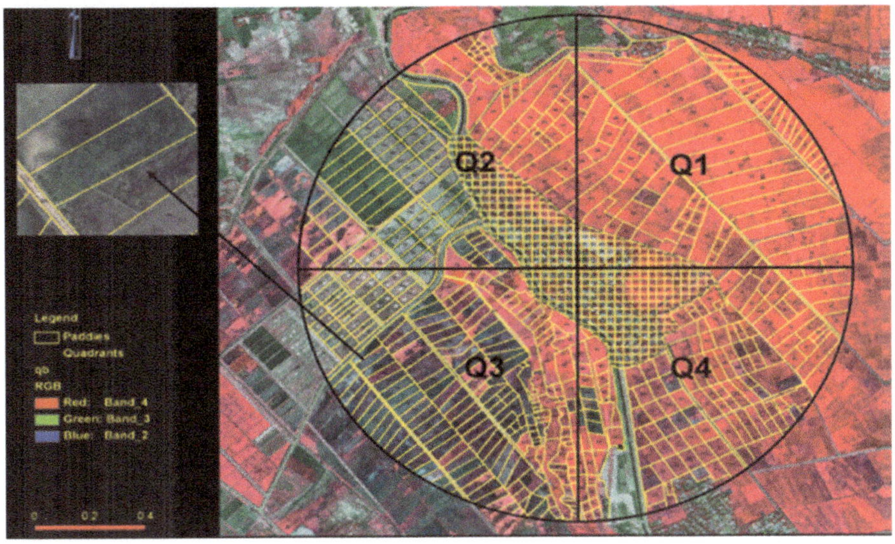

Figure 2. A stratified digitized gridded matrix overlaid onto a QuickBird visible and near infra-red of the spatially targeting quadrants for insecticide treatment at the Karima epidemiological study site.

Error coefficients were calculated for each LULC class and for the overall mapped area. Land use/cover types cannot be effectively extracted when different land cover types bear similar spectral characteristics or the same land cover classification has different spectral responses [22]. Importantly, there are many sources of both conservative and optimistic bias in seasonal mosquito land cover classification accuracy assessment, many of which are impossible to avoid [2]. Bias occurs when a classification estimate is optimistic or conservative [5]. Interestingly, there are at least three significant sources of conservative bias in time series vector arthropod-related image processing: errors in reference data, positional errors, and minimum mapping unit of reference grid [2]. Further, there are at least three significant sources of optimistic bias: use of training data for accuracy assessment, restriction of reference data sampling to homogeneous areas, and sampling of reference data not independent of training data. The magnitude and direction of bias in seasonal mosquito related classification accuracy estimates depends on the methods used for classification and reference data sampling [2]. Therefore, simply reporting a geo-spatiotemporal predictive error matrix and associated classification accuracy estimates we assumed was not enough for accurate seasonal malarial mosquito risk mapping the Karima epidemiological study site. The error matrix is practically meaningless unless methods are reported in sufficient detail to enable readers to assess the potential for bias in the classification accuracy estimates [5].

In this research, we evaluated the results using pixel-based contingency matrices. Change maps for identifying seasonal malaria-related larval habitat 'hot spots" generated from satellite data, and change overlays created from traditional stand maps updated with aerial photography may be site-specifically cross-verified using contingency time series error matrices and GIS modeling [2]. It has been demonstrated in GIS literature clearly that a majority of the so-called digital change classification inconsistencies are not errors, but embody instead a powerful source of sub-stand information with the potential to significantly impact the sustainability of the resource management [5]. While our QuickBird pixel approach did not directly encompass the stand concept (stand representing the actual field sampled georeferenced An. arabiensis aquatic larval habitat), qualitative and quantitative seasonal malaria-related mosquito larval habitat data on change at the stand level was readily extracted from the satellite imagery [2]. Error matrices were then generated for our larval habitat classifications with the purpose of evaluating the map product for thematic accuracy. The associated error matrix provides the user with values indicating how true the map product is with respect to actual conditions on the ground [23] [24]. Errors of omission and commission can be calculated to update and refine the LULC classification for riceland environments or for estimating confidence on a per-class basis [25].

In this research overall accuracy and class-specific user and producer accuracies were calculated for each of the resultant land cover classes. For each mapping region, time series stratified sampling formulas were applied to estimate the error matrix cell proportions [22], and consequently, the estimates of overall and class-specific user's and producer's accuracy [23]. An overall accuracy and a producer and user accuracies for each LULC class was calculated using these methods. The overall accuracy was calculated by dividing the number of pixels correctly classified (i.e., the sum of the diagonal axis of the matrix) by the total number of pixels included in the evaluation process. The producer accuracy is a measure of the omission error and indicates the percentage of pixels of a given land cover type that are correctly classified [5]. The user accuracy is a measure of the commission errors and indicates the probability that a pixel classified into a given class actually represents that class on the ground [5]. These validation techniques were calculated by dividing the number of pixels of the ith class correctly classified by the total number of pixels classified as the ith class.

In this research, the producer's accuracy related to the probability that a land cover class was correctly mapped and measured using the errors of omission (1-producer's accuracy). In contrast, the user's accuracy indicated the probability that a geo-spatiotemporal sampled An. arabiensis aquatic larval habitat from the land cover map actually matched the information from the georeferenced empirical datasets by measuring the error of commission (1-user's accuracy). Accuracy results were computed by weighting the proportions of each land cover, within the Karima epidemiological study site, against total land cover area used in the sampling frame. Specifically, the overall accuracy $(\hat{P})$ and producer's accuracy $(\hat{P}_{Aj})$ were estimated using post-stratified formulas [22]. Post-stratified estimators use the known pixel totals for each land-cover class (Ni+), treating the sample as a stratified random sample of ni+ pixels from the Ni+ pixels in that class [5]. In this research the user's accuracy $(\hat{P}_{Ui})$ was based on the simple random sampling formulas:

- $$\hat{P} = \frac{1}{N}\sum_{k=1}^{q}\frac{N_{k+}}{n_{k+}}n_{kk}$$
- $$\hat{P}_{Ui} = n_{ii}/n_{i+}$$
- $$\hat{P}_{Aj} = \frac{\left(N_{j+}/n_{j+}\right)n_{jj}}{\sum_{k=1}^{q}\left(N_{k+}/n_{k+}\right)n_{kj}}$$

The spatial distribution of the sampled georeferenced aquatic larval habitats were created as a layer and overlaid on the land cover layer, and the number of An. arabiensis aquatic habitats in each land cover class was then calculated using Zonal stat, an avenue script in ArcGIS® [6]. Jacob et al. [9] created a 2 km buffer zones around Karima villages and then uses zonal statistics to estimate the mean of normalized difference vegetation index (NDVI), average precipitation, temperature, elevation and slope within each buffer zone to determine which factors are associated with high mosquito densities and high rates of malaria. In this research, the chi-square analysis for the sampled field and remote predictor variables was used to examine whether there were significant differences in proportions of positive and negative sampled aquatic habitats in each grid cell located in different LULC classes. Tukey style multiple comparison of proportions were used for post-hoc analyses. Each predicted estimate in the LULC classification was ground verified which revealed all classes were within correct statistical limitations. Unless each and every minimum mapping unit, which may be a pixel, polygon, or any other feature with spatial properties, has been verified on the ground, there is still unknown and un-quantified error [5].

Object Oriented Classification

ENVI® spectral tools were used to analyze the QuickBird visible and NIR data, in order to confirm the location of varying states of rice crops. In ENVI 4.6®, a spectrum plot, known as a Z-profile, of the pixel under the cursor, was run through all bands of the QuickBird image of the An. arabiensis aquatic habitats in the Karima agro-vil- lage complex epidemiological study site. The basic workflow involved importing the data collected in the field from the study site into a spectral library. The content of the spectral library was then used in the Endmember Collection workflow to perform a supervised classification, based on the empirical ecological covariates sampled at each individual georeferenced aquatic larval habitat. Binary Encoding, Spectral Angle Mapper (SAM), and spectral feature fitting were used to rank and match any unknown spectrum to the materials in the spectral library. SAM is a method to compare known spectral target properties to every pixel in the scene [23]. Land cover class representative pixels were selected and compared to a reference dataset.

After pixels in a raster dataset have been assigned to categories, post-classification methods were used to clean up the resulting raster class image in preparation for conversion to vector data. Raster classification of riceland An. arabiensis aquatic habitat results often contain scattered individual pixels of one LULC class surrounded by a larger area of another LULC class (i.e., islands) [25]. ENVI used sieving (i.e., removal of isolated aquatic larval habitat pixels from our QuickBird image) and clumping (clumping adjacent, similar larval habitat pixels together) to minimize database clutter in our GIS analyses.

During the segmentation procedure, image objects from the QuickBird scenes were generated based on color, shape, and texture. Optimum segmentation parameters were generated for extracting different land cover classes associated with the sampled An. arabiensis aquatic larval habitats. Image segmentation is the key technology from image processing to image analysis, which divides an image into non-overlapping homogeneous regions and the combined heterogeneous adjacent regions [22]. In this research, image classification was done using FastLine-of-sight Atmospheric Analysis of Spectral Hypercubes (FLAASH™) using a spatial model (GMD file) that converted the image's digital numbers (DN) associated to An. arabiensis aquatic habitat surface components to at-sensor radiance and computed at-sensor reflectance while normalizing solar elevation angle.

Interestingly, FLAASH is a first-principles atmospheric correction tool that corrects wavelengths in the visible through near-infrared and shortwave infrared regions, up tμm (http://www.exelisvis.com/docs/FLAASH.html). FLAASH works with most hyperspectral and multispectral sensors. Water vapor and aerosol retrieval are only possible when the image contains bands in appropriate wavelength positions. FLAASH can correct images collected in either vertical (nadir) or slant-viewing geometries (www.esri.com). Previously Yuan et al. [22] employed the capability of FLAASH in ENVI software to make atmospheric correction for EO-1 Hyperion hyperspectral image. Hyperion hyperspecreal image of Zhangye city in Heihe River valley of Gansu province, China was acquired on September 10, 2007. Canopy spectra, biochemical component and GPS data of 41 plots were measured in near real-time during the satellite overpass. Hyperion hyperspectral image was geometrically corrected using Lansat-7 ETM+ image. Thereafter, the digital numerical coefficients were transformed to radiance and apparent reflectance, and atmospheric correction of Hyperion image was made using FLAASH. The resulting radiance, apparent reflectance and reflectance after FLAASH of four typical objects, including corn, water body, desert and building, were compared. ASD spectra of corn were resampled to Hyperion corresponding bands using Gaussian filter function. The comparison between ASD resampled spectra and Hyperion spectra after FLAASH demonstrated that the atmospheric correction using FLAASH is very effective and these two spectra are consistent with each other and the correlation coefficient reached 0.987. In this research data pre-processing involved converting digital numbers to radiance, employing FLAASH atmospheric corrections and the equation:

$$\rho_{BandN} = \frac{\pi\left(L_{BandN} \times Gain_{BandN} + Bias_{BandN}\right) \times D^2}{E_{BandN} \times \left(\cos\left((90-\theta) \times \pi/180\right)\right)}$$

where,

r_{BandN} = Reflectance for Band N;

L_{BandN} = Digital Number for Band N;

D = Normalized Earth-Sun Distance;

E_{BandN} = Solar Irradiance for Band N.

Regression Analyses

A Poisson regression with a 95% confidence level was then constructed in PROC NLMIXED MOD using the spatiotemporal sampled field and remote specified An. arabiensis larval habitat predictors. In actuarial literature, researchers have suggested various Poisson regression model, which is also known as the Generalized Linear Model (GLM) with Poisson error structure [6] We expected the seasonal sampled An. arabiensis aquatic habitat larval/pupal count in the Karima study site to follow a Poisson distribution, as was the case in previous research in irrigated riceland areas. Poisson process is a commonly used starting point for modeling stochastic variation of ecological count data around a theoretical expectation [7]. In this research, PROC GENMOD used a class statement for specifying our sampled categorical indicator variables. The parameter $\lambda_i(X_i)$ is both the mean and the variance of the Poisson distribution for a sampled anopheline larval habitat i [2]. The dependent variable was the total An. arabiensis larval count in the sampled larval habitat.

We generated regression models for comparing transmission intensity at the Karima epidemiological study site. The regression analyses assumed independent counts (i.e., n_i), taken at the sampled larval habitat locations $i = 1,2,L\ ,n$, where each of the explanatory predictor covariates represented in the linear framework was from a Poisson distribution. Our Poisson regression assumed the response variable Y had a Poisson distribution and assumed the logarithm of its expected value was modeled by a linear combination of the sampled larval habitat sampled parameter estimates. This expression was written more compactly as $\log\left(E\left(Y|x\right)\right)$ where x was an n + 1-dimensional vector consisting of n independent variables concatenated to 1 and θ, which in this research was simply expressed as a linearly linked to b. Thus, in our Poisson regression model θ was an input vector x and, the predicted mean of the associated Poisson distribution was given by $E\left(Y|x\right) = a^{\theta' x}$, if x Î R^n was a vector of the sampled independent variables. Thereafter, the Poisson models took the from $\log\left(E\left(Y|x\right)\right) = a'x + b$ where a Î R^n and b Î R. Positing salient estimators using the maximization of an auto-Gaussian log-likelihood function and a set of eigenvectors where lambda is a subspace of R^n can identify important explanatory predictor covariates associated to productive habitats based on spatiotemporal field-sampled count data [1] - [5].

After the regression, we attempted to ascertain whether the proportions of the sampled An. arabiensis immature data differed by digitized grid cells in the study site. The model assumed a random sample between Y_i, (i.e., sampled habitat total larval count density data), and the regressors $X_{i1},L\ ,X_{ip}$. A disturbance term, ε_i, which was a random variable, was added to this assumed relationship to capture the influence of all the parameters estimators sampled on Y_i other than $X_{i1},L\ ,X_{ip}$. The random error term, ε, in the regression models was assumed to be normally distributed with mean zero and variance σ^2. Statistical characteristics of the sampled data were examined in PROC UNIVARIATE. The PLOT option in the PROC UNIVARAITE statement generated histograms and boxplots. The NORMAL option was used to test whether the field and remote-sampled explanatory field and remote specified endemic transmission oriented covariate coefficients had a normal distribution. The regression equation for the study sites was: $Y_i = \beta_0 + \beta_1 X_{i1} + \beta_1 X_{2i1}$, $i = 1,2,L\ ,n$.

It was important to distinguish the geo-spatiotemporal-sampled An. arabiensis aquatic larval habitat models in terms of random variables and the observed values of the random variables. Thus, we determined p + 1 parameters $\beta_0,L\ ,\beta_p$. In order to estimate the sampled parameters, it was also useful to employ the matrix notation $Y = X\beta + \varepsilon$, where Y was a column vector that included the geo-spatiotemporal-sampled An. arabiensis aquatic larval habitat count values of $Y_1,L\ ,Y_n$, sampled in the study site, which in this research included the unobserved stochastic components $\varepsilon_1,\text{L},\varepsilon_n$ and the matrix X. This matrix was the observed larval habitat parameter values of the regressors expressed as: $X=\begin{pmatrix}1 & x_{1i} & \text{L} & x_{1p}\\ 1 & x_{2i} & \text{L} & x_{2p}\\ \text{M} & \text{M} & \text{O} & \text{M}\\ 1 & x_{n1} & \text{L} & x_{np}\end{pmatrix}$.

Thereafter, the sampled explanatory predictor covariate coefficient values were log-transformed to normalize the distribution and minimize standard error in the residual forecasts. Multicollinearity diagnostics from the COLLIN option in SAS were estimated. Residual-based diagnostics for univariate and multivariate conditional heteroscedastic models previously constructed from clustering field and remote-sampled vector insect habitat parameter estimates have revealed that errors in variance uncertainty estimation can substantially alter numerical predictions models due to multicollinearity [5]. The SAS COLLIN option produced eigenvalues and condition index, as well as proportions of variances with respect to individual-sampled An. arabiensis aquatic larval habitat explanatory predictor covariates in the models. The conditional index scores indicated no significant multicollinearity in either model output.

In this research extra-Poisson variation was detected in the residual variance estimates of both An. arabiensis aquatic larval habitat models. Extra-binomial (i.e., extra-Poisson) variation occurs when discrete data comes in the form of counts or proportion that display greater variability than would be predicted when fitting a model [7]. When sampled vector mosquito aquatic larval habitat data are overdispersed, the square root and logarithmic transformations may be less effective at making the mean and variance independent [1] - [5]. As such, we employed a negative binomial regression with a gamma distributed non-homogenous mean constructed in PROC GENMOD to account for the overdispersion in both the geo-spatiotemporal An. arabiensis larval habitat models. The negative binomial distribution arises as a continuous mixture of Poisson distributions in a vector mosquito larval habitat distribution model where the mixing distribution of the Poisson rate is a gamma distribution [2]. The negative binomial model is a quadratic function of the mean and the variance which commonly affect the weights in the iteratively weighted least-squares algorithm for fitting spatiotemporal-sampled vector mosquito aquatic larval habitat data [4]. We used the negative binomial regression model with a non-homogenous gamma distributed mean to linearly adjust the sampled data in the Karima rice agro-village complex epidemiological study site. In this research, the negative binomial distribution [i.e., NB(r, p)] was represented as a compound Poisson distribution by letting Y_n, $n \in N_0$ denote a sequence of the independent and identically distributed An. arabiensis aquatic larval habitat randomized variables, each one having the logarithmic distribution Log(p). The negative binomial for a predictive linear vector mosquito aquatic larval habitat model is a Poisson (λ) distribution, where λ is itself a random variable, distributed according to Gamma(r, p/(1 − p)) [2].

Steps in constructing the negative binomial models in SAS was similar to our description of the Poisson An. arabiensis larval habitat time series predictive regression models, that is, the log of the mean, m, was a linear function of the independent variables, $\log(\mu)=\text{intercept}+b_1\times X_1+b_2\times X_2+\text{L}+b_3\times X_m$, which implied that m was the exponential function of independent variables, $\mu=\exp(\text{intercept}+b_1\times X_1+b_2\times X_2+\text{L}+b_3\times X_m)$. Instead of assuming as before that the distribution of Y, (i.e., sampled An. arabiensis aquatic larval habitat parameters) were Poisson, we now assumed that Y had a negative binomial distribution. We relaxed the assumption about equality of mean and variance (i.e., Poisson distribution property), since in our models the variance of negative binomial was equal to m + km², when k ³ 0 was a dispersion parameter. The maximum likelihood estimation method was used to estimate k as well as the parameters of the regression models for log(m). Since the Poisson regression model can be generalized by introducing an unobserved heterogeneity term for observation *i* [5], individual sampled An. arabiensis larval habitat data at the Karima study site were assumed to differ randomly in a manner that was not fully accounted for by the observed covariate coefficient indicator values. In this research this was formulated where the unobserved heterogeneity term $\tau_i=e^{\varepsilon_i}$. I was independent of the vector of regressors X_i. Then the distribution of y_i conditional on x_i and τ_i was Poisson with conditional mean and conditional variance (i.e., $\mu_i\tau_i$) which in this research was expressed as $f(y_i|X_i,\tau_i)=\dfrac{\exp(-\mu_i\tau_i)(\mu_i\tau_i)^{y_i}}{y_i!}$.

We then let $g(\tau_i)$ be the probability density function of τ_i in the linear vector time series explanatory field and remote specified endemic transmission oriented An. arabiensis aquatic larval habitat models. Then, the distribution $f(y_i|X_i)$ was no longer conditional on τ_i, which was obtained by $f(y_i|X_i)$, τ_i with respect to τ_i: $f(y_i|X_i)=\int_0^\infty f(y_i|X_i,\tau_i)g(\tau_i)d\tau_i$ in both larval habitat models. Our autoregressive algorithms had an analytical solution to this integral which existed when τ_i was assumed to follow a gamma distribution. This solution was the negative binomial distribution. When the model contains a constant term, it is necessary to assume that $E(e^{\varepsilon_i})=E(\tau_i)=1$, in order to identify the mean of the distribution [7]. Thus, it was assumed that τ_i generated from the An. arabiensis aquatic larval habitat explanatory predictor covariates sampled in the Karima epidemiological rice agro-village complex study site followed a gamma (θ, θ) distribution with $E(\tau_i)=1$ and $V(\tau_i)=1/\theta$:

$$g(\tau_i)=\frac{\theta^\theta}{\Gamma(\theta)}\tau_i^{\theta-1}\exp(-\theta\tau_i)$$

where $\Gamma(\theta)=\int_0^\infty Z^{x-1}\exp(-z)dz$ was the gamma function and θ was a sampled field and/or remote sampled endemic transmission oriented parameter estimates. Then, the density of given X_i was derived as $f(y_i|X_i)$ which rendered $\int_0^\infty f(y_i|X_i,\tau_i)g(\tau_i)d\tau_i$

$\frac{\theta^{\theta}\mu_i^{y_i}}{y_i!\Gamma(\theta)}\int_0^{\infty} e^{-(\mu_i+\theta)\tau_i}\tau_i^{\theta+y_i-1}d\tau_i$, $\frac{\theta^{\theta}\mu_i^{y_i}\Gamma(y_i+\theta)}{y_i!\Gamma(\theta)(\theta+\mu_i)^{\theta+y_i}}$, and $\frac{\Gamma(y_i+\theta)}{y_i!\Gamma(\theta)}\left(\frac{\theta}{\theta+\mu_i}\right)^{\theta}\left(\frac{\mu_i}{\theta+\mu_i}\right)^{y_i}$. Making the substitution $\alpha=\frac{1}{\theta}(\alpha>0)$, the distribution in both larval habitat models was then rewritten as $f(y_i|X_i)=\frac{\Gamma(y_i+\alpha^{-1})}{y_i!\Gamma(\alpha^{-1})}\left(\frac{\alpha^{-1}}{\alpha^{-1}+\mu_i}\right)^{\alpha}\left(\frac{\mu_i}{\alpha^{-1}+\mu_i}\right)^{y_i}$, $y_i=0,1,2,L$

Thus, the negative binomial distributions of the sampled aquatic larval habitats in the Karima study site were derived as a gamma mixture of Poisson random variables. The models had a conditional mean $E(y_i|X_i)=\mu_i=e^{x_i'\beta}$ and conditional variance $V(y_i|X_i)=\mu_i\left[1+\frac{1}{\theta}\mu_i\right]=\mu_i[1+\alpha\mu_i]>E(y_i|X_i)$. The conditional variance of the negative binomial distribution exceeded the conditional mean. Overdispersion in linear vector mosquito aquatic larval habitat models results from neglected unobserved heterogeneity [7]. The negative binomial model with variance function (i.e., $V(y_i|X_i)=\mu_i+\alpha\mu_i^2$) is quadratic in the mean, which is referred to as the NEGBIN2 model [7]. To estimate this model in this research, we specified DIST = NEGBIN (p = 2) in the MODEL statements. A test of the Poisson distribution was carried out by testing the hypothesis that $\alpha=\frac{1}{\theta_i}=0$. The Poisson distribution is a special case of the negative binomial distribution where $\alpha=0$ [5]. A Wald test of this hypothesis was provided. In this research this test was the reported statistic for the estimated in the linear study site models. The log-likelihood function of the models was then given by

$$\sum_{1\,i=1}^{N}\left\{\sum_{j=0}^{y_i-1}\ln(j+\alpha^{-1})-\ln(y_i!)-(y_i!)-(y_i+\alpha^{-1})\ln(1+\alpha\exp(X_i'\beta))+y_i\ln(\alpha)+y_iX_i'\beta\right\}$$

and the equation $\Gamma(y+a)/\Gamma(a)=\prod_{j=0}^{y-1}(j+a)$. The gradient in both our time series explanatory field and remote specified endemic transmission oriented An. arabiensis aquatic larval habitat pre models was

$$\frac{\partial l}{\partial\beta}=\sum_{i=1}^{N}\frac{y_i-\mu_i}{1+\alpha\mu_i}X_i$$

and

$$\frac{\partial l}{\partial\alpha}=\sum_{i=1}^{N}\left\{-\alpha^{-2}\sum_{j=0}^{y_i-1}\frac{1}{(j+\alpha^{-1})}+\alpha^{-2}\ln(1+\alpha\mu_i)+\frac{y_i-\mu_i}{\alpha(1+\alpha\mu_i)}\right\}.$$

Thereafter, we used the seasonal sampled immature count data and standard deviations of the log number of larval/pupal counts collected in the Karima study site to determine sample size requirements. We applied a sampling intensity formula for determining the number of immature Anopheles samples to collect from a finite population where $n=(t_s/E)^2$ where t = t value (t ≈ 2), s = the standard deviation of log-larval/pupal count values observed in the study site(s = 0.889) and E was the desired half-width of the confidence interval around the mean expressed in same units as standard deviation where E = ln(1.25). Applying this formula and assuming that the geo-spatiotemporal larval/pupal production was similar for the sampled An. arabiensis aquatic larval habitats in the study site, we determined 152 samples were required. We overlaid vector images of the sampling scheme with the raster image of the study site to identify AOI's within the sampling frame for field crews to visit. All potential aquatic georeferenced aquatic larval habitat sites were identified and data relative to species composition and abundance, predators, water quality and environmental parameters were collected longitudinally by the field crews.

Field data parameters were entered in Microsoft Excel files and analyzed using and SAS 9.1.3®(SAS Inc. Carey, NC, USA). Before analyses the data was tested for collinearity using design matrix from a Poisson regression model and run through SAS PROCREG/VIF (Variant Inflation Factor) procedure which indicated the absence of problematic correlation among the predictors. ANOVA test was used to compare the differences in the An. arabiensis larval/pupal abundance among the rice stages. The analyses revealed that post-transplanting was the rice stage associated with the highest larval/pupal count throughout the rice cycle. Differences in immature count between the An. arabiensis aquatic habitats were compared using Student's t-test. Forward multiple regression analysis was then used to obtain the best predictor variables explaining the abundance of mosquito immatures. In probability and statistics, Student's t-distribution (or simply the t-distribution) is a family of continuous probability distributions that arise when estimating the mean of a normally distributed population in situations where the sample size is small and population standard deviation is unknown [5]. A widely used statistical analyses in time series vector mosquito larval habitat endemic transmission oriented risk modeling is the Student's t-test for assessing the statistical significance of the difference between two sample means, the construction of confidence intervalsand for parsimoniously quantitating the difference between two linearized population means [2]. The relative abundance of mosquitoes was expressed as the number of mosquito larvae/pupae per 20 dips because the number of larvae/pupae sampled was low. Statistical analyses was done using log-trans- formed ($\log_{10}$ n + 1) larval/pupal counts to normalize the data. Results were considered significant at $P < 0.05$.

RESULTS

The overall accuracy of the land cover classification from the QuickBird image was Kappa statistic, 0.91 for the non-riceland site and 0.94 for the riceland epidemiological study site. Generally, the frequency of confusion of the LULC classification was low (Table 2). Overall, instances of confusion were minimal and did not affect user's classification accuracy for the study sites. The user's accuracy ranged between 92% and 97%, with relatively low errors of commission (excesses) and the producer's accuracy ranged between 91% and 98%.

All QuickBird polygons generated were conformed to larval habitat boundaries within the 1 km buffer of the study site. Multiple data layers of spatial and attribute data were created, using different coded values for various rice field attributes, which were linked to the same grid cell. This digitized grid-based algorithm allowed retrieval and transformation of seasonal larval habitat data, regardless of spatial dimensionality of a riceland An. arabiensis aquatic habitat, in the study site.

Table 2. Producer's and user's accuracy levels of the QuickBird imagery of land cover types in the Karima study site.

Land Cover Type	Producer's Accuracy (%)	Omission Errors (%)	User's Accuracy (%)	Commission Errors (%)
Plowing	91	3	96	3
Flooding	96	5	97	9
Post-harvested	98	2	94	6
Tillering	91	8	92	5

In this research, the relationship between prevalence of each individual potential predictor variable sampled in the epidemiological study site was investigated by single variable regression analysis in PROC NL MIXED. We used the regression line

$(y_i - \bar{y}) = (\hat{y}_i - \bar{y}) + (y_i - \hat{y}_i)$ to generate a pseudo R^2 value where the first term was the total variation in the response y (i.e., total density larval count of An. arabiensis aquatic habitats) and the second term was the variation in mean response based on the sampled parameters. The third term was the residual value in the risk model estimates. Squaring each of these terms and adding over all of the sampled observations generated the equation $\sum (y_i - \bar{y})^2 = \sum (\hat{y}_i - \bar{y})^2 + \sum (y_i - \hat{y}_i)^2$. This equation was then written as SST = SSM + SSE, where SS was notation for sum of squares and T, M, and E were the notation for total, model, and error, respectively. The square of the sample correlation was equal to the ratio of the estimates while the sum of squares was related to the total sum of squares: r^2 = SSM/SST. This formalized the interpretation of R^2 as explaining the fraction of variability in the sampled An. arabiensis aquatic habitat data explained by the regression model. The sample variance s_y^2 was equal to $\sum (y_i - \bar{y})^2 / (n-1)$, which in turn was equal to the SST/df, the total sum of squares divided by the total df. A regression equation was then constructed using the mean square model (i.e., MSM) = $\sum (\hat{y}_i - \bar{y})^2 / (l)$, which in this research was equal to the SSM/df. The corresponding mean square error (i.e., MSE) was $\sum (y_i - \hat{y}_i)^2 / (n-2)$, then was equal to SSE/df and the estimate of the variance about the regression line (i.e., σ^2). The MSE is an estimate of σ^2 for determining whether or not the null hypothesis is true [24]. For the geo-spatiotemporal sampled explanatory field and remote specified endemic transmission oriented predictive variables, (p) the An. arabiensis aquatic larval habitat modeled the DFM which in this research was equal to p, and the error degrees of freedom (dfe) which was equal to (n □ p □ 1), and the total degrees of freedom (dft) which was equal to (n □ 1), the sum of DFM and DFE. Explanatory and response variables were numeric. The relationship between the mean of the response variable (i.e. larval count) and the level of the sampled explanatory predictor covariates in the regression equation were assumed to be approximately linear (i.e., straight line). The corresponding table generated classified each the field and remote-sam- pled An. arabiensis aquatic habitat larval habitat parameters in SAS as in Table 3.

In the multiple regression analyses, the test statistic MSM/MSE had an F(p, n – p – 1) distribution. In this research, the null hypothesis was $\beta_1 = \beta_2 = L = \beta_p = 0$, and the alternative hypothesis was at least one of the sampled An. arabiensis aquatic habitat larval habitat parameters $\beta_j \neq 0$, $j = 1, 2, L, p$. The F test did not indicate which of the parameters $\beta_j \neq$ was not equal to zero, only that at least one of them was linearly related to the response variable. The ratio SSM/SST = R^2 (i.e., squared multiple correlation coefficient) was the proportion of the variation in the response variable that was explained by the immature sampled habitat data. The square root of R^2 (i.e., the multiple correlation coefficient) was the correlation between the sampled An. arabiensis aquatic larval habitat observations (i.e., y_i) and the fitted values (i.e., $\hat{y}_i$). Additionally, from the sampling distribution, generated from the sampled t parameters, the probability of obtaining an F was large or larger than the one was calculated. We noted that, the t-test and the F-test were equivalent; the relation between ANOVA and t was given by F = t^2. Significant differences by ANOVA were noted for mean numbers of the sampled An. arabiensis aquatic larval habitat collected throughout the sampling frame (F = 44.7, df = 1).

A Poisson regression analyses was then constructed in PROC NL MIXED to determine the relationship between the sampled An. arabiensis aquatic habitat larval count data and the sampled habitat characteristics. The Poisson models were built using the time series field and remote-sampled larval habitat data. A negative binomial regression had to be employed, however, as an examination of the data indicated that over dispersion was a significant problem in the Poisson model.

Table 3. The An. arabiensis aquatic habitat larval habitat regression-based model parameters.

Source	Degrees of Freedom	Sum of Squares	Mean S

Model	p MSM/MSE	$\sum(\hat{y}_i - \bar{y})^2$	SSM/DFM
Error	n – p – 1	$\sum(y_i - \hat{y}_i)^2$	SSE/DFE
Total	n – 1	$\sum(y_i - \bar{y})^2$	SST/DFT

The Poisson distribution is a special case of the negative binomial distribution, where the mean approximates the standard deviation [24]. We assumed that the log of the mean, m, was a linear function of independent variables, $\log(\mu) = \text{intercept} + b_1 \times X_1 + b_2 \times X_2 + \text{L} + b_3 \times X_m$ in the model which implied that m was the exponential function of independent variables when multiple independent geo- spatiotemporal $\mu = \exp(\text{intercept} + b_1 \times X_1 + b_2 \times X_2 + \text{L} + b_3 \times X_m)$ [25]. Therefore, instead of assuming that the distribution of the sampled An. arabiensis aquatic habitat larval habitats parameter estimates (i.e., Y) was Poisson, we were able to assume that Y had a negative binomial distribution. We relaxed the assumption about equality of mean and variance (i.e., Poisson distribution property), since the variance of negative binomial was equal to m + km^2, where k ³ 0 was a dispersion parameter. The maximum likelihood method was used to estimate k, as well as the sampled larval habitat parameters of the regression model for log(m). For the negative binomial distribution, the variance was equal to the mean + k mean^2 (i.e., k ³ 0) as the negative binomial distribution reduced to Poisson when k was 0.

In the regression analyses the null hypothesis was: H_0: k=0 and the alternative hypothesis was: H_a: k > 0. We recorded the log-likelihood (i.e., LL) for the time series field and remote specified endemic transmission oriented An. arabiensis aquatic habitat larval habitat models. We used the likelihood ratio (LR) test to compute the LR statistic using -2(LL) (Poisson) and the LL (i.e., negative binomial). The asymptotic distribution of the LR statistic had probability mass of one half at zero and one half-chi-square distribution with 1 df. To test the null hypothesis at the significance level a, we employed the critical value of chi-square distribution corresponding to significance level 2a, that was rejection of H_0, if LR statistic > $\chi^2_{(1-2\alpha,1\,\text{df})}$. We generated the log of the mean, m, which in this research was a linear function of independent variables, $\log(\mu) = \text{intercept} + b_1 \times X_1 + b_2 \times X_2 + \text{L} + b_3 \times X_m$, in the larval habitat model which implied that m was the exponential function of the independent variables when $\mu = \exp(\text{intercept} + b_1 \times X_1 + b_2 \times X_2 + \text{L} + b_3 \times X_m)$.

Table 4 lists the dependent and independent variables collected in the Karima epidemiological riceland agro-village complex study site. The count of An. arabiensis larvae/pupae collected at a habitat had a mean of 4.7, with a standard deviation of 5.6. The median count was 5 Anopheles larvae/pupae and the count range from 0 to 41 Anopheles larvae. The distribution was right skewed with 75% of the sampled aquatic habitats having 4 or less larvae. Significant correlations existed between some of the independent variables including study site and number of tillers, (r = –0.23, p = 0.004). The table reveals the abundance of riceland *Anopheles* larvae/20 dips collected in the paddy and canal habitats at the Karima study site. In the study site, the difference in the abundance of pupae and 1st, 2nd and 3rd instars larvae collected in paddy and canal habitats was not significant ($P > 0.05$), while that of 4th instars larvae was significantly higher in the paddy habitats than in the canal habitats. (t = 5.19, df 179, $P < 0.05$).

The regression analyses identified rice height, levels of turbidity and number of tillers as significantly influencing the count of An. arabiensis mosquitoes in the study site. Muturi et al. [8] reported a positive association between turbidity and An. arabiensis in the Mwea rice fields. In the rice growing cycle, rice plants increase in height and tiller numbers and affect the microhabitat conditions of mosquito larval habitats [2].

The Feature Extraction Module in ENVI allows quick and accurate extraction of an An. arabiensis aquatic habitat from the QuickBird visible and NIR data. All post-transplanting habitat pixels were isolated from a LULC map. The endmember determination estimated the set of distinct spectra that comprised the mixed QuickBird pixels in the riceland scene. The produced abundance planes provided estimates of the fractional abundances for the endmembers in each QuickBird visible and NIR pixel.

Table 4.The mean number of *An. arabiensis* larvae/pupae collected (mean ± SE) per 20 dips in paddies containing different stages of rice growth using field sampled and QuickBird 0.61 m visible and near infra-red (NIR) data in Karima rice-village complex.

Paddy category	Number of habitats	1st in-stars	2nd un-stars	3rd in-stars	4th in-stars	Pupae
Ploughed	25	1.41	0.95	0.09	0.00	0.36
Flooded	23	1.67	1.10	0.30	0.00	0.36
Post-transplanting	30	8.02	9.00	8.89	3.22	0.99
Tillering	28	2.00	6.67	2.00	1.20	0.67
Fallow	27	1.00	0.67	0.00	0.00	0.01

The ENVI Feature Extraction Module had the ability to extract a variety of An. arabiensis aquatic habitat features. The feature extraction workflow guides you through the process of segmenting the image by dividing the image into discreet, real-world, objects through the identification of edges and the grouping of pixels into regions (http://www.exelisvis.co.uk/). Prior to the final classification and extraction, the ENVI preview portal window allowed us view the preliminary results, so that we could make adjustments to the analysis before processing the entire image data. ENVI also provided us tools to polish and refine our extraction results. We chose to annotate the results, count individual features, edit or smooth vectors, assess classification accuracy, and generate summary statistics about extracted features. Following image segmentation, the workflow provided us with several options for defining the time series An. arabiensis aquatic larval habitat remote features that need to be extracted We chose example segments in the larval habitat image that were representative of each habitat feature class or create rule sets that must be met in order to assign segments to a feature class.

Vector map layers of rice fields were outlined as polygons, using ESRI shape files and MapInfo tab-file formats. By tapping on the PDA screen of a rice paddy polygon layer in the Karima study site, the end user had the ability to toggle a particular polygon as ON/OFF or TREATED/NOT TREATED, by changing the polygon line color, fill pattern, or both. The Recon X® supported both the raster image layer display and a vector-based data layer used for mapping rice paddy polygons. There was a minimal requirement to be able to output a simple text file of information indicating the paddy had been treated versus not treated (Figure 3). Spatial data sets consisting of a raster image layer and vector line work describing the QuickBird polygons were stored and displayed on the Recon X®. When uploaded into a Recon X, the polygons displayed all treated and untreated paddies within a 1 km buffer of the study site.

DISCUSSION

Initially, Poisson regression was used to model the geo-spatiotemporal larval productivity of the spatiotemporal-sampled An. arabiensis aquatic larval habitats in the Karima agro-village riceland epidemiological study site in our customized real-time bidirectional PDA-GIS-DGPS-RS cyber-environment. An examination of the data, however, indicated that significant overdispersion was present. Therefore, a negative binomial was used to model the overdispersed Poisson larval count data. Negative binomial regression models estimate a dispersion parameter that is used to remove the effects of overdispersion and provide more accurate estimates of standard error [5]. The negative binomial was derived as a Poisson-gamma mixture and as a GLM. PROC NL MIXED expressed the variance of the response for the negative binomial as variance $(y) = \mu + k\mu^2$ as opposed to the more common notation $(y) = \mu + \mu^2/v\mu$ [2].

Figure 3. QuickBird visible and near infra-red data of treated and untreated paddy habitats Karima epidemiological study site displayed in the Trimble Recon X, 400 MHz Intel PXA255 Xscale CPU.

The difference in notation was trivial ($k = 1/v$). The straightforward derivation of the models, from the negative binomial probability distribution function, did not, however, equate with the Poisson–gamma mixture-based version of the negative binomial. Rather, canonical link and inverse canonical link were converted to log form in the real-time PDA-GIS-DGPS-RS cyber-environment. A GLM-based negative binomial was produced that yielded identical parameter estimates of An. arabiensis aquatic larval habitats to those calculated by the mixture-based model for the Karima epidemiological study site.

As a non-canonical linked model, however, the standard errors did differ slightly from the mixture model. A maximum likelihood estimator was employed by default as an observed information matrix, to produce standard errors in the cyber-environment. The GLM algorithm produced standard errors based on the expected information matrix using the difference in standard errors between the two versions of negative binomial analyses. The GLM negative binomial algorithm was amended to allow production of standard errors based on the field-sampled larval count data of An. arabiensis aquatic larval habitats at the epidemiological study site. The amended GLM-based negative binomial rendered identical estimates and standard errors to that of the mixture-based negative binomial analyses. The log-negative binomial data was then imported into an SAS database within our real-time bidirectional PDA-GIS-DGPS-RS cyber-environment.

A time series stratified digitized grid-based algorithm in the real-time PDA-GIS-DGPS-RS cyber-environ- ment identified the field and remote-sampled An. arabiensis aquatic habitats in the Karima epidemiological rice agro-village complex study site. By digitally tracing riceland An. arabiensis aquatic habitats using QuickBird visible and NIR data in a GeoGrid database, polygons were generated that approximated the curvature of the habitat boundaries. All visible and NIR polygons were conformed to the sampled aquatic habitat boundaries within the 2 km buffer of the epidemiological study site. Field attribute tables were then associated to the polygons.

Building a digitized grid geodatabase PDA-GIS-DGPS-RS cyber-tools can identify relationships between riceland mosquito aquatic habitats, rice plant stage of development and agro-village-complex within the framework of relational database technology. Thus, by digitally tracing riceland*An. arabiensis* aquatic habitats using QuickBird visible and NIR data in a PDA-GIS-DGPS-RS cyber-environment; a vector biologist, an epidemiologist, or other data analysts can generate polygons that approximate the curvature of the paddy and canal habitat boundaries. Field attribute tables can then be associated to the polygons. Multiple data layers can be created thereafter using different coded values for geographically representing various field attributes to the same digitized grid cell which can allow for multiple interactions between compatible ecological geodatabases enabling retrieval and transformation of seasonal larval habitat data efficiently regardless of spatial dimensionality of a georeferenced riceland agro ecosystem An. arabiensis aquatic larval habitat. Multiple data layers can then be created PDA-GIS-DGPS-RS cyber-environment using different coded values for various field attributes for the same digitized grid cell which can allow for multiple interactions between compatible ecological databases while enabling retrieval and transformation of seasonal collected habitat data efficiently regardless of spatial dimensionality of a sampled habitat.

In this research, ENVI software in the real-time bidirectional PDA-GIS-DGPS-RS cyber-environment automatically categorized individual sampled An. arabiensis aquatic habitats pixels into separate spectral classes, converted remotely sensed raster layers to vector coverage and classified the layers as ESRI shape files. In ENVI the sampled habitat data and corresponding LULC classification generated in ArcGIS was evaluated using spectral and spatial browsing, and using color composites to characterize spectral variability of the sampled An. arabiensis aquatic habitat explanatory covariates. ENVI's built-in spectral libraries provided quick reference to the QuickBird visible and NIR data, giving well-characterized reference sets of habitat spectral values based on the LULC classifications. This was important as there was significant emissivity spatial heterogeneity within the georeferenced empirical ecological dataset based on individual sampled habitat reflectance estimates. Therefore several LULC feature data corresponding to the sampled An. arabiensis aquatic habitats had to be included in the spectral library. ENVI's spectral analysis tools also used an adjacency correction method to classify the images based on LULC reflectance spectral estimates (http://www.ittvis.com). For example, the MODTRAN- based Fast Line-of-Sight Atmospheric Analysis of Spectral Hypercubes (FLAASH) technique in ENVI removed most of the solar and atmospheric effects, transforming the georeferenced LULC classified An. arabiensis aquatic larval habitat data from radiance to surface reflectance values. However, the most important tool in the PDA- GIS-DGPS-RS cyber-environment was the Feature Extraction module which helped automate the process of performing accurate segmentations for image classification of the sampled An. arabiensis aquatic larval habitats by LULC, thus accounting for larval/pupal development by rice stage. For instance, the FLAASH™ model in- cluded a method for retrieving an estimated haze amount from selected "dark" land covers pixels of sampled An. arabiensis aquatic larval habitats in the QuickBird scene. When looking for a single target, the result is a grayscale image where the pixels that best match the target properties are the darkest (www.ittvis.com). ENVI computed the texture features in the PDA-GIS-DGPS-RS cyber-environment of each individual sampled An. arabiensis aquatic larval habitat attribute in the epidemiological riceland agro-village epidemiological study site and converted raw texture descriptors into spectrally quantifiable data.

The spatiotemporal-sampled georeferenced explanatory field and remote specified An. arabiensis aquatic larval habitat data from ENVI was entered into the Recon X® in the PDA-GIS-DGPS-RS cyber-environment. The Recon X® utilized a touch screen for entering spatiotemporal explanatory field and remote data of An. arabiensis aquatic habitats. All map and data storage in the cyber-environment of the sampled data was done on removable Compact Flash (CF) storage cards. The PDA was configured with 64M RAM, 256M Flash memory, and a 2 GB SanDisk Extreme IV CF card. This process provided adequate backup protection to reload programs using the ecological time series sampled An. arabiensis aquatic habitat data. A CF card reader PDA-GIS-DGPS-RS cyber-environment was then specified to assist the field research team in quickly populating the CF cards with project maps and to transfer data with laptop PCs.

Live data entry pertaining to riceland agroecosystem treatment data of the sampled An. arabiensis aquatic habitats in our bidirectional PDA-GIS-DGPS-RS cyber-environment was performed on the Recon X® and, was transferable to a laptop computer via a wired (or optional wireless) connection. For this study, transfers were done using ActiveSync software that came with the Windows Mobile powered Recon X®. The Mobile VCMS™ software was used to record field data on the Trimble Recon X® which was synchronized automatically with a centralized VCMS database in the PDA-GIS-DGPS-RS cyber-environment running on a laptop computer. For this research project, some of the PDAs also supported an integrated digital camera, allowing the field staff to geocode, map and take pictures of An. arabiensis aquatic larval habitat locations in the study site. Interestingly, the Recon X® included an optional wireless LAN connection feature that we used with Trim Pix™ image technology to connect a Wi-Fi capable Nikon digital camera for automated capturing and geocoding of digital images of the sampled aquatic larval habitats in the PDA-GIS-DGPS-RS cyber-environment. Local area network (LAN) supplies networking capability to a group of computers in close proximity to each other [5]. A LAN in turn connected to the Internet.

Ecological sampled information stored on a Recon X® in the PDA-GIS-DGPS-RS architecture then directly forwarded and/or synchronized with other existing information systems to support further time series An. arabiensis aquatic larval habitat data analyses. A dedicated server was used as the back-end data repository which in this research we used to store collected ecological sampled habitat data, as it is aggregated from multiple Recon PDAs. The back-end dedicated server scenario supported the dissemination of the larval habitat information within the PDA-GIS-DGPS-RS cyber-environment via web-based reporting systems when connected to the Internet.

Web-based technologies have made anonymous, automated, and integrated reporting systems a reality [5].

Moreover, the use of Web-based technologies for time series endemic transmission oriented An. arabiensis aquatic larval habitat data in a robust PDA-GIS-DGPS-RS cyber-environment ensures reporting consistency and data quality, as well as the ability to survey and detect adverse events. This, in turn, provides data that may be measured quantitatively and, analyzed for trends, then used to generate feedback for the quality assurance pro- cess. Thus, web-based technology has the potential to improve workflow and the fluid transmission of vital information on errors that surface from a system perspective, without placing blame on fallible individuals and institutions. Safety may be improved in an ongoing manner with the use of technology, particularly as the effects of systemic changes are being assessed. The use of information technology for oriented time series An. arabiensis aquatic larval habitat data PDA-GIS-DGPS-RS cyber-environment can reduce errors in three ways: (1) adverse events (e.g., flooding, droughts) and errors associated with these events may be prevented before they can occur; (2) the response time needed to resolve the root cause of adverse events can be decreased when implementing gridded sub-meter resolution sampled data feature attributes, thus preventing uncertainty reoccurrence; and (3) LULC trends can be tracked and pertinent feedback about errors and adverse events then can be disseminated to other vector biologists, epidemiologists or research collaborators. Remote monitoring of An. arabiensis aquatic larval habitat data can help simplify therapeutic procedures through a user-friendly mobile device supplemented with interactive audiovisual cues which can help keep track of exchange details using electronic records [8].

While time series web-based reporting systems within PDA-GIS-DGPS-RS cyber-environment are becoming more common in public health for disease surveillance and for tracking interventions such as field sampled An. arabiensis aquatic habitat data, their use for program evaluation is relatively new. Alternatively, a broadband satellite access point could be employed to wirelessly link multiple mobile field data collection devices to a server, via the Field Bridge middleware software. For example, Hughes Network Systems' (HNS) 9201 Broadband Satellite IP Modem and Wi-Fi Access Point can be utilized as part of the field data communications within a robust PDA-GIS-DGPS-RS architecture for data transfer of seasonal-sampled field and/or remote An. arabiensis aquatic larval habitat endemic transmission oriented data to any Internet connected, back-end VCMS database server. The HNS9201 can support multiple field teams to simultaneously send and receive IP packet data of sampled An. arabiensis aquatic habitat data in PDA-GIS-DGPS-RS cyber-environment via USB, Ethernet, ISDN and Wi-Fi interfaces, using the Inmarsat BGAN satellite network and the Recon X® (Figure 4). This type of satellite link can provide 432 kbps IP data (transmit and receive) capability, Integrated Services Digital Network (ISDN) voice (4kbps) ISDN data (64 kbps) and Wi-Fi access point with multi-user capability. Firewall protection and data encryption techniques (e.g., 128 bit Advanced Encryption Standard (AES) Encryption) can prevent hackers from accessing the sampled field/remote endemic transmission oriented data, while firewall and password protection/encryption measures, at the database level can protect the transmission and storage of sampled An. arabiensis aquatic habitat data PDA-GIS-DGPS-RS cyber-environment. All the data managed by such a remote surveillance system can be spatially displayed via the use of GIS software, both as an embedded component of the VCMS database, or in a standalone desktop GIS using industry standard shape files. Traditional shapefiles contain cryptic field names, no rendering information and no metadata [5].

By using a PDA-GIS-DGPS-RS web-based reporting system connected to the Internet, the Trimble Recon X® enables the export or import of data layers completely-with labels, colors, rendering and metadata. Compatible geospatial software such as ESRI ArcGIS Interoperability tools, allows for a seamless transfer of spatiotemporally sampled field and remote specified explanatory An. arabiensis larval habitat endemic transmission oriented information between users of the software. Therefore, a Trimble Recon X® web-based reporting system can easily import complete packages of data and meta-data while viewing all field information sampled. This system will allow a field manager to easily distribute their data in a single file with meaningful variable names and descriptions; metadata and full map color and symbol rendering.

Furthermore, add-in technology can extend the functionality of a remote bidirectional PDA web-based reporting system by allowing vector biologists, epidemiologists and other\data analysts to have the ability to develop and run their own custom add-ins modules while providing a suitable environment to run complex spatial predictive eco-epidemiological risk models. For example, an empirical ecological dataset of georeferenced explanatory spatiotemporal sampled field and remote specified endemic transmission-oriented georeferenced An arabiensis aquatic larval habitat covariate coefficients values in a PDA-GIS-DGPS-RS cyber-environment can be exported from different GIS compatible software, such as spdep from R®, GeoDa®, D2K® Python® and from workflow environments from the National Center for Supercomputing Applications (NCSA)'s Cyber-Integrator which can then be analyzed and imported into a compatible field databases.

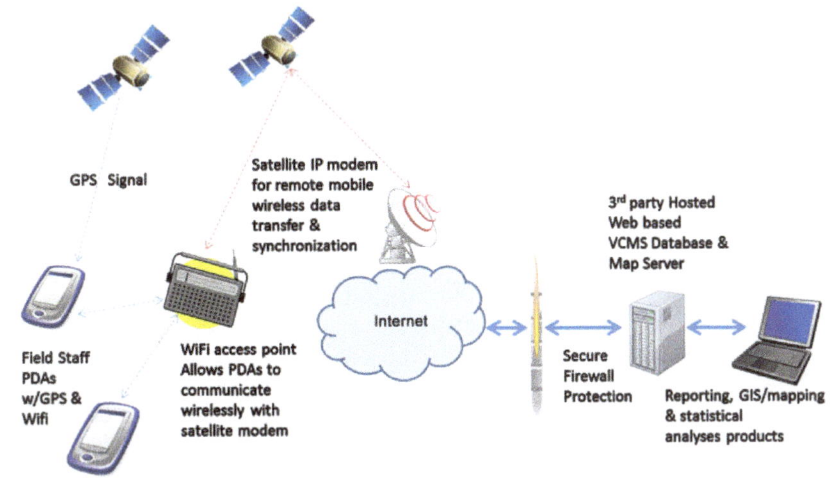

Figure 4. A bidirectional PDA web-based reporting system using a broadband satellite access point and the Internet.

A web-based joint database can then combine the spatiotemporal field and remote-sampled An. arabiensis aquatic habitat data and relevant programmatic information can be used to facilitate evidence-based decision-making, by geographically/non geographically spatiotemporally modeling key covariate coefficient indicators in robust PDA-GIS-DGPS-RS cyber-environments (e.g., parameter estimators of prolific larval habitats based on field and/or remote seasonal sampled data). The key advantages of a web-based joint database within a PDA-GIS-DGPS-RS cyber-infra- structure is: 1) the ability to integrate data (demographic, social, economic weather, etc.) with data that are critical to a project, 2) Ability to share data easily with others by sending with the specific An arabiensis aquatic larval habitat risk map view, the symbology and all data in one file—so they can quickly understand the issue under discussion, 3) Ability for data recipients to repurpose, redisplay and remix the data and, 4) All-time series field and remote data stored and analyzed is in GIS shape file format.

In the foreseeable future, GIS software in a PDA-GIS-DGPS-RS cyber-environment will continue to play essential roles for breaking through scientific challenges in numerous spatiotemporal explanatory georeferencable field and remote sampling frames for parsimoniously quantitating An. arabiensis aquatic larval habitat field and remote sampled endemic transmission oriented data for improving decision-making practices with broad societal impacts (e.g., implementation of IVM). Cyber GIS is fundamentally new software framework under development at Georgia Tech that can provide seamless integration of cyber-infrastructure for GIS, and spatial analysis and malarial time series risk modeling of geo-spatiotemporal sampled An. arabiensis data in a PDA-GIS-DGPS-RS cyber-environment.

Critical pieces of the cyber-infrastructure include supercomputers, high-capacity, mass-storage systems, system software suites and programming environments, scalable, interactive, visualization tools, productivity-enhancing software libraries and tools, large-scale data repositories and digitized scientific data management systems [5], However, fulfilling such roles in a PDA-GIS-DGPS-RS cyber-environment for efficient predictive risk modeling empirical sampled georeferenced field and remote specified An. arabiensis larval habitat explanatory data sets would be dependent on the ability to handle very large geo-spatiotemporal and complex analysis software based on synthesizing computational and spatial thinking enabled by the cyber-infrastructure, which conventional GIS-based software approaches do not provide.

An epidemiologist, vector biologist or other data analyst may establish CyberGIS as a fundamentally new software framework comprising a seamless integration of GIS, and spatial analysis/risk modeling capabilities in the PDA-GIS-DGPS-RS cyber-infrastructure. Specifically, such a project may robustly spatially target seasonally prolific georeferenced field and remote specified An. arabiensis larval habitat explanatory covariate coefficients in the cyber-environment by: 1) engaging in multidisciplinary tactics [e.g., field and remote GIS science, validation metrics derived from real time geostatistical non-linear algorithm] employing evolving CyberGIS software requirements; 2) integrating and sustaining a core set of composable, interoperable, manageable, and reusable CyberGIS software elements based on epidemiological interventional study site -driven and open source strategies; 3) empowering high-performance and scalable CyberGIS by exploiting spatial characteristics of the sampled endemic transmission oriented data and analytical operations for achieving unprecedented capabilities for geospatial risk malarial risk mapping; 4) enhancing an online geospatial problem solving environment to allow for the contribution, sharing, and learning of CyberGIS software by numerous users (e.g., vector biologists, epidemiologists) for fostering the development of education, outreach, and IVM crosscutting multiple disciplines; 5) deploy and test CyberGIS software by linking with national and international cyber-infrastructure to achieve scalability to significant sizes of geospatial problems, and vector malarial mosquito cyber infrastructure resources, and user communities; and 6) evaluating and improving a time series endemic transmission oriented CyberGIS framework through domain science applications and vibrant partnerships to gain better understanding of the complexity of coupled natural and anthropogenic influenced epidemiological ecosystems.

A time series endemic transmission oriented explanatory field and remote specified Cyber GIS software framework in a robust PDA-GIS-DGPS-RS cyber-environment will shift the current paradigm of GIS and associated spatial analysis/ modeling software to create further scalable and sustainable software while achieving groundbreaking scientific advances in understanding human-natural ecosystems associated to immature An. arabiensis that would be impossible otherwise. These advances will dramatically advance the understanding of flooding preparedness and response and impacts for droughts. This framework will empower high-performance and collaborative geospatial problem solving and serve as a key driver for the interoperability of international cyber-infrastructure based on broad engagement of user communities related to GIS for both research and education purposes for risk mapping seasonal An. arabienesis larval habitat field and remote specified endemic transmission oriented data.

In our PDA-GIS-DGPS-RS cyber-environment, the Trimble Recon handheld features a high-performance 400 MHz processor, built-in Bluetooth wireless technology, and built-in wireless LAN. The system provided two Compact Flash (CF) slots, letting us add peripherals such as GNSS cards, barcode scanners, or memory cards. We also had the ability to employ Bluetooth to connect wirelessly to other devices such as a laser rangefinder, a mobile phone for connection to the Internet, or Trimble's GPS Pathfinder® receivers. And since we were within range of a Wi-Fi network, the built-in wireless LAN in the Trimble Recon handheld in the cyber-environment made it very efficient to send and receive spatiotemporal-sampled field and remote sampled An. arabiensis s.s. larval m data. As soon as a Wi-Fi hotspot was discovered we could quickly and securely transfer large amounts of time series dependent endemic transmission oriented field and/or remote geosampled aquatic larval habitat data into the network.

Developing and implementing streamlined data collection, aggregation and reporting methodologies, employing a PDA-GIS-DGPS-RS cyber-environment can provide geographically detailed, real-time information which can lower overall seasonal An. arabiensis aquatic habitat treatment costs. For example, a bidirectional PDA-GIS-DGPS-RS web-based reporting system using a broadband satellite access point and the Internet can provide optimally efficient and timely amounts of relevant field level information. This is important, because riceland An. arabiensis aquatic habitat treatment and management information has the most value when it is quick, clear, easy to understand, and relevant to decisions that need to be made immediately. An adaptable modular information surveillance system in the cyberinfrastrure can help insure that the right decisions are made to reduce the parasite, the vector or both in riceland areas. This system can support dissemination of information such as data about field activities and costs as they are compiled to provide an expedited understanding of how much is being spent in riceland An. arabiensis aquatic habitat treatment, and to determine whether the intended outcomes are being realized. This will allow both field managers and health ministries timely and optimal information to make field operational adjustments and maintain the most efficient and economically feasible pressure on riceland An. arabiensis aquatic habitat populations. For example, once patterns and correlations of spatiotemporal sampled An. arabiensis aquatic habitats are elucidated, field management practices can be modified to optimize applications of fertilizers and pesticides yielding lower overall costs, and minimizing environmental impacts caused by excessive insecticide applications. In addition, a PDA-GIS-DGPS-RS cyber-environment web-based reporting system can provide a module that is dedicated to measuring the economic status of the riceland community and clusters of communities, in order to monitor and measure the impacts of malaria and the economic benefits of malaria control interventions. These optimizations are a critical factor for malaria eradication.

Globally accessible, PDA-GIS-DGPS-RS cyberinfrastructural web-based An. arabiensis aquatic larval habitat maps, that are generated from near real-time map and field data updates, can be incorporated into other ecological sampled riceland datasets for fast and timely analyses. Although this project did not fully utilize web-based data dissemination techniques, the capability are there using the same off-the-shelf technologies employed for our field studies. The logical extension of this field research is to continue to expand the use of mobile field data collection and wireless communication technologies in a PDA-GIS-DGPS-RS cyber-environment to provide a single, seamless and cost effective information clearing house for IVM. Such a system can leverage GIS-mapping functionality, wireless communications and web-based publications. This study provides an important starting point for the creation of standardized techniques for mobile field data collection, global data transmission, data aggregation and display, and analyses of multivariate An. arabiensis aquatic larval habitat models within a robust cyber-environment. The standardization of a logical geodatabase design for implementing IVM will ultimately support linkage with other relational geodatabases, statistical packages and software applications.

The PDA-GIS-DGPS-RS cyberinfrastructural architecture, deployed as part of this project, is the first step in seamlessly sharing geospatially enabled An. arabiensis aquatic habitat data between end users, including field teams, project managers and other researchers. The envisioned web application of this technology would enable: the sourcing and integration of time series An. arabiensis aquatic larval habitat data, supporting the creation of new data sets, and the coordinated delivery of focused decision-supported information for generating reports, charts, maps, tables, using common data exchange formats (e.g., Excel, PowerPoint, SPSS, SAS). Additionally, revisal software can also be installed in the Trimble Recon X® in a PDA-GIS-DGPS-RS cyber-environment which can create An. arabiensis aquatic habitat maps from spreadsheet or database data which can be integrated with data from weather and other scientific studies, i.e. Virtual Earth™. The initial deployment of the custom application described here can ensure rapid deployment and proper site configuration needed to support local decision-making and malaria control efforts. A web-based shared database project PDA-GIS-DGPS-RS cyber- environment can combine a variety of sampled field and remote-sampled mosquito data and relevant program- matic information, in order to facilitate evidence based decision-making for control operations in a real-time or near real-time environment.

Finally, using spatial modeled epidemiological time series data, acquired through a PDA-GIS-DGPS-RS cyberinfrastructural remote bidirectional surveillance system, can simulate disease transmission among An. arabiensis aquatic habitat based on larval/pupal productivity. For instance, Markov simulations models can be created from field sampled data using WinBUGS and its spatial module GeoBUGS which can be facilitated as part of the end-user repository database. Error propagation in these non-linear model estimates can also be spatially quantified using an autocovariate matrix [25]. The first prototype implementation of the Java virtual machine, done at Sun Microsystems, Inc., emulated the Java virtual machine instruction set in software hosted by a handheld device that resembled a contemporary PDA. Thus, accurate models can be developed by sampled immature An. arabiensis aquatic habitat data which can define a set of probabilistic environmental interactions that may affect transitions in aquatic habitats in riceland areas. These riceland An. arabiensis aquatic habitat models in a robust PDA-GIS-DGPS-RS cyberinfrastructural can be made to analysis and display data for predicting habitat locations and for testing and improving various research hypotheses. These models can also provide information about realistic sampling grids, ecological covariates that are involved in high larval/pupal population levels and other transmission dynamics from simple, transparent simulated representations. Geostatistical models of An. arabiensis aquatic habitats derived from PDA-GIS-DGPS-RS cyberinfrastructural data cannot replace mental, qualitative intuition and other intervention techniques, but models can expand into more formal and quantitative realms in which field components and their environmental interactions with productive aquatic habitats are made more specific which can help reduce the overall cost of implementing mosquito control strategies in riceland environments [25].

Geographic information science for seasonal malarial risk mapping in a PDA-GIS-DGPS-RS cyberinfra- structural is a dynamic area in which there are continuous innovations. In remote sensing, for example, increas- ing use of high resolution imagery (e.g., panchromatic QuickBird), and time-series data present both opportuni- ties (e.g., improved characterization of land cover) and challenges (e.g., handling large data volumes, image un- derstanding) for data fusion and integrated data analysis. Enhancements in internet, wireless and satellite commu- nications, and innovations in in-situ sensors, for PDA-GIS-DGPS-RS cyber-infrastructures are paving the way for increasingly robust "real time" applications of remote sensing and GIS, (i.e. telegeoprocessing) for accurate seasonal malarial predictive risk mapping. Web-based tools, such as Google Earth (http://earth.google.com/) and Internet Map Service (IMS) applications PDA-GIS-DGPS-RS cyber-environments, now provide an increasingly larger audience of vector ecologists, epidemiologists and other research collaborators with ready access to geo- spatial malaria-related data while allowing elementary integration of imagery and graphics in PDAs, but more sophisticated implementation of web-based integrated geospatial analysis will require resolution of issues related to metadata standards, data transmission formats, client/server computation and communication protocols for advanced quantification of seasonal sampled malaria-related explanatory covariates.

In conclusion, QuickBird visible and NIR image datasets in a customized PDA-GIS-DGPS-RS cyber-envi- ronment containing digitized habitat boundaries, with unique identifiers displayed all sampled habitats within a 1 km buffer of the study site. An object-oriented classification in ENVI technology separated the sampled habi- tats by larval/pupal productivity. Thereafter, a regression analyses revealed significantly higher An. arabiensis larval/pupal counts in the post-transplanting stage of rice development in the cyber-environment. All field and remote-sampled explanatory variables were entered into the PDA-GIS-DGPS-RS cyberinfrastructural architec- ture. The Recon X displayed all treated and untreated An. arabiensis aquatic habitats in the study site. Recon X integrated solutions can enable field-workers to capture observation and location data at the same time. GIS/ GPS/and Recon X technological integration may be used to build new databases, maintain database accuracy, and conduct spatial analysis in the field. A Recon X unit can be attached to a laptop for direct data input into a PDA-GIS-DGPS-RS cyberinfrastructural server. The solution can be further synchronized with applications that include digital photography and date and time of data capture. A web-based joint database within a robust PDA- GIS-GPS-RS cyber-environment can ensure timely deliverable of explanatory seasonal mosquito sampled data. This platform lends itself to a larger public health capability and has the ability to scale-in data, providing an easily accessible tool linking data from handheld to desktop to web. This database extends current field sam- pling technology by providing the ability to transfer geographic information from either a centralized repository or directly between users. Data-based location intelligence aggregated and analyzed in PDA-GIS-GPS-RS cy- ber-environment servers can provide effective dissemination of information which will lead to improved pro- gram management and therefore impact.

Benjamin G. Jacob, Robert J. Novak (e.g., roads, rivers, towns, political units) in a seamless, stand-alone Windows-based application or into a web-based environment that would express data in simple and engaging ways across the malaria sector. Unlike simple habitat mapping that just plots points on a map; a PDA-GIS- DGPS-RS cyberinfrastructural web-based reporting system can create dynamic maps that are interactive. Trimble Recon X® technology can plot results-based information on a collective mapping system designed to help malaria researchers assess and plan improved programs. Mobile field data collection and reporting, using PDA-based technologies, should be part of a tailored approach for reducing the immature population of An. arabiensis aquatic larval habitats through well-timed/targeted spraying. Off-the-shelf mobile field computing technologies, such as the Windows Mobile powered Trimble Recon X® configured with user-friendly data entry software and spatial data such as QuickBird visible and NIR data can play an important role in supporting malaria control field activities.

REFERENCES

1. Sattler, M.A., Mtasiwa, D., Kiama, M., Premji, Z., Tanner, M., Killeen, G.F. and Lengeler, C. (2005) Habitat Characterization and Spatial Distribution of Anopheles sp Mosquito Larvae in Dar es Salaam (Tanzania) during an Extended Dry Period. Malaria Journal, 4, 4-11. http://dx.doi.org/10.1186/1475-2875-4-4 [Paper reference:19]

2. Jacob, B.G., Muturi, E.J., Funes, J., Shililu, J., Githure, J.I., Regens, J.L. and Novak, R.J. (2007) Using Imaging Technologies to Control Malaria. Imaging Notes, 22, 17-22. [Paper reference:67]

3. Gu, W. and Novak, R.J. (2005) Habitat-Based Modeling of Impacts of Mosquito Larval Interventions on Entomological Inoculation Rates, Incidence, and

Prevalence of Malaria. American Journal of Tropical Medicine and Hygiene, 11, 546-552. [Paper reference:17]

4. CDC (Centers for Disease Control and Prevention). The President's Malaria Initiative (IPM) Rolls into Action. http://www.cdc.gov. [Paper reference:5]

5. Jensen, J.R. (2005) Introductory Digital Image Processing: A Remote Sensing Perspective, 3rd Edition, Prentice-Hall, Upper Saddle River. [Paper reference:40]

6. Environmental Systems Research Institute. http://www.esri.com [Paper reference:12]

7. Digital Globe LTD. http://digitalglobe.com [Paper reference:13]

8. Muturi, E.J., Mwangangi, J., Shililu, J., Muriu, S.M., Jacob, B.G., Kabiru, E.W., Gu, W., Mbogo, C.M., Githure, J. and Novak, R.J. (2007) Mosquito Species Succession and Physicochemical Factors Affecting Their Abundance in Rice Fields in Mwea, Kenya. Journal of Medical Entomology, 44, 336-344. http://dx.doi.org/10.1603/0022-2585(2007)44[336:MSSAPF]2.0.CO;2 [Paper reference:7]

9. Jacob, B.G., Nelson, P., Lampman, R., Morris, J., Raim, A., Funes, J., LaPointe, C. and Novak, R.J. (2006) Comparing GPS Technology for Identifying Spatial Ecological Variation for Urban Mosquito Management. Wing Beats, 16, 30- 33. [Paper reference:6]

10. Gillies, M.T. and Coetzee, M. (1987) A Supplement to the Anophelinae of Africa South of the Sahara. South African Institute for Medical Research, 55, 1-143. [Paper reference:2]

11. Service, M.W. (1993) Mosquito Ecology: Field Sampling Methods. 2nd Edition, Elsevier Publishers, Essex. [Paper reference:2]

12. Edwards, F.W. (1941) Notes on British Mycetophilidae. [Paper reference:1]

13. Bioequip Products Inc. www.bioquip.com [Paper reference:1]

14. Meritt, R.W. and Cummins, K.W. (1996) An Introduction to the Aquatic Insects of North America. 3rd Edition, Kendall Hunt Editor, Dubuque. [Paper reference:1]

15. ERDAS. http://geospatial.intergraph.com/Homepage.aspx [Paper reference:1]

16. Rejmankova, E., Roberts, D., Pawley, A., Manguin, A. and Polanco, S. (1995) Predictions of Adult Anopheles albimanus Densities in Villages Based on Distances to Remotely Sensed Larval Habitats. American Journal of Tropical Medicine and Hygiene, 53, 482-491. [Paper reference:1]

17. Beck, L.R., Rodriguez, M.H., Dister, S.W., Rodriguez, A.D., Rejmankova, E., Ulloa, A., Meza, R.A., Roberts, D.R., Paris, J.F. and Spanner, M.A. (1994) Remote Sensing as a Landscape Epidemiologic Tool to Identify Villages at High risk Malaria Transmission. American Journal of Tropical Medicine and Hygiene, 51, 271-280.

18. Wood, B.R., Washino, R., Beck, L., Hibbard, K., Pitcairn, M., Roberts, D., Rejmankova, E., Paris, J., Hacker, C., Salute, J., Sebesta, P. and Legters, L. (1991) Distinguishing High and Low Anopheline-Producing Rice Fields Using Remote Sensing and GIS Technologies. Preventive Veterinary Medicine, 11, 277-288. http://dx.doi.org/10.1016/S0167-5877(05)80014-5 [Paper reference:1]

19. Clarke Mosquito Control Products, Inc., Roselle, IL for the VCMS, Mobile VCMS and FieldBridge Software. https://www.clarkemosquito.com/vcms_tech_support [Paper reference:1]

20. Jacob, B.G., Muturi, E.J., Funes, J.E., Shililu, J.I., Githure, J.I., Kakoma, I.I. and Novak, R.J. (2006) A Grid-Based Infrastructure for Ecological Forecasting of Rice Land Anopheles arabiensis Aquatic Larval Habitats. Malaria Journal, 5, 941-947.http://dx.doi.org/10.1186/1475-2875-5-91 [Paper reference:1]

21. Dolo, G., Briët, O.J., Dao, A., Traoré, S.F., Bouaré, M., Sogoba, N. and Touré, Y.T. (2004). Malaria Transmission in Relation to Rice Cultivation in the Irrigated Sahel of Mali. Acta Tropica, 89, 147-159. http://dx.doi.org/10.1016/j.actatropica.2003.10.014 [Paper reference:1]

22. Green, K., Kempka, D. and Lackley, L. (1994) Using Remote Sensing to Detect and Monitor Land Cover and Land Use Changes. Photogrammetric Engineering and Remote Sensing, 60, 331-337. [Paper reference:5]

23. Congalton, R.G. (1991) A Review of Assessing the Accuracy of Classifications of Remotely Sensed Data. Remote Sensing of Environment, 37, 35-46.http://dx.doi.org/10.1016/0034-4257(91)90048-B [Paper reference:3]

24. Arbia, G. (1988) Spatial Data Configuration in the Statistical Analysis of Regional Economics and Related Problems. Kluwer, Dordrecht. [Paper reference:3]

25. Jacob, B.G., Griffith, D., Gunter, J., Muturi, E.J., Caamano, E., Shililu, J., Guthure, J., Regens, J. and Novak, R.J. (2009) A Spatial Filtering Specification for an Auto-Negative Binomial Model of Anopheles arabiensis Aquatic Habitats. Transactions in GIS, 12, 515-539. http://dx.doi.org/10.1111/j.1467-9671.2008.01110.x [Paper reference:6]

CHAPTER 14

SHIFTED LINEAR INTERPOLATION FILTER

H. Olkkonen and J. T. Olkkonen

Keywords: Linear Interpolation, Cubic Convolution, B-splines, VLSI

INTRODUCTION

Linear interpolation plays an important role in modern signal and image processing applications due to its simple implementation and low computational cost. Plonka [1,2] has extensively studied the effect of the shifted interpolation nodes on the approximation error in the B-spline interpolation. According to the theoretical analysis the shift $\tau = 0$ yields the minimal error norm. Blu *et al.* [3] applied the same idea to the piecewise-linear interpolation. On the contrary, they observed that interpolation error diminishes if the sampling knots are shifted by a fixed amount. Through the theoretical considerations Blu *et al.* [3] determined the optimal shift of the knots is close to $\tau = 1/5$ of the distance between the knots.

In this work we consider a modification of the linear interpolation. In standard linear interpolation the kernel is the second order B-spline $\beta_2(t)$. Instead of shifting the sampling knots we introduce the shifted linear interpolation (SLI), where the B- spline kernel has been time-shifted by a fraction D of the sampling period. We describe the z-transform SLI filter for efficient implementation of the shifted B-spline kernel $\beta_2(t - D)$ especially for VLSI and microprocessor circuits.

THEORETICAL CONSIDERATIONS

B-Spline Signal Processing

B-spline approximation of the continuous signal $x(t)$ is based on the summation [4]

$$x(t) = \sum_k c[k]\beta_p(t-k) \tag{1}$$

where $c[k]$ are the scale coefficients. The B-spline $\beta_p(t)$ is constructed by convolving an indicator function $\beta[t]$ by itself

$$\beta_p(t) = \underbrace{\beta(t) * \beta(t) * \cdots \beta(t)}_{\text{p times}} \tag{2}$$

where

$$\beta(t) = \begin{cases} 1 & 0 \le t \le 1 \\ 0 & \text{elsewhere} \end{cases} \tag{3}$$

The linear interpolation is obtained in the case $p = 2$

$$x(t) = \sum_k c[k]\beta_2(t-k) \tag{4}$$

where $\beta_2(t)$ is the triangle function Figure 1 defined as

$$\beta_2(t) = \begin{cases} t & \text{for } 0 \le \text{t} < 1t \\ 2-t & \text{for } 1 \le \text{t} \le 2 \\ 0 & \text{elsewhere} \end{cases} \tag{5}$$

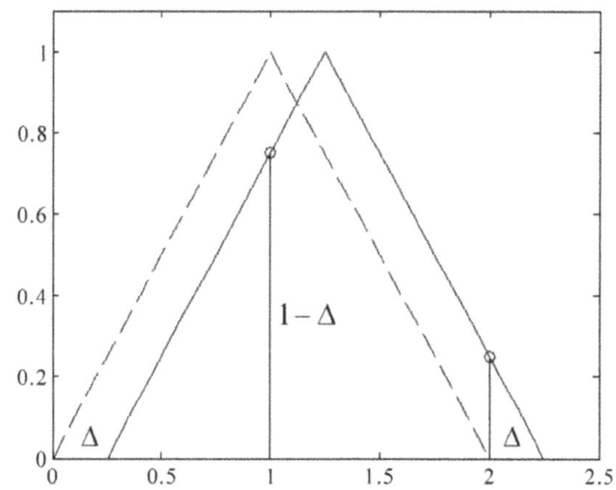

Figure 1. The shifted B-spline $\beta_2(t - D)$.

Citation: H. Olkkonen and J. Olkkonen. "Shifted Linear Interpolation Filter," Journal of Signal and Information Processing, Vol. 1 No. 1, 2010, pp. 44-49. DOI: 10.4236/jsip.2010.11005.

Shifted Linear Interpolation

Let us consider the approximation of the continuous-time signal $x(t)$ by a summation of the shifted B-spline $\beta_2(t - D)$

$$x(t)=\sum_k c[\Delta,k]\beta_2(t-\Delta-k) \tag{6}$$

where c[D, k] D Î [0,1] is the scale sequence for the time-shifted kernel. In the sequel we call (6) the shifted linear interpolation (SLI). Usually the data points are known in discrete time intervals $t = nT$ (n = 0,1,2,···). In the z-transform domain we obtain

$$Z\{x[n]\}=X(z)=C(\Delta,z)Z\{\beta_2(t+\Delta)\}$$
$$=C(\Delta,z)\left[(1-\Delta)+\Delta z^{-1}\right] \tag{7}$$

The SLI filter is now obtained from (6) as

$$Z\{x[n-d]\}=C(\Delta,z)\left[(1-\Delta-d)+(\Delta+d)z^{-1}\right] \tag{8}$$

By inserting the scale sequence C(D, z) from (7) we finally have

$$Z\{x[n-d]\}=\frac{(1-\Delta-d)+(\Delta+d)z^{-1}}{1-\Delta+\Delta z^{-1}}X(z)$$
$$=SLI(d,\Delta,z)X(z) \tag{9}$$

where SLI(d, D, z) denotes the SLI filter.

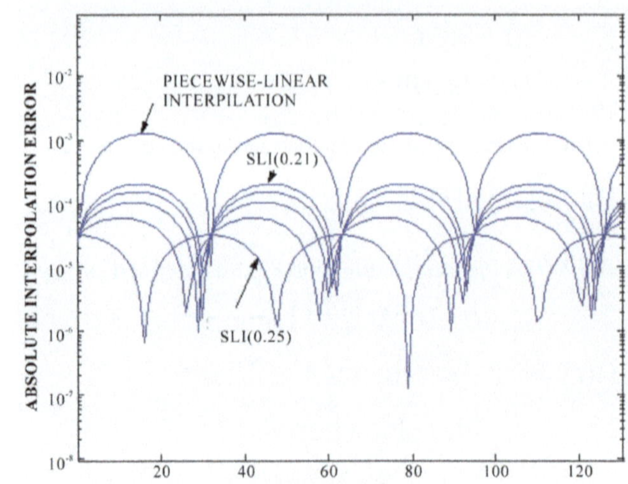

Figure 2. Interpolation error of function sin (0.1 n) in the case d = 0.5 for piecewise-linear interpolation (Δ = 0) and for SLI filter from Δ = 0.21 to Δ = 0.25.

Inverse SLI Filter

In some applications it is required that the original signal has to be reconstructed from the delayed one. In the general case the SLI filter (9) has no stable inverse since the roots of the nominator may lie outside the unit circle. For D = 0.25 the roots are inside the unit circle in the range 0 £ d £ 0.25. The inverse SLI filter can be realized by the inverse filtering procedure described in Appendix I.

EXPERIMENTAL

Experimental Verification of the SLI Algorithm

We tested experimentally the quality of the SLI filter for a variety of synthetic signals with varying frequency components. The interpolation error was defined as the absolute difference between the analytically calculated and interpolated signal: $e(t)=|x(t)-x_{\text{int}}(t)|$. The timeshift of the interpolation kernel was altered in the range 0 £ D £ 1. The interpolation error was strongly related to the time-shift. Irrespective of the type of the signal the minimum interpolation error was obtained in every case at D = 0.25. Figure 2 shows an example of the interpolation error versus D for signal sin(0.1 t), when d = 0.5. The worst case is the linear interpolation (D = 0). Then the interpolation error diminishes from D = 0.21 to D = 0.25. At bigger D values the interpolation error starts to increase.

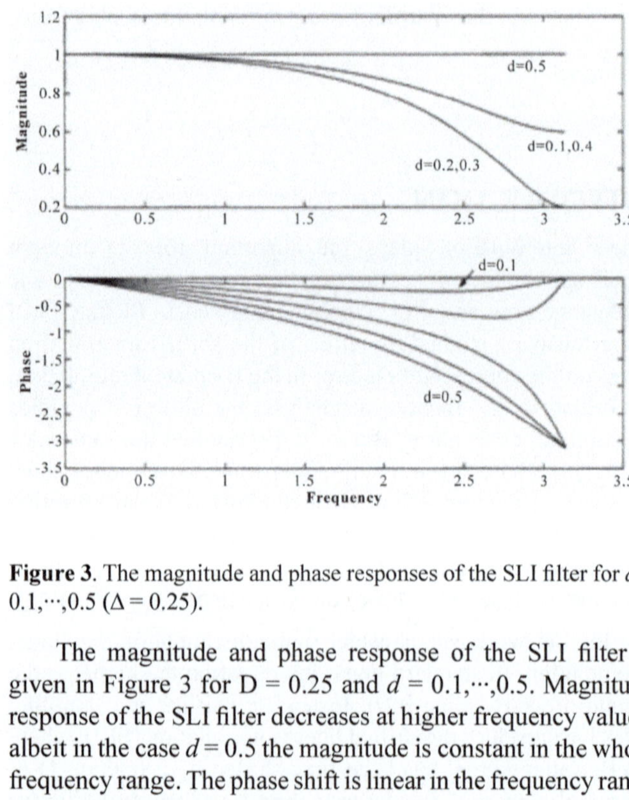

Figure 3. The magnitude and phase responses of the SLI filter for d = 0.1,···,0.5 (Δ = 0.25).

The magnitude and phase response of the SLI filter is given in Figure 3 for D = 0.25 and d = 0.1,···,0.5. Magnitude response of the SLI filter decreases at higher frequency values, albeit in the case d = 0.5 the magnitude is constant in the whole frequency range. The phase shift is linear in the frequency range 0 £ w £ 1. For d > 0.5 and D = 0.25 the magnitude response of the SLI filter turns to high-pass and tends to intensify the noise occurring in high frequency range. As a practical solution the interpolated signal can be time-reversed and 1- d can be inserted instead of d in the SLI algorithm.

Half Delay Filter

An interesting application of the SLI is the half delay filter, which delays the signal half of the sampling period and interpolates the intermediate points between the evenly spaced knots. For D = 0.25 and d = 0.5 we have

$$SLI(0.5,0.25,z)=\frac{(1/3)+z^{-1}}{1+(1/3)z^{-1}} \tag{10}$$

which is an all-pass filter having the unity magnitude response. The interpolation error of the half delayed filter (11) is especially low Figure 2. Figure 4 shows the phase response and the phase error compared with the ideal phase response.

Comparison with the Cubic Convolution Interpolation

The absolute interpolation error using the cubic convolution method [5] and the SLI-algorithm for signals $x[n] = \sin(n)$ and $x[n] = \sin(2n)$ are given in Figures 5 and 6 for $d = 0.5$ and D = 0.25. At $w = 1$ the cubic convolution interpolation performs slightly better, but at $w = 2$ the quality of the SLI algorithm equals the cubic convolution.

DISCUSSION

The linear interpolation has been used in many microprocessor and VLSI applications due to its simple and fast implementation. The new approaches based on higher degree B-spline interpolation kernels [6-8] suffer from computational complexity and usually require the time reversed inverse filtering procedure, which is not possible to implement in real-time. In this work we examined the shifted linear interpolation, which outperforms the standard linear interpolation Figure 2. The SLI algorithm can be implemented via the z-transform filter (9) with integer coefficients. The interpolation error of the SLI filter is comparable to the more elaborate cubic convolution interpolation Figures 5,6, which is a standard image and signal processing tool. The computational complexity of the cubic convolution method is significantly higher compared with the SLI algorithm.

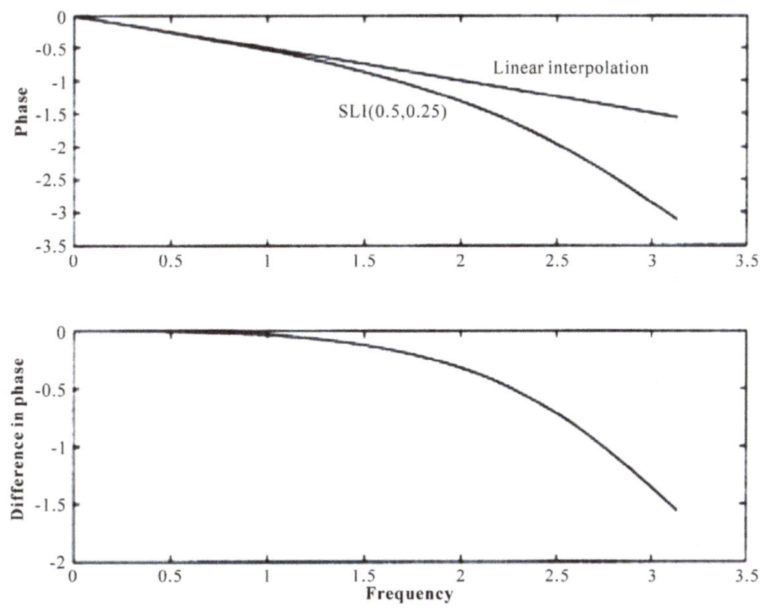

Figure 4. Comparison of the phase responses of the piecewise-linear interpolation and the SLI filter(d = 0.5, D= 0.25).

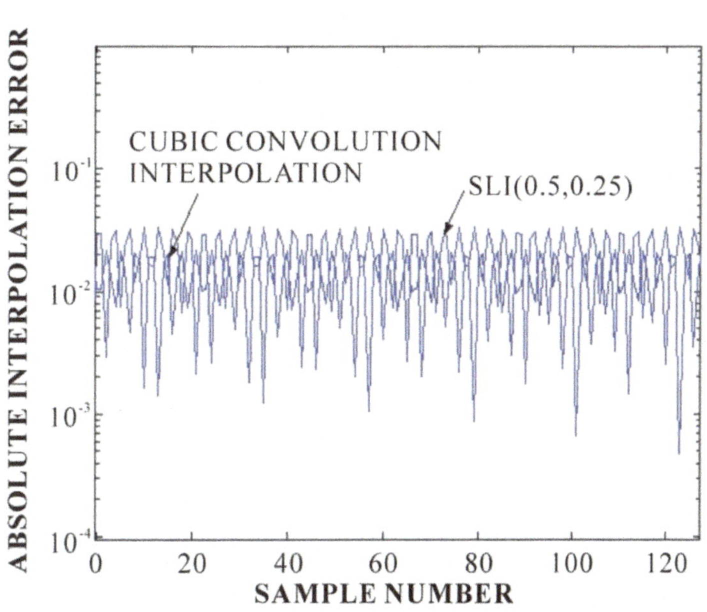

Figure 5. The absolute interpolation error yielded by the SLI algorithm ($d = 0.5$, D = 0.25) and the cubic convolution interpolation for the sinusoidal signal $x[n] = sin(n)$.

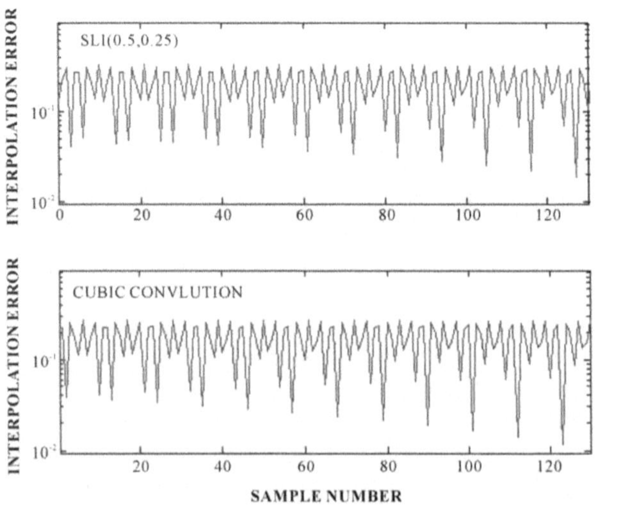

Figure 6. The absolute interpolation error yielded by the SLI algorithm (d = 0.5, Δ = 0.25) and the cubic convolution interpolation for the sinusoidal signal $x[n]$ = $sin(2n)$.

The interpolation function used as a kernel in the SLI algorithm correspond the B-spline order p = 2 Figure 1. The B-spline $\beta_2(t)$ has the z-transform (1 - D) + Δz^{-1}. Probably the main reason for the effectiveness of the SLI comes from the two term z-transform, which allows the solution of the scale coefficient sequence C(D, z), in (7). This leads to the definition of the SLI filter (9). As a surprising experimental observation of the SLI algorithm is the strong dependence of the interpolation error on the time shift D. Our comparisons using a variety of functions with different waveforms and frequency content showed that every time the interpolation error attains its minimum at D = 0.25. Previously Plonka [1,2] has studied the B-spline interpolation with shifted nodes. In the theoretical analysis the shift parameter t = 0 yielded the minimal error norm. Blu et al. [3] conducted theoretically the piecewise-linear interpolation error versus the shift t in the measurement knots. Their analysis predicted the optimal choice t = 0.21. Blu et al. [3] used a pre-filter to shift the measurement knots and then the standard piecewise linear interpolation for computation of the interpolated signal. An advantage in this procedure is that the prefiltered signal can be directly implemented by any existing linear interpolation hardware, while the implementation of the SLI filter needs a new hardware design.

An essential observation in this work is that the SLI filter has low-pass frequency response characteristics only in the range 0 £ d £ 0.5. For d > 0.5 the frequency response of the SLI filter turns to high-pass and intensifies the noise occurring in high frequency range. The noise interference for d > 0.5 can be avoided by time reversing the interpolated signal. Then 1- d can be inserted instead of d in the SLI algorithm.

A shortcoming in the SLI algorithm is the phase error occurring in the high frequency range compared with the linear interpolation. Fortunately, in most of the signal processing applications the signal is band limited and the high frequency band 2 £ w £ p contains only some noise components.

An interesting observation is that for d = 0.5 and D = 0.25 the half delay SLI algorithm reduces to the simple all-pass filter (10), which has the unity magnitude response and the phase response is close to the ideal phase response in the range 0 £ w £ 1. If the signal bandwidth is limited to this range the SLI algorithm yields interpolation precision and accuracy, which is comparable to the more elaborate interpolation methods. The interpolation of intermediate points between the uniformly distributed knots is probably the most frequently used operation is image and signal processing. The half delay filters are also important tools in construction of shift invariant wavelet transform [9]. The half delay SLI filter (10) can be readily applied to replace the more elaborate half delay filters based on the Thiran filters [10]. For 0 £ d £ 0.25 and D = 0.25 the SLI filter has a stable inverse. If the original data points have to be reconstructed from the interpolated ones, the inverse SLI filter for higher values of d can be realized by using the inverse filtering procedure (Appendix I).

CONCLUSIONS

In signal and image processing applications linear interpolation has been frequently used due to its simple implementation and low computational cost. In standard linear interpolation the kernel is the second order B-spline $\beta_2(t)$. We show in this work that the interpolation error can be remarkably reduced by adapting the time-shifted B-spline $\beta_2(t$ - D) as an interpolation kernel. By experimental tests we verify that the optimal shift is D = 0.25. The shifted linear interpolation (SLI) algorithm can be effectively implemented by the z-transform filter in VLSI and microprocessor applications. The interpolation error of the SLI filter is comparable to the more elaborate cubic convolution interpolation. The future research goal is to extend the time-shifted interpolation kernel to higher order B-splines.

APPENDIX I

Implementation of the inverse filter $SLI^{-1}(d$, D, $z)$ for D = 0.25 and d > 0.25.

By denoting the interpolated signal by $Y(z) = Z\{x[n-d]\}$ the inverse filter $SLI^{-1}(d$, D, $z)$ is

$$SLI^{-1}(d,\Delta,z)=\frac{1-\Delta+\Delta z^{-1}}{(1-\Delta-d)+(\Delta+d)z^{-1}}=\frac{X(z)}{Y(z)} \tag{11}$$

The inverse filter $SLI^{-1}(d$, t, $z)$ is not stable which is not stable for D = 0.25 and d > 0.25, since the root in the denominator is outside the unit circle. However, the inverse filter can be implemented by the following inverse filtering procedure. First we replace z by z^{-1}

$$SLI^{-1}(d,\Delta,z^{-1})=\frac{X(z^{-1})}{Y(z^{-1})}=\frac{\Delta+(1-\Delta)z^{-1}}{(\Delta+d)+(1-\Delta-d)z^{-1}} \tag{12}$$

The inverse filter $SLI^{-1}(d$, D, $z^{-1})$ is stable for D = 0.25 and d > 0.25. The $X(z^{-1})$ and ( ) 1 Y z- are the z-transforms of the time reversed signal $x[N - n]$ and the interpolated signal $y[N - n]$. The following Matlab program ISLI.m demonstrates the computation procedure.

```
function x = ISLI(y,d,delta)
y = y(end:-1:1);

x = filter([delta 1-delta],[delta+d 1-delta-d],y);
x = x(end:-1:1);
```

REFERENCES

1. G. Plonka, "Periodic Spline Interpolation with Shifted Nodes," *Journal of Approximation Theory*, Vol. 76, No. 1, 1994, pp. 1-20.
2. G. Plonka, "Optimal Shift Parameters for Periodic Spline Interpolation," *Numerical Algorithms*, Vol. 6, No. 2, 1994, 297-316.
3. I. J. Schoenberg, "Contribution to the Problem of Approximation of Equidistant Data by Analytic Functions," *Quarterly of Applied Mathematics*, Vol. 4, No. 1, 1946, 45-99, 112-141.
4. T. Blu, P. Thevenaz and M. Unser, "Linear interpolation revitalized", IEEE Trans. Image Process. Vol. 13, No. 5, 710-719, May 2004.
5. R. G. Kay, "Cubic Convolution Interpolation for Digital Image Processing," *IEEE Transactions on Acoustics, Speech, Signal Processing*, Vol. 29, No. 6, December 1981, pp. 1153-1160.
6. P. Thevenaz, T. Blu and M. Unser, "Interpolation Revisited," *IEEE Transactions on Medical Imaging*, Vol. 19, No. 7, July 2000, pp. 739-758.
7. J. T. Olkkonen and H. Olkkonen, "Fractional Delay Filter Based on B-Spline Transform," *IEEE Signal Processing Letters*, Vol. 14, No. 2, February 2007, pp. 97-100.
8. J. T. Olkkonen and H. Olkkonen, "Fractional Time-Shift B-Spline Filter," *IEEE Signal Processing Letters*, Vol. 14, No. 10, October 2007, pp. 688-691.
9. H. Olkkonen and J. T. Olkkonen, "Half Delay B-Spline Filter for Construction of Shift Invariant Wavelet Transform," *IEEE Transactions on Circuits and Systems II*, Vol. 54, No. 7, July 2007, pp. 611-615.
10. I. W. Selesnick, "The Design of Approximate Hilbert Transform Pairs of Wavelet Bases," *IEEE Transactions on Signal Processing*, Vol. 50, No. 5, May 2002, pp. 1144-1152.

CHAPTER 15

EMBEDDED UHF RFID TAG FOR DURABILITY MONITORING IN CONCRETE

Jean-Marc Laheurte, Aladdin Kabalan, Houssam Retima, Eric Piedallu, Fulvio Michelis, and Bérengère Lebental

Keywords: UHF RFID Technologies, Sensing RFID, UHF Tag Antennas, Concrete Monitoring

INTRODUCTION

Structural deficiencies in concrete structures often originate at their core from micro scale defects, whose detection is the key to predict structural ageing. For this purpose, a flexible strain gauge based on carbon nanotubes on Ethylene Tetrafluoroethylene (ETFE) was recently proposed by the authors showing highly reproducible, hysteresis-free performances [1]. However, the in-situ, real-time detection of such defects and more generally, the embedded monitoring of concrete structures remain a major technological challenge.

Recent researches demonstrated the feasibility of wirelessly powering nodes embedded in challenging environment such as soil and concrete [2] - [5]. UHF RFID is particularly appealing in this regard as embedded RFID tags can either be used in their battery-less mode or require very low energy in their battery-assisted mode. In [6], a dedicated RFID tag design was proposed where the upper part of the antenna was not inside but at the surface of a concrete soil. The tag reading was possible up to 12.8 meters for a −15.3 dB·m chip sensitivity but no sensing capabilities were associated to the tag. Meanwhile, chip manufacturers have developed a new generation of recently commercialized battery-assisted UHF RFID chip integrating communication and sensing capabilities [7] [8]. This paper presents the first description of the wireless communication part of a strain sensor node buried inside a concrete structure and built around one of these chips. It includes the design and fabrication issues as well as an experimental validation and a sensitivity analysis upon the variability of the concrete characteristics.

Citation: Laheurte, J., Kabalan, A., Retima, H., Piedallu, E., Michelis, F. and Lebental, B. (2016). Embedded UHF RFID Tag for Durability Monitoring in Concrete. *Wireless Sensor Network*, 8, 137-144. DOI: 10.4236/wsn.2016.87012.

DESCRIPTION OF THE UHF RFID TAG

The tag shown in Figure 1 has been optimized for its embedding into mortar. It is built around the battery-as- sisted SL900A chip by AMS [8] showing a 31 μW sensitivity in its battery mode. This chip has a fully integrated temperature sensor and an interface compliant for two additional external sensors including an Analog- to-Digital Converter (ADC). The SL900A does not provide protection against DC short circuits between the antenna pads. In particular, the matching circuit included in the antenna design should not be terminated with a short circuit, but with an open one.

The tag also includes a microstrip dipole antenna and a low-power signal conditioning circuit [9] used to adjust the sensor output to the input requirements of the ADC (Figure 1). Power is supplied by a 3 V button cell battery (CR1225) offering a good compromise between size, capacity and lifetime. The use of a battery is acceptable for this application as sensor interrogation is required only weekly. The antenna, the conditioning circuit and the RFID chip are etched or soldered on the same low-cost dual-sided PCB with 35 μm thick copper metallization and FR4 dielectric ($\varepsilon_r = 4.25$, $\tan\delta = 0.02$, thickness = 1.54 mm).

The circuitry is protected by a 53 mm × 53 mm Teflon® housing ($\varepsilon_r = 1.78$, $\tan\delta = 0.08$, thickness = 3 mm) to insulate the device from the harsh concrete environment. The housing includes a 5.5 mm air layer to further improve the radiation efficiency of the antenna. The housing has an opening on its side for the passage of the wires to connect an external strain gauge exposed to concrete outside. This opening is sealed with silicone for water- proofing.

The tag can communicate with any reader using the proprietary protocol EPC Gen2 Class 3 [8], an evolution of the EPC Gen2 protocol dedicated to UHF RFID sensor tags.

ANTENNA DESIGN IN A MORTAR BLOCK

Like mortar, concrete is a mixture of sand, cement, and water, but it also contains rock chippings or gravel which makes it much stronger and more durable than mortar. Therefore, mortar and concrete might have slightly different electrical properties depending on the aggregates type and density. Mortar is used instead of concrete in the experimental characterization because its homogeneity is controlled and its electrical characteristics are known.

For the simulations and the measurements, a (12 cm × 12 cm × 16 cm) block of mortar is considered. The upper side of the tag housing is located 5 cm below the block interface. The tag is centered with respect to the mortar surface.

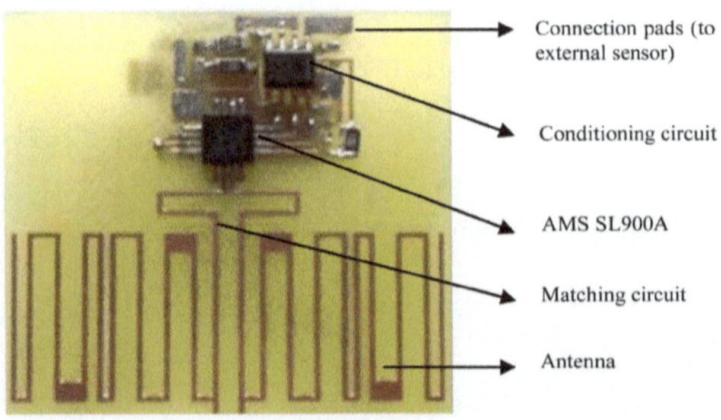

Figure 1. Assembled UHF RFID tag.

To avoid the disruption that the device may cause in the concrete structure in which it is embedded, a maximum device footprint of 5 cm × 5 cm is set as the maximum diameter of the aggregates regularly used in the construction industry is typically around 5 cm [10] [11].

To match the chip internal series impedance of (31-j320) Ω around 865 MHz [8], a T-matching network is associated to a meandered dipole using classical design rules [12] for an optimized bandwidth over a limited surface (Figure 2). The antenna parameters are optimized in the European UHF band (865 - 868 MHz) with ANSYS HFSS 14 [13] including the Teflon box and the block of mortar. The values of all tag dimensions are summarized in the legend of Figure 2. The relative permittivity $\varepsilon_r = 5$ and conductivity $\sigma = 0.01$ S/m of mortar correspond to the properties of a dry mortar at 900 MHz extracted from the literature [14].

READING RANGE FORMULATION

The maximum Reading Range (RR) of a tag is calculated from the following formula [15] derived from the Friis equation:

$$RR = \frac{\lambda}{4\pi}\sqrt{\frac{P_{EIRP}G_{tag}\left(1-|\Gamma|^2\right)}{P_{th}}} \tag{1}$$

$P_{EIRP} = 3.28$W is the maximum regulated effective isotropic power radiated by the reader's antenna. P_{th} is the chip sensitivity (31 μW for the SL900A in its battery mode). G_{tag} is the gain of the tag antenna. Γ is defined as:

$$\Gamma = \frac{Z_A - Z_{chip}^*}{Z_A + Z_{chip}} \tag{2}$$

where Z_{chip} is the internal chip impedance and Z_A the complex impedance of the tag antenna.

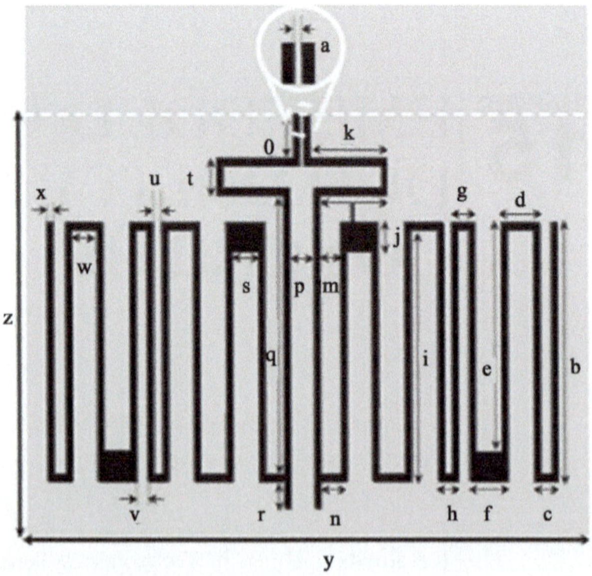

Figure 2. Schematics of the tag antenna. All line widths are 5 mm except f = 1.5 mm. Line lengths: a = 0.3 mm, b = 17.15 mm, c = 2 mm, d = 3.15 mm, e = 15.15 mm, f = 3.15 mm, g = 1.98 mm, h = 1.67 mm, i = 16.65 mm, j = 1.5 mm, k = 6.35 mm, l = 5.5 mm, m = 1.6 mm, n = 2.1 mm, o = 2.9 mm, p = 2 mm, q = 18.38 mm, r = 1.72 mm, s = 2.15 mm, t = 2.5 mm, u = 0.67 mm, v = 0.98 mm, w = 2.15 mm, x = 0.5 mm, y = 43 mm, z = 26.5 mm.

In order to deliver the maximum power to the antenna, the reflection coefficient Γ has to be minimized, i.e., $Z_A = Z_{chip}^*$ (where * indicates the complex conjugate).

In this work, the theoretical RR is estimated by using the simulated values of Z_A and G_{tag} provided by HFSS. The experimental RR is obtained experimentally with the Tagformance UHF RFID measurement system commercialized by Voyantic [16]. Operated at a pre-set distance of 30 cm between the reader antenna and the tag, the Tagformance system measures the minimum effective isotropic power P_{EIRP}^{min} required to wake-up and read the tag. Using Friis equation, one finds:

$$0.3 = \frac{\lambda}{4\pi}\sqrt{\frac{P_{EIRP}^{min}G_{tag}\left(1-|\Gamma|^2\right)}{P_{th}}} \tag{3}$$

The 30 cm distance is chosen so that the measurements are carried out in far field for the frequency range of interest. It is achieved using a foam cube (30 cm ×30 cm) whose electromagnetic characteristics are close to air. The experimental RR provided by the measurement system is finally obtained by combining Equation (1) and Equation (3):

$$RR = 0.3\sqrt{P_{EIRP}/P_{EIRP}^{min}} \tag{4}$$

SENSITIVITY ANALYSIS OF THE READING RANGE UPON THE VARIABILITY OF THE CONCRETE CHARACTERISTICS

Figure 3 and Figure 4 emphasize respectively the influence of the mortar relative permittivity (ε_r) and conductivity (σ) on RR.

It can be used to anticipate the behavior of RR during the drying process of concrete. Indeed, ε_r and σ values [14] increase with the percentage of water in concrete, which decreases with time. Values such as $\varepsilon_r = 8$ and $\sigma = 0.08$ S/m correspond to the early stage of mortar with a 12% water ratio; $\varepsilon_r = 5$ and $\sigma = 0.01$ S/m correspond to an advanced stage in the drying process with a water percentage of 2.8%. RR is normalized to its maximum value (1.55 m for Figure 3 and 1.3 m for Figure 4).

As shown in Figure 3, the antenna is well tuned and sufficiently wide band to accommodate the resonance shift introduced by the change in permittivity (ε). An increased conductivity (σ) of concrete enhances losses and affects the antenna efficiency. As a consequence, RR decreases for increasing values of (σ) in Figure 4.

INFLUENCE OF THE PROTECTIVE BOX

In order to prove that the protective box reduces the influence of concrete and improves the performances of the antenna, the reference tag is simulated within a protective box featuring no air layer between antenna and box, and its performances are compared with those of the reference geometry.

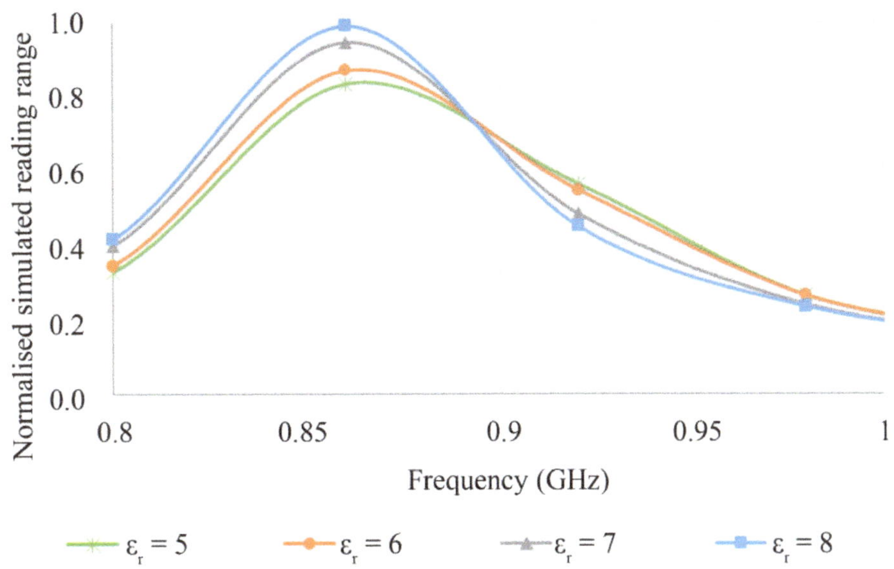

Figure 3. Normalized reading range of the antenna as a function of the frequency and the mortar relative permittivity (HFSS simulations) for a fixed conductivity $\sigma = 0.02$ S/m. The maximum RR is 1.55 m, obtained at 860 MHz for $\varepsilon_r = 8$.

In Figure 5, the relative shift of resonant frequency with respect to the reference tag is given as a function of mortar permittivity for a fixed mortar conductivity $\sigma = 0.01$ S/m. The resonant frequency of the antenna protected by the thinner box is much more sensitive to the changes in mortar permittivity.

In Figure 5, the relative decrease of RR with respect to the reference tag is also given as a function of mortar conductivity for a fixed mortar permittivity $\varepsilon_r = 5$. In the absence of air layer, RR decreases more rapidly when the mortar conductivity is increased, which confirms that the air layer around the antenna is highly beneficial for improving the radiation efficiency of the antenna. However, one can observe that this effect is attenuated when the conductivity reaches value such as $\sigma \leq 0.01$ S/m, i.e. at low water percentage when the mortar is fully dried.

EVALUATION OF THE EXPERIMENTAL READING RANGE

RR measurements are performed between 800 and 1000 MHz using the UHF RFID measurement system [16] and the test bench described in Figure 6. The device is first evaluated in air without and with housing. Then, the device is placed in a 2-litre mould of mortar. Measurements are carried out three days after casting the mortar, so that it is expected to be fully hydrated. The maximum RR = 270 cm occurs at 934 MHz in air without housing (Figure 7). With housing, the maximum RR = 240 cm is shifted to 897 MHz. The 140 cm RR in the European band shows that the tag in its housing can still operate in free space even though it has been designed to operate in concrete.

The experimental Q-factor and the experimental RR are significantly lowered in mortar as indicated by the smooth variation of the RR response. Within mortar, the maximum RR drops to 55 cm at 915 MHz and 50 cm at 865 MHz. It is likely that the variation of water content and drying conditions from one type of mortar to another and from one preparation method to another alter the maximum RR which was expected to reach 95 cm in HFSS simulations for the European UHF band. Also, the conditioning circuit was roughly modeled by metallic pads in HFSS simulations. Therefore, its coupling with the antenna and its influence on Z_A was probably underestimated

CONCLUSIONS

This paper described a UHF RFID tag embedded in mortar, which was the first published prototype built around a battery-assisted chip integrating sensing capabilities. A protective case sheltered the device from the aggressive environment in concrete. The overall tag size is comparable with the volume of the largest aggregates used in concrete industry.

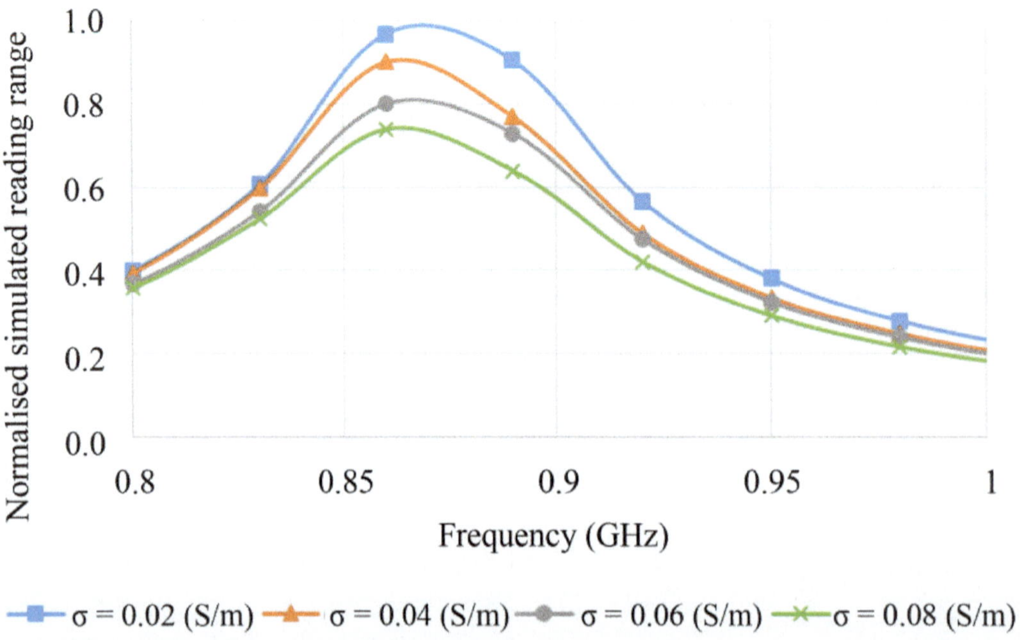

Figure 4. Normalized reading range as a function of the frequency and the mortar conductivity (HFSS simulations) for a fixed relative permittivity $\varepsilon_r = 5$. The maximum RR is 1.3 m, obtained at 870 MHz for $\sigma = 0.02$ S/m.

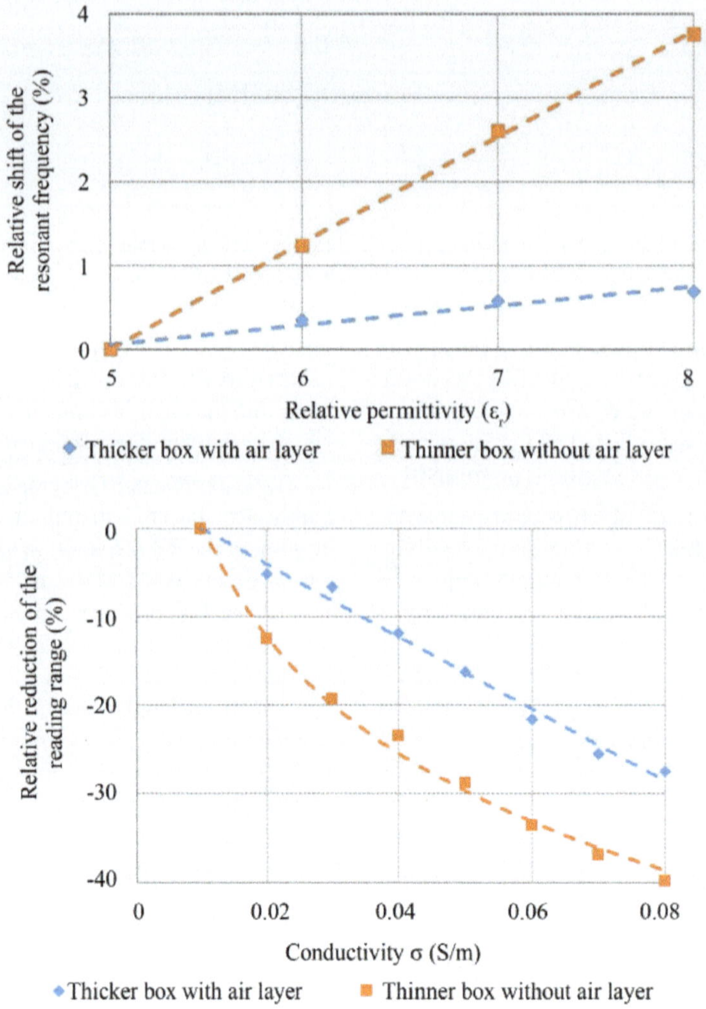

Figure 5. Relative downshift of the resonance frequency (upper) and relative reduction of the reading range (bottom) with increasing relative permittivity for a device with or without air layer between the antenna and the Teflon® box. The shift is expressed with respect to the resonance frequency (upper) and reading range (bottom) of the reference geometry with or without air layer for the lowest conductivity $\sigma = 0.01$ S/m and dielectric permittivity $\varepsilon_r = 5$.

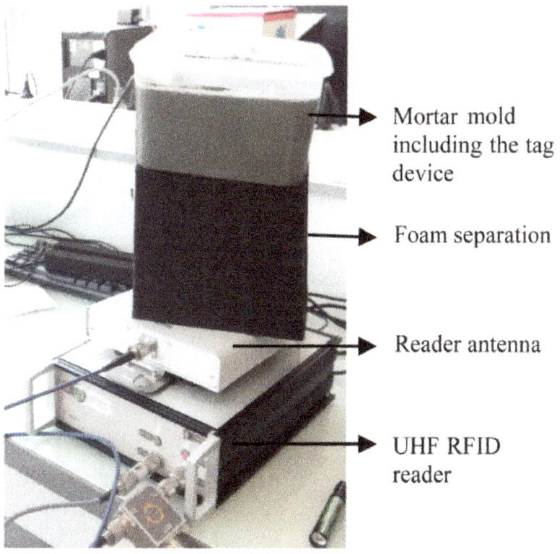

Figure 6. Test bench including the Tagformance measurement system. Tag housing centered in the (12 cm × 12 cm × 16 cm) block of mortar, 5 cm below its surface.

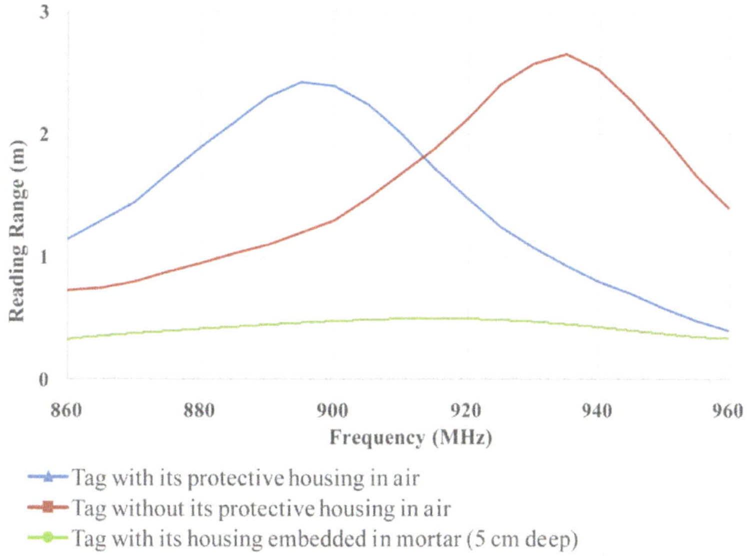

Figure 7. Experimental read range as a function of frequency for the UHF tag in air with and without housing and embedded in the mortar block.

The price of miniaturization is the decrease of antenna bandwidth of the antenna, but according to the HFSS simulations, the RR bandwidth is sufficient to accommodate some variability in the concrete characteristics. The sensitivity of the tag to concrete variability was strongly reduced by implementing a 5mm-thin air layer between the protective casing and the antenna.

The performances of the device were evaluated experimentally for a 5 cm burying depth in a mortar block. The reading distance reached up to 50 cm while the final volume of the assembled device was more than 3 times smaller than the state of the art embedded wireless nodes for concrete [17]. A 50 cm reading distance is sufficient to enable the monitoring of a concrete structure with a handheld RFID reader, especially for the transfer of strain gauge data.

The reading distance could be further improved following two approaches: firstly a more accurate evaluation of the conductivity and permittivity of the concrete to be monitored, secondly a better decoupling between the conditioning electronics and the antenna.

REFERENCES

1. Michelis, F., Bodelot, L., Bonnassieux, Y. and Lebental, B. (2015) Highly Reproducible, Hysteresis-Free, Flexible Strain Sensors by Inkjet Printing of Carbon Nanotubes. Carbon, 95, 1020-1026. http://dx.doi.org/10.1016/j.carbon.2015.08.103
2. Jiang, S. (2011) Optimum Wireless Power Transmission for Sensors Embedded in Concrete. Florida International University, University Park.
3. Shams, K. and Ali, M. (2007) Wireless Power Transmission to a Buried Sensor in Concrete. IEEE Sensors Journal, 7, 1573-1577. http://dx.doi.org/10.1109/JSEN.2007.908230
4. Ishida, H. and Furukawa, H. (2015) Wireless Power Transmission through Concrete Using Circuits Resonating at Utility Frequency of 60 Hz. IEEE Transactions on Power Electronics, 30, 1220-1229. http://dx.doi.org/10.1109/TPEL.2014.2322876

5. Jonah, O. and Georgakopoulos, S.V. (2013) Wireless Power Transfer in Concrete via Strongly Coupled Magnetic Resonance. IEEE Transactions on Antennas and Propagation, 61, 1378-1384. http://dx.doi.org/10.1109/TAP.2012.2227924

6. Jeong, S.H. and Son, H.W. (2011) UHF RFID Tag Antenna for Embedded Use in a Concrete Floor. IEEE Antennas and Wireless Propagation Papers, 10, 1158-1161. http://dx.doi.org/10.1109/LAWP.2011.2172179

7. EM Microelectronic (2012) EM4325 Datasheet [Online]. http://www.emmicroelectronic.com/

8. AMS (2013) SL900A Datasheet [Online]. http://ams.com/eng

9. Michelis, F., et al. (2014) Wireless Flexible Strain Sensor Based on Carbon Nanotube Piezoresistive Networks for Embedded Measurement of Strain in Concrete. EWSHM, 7th European Workshop on Structural Health Monitoring, Nantes, 8-11 July 2014, 1780-1787.https://hal.inria.fr/hal-01022026/document

10. America's Cement Manufacturers (2015) Aggregates. http://www.cement.org/cement-concrete-basics/concrete-materials/aggregates

11. (2015) Aggregates for Concrete. University of California, Berkeley.

12. Laheurte, J.-M. (2014) UHF RFID Tags: Design and Technologies. John Wiley & Sons, Inc., Hoboken, 84-85.

13. [Online]. http://www.ansys.com/Products/

14. Soutsos, M., Bungey, J., Millard, S., Shaw, M.R. and Patterson, A. (2001) Dielectric Properties of Concrete and Their Influence on Radar Testing. NDT & E International, 34, 419-425. http://dx.doi.org/10.1016/S0963-8695(01)00009-3 [Paper reference:2]

15. Nikitin, P.V., Rao, K.V.S., Lam, S.F., Pillai, V., Martinez, R. and Heinrich, H. (2005) Power Reflection Coefficient Analysis for Complex Impedances in RFID Tag Design. IEEE Transactions on Microwave Theory and Techniques, 53, 2721-2724. http://dx.doi.org/10.1109/TMTT.2005.854191

16. Voyantic Ltd. Kutojantie 11, FI-02630 Espoo, Finland. [Paper reference:2]

17. Chang, C.-Y. and Hung, S.-S. (2012) Implementing RFIC and Sensor Technology to Measure Temperature and Humidity inside Concrete Structures. Construction and Building Materials, 26, 628-637. http://dx.doi.org/10.1016/j.conbuildmat.2011.06.066

CHAPTER 16

GENERALIZED ELECTROMAGNETIC POLLUTION MONITORING USING WSN

Sara Nouh, Nada Elgaml, Nora Ali, Ahmed Khattab, Ramez Daoud, and Hassanein Amer

Keywords: WSN, EM Pollution, Monitoring, Random Distribution

INTRODUCTION

Recently Wireless Sensor Networks have been an attractive field, since they could be implemented in a huge number of applications. Different examples of these applications could be alarm systems, office and home automation, traffic control, civil infrastructure, environmental monitoring, personal health and many others [1]. Nowadays electromagnetic (EM) radiations are increasing tremendously due to the wide expansion of smartphones and wireless devices [2]. Moreover, people are exposed to other EM radiations due to the existence of several wireless networks, such as Global System for Mobile communications (GSM), Universal Mobile Telecommunications System (UMTS) and Wi-Fi, and the existence of FM Radio, TV, power systems and transmission powers [3]. Nevertheless, there is an increased demand of building wireless infrastructures, in order to provide higher data bandwidth and better coverage [4]. Therefore, concerns have risen due to the fact that these radiations could affect the personal health and hence, the need for monitoring these radiations has grown [4]. Possible hazardous effects of EM pollution could be cancer, leukemia and neuropsychological disorders [5].

This paper proposes a WSN-based framework to monitor these emissions and report any security violation; still the aim of this paper is not only to monitor the excessive electromagnetic waves, but also to prolong the WSN network's lifetime. The framework presented here is more general and more flexible than the one presented in [6], several parameters are modeled as random variables instead of constants. Furthermore, several case studies are analyzed to illustrate the use of the generalized framework.

The rest of the paper is organized as follows. Section 2 presents some background information regarding the system presented in [6]. Section 3 introduces the proposed framework and the main three parameters that affect the network's lifetime. Section 4 examines the use of different random distributions for the three main parameters. It is worth mentioning that the scope of this paper is only limited to measuring the EM pollution. Finally, Section 5 concludes the paper.

RELEVANT BACKGROUND AND RELATED WORK

As previously mentioned, there is always the desire of increasing the wireless infrastructure due to the need of a better coverage and higher bandwidth. A simple example of that is the mobile base stations of different service providers, where multiple signals could overlap together causing higher EM exposure [2]. Additionally, some countries do not have regulations to limit the placement of antennas on offices or residential buildings [7]. Hence, this paper presents a simple wireless sensor network model based on [6] that can monitor the electromagnetic violation of four base stations from four different service providers. It is assumed that the frequency polluters (F1, F2, F3 and F4) are placed on each side of a 100×100 m^2 area as illustrated in Figure 1. The sensors are distributed uniformly in the 100×100 m^2 area; in order to make sure that the whole area is covered and also match the most commonly used applications.

There are 100 narrow band sensors inside this area and 25 sensors are associated to each frequency polluter as in Figure 1. It is assumed that in the case of violation only 11 sensors out of 25 that are close to their frequency polluter are the ones who

Citation: Nouh, S., Elgaml, N., Ali, N., Khattab, A., Daoud, R. and Amer, H. (2016). Generalized Electromagnetic Pollution Monitoring Using WSN. Wireless Sensor Network, 8, 85-92. DOI: 10.4236/wsn.2016.86009.

sense the violation. In general, the 25 sensors should cover the whole area and report any violation. However, for simplicity and due to the fixedness of the polluter's transmitting power, only the closest sensors to the frequency polluter are allocated to it. Hence, 11sensors have been chosen to report any violation. However, if the transmitting power is varied, the number of associated sensors could be changed accordingly.

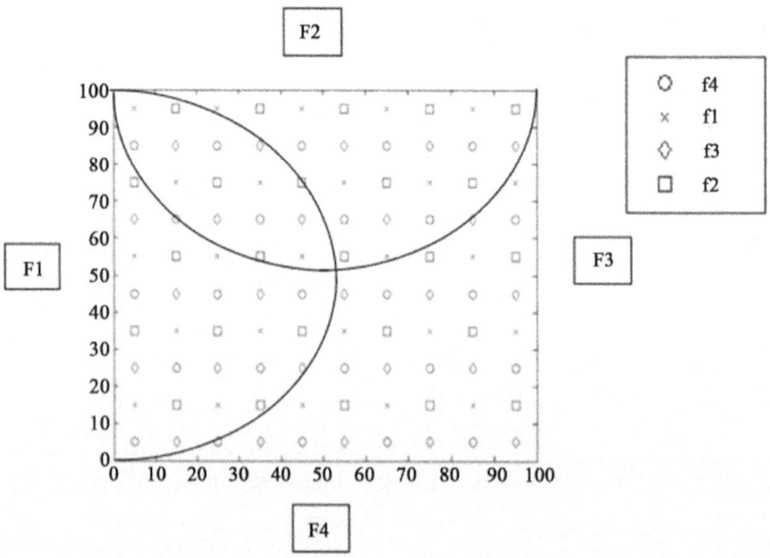

Figure 1. Placement of wireless sensors that correspond to each frequency.

The sensors in this model could either be active sensors, which sense the violation, or network masters (NM), which gather the data from the sensors and send it to the sink. Having a network master, also known as a Cluster Head (CH) in wireless sensor networks, has been introduced in LEACH and LEACH C [8] [9], in order to optimize the utilized energy in the network. This idea was also improved further in [10] by first having an optimized number of cycles of each NM, where each sensor could be an NM several times during the network's lifetime. The second algorithm was to calculate an optimized number of cycles for each NM, but each sensor could be NM only once. In [6] [11], this idea was developed further, so that each sensor, starting from the center and going in a circular order, could be an NM depending on a predefined threshold. If the sensor has enough energy above this threshold, it could act as an NM for several cycles, as long as it has enough energy to do so. When it depletes its required energy to act as an NM, the next sensor in row will be examined if it has enough remaining energy above the threshold that could enable it to act as an NM. The threshold for choosing the NM is based on a maximum threshold technique that was studied in [6] with several other techniques. It was then derived that the maximum threshold is the most practical threshold technique that could be used in this model.

While the NM selection is occurring, the sensors are continuously monitoring the power levels of their frequency polluters. This is done using the watchdog algorithm, where the sensors only send their data to the NM, when they detect violation. Other than that, when there is no violation, each sensor should send an "alive" data packet, every predefined period, in order for the sink to detect the network's failure by the death of the first node. The predefined period in this paper is chosen as in [6], which is every three cycles, where one cycle is equivalent to one hour.

GENERALIZED FRAMEWORK

The event by event algorithm used in [6] is developed in order to monitor frequency pollution. Four different frequencies as previously mentionedF1, F2, F3 and F4 are introduced and each one of them is assumed to be an operator that causes violation. The frequency polluter was placed at each side of the 100 × 100 m^2 network area resulting in a total number of four polluters. There was a specified schedule for this system, meaning that the first frequency polluter violates during the last six hours of the day and on the next day the second polluter violates during the last six hours of the second day and so on. This process repeats itself every four days and upon that is the lifetime of the total network measured. Such an event by event algorithm is not very flexible to capture any diverse and dynamic network change. Therefore, in this paper, a new generic framework is developed in order to add flexibility to [6] and make it more applicable for diverse situations. The same parameters that were used in [6] [8] - [11] are going to be used in this proposed generalized framework and they are listed below:

1. Network size: 100 ′ 100 m^2
2. Number of Sensors (N): 100 Sensors
3. Initial Energy: 2J
4. Transmitter/Receiver Electronics (E_{elec}): 50 nJ/bit
5. Transmitter Amplifier (E_{amp}): 100 pJ/bit/m2
6. Path Loss factor (n): 2
7. Aggregation Energy (E_{agg}): 5 nJ/bit/Signal
8. Data packet size sent by active nodes to NM (K): 64 bits
9. Data packet size sent by the NM to the sink (K1): 512 bits
10. Data packet size equivalent to sensing power levels (K2): 1 bit

11. Sink location: field center
12. Distribution: Homogeneous Density (Figure 1)

The event by event algorithm in [6] is not a generic system, since the main parameters used are fixed parameters. Those parameters are the starting time of the violation, the duration of the violation and the number of polluters per day. Random variables are going to be used for each parameter in the suggested framework. Each one of them will be described separately as follows.

Starting Time

In the event by event algorithm, the staring time of the violation was at 6pm everyday and it lasted for six hours. This has been modified in the proposed algorithm to be a random starting time between 12 am, which is the beginning of the day, until 6 pm. Since the violation duration is 6 hours, the last starting time has to be 6 pm, so that when the violation starts at that time it will not extend over the following day.

Matlab [12] functions are used here to generate a stream of random numbers between 0 and 18 for each frequency polluter using different random distributions, such as uniform, Gaussian, and exponential distributions.

Violation Duration

The violation duration for each polluter was six hours as stated before. Using Matlab [12] the duration period becomes a random number between one and six hours. Thus, a violator could violate for one hour or more, but the maximum violation period is tied by the previous assumption in the event by event algorithm, which is six hours. However, the flexibility of this new generic framework allows the possibility of changing this assumption and replacing it by the desired violation period.

Number of Polluters Per Day

Previously, in the event by event algorithm it was assumed that each violator will violate on a separate day and each will repeat violation every four days. For instance F1 will violate on day one, F2 on day two, F3 on day three, F4 on day four, and the cycle repeats again,F1 on day five and so on. Here, the number of violators per day became unknown. This means that a minimum of one violator and a maximum of four violators can violate on the same day. Looking at the example of having two polluters violating on the same day, these polluters can be F1 and F2, F1 and F3, F1 and F4, F2 and F3, F2 and F4 or F3 and F4. The total number of combinations here will be:

$$C_r^n = C_2^4 = 6 \text{ combinations} \tag{1}$$

where n is the total number of polluters and r is the number of polluters violating on the same day. Consequently if three violators will violate on the same day then we will have $C_3^4 = 4$ possible combinations of random polluters per day.

Changing the above three parameters that were used as fixed assumptions in the algorithm used in [6] and making them random makes the electromagnetic pollution monitoring technique more flexible and more suitable for different situations and applications.

Combining the Three Random Variables together Using Different Distributions

In order to add more flexibility to the newly proposed framework, it is designed to accommodate the randomness of the three above-mentioned parameters at the same time. Meaning that according to preference, one can choose to make for example the starting time and the violation duration random at the same time, or have all three parameters be random simultaneously. Different combinations can be chosen to simulate real life examples.

Additionally, different distributions will be introduced to the framework in order to expand its flexibility and meet the requirements of different applications. For simplicity the uniform distribution will be used as the default random distribution. However, one can also choose between the Gaussian and the exponential distribution according to the desired scenario. Certainly it is also possible for every parameter to have a different random distribution than the other or to choose the same distribution for all parameters. Figure 2 shows an overview of the Generalized Framework. Next different scenarios will be examined to show the usage of the new algorithm.

EXAMINED SCENARIOS

In this section, several scenarios are examined to showcase the capabilities of the proposed algorithm. When monitoring EM pollution different situations can occur. Using this framework, it is easier to simulate these situations, since the main aim of proposing this framework is to make [6] generic and applicable for many scenarios and applications. Most of the assumptions are the same as in [6], meaning that there are four violators; each of them is placed on one side of the network area. As mentioned before, these violators have fixed criteria when they breach, where each one of them violates only for six hours per day starting with F1 on the first day, then F2 on the second and so on and they repeat every four days. These six hours used to be the last six hours within the day. This scenario yielded a lifetime of 162,296 cycles and for further use, this lifetime will be referred to as the default lifetime.

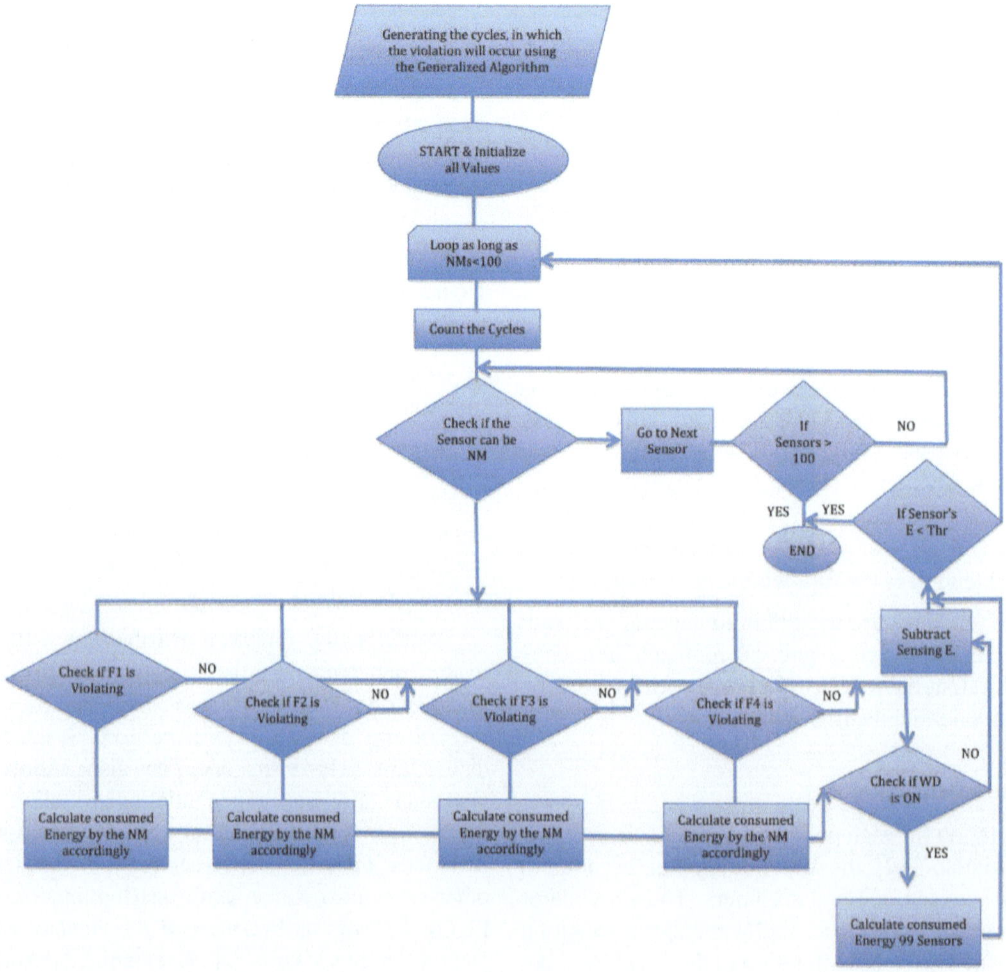

Figure 2. Flowchart of the generalized framework.

One of the common scenarios that one can compare to the previous default example is the effect of having four violators breaching at the same time. It is not possible to examine such scenario with the system in [6], however using the proposed framework it can be easily tested. It will be assumed that F1, F2, F3 and F4 are violating everyday, while all other parameters are fixed as follows: starting time = 6 pm, duration = six hours. This will result in a total lifetime of 122,977 cycles, which is a decrease by 24.227% compared to the default scenario. One would expect that having four violators per day compared to one violator, would automatically decrease the lifetime by 75%. However, this is not the case here, because of using the watchdog technique, where every three cycles all sensors should send to the network master a packet indicating that they are alive. Additionally, the number of network masters will differ in both cases, as on one day four polluters violating and exploiting from the sensors more energy than in the case of having one polluter per day. Accordingly, the calculated consumed energy using the distances between each network master and each sensor will be different in both scenarios. All these factors contribute to the 24.227% decrease in lifetime compared to the expected 75% with four violators per day,

Next the effect of randomness on lifetime is going to be obtained by removing all fixed assumptions and considering each parameter as a random variable with different distributions.

Effect of Starting Time Randomness on Lifetime

There are three main parameters in the algorithm in [6] that can be changed from fixed variables to random variables using the new proposed framework. These parameters are the starting time and the duration of the violation and at last the number of polluters violating per day. Therefore their effect on lifetime after being random variables needs to be deeply investigated in order to identify the dominant parameter that affects the lifetime mostly.

The first parameter, which is the starting time, will be examined separately. The starting time will be a random variable with uniform distribution from (1 - 19), which is equivalent to 1 am until 6 pm. The limit is at 6pm because the maximum duration violation (six hours) should not spill into the next day. The other two parameters will remain the same as the default case; the number of violators per day will be one starting with F1, F2 etc. and repeating every four days. Also the violation duration will remain six hours. This will result in a lifetime of 162,304 cycles, which has increased the default lifetime by only eight cycles. Since the 0.0049% increase of lifetime is very small, this shows that changing the starting time parameter to a random variable will result in an insignificant increase of the network's the lifetime.

Effect of Number of Polluters vs. the Duration Randomness on Lifetime

Since it was shown in the previous section that the starting time, as a random variable, does not have a significant effect on lifetime, it will remain as a fixed variable. Furthermore, the other two parameters the duration and the number of polluters per day, will be investigated further to identify the most effective variable on lifetime. A simple experiment will be simulated, where in scenario (a) it will be assumed that the number of polluters per day is fixed in each case, having F1 fixed everyday and hence there are four cases to simulate all possible situations. Meanwhile, the duration is a random variable uniformly distributed between (1,4). Only four hours are considered here in order to have a fair comparison between both parameters. The results of this experiment are shown in Table 1.

The results show that changing the number of polluters from 1 - 4 does not show a significant change on lifetime. The lifetime only differs by 0.01% to 2.2%. The same experiment will be repeated, but this time the duration will be fixed, while the number of polluters will be a random variable uniformly distributed form (1,4). This will yield to the following results in Table 2.

Table 1. Scenario (a)

No. of Polluters Fixed vs. Duration Random		
		Duration
Cases	**No. of Polluters per Day**	**U (1,4)**
A	F1	162.470
B	F1, F2	160.973
C	F1, F2, F3	160.373
D	F1, F2, F3, F4	158.849

Table 2. Scenario (b)

Duration Fixed vs. No. of Polluters Random		
		Duration
Cases	**Duration per Cycle**	**U (1,4)**
A	1	181.799
B	2	168.117
C	3	150.278
D	4	150.242

Scenario (b) shows that the duration variable has a huge effect on the lifetime. When 1-4 polluters violate for only one hour per day, the default lifetime will increase by 12.02%. Also when comparing the different cases A, B, C, and D, one can obtain a significant change in lifetime between 7.53% and 17.36%. Hence, the duration parameter has the most notable effect on lifetime.

Effect of Changing Random Distribution on Lifetime

Through the previous experiments, the duration parameter has proven to yield a significant effect on lifetime when it is changed. Therefore, different distributions will be investigated next with the duration parameter as a random variable. Thus, the same scenario (b) will be implemented again but using Gaussian and exponential distributions, in addition to the existing uniform distribution. The results of this attempt will be as shown in Table 3.

In order to have a fair comparison, the same mean is used in all three distributions. The Gaussian distribution shows a better outcome than the uniform distribution, while the exponential distribution yields a higher lifetime compared to the Gaussian and the uniform distribution. Comparing the exponential distribution to the uniform distribution, it has increased the lifetime by a factor of 3.97% to 5.57%. Nevertheless this increase is not very significant and this is due to the low range of samples, since the random variable only varies between 1 and 4. Therefore, a wider range will be investigated next.

Effect of Changing Random Distributions on Lifetime with a Wider Range of Variables

In this example, case D in Table 3 will be examined more thoroughly and will be called scenario (d); four violators are violating everyday and the starting time is at 1 am, in order to be able to have a wider range of duration. Henceforth, the duration will be a random variable from (1,23). Using the same distributions that were used before, Table 4 shows the results of this experiment:

It is clear that the results follow the same trend like scenario (c). The uniform distribution has the least lifetime, while the exponential distribution has the highest lifetime. The Gaussian distribution lies between both of them, but is closer to the uniform distribution lifetime. The difference between the uniform and Gaussian distribution is about 3.4%, while the difference between the uniform and exponential distribution is 13.08%. This is more than a double increase compared to scenario (c). The reason for that is that the wide range of random variable has revealed the real effect of the random distribution on the duration parameter.

Table 3. Scenario (c)

Different Distribution for Duration Random Variable				
		Duration		
Cases	**No. of Polluters per Day**	**U (1,4)**	**N(2,0.5)**	**Exp(2)**
A	F1	162.396	167.245	171.968
B	F1, F2	160.973	167.297	169.991
C	F1, F2, F3	160.373	163.749	166.271
D	F1, F2, F3, F4	158.849	162.344	165.410

Table 4. Scenario (d)

Different Distribution for Duration Random Variable				
		Duration		
Cases	**No. of Polluters per Day**	**U (1,23)**	**N (11, 0.6)**	**Exp (11)**
D	F1, F2, F3, F4	92,111	95,278	105,972

CONCLUSIONS

Wireless Sensor Networks are used in several applications that involve monitoring, controlling and tracking. In this paper a wireless sensor network-based framework was developed to monitor the EM emissions transmitted from frequency polluters. The framework developed here is more flexible and generic compared to previous systems in the literature; since all main parameters are treated as random variables. The main parameters are the starting time of the violation, the duration of the violation and at last the number of polluters violating per day.

To illustrate the use of the proposed framework, several case studies were investigated. It was identified through several simulations that the duration parameter is the most affective parameter and its variation can increase the network's lifetime between 7.53% and 17.36%. Moreover, the change of random distributions used for the parameter's random variables was investigated further through the examined scenarios. The outcome was that using the exponential distribution compared to the uniform distribution could prolong the network's lifetime by 13.08%. Finally, it is very important to note that using a wider range for each random variable will always yield better results, when comparing parameters together.

REFERENCES

1. Mikhaylov, K., Tervonen, J., Heikkil, J. and Knskoski, J. (2012) Wireless Sensor Networks in Industrial Environment: Real-Life Evaluation Results. 2012 2nd Baltic Congress on Future Internet Communications (BCFIC), Vilnius, 25-27 April 2012, 1-7. http://dx.doi.org/10.1109/bcfic.2012.6217971
2. Viani, F., Donelli, M., Oliveri, G., Trinchero, D. and Massa, A. (2011) A WSN-Based System for Real-Time Electromagnetic Monitoring. 2011 IEEE International Symposium on Antennas and Propagation (APSURSI), Spokane, 3-8 July 2011, 3129-3132. http://dx.doi.org/10.1109/aps.2011.5997195
3. Djuric, N., Prsa, M., Kasas-Lazetic, K. (2011) Serbian Remote Monitoring System for Electromagnetic Environmental Pollution. Acta Electrotehnica, Proceedings of the 4th International Conference on modern Power Systems MPS 2011, Cluj-Napoca, 17-20 May 2011, 140-142.
4. Derr, K. and Manic, M. (2015) Wireless Sensor Networks-Node Localization for Various Industry Problems. IEEE Transactions on Industrial Informatics, 11, 752-762. http://dx.doi.org/10.1109/TII.2015.2396007
5. Das, H., Gogoi, K. and Chatterjee, S. (2015) Analysis of the Effect of Electric Field Due to High Voltage Transmission Lines on Humans. 2015 1st Conference on Power, Dielectric and Energy Management at NERIST (ICPDEN), Itanagar, 10-11 January 2015, 1-4. http://dx.doi.org/10.1109/icpden.2015.7084491
6. Abou El Seoud, D., Nouh, S., Abbas, R.A., Ali, N.A., Daoud, R.M., Amer, H.H. and El Sayed, H.M. (2010) Monitoring Electromagnetic Pollution Using Wireless Sensor Networks. Proceedings of the 15th International Conference on Emerging Technologies and Factory Automation ETFA, Bilbao, September 2010.
7. Stacenko, L. and Anna, A. (2015) Monitoring of Levels of Electromagnetic Radiation from Base Station Antennas. 2015 International Siberian Conference on Control and

Communications (SIBCON), Omsk, 21-23 May 2015, 1-3. http://dx.doi.org/10.1109/sibcon.2015.7147299

8. Heinzelman, W., Chandrakasan, A. and Balakrishnan, H. (2000) Energy-Efficient Routing Protocols for Wireless Microsensor Networks. Proceedings of the 33rd Hawaii International Conference on System Sciences (HICSS), Maui, 4-7 January 2000. http://dx.doi.org/10.1109/HICSS.2000.926982

9. Heinzelman, W., Chandrakasan, A. and Balakrishnan, H. (2002) An Application Specific Protocol Architecture for Wireless Micro sensor Networks. IEEE Transactions on Wireless Communications, 1, 660-670. http://dx.doi.org/10.1109/TWC.2002.804190

10. Botros, S., El Sayed, H.M., Amer, H.H. and El-Soudani, M.S. (2009) Lifetime Optimization in Hierarchical Wireless Sensor Networks. Proceedings of the 14th International Conference on Emerging Technologies and Factory Automation, ETFA, Mallorca, 22-25 September 2009, 1-8. http://dx.doi.org/10.1109/etfa.2009.5347016

11. Nouh, S., Abbas, R.A., Abou El Seoud, D., Ali, N.A., Daoud, R.M., Amer, H.H. and El Sayed, H.M. (2010) Effect of Node Distributions on Lifetime of Wireless Sensor Networks. Proceedings of the IEEE International Symposium on Industrial Electronics ISIE, Bari, July 2010.

12. Official Site for MATLAB. http://www.mathworks.com

CHAPTER 17 ANALYSIS OF GEO POSITIONING SYSTEM USING RF & DA

Jaspreet Kaur Kohli and Amol Potgantwar

Keywords: Global Positioning System (GPS), Indoor Positioning System (IPS), Radio Frequency Identification (RFID), Received Signal Strength (RSS), Directional Antenna (DA)

INTRODUCTION

The idea of RFID based navigation in indoor environment is becoming very popular in today's era because of its low cost and less consumption of energy. Moreover, it is even gaining a lot of interests in industry and academics [1]. The wide usage of wireless technologies is an attraction for researchers to develop an application for location tracking in indoor environment and to deploy these systems on handheld devices such as mobile phones, tablets, etc. The idea here is to propose a system for portable devices with Radio Frequency and Directional Antenna to provide flexibility and dynamic solutions for the realistic and small-scale indoor areas such as some organizations (college, industries, etc.). This proposed work of the system will initiate the mobile phone device that connects to the server and accordingly gives the response to the user of their locations and command to take accurate to find the destination.

Indoor positioning system is generally used for locating required objects or places or people in buildings i.e. in either small or large closed environments. Global Positioning System (GPS) [2] is a very commonly used localization system in today's era. GPS is an existing system whose working depends on satellites and its signal. Due to the attenuation of the satellite signals GPS does not work when the respective object is detected in an indoor/closed environment [3]. So, to overcome the disadvantage of working GPS in closed area and locating objects accurately, a varied number of researchers developed indoor positioning systems such as RADAR system which was developed by Microsoft's researcher's laboratory; this system's working is based on Wi-Fi signal strength and location fingerprinting techniques [4]. Another indoor location sensing system developed by AT & T Cambridge was Active Badges [5] ; this developed system used diffuse infrared technology to realize indoor objects or locations. Crickets [6], Active Bats [7] and Dolphin [8] are some more IPS based on ultrasonic positioning systems. The computer vision system [9] and the cell phone positioning system are yet another development of indoor positioning system. The system, Spot ON RFID [10] sensing in the year 2013, consists of RFID tags, RFID readers and the signal communicating between the tags and readers. The foundation of this system is the radio signal strength information from several tags to reader and then uses the classifier algorithm to describe the user's location. The high end terminologies used in the proposed idea are as follows:

RFID Technology

RFID is a tracking technology which is used to authenticate and identify tags that are applied to any object or individuals. Radio Frequency Identification and Detection is a generalized term used for technologies that makes use of radio waves to identify objects and people. The RFID uses power from the reader's signal to reflect this signal back. Some tags contain build in power source and some needs power from external battery. Tags may modulate the signal to send information, such as an ID no. back to the reader. In short, in a typical RFID System, tags are attached to objects to be technology that make use of radio waves to automatically identify individual objects. An RFID system is composed of three main elements as shown in Figure 1.

Citation: Kohli, J. and Potgantwar, A. (2016). Analysis of Geo Positioning System Using RF & DA. Wireless Sensor Network, 8, 69-75. DOI: 10.4236/wsn.2016.84007.

RFID Tag

The proposed idea of the system will be using RFID tag to guide the visitor to their destination. RFID tag will contain information about the user like stag id, visitor id etc. The tag's chip or integrated circuit (IC) delivers performance, memory and extended features to the tag. The chip is pre-programmed with a tag identifier (TID), a unique serial number assigned by the chip manufacturer, and includes a memory bank to store the items' unique tracking identifier (called an electronic product code or EPC). The basic types of tags are Active and Passive tags. Passive tags basically do not contain internal battery. That means passive tags take their operating power from the field generated by the person who reads. Active tags have a transmitter and their internal battery (power).

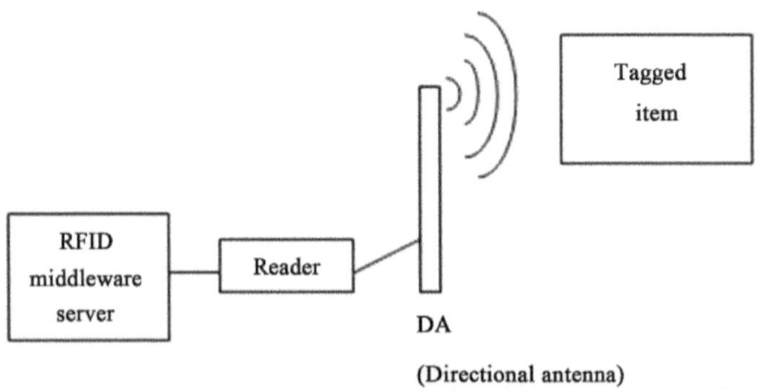

Figure 1. RFID grid.

RFID Reader

System uses RFID reader for reading the RFID tags and sends the current location of the visitor to the server. An RFID reader, also known as an interrogator, is a device that provides the connection between the tag data and the server. The reader communicates with tags that are within its field of operation, performing number of tasks such as encoding, filtering etc.

Directional Antenna

Directional Antenna is basically used in order to send the tag information of visitor to the server, this information will be transferred wirelessly. Tag antennas collect energy and channel it to the chip to turn it on. Generally, the larger the tag antenna's area, the more energy it will be able to collect and channel toward the tag chip, and the further read range the tag will have.

Received Signal Strength

It is a measurement of the energy that is available in the received radio signal. In many of the existing IPS, based on radio waves uses RSS on a great scale. Received signal strength determines the characteristics and nature of location fingerprinting. The generalized formula used for calculating RSS is as given in the Equation (1).

$$\mathrm{RSS} = P_t - PL(d_0) - 10\alpha \log_{10}\frac{d}{d_0} + X_{\sigma RSS} \quad (1)$$

where,

RSS: Received Signal Strength measured in dBm

Variable d: Distance from the sender to the receiver

α: Path loss component

P_t: Transmitted energy of the sender

PL (d_0): Loss in energy at distance d_0

X: Random variable representing the noise with variance σRSS

LITERATURE SURVEY

Geographical navigation by using various GPS tracking system is become most important part in day to day activity, where people have to visit various unknown locations for various purposes [11]. During these activities wastage of time due to unknown location or unawareness of accurate address may cause delay in meeting etc. [12].

GPS navigation by using current available devices in market is become much more impressive option to avoid various security, misguide in travelling, delay etc. problems [13]. Likewise same problems are being faced by users some sort of big organization such as college, industry or any equivalent places [5] [14] [15]. So to facilitate such activity the new device idea is being proposed which will interact with central server of concerned organization & visitor's mobile with the help of RFID, DA (Directional Antenna). During navigation various input& output parameters such as accuracy, scalability, energy consumption, efficiency and division multiple access (GSM/CDMA) mobile cellular network to estimate the location of outdoor mobile clients. However, the accuracy of the method using cell-ID or enhanced observed time difference (E-OTD) is generally low (in the range of 50 - 200 m), depending on the cell size [16] [17]. Environmental factors have to be monitored for the purpose of performance contribution & to obtain efficiency & accuracy of the system [18]. A number of systems have used global system of mobile/ code. Table 1 shows various Indoor positioning technologies [5] used.

PROPOSED SYSTEM IDEA

For the purpose of avoiding the drawbacks of existing system this paper propose the new system with extended ideas. Proposed system will implement a navigation system for robot application. Here system will use RFID tags for the purpose of giving instructions to the robot. System will load some tag numbers and direction instructions in programmer. When robot system will be started to move, it will detect the path for navigation and start navigation on given path. At some point in given specified area, user has to put RF tag. This tag will be read by RF reader which is placed on robot. Every tag has one unique no. System will scan the number and its instruction for direction. As per direction instruction, robot will change the direction and reach to destination.

Table 1. Various indoor positioning techniques.

Technologies	Advantages	Disadvantages	Accuracy
Ultrasound	-Simple and cheap equipment -Precision measurement possibilities	-The necessity of the establishment of the recipients in every room -Negative influence a high density-tone	-A few centimeters when it is placed frequent enough
RFID	-The transmitter and receiver don't need to see each other directly. -Opportunity to work in all types of environments -Fast interactions	-Passive tags have low precision than active tags. -RFID Readers are expensive	1 - 10 meters (extendable)
Bluetooth	-Any kind of Bluetooth device can be monitored (cell phone, mp3 player etc.) -Variable reading distance (1 m vicinity)	-Cost effective for large scale installation -Limited bandwidth -Transmission delay. -The possibly of a maximum of seven sub-links corresponds to main connection	2 - 10 meters
WLAN (Wi-Fi)	-The possibility of use the infrastructure of IEEE 802.11 -Low cost	-Poor performance for multilayered and very dense areas. -Due to signal reflection and dynamic network structure variable signal strength.	1 - 3 meters (in range of 50 meters)
Infrared	-Compact -Low power consumption	-Sensitive to daylight -Receiver and transmitter needs to see each other directly	5 - 10 meters

Here system requires multiple tags to create random path. Also as per the movement, robot will send its GPS location to machine or personal computer (PC) and it will display position of the RFID tag on it. Here system will give forward, backward, left and right and stop instructions to robot.

Segment Identification

A particular block of signal in terms of time interval will be taken for sampling and analysis purpose. The processed segments further will be transferred to the range estimation block. In this proposed idea the segment will be transferred via directional antenna to RFID middleware connected to the server.

Range Estimation

The proposed range estimation algorithm calculates range of the registered users depending on the scanning of RFID tags on previous reader nodes and the current reader node. Then the location of the user's mobile is calculated by using RSS values and eradicating the angle dependent loss errors. The noise in the signal i.e. the error in the signal will be reduced from it by using Kalman filter technique. After the unwanted part of the signal is cleared the resultant signal will give a better output strength.

Angle Dependent Loss Component

The angle loss within the tracking area is calculated by using the backscattered signals from RFID devices. After identification of the reader's location the angles will be calculated. It is usually calculated from the specific antenna radiation pattern.

METHODOLOGY

The system consist multiple components as shown in the Figure 2. There will be a visitor and an admin who directly interact with the system. Visitors have to register first. After registration visitors get the RFID tags. The tags contain the information about the visitor like visitor id, name, destination, purpose etc. The readers will be installed on each corner of the campus. Visitors need to swap the tag to reader in order to get the direction of destination. As soon as visitor swaps the tag, the RF reader makes contact with the server. Server gets the visitor id, name, destination, tag id and start finding the nearest path to the destination from the current position of the visitor. Then server sends the information of nearest path with direction notification form onto the visitors' mobile device. After getting direction visitors can move forward to next reader and swap card, and can get the destination in less time and more accurately.

EXPERIMENTAL SETUP

The implementation of this system will be performed using java jdk1.8 & Net Beans 8.0 on windows OS with 1 GB RAM & 500 GB hard disk. Hardware module will be integrated with RS232 cable using comp port. This system access location parameters from all the RF tags. The major hardware components of the system consist of ARM 7 controller, RFID tags, RFID readers interconnected with each other. Hardware Layout is presented in Figure 3.

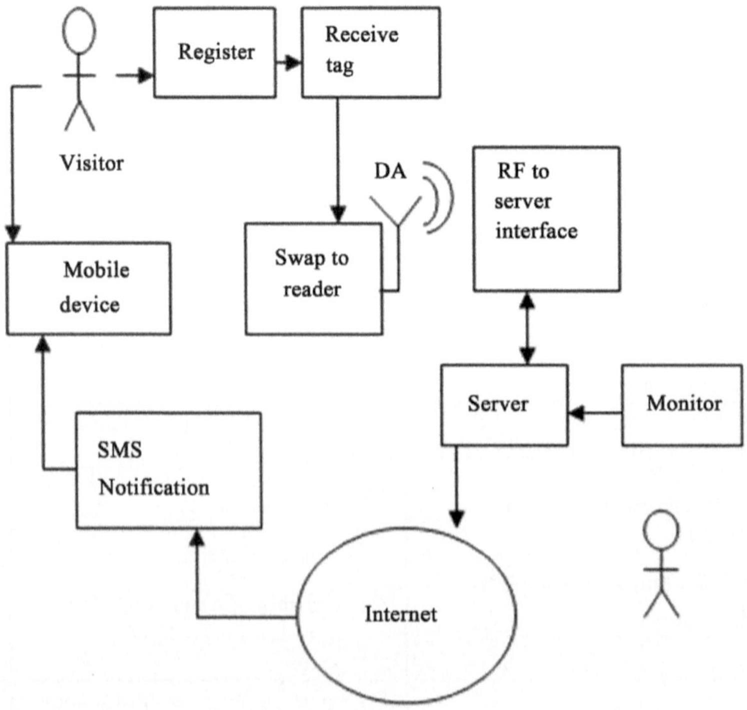

Figure 2. Block diagram of proposed system idea.

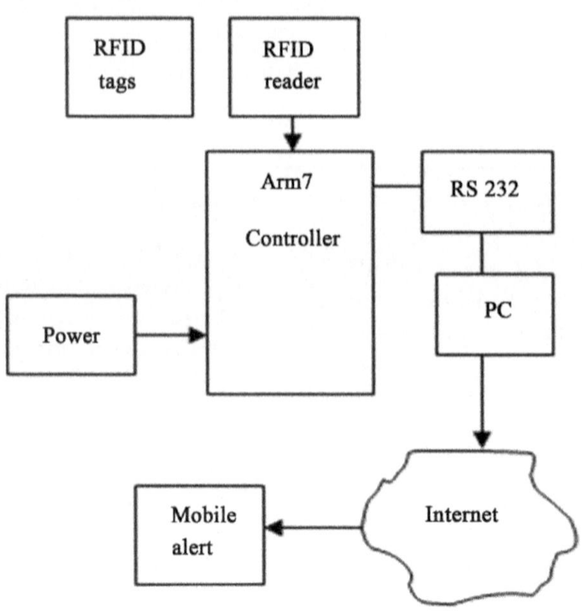

Figure 3. Block diagram of hardware layout.

CONCLUSION

In this implementation the study of effects of real time system with RFID based positioning system is processed. The comparison of various parameters such as signal strength, geographical position & radiation pattern etc. is performed along with the analysis of various parameters such as accuracy, scalability, precision, delay, etc. A study on RFID based indoor positioning system with directional antenna has been introduced. Usages and performances of various types of readers and tags have been investigated and the best combination has been selected. This idea of proposed system can be merged with many existing systems such as visitor guidance, information kiosk in organization, large and geographically distributed organization, etc. This work can be further expanded to quality and reliability study.

REFERENCES

1. Qiu, R.G. (2007) RFID-Enabled Automation in Support of Factory Integration. Robotics and Computer-Integrated Manufacturing, 23, 677-68. http://dx.doi.org/10.1016/j.rcim.2007.02.002
2. Department of Defence United States of America and GPS Navstar (2008) Global Positioning System Standard Positioning Service Performance Standards. 4th Edition. Want, R., Hopper, A., Falcao, V. and Gibbons, J. (1992) The Active Badge Location System. ACM Transactions on Information Systems, 40, 91-102. http://dx.doi.org/10.1145/128756.128759
3. Garmin Corporation (2001) About GPS. Website. http://www.garmin.com/aboutGPS/
4. Pathak, O., Palaskar, P., Palkar, R. and Tawari, M. (2014) Wi-Fi Indoor Positioning System Based on RSSI Measurements from Wi-Fi Access Points—A Trilateration Approach. International Journal of Scientific and Engineering Research, 5, 1234-1238.
5. Want, R., Hopper, A., Falcao, V. and Gibbons, J. (1992) The Active Badge Location System. ACM Transactions on Information System, 40, 91-102. http://dx.doi.org/10.1145/128756.128759 [Paper reference:3]
6. Priyantha, N.B., Chakraborty, A. and Balakrishnan. H. (2000) The Cricket Location-Support System. Proceeding of MOBICOM 2000, Boston, August 2000, 32-43. http://dx.doi.org/10.1145/345910.345917
7. Harter, A., Hopper, A., Steggles, P., Ward, A. and Webster, P. (1999) The Anatomy of a Contextaware Application. Proceeding of the 5th Annual ACM/IEEE International Conference on Mobile Computing and Networking, Seattle, August 1999, 59-68. http://dx.doi.org/10.1145/313451.313476
8. Minami, M., Fukuju, Y., Hirasawa, K., Yokoyama, S., Mizumachi, M., Morikawa, H. and Aoyama, T. (2004) Dolphin: A Practical Approach for Implementing a Fully Distributed Indoor Ultrasonic Positioning System. Ubicomp, 347-365.http://dx.doi.org/10.1007/978-3-540-30119-6_21
9. Krumm, J., Harris, S., Meyers, B., Brumitt, B., Hale, M. and Shafer, S. (2000) Multi Camera Multi-Person Tracking Easy Living. Third IEEE International Workshop on Visual Surveillance, Dublin, 1 July 2000, 1-8. http://dx.doi.org/10.1109/VS.2000.856852
10. Hightower, J., Want, R. and Borrielo, G. (2000) SpotON: An Indoor 3D Location Sensing Technology Based on RF Signal Strength. Technical Report, University of Washington, Seattle, Wash.
11. http://www.safecomprogram.gov/NR/rdonlyres/D8BFF5A9-47A7-4496-AB93-611DD043C3E0/0/emerging_wireless_technologies_enhanced_911.pdf
12. Lee, C., Chang, Y., Park, G., et al. (2014) Indoor Positioning System Based on Incident Angles of Infrared Emitters. Proceedings of the 30th Annual Conference of IEEE Industrial Electronics Society (IEOCON'04), 3, 2218-2222.
13. Rodriguez, M., Pece, J. and Escudero, C. (2005) In-Building Location Using Bluetooth. International Workshop on Wireless Ad-Hoc Networks.
14. Dong, Z., Wu, Y. and Sum, D. (2013) Data Fusion of the Real-Time Positioning System Based on RSSI and TOF. 5th International Conference on Intelligent Human-Machine Systems and Cybernetics, 2, 503-506.http://dx.doi.org/10.1109/ihmsc.2013.267
15. Torteeka, P., Xiu, C. and Yang, D.K. (2014) Hybrid Technique for Indoor Positioning System Based on Wi-Fi Received Signal Strength Indication. 2014 International Conference on Indoor Positioning and Indoor Navigation (IPIN), Busan, 27-30 October 2014, 48-57.
16. Ni, L.M., Liu, Y., Lau, Y.C. and Patil, A.P. (2003) LANDMARC: Indoor Location Sensing Using Active RFID. Proceedings of the 1st IEEE International Conference on Pervasive Computing and Communications (PerCom'03), Fort Worth, 23-26 March 2003, 407-415.
17. UbiSense Company. http://www.ubisense.net
18. Aitenbichler, E. and Muhlhauser, M. (2003) An IR Local Positioning System for Smart Items and Devices. Proceedings of the 23rd International Conference on Distributed Computing Systems Workshops, Providence, 19-22 May 2003, 334-339. http://dx.doi.org/10.1109/icdcsw.2003.1203576

CHAPTER 18

INTERFERENCE MITIGATION TECHNIQUES USING RECEIVER PROCESSING AND RESOURCE ALLOCATION

Jesvin Veancy Boaz and Yogesh Palanichamy

Keywords: Cellular Network, Orthogonal Frequency Division Multiplexing, Interference, Scheduling, Post Processing Signal to Noise Ratio

INTRODUCTION

The mobile broadband traffic is growing exponentially during the recent years due to the evolution of the capabilities of the terminals [1]. The mobile service providers are facing severe challenges to meet the growing demand of the very high data rate applications. Spectrum is a scarce resource and various techniques are required to increase the network capacity and meet the growing needs of data rate [2]. Adding more antennas to enhance the capacity will increase the signal processing between the transmitter and receiver. Shrinking of the macro cell size in urban and dense areas by deploying more base stations a.k.a. eNodeBs (eNBs) by operators is also done to increase the capacity [1] - [6]. This will increase the interference to the system. The problem of this method is that it is very expensive to add more eNBs due to the need for expensive drive tests in finding suitable locations for deployment, and also the high real estate cost to put the towers. The alternative approach is to introduce low power and less planned pico eNBs overlaid on the macro network. This is called as heterogeneous networks (HetNet) [1] [4]. The pico nodes are added depending on the local capacity needs [2]. However, the interference due to addition of the pico nodes is more severe due to the fact that the macro and pico transmit at different power. An example of a HetNet that consists of one pico eNB overlaid on a macro eNB coverage area is shown in Figure 1.

Citation: Boaz, J.V. and Palanichamy, Y. (2016). Interference Mitigation Techniques Using Receiver Processing and Resource Allocation. *Circuits and Systems*, 7, 2893-2903. DOI: 10.4236/cs.2016.710248.

The coverage of the pico eNB is quite limited due to its low transmit power, and the strong interference that user equipment (UE) receives from the macro eNB. In this scenario, the equal signal to interference plus noise (SINR) boundary where the downlink (DL) SINR observed from the macro eNB and the pico eNB is the same is very closer to the pico cell. Therefore, only a small percentage of UEs are associated to the pico eNB [6]. Due to lack of sufficient number of UEs, the spectrum is not fully utilized by the pico eNBs. Therefore, new techniques are required to increase the network efficiency of HetNets by offloading macro cell traffic to pico cells [6]. The performance of the network is significantly improved when more UEs are allowed to attach to the pico eNB by further reusing the frequency resources enabling additional cell-splitting gains [2]. More UEs are associated to the pico eNB by applying a positive bias to the SINR observed with respect to the pico eNB, and the UEs that are thus associated to the pico eNB are called as cell range extension or expansion (CRE) UEs, and the region is called as CRE region [2]. However, biasing makes the interference further worse for the UEs served by the pico eNB at the boundary between the macro and pico cells. Therefore, efficient interference mitigation techniques are required to achieve the real benefits of adding the low power picos [2] - [4] [7]. The heterogeneous network that illustrates the CRE region is shown in Figure 2.

A single technique cannot fully mitigate the interference, and therefore, more than one techniques are required to mitigate the effect of interference in a cellular system. In this paper, one such interference mitigation technique based on receiver processing along with resource allocation using appropriated scheduling method is proposed to improve the throughput, average packet delay, and packet delivery ratio performance of the system. The performance is compared and analyzed with the round robin (RR) [8] - [10], proportional fair (PF) [9] [10], and modified largest weighted delay first (MLWDF) [10] algorithm along with interference suppression receiver to decide on resource allocation. It is found from simulation results that the MLWDF algorithm achieves the highest throughput with minimum average delay of packets with the best delivery ratio compared to the PF and RR schedulers.

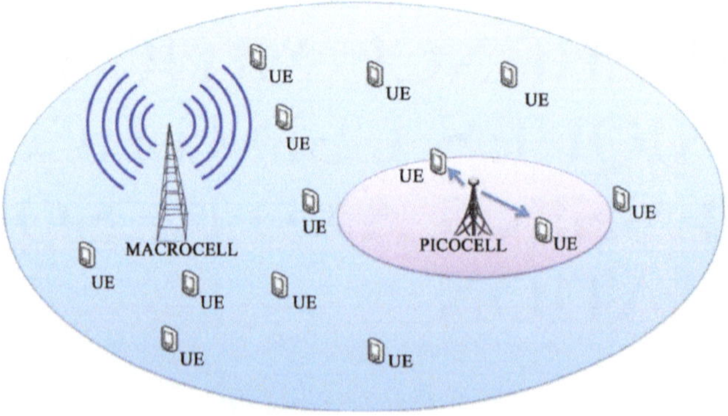

Figure 1. Example of a HetNet with one macro and a pico eNB.

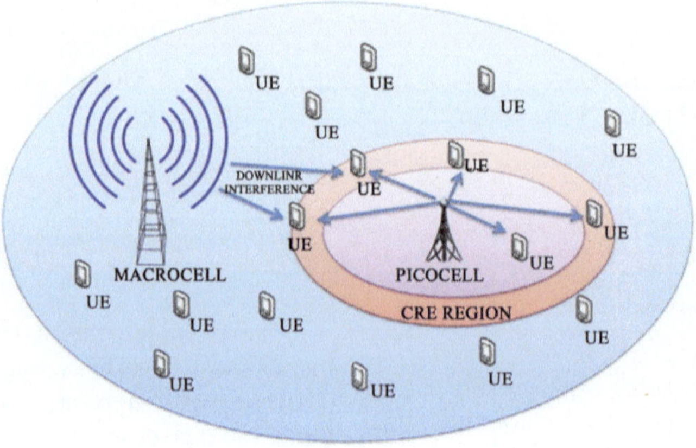

Figure 2. Downlink interference from macro and range extension.

SYSTEM MODEL

The fourth generation (4G) long term evolution (LTE) technology is considered to simulate the proposed interference mitigation methods. The DL transmission of LTE is based on orthogonal frequency division multiplexing access (OFDMA) technique for providing services to multiple UEs. OFDMA divides the total available bandwidth into multiple parallel narrow band carriers called subcarriers. The resource to be allocated for data transmission consists of 12 such subcarriers over six or seven OFDM symbols a.k.a resource block (RB). The numbers of RBs are 25, 50, and 100 for the bandwidth of 5, 10, and 20 MHz, respectively. The UEs depending on the data rate requirement will be allocated with a RB or group of RBs [3] [4] [9] [11] [12].

The system consists of L eNBs, each having N_t antennas, and Q UEs, each having N_r antennas. The total transmit power of the desired eNB is P_s and interfering eNB is P_i, i = 1,2,⋯,L–1, Therefore, the $N_r \times 1$. vector $r(n, k)$ received by the N_r antennas at the UE for the $N_t \times 1$ vector $s(n, k)$ transmitted by the N_t antennas at the eNB is given by the following equation

$$\boldsymbol{r}(\boldsymbol{n},\boldsymbol{k})=\sqrt{P_s}\boldsymbol{H}(\boldsymbol{n},\boldsymbol{k})\boldsymbol{s}(\boldsymbol{n},\boldsymbol{k})+\sum_{i=1}^{L-1}\sqrt{P_i}\boldsymbol{I}_i(\boldsymbol{n},\boldsymbol{k})\boldsymbol{u}_i(\boldsymbol{n},\boldsymbol{k})+\boldsymbol{v}(\boldsymbol{n},\boldsymbol{k}) \quad (1)$$

where, $H(n, k)$ and $I_i(n, k)$ are the $N_r \times N_t$ desired and interference channel matrix from the ith eNB, respectively on the subcarrier at time instant in the frequency domain, and they are given by

$$\boldsymbol{H}(\boldsymbol{n},\boldsymbol{k})=\begin{bmatrix} h_{11} & h_{12} & \cdots & h_{1N_t} \\ h_{21} & h_{22} & \cdots & h_{2N_t} \\ \vdots & \vdots & \ddots & \vdots \\ h_{N_r1} & h_{N_r1} & \cdots & h_{N_rN_t} \end{bmatrix}, \quad \boldsymbol{I}_i(\boldsymbol{n},\boldsymbol{k})=\begin{bmatrix} i_{11} & i_{12} & \cdots & i_{1N_t} \\ i_{21} & i_{22} & \cdots & i_{2N_t} \\ \vdots & \vdots & \ddots & \vdots \\ i_{N_r1} & i_{N_r1} & \cdots & i_{N_rN_t} \end{bmatrix}. \quad (2)$$

The entries h_{jm} of $H(n, k)$ and i_{jm} of $I_i(n, k)$ are the coefficients of desired and interference channels, respectively from the mth antenna at eNB to the jth antenna at UE. They are obtained using complex Gaussian random variable with zero mean and unit variance for each tap of a frequency selective multipath channel. The variance of each tap is appropriately scaled based on the 3GPP defined power delay profile [13], and the frequency domain channel is obtained using fast Fourier transform (FFT). The $N_t \times 1$ vector $u_i(n, k)$ is the interfering signal transmitted from the ith eNB. The entries of the $N_r \times 1$ noise vector $v(n, k)$ are obtained using complex Gaussian random variable with zero mean and N_0 variance. Then, the preprocessing SINR is given by

$$SINR_{pre} = \frac{P_s}{N_0 + \sum_{i=1}^{L-1} P_i} . \quad (3)$$

The signal to interference ratio (SIR) for a system with L eNBs is defined as

$$SIR = 10\log\left(\frac{P_s}{\sum_{i=1}^{L-1} P_i}\right) .$$

The signal to noise ratio (SNR) for various SIR is defined as SNR = P_s/N_0.

The SNR is varied by varying the noise variance N_0 for link level simulations.

INTERFERENCE MITIGATION TECHNIQUES

The extra antennas at the receiver could be used to cancel or suppress the interference at the receiver. It is well known that a system with N_r receive antennas can cancel up to $N_r - 1$ interferers [7]. Maximum likelihood (ML) receivers are used to exactly cancel the interferers [14] with full knowledge of the desired and all the interference channels. They are computationally complex as it requires to jointly decoding the desired and interference data. However, there is another class of receivers, which may not be able to fully cancel the interference, but can suppress the effect of interference to a minimum, and are called as minimum mean square error (MMSE) receivers [15] - [17]. This is a less complex receiver as it requires knowing only the interference covariance, and not the exact interfering channels. The performance of these receivers is enhanced with a slight increase in complexity when some of interfering channels are known, and are called as eMMSE receivers [16] [17].

MMSE Receiver

The receiver operation is performed on every subcarrier at each time instant in the OFDM system. Therefore, without loss of generality, the time and frequency indices can be dropped from (1), and considering $P_s = P_i = 1$ the received signal equation is written as

$$\boldsymbol{r} = \boldsymbol{Hs} + \sum_{i=1}^{L-1} \boldsymbol{I}_i \boldsymbol{u}_i + \boldsymbol{v} . \quad (4)$$

It can be rewritten as

$$\boldsymbol{r} = \boldsymbol{Hs} + \boldsymbol{w} .$$

where, w is the total noise plus interference vector of size $N_r \times 1$. The low complexity receiver weights for the MMSE solution is given by

$$\boldsymbol{B}_{MMSE} = \boldsymbol{H}^H \boldsymbol{R}_{MMSE}^{-1} .$$

where, R_{MMSE} is the total received signal covariance matrix estimate, and is given by

$$\boldsymbol{R}_{MMSE} = \boldsymbol{HH}^H + \boldsymbol{R}_{ww} .$$

where, R_{ww} is the covariance matrix of inter-cell interference and white noise, and is given by

$$\boldsymbol{R}_{ww} = \boldsymbol{ww}^H .$$

Therefore, the estimate of S is given by

$$\hat{\boldsymbol{s}} = \boldsymbol{B}_{MMSE}\boldsymbol{r} . \quad (5)$$

eMMSE Receiver

When one of the dominant interference is known, (4) can be written as

$$\boldsymbol{r} = \boldsymbol{Hs} + \boldsymbol{I}_1 \boldsymbol{u}_1 + \sum_{i=2}^{L-1} \boldsymbol{I}_i \boldsymbol{u}_i + \boldsymbol{v} . \quad (6)$$

It can be rewritten as

$$\boldsymbol{r} = \boldsymbol{Hs} + \boldsymbol{I}_1 \boldsymbol{u}_1 + \boldsymbol{x} .$$

where, x is sum of the noise plus rest of the interference vector of size $N_r \times 1$. The low complexity receiver weights for the MMSE solution is given by

$$\boldsymbol{B}_{eMMSE} = \boldsymbol{H}^H \boldsymbol{R}_{eMMSE}^{-1} .$$

where, R_{eMMSE} is the total received signal covariance matrix estimate, and is given by

$$\boldsymbol{R}_{eMMSE} = \boldsymbol{HH}^H + \boldsymbol{I}_1 \boldsymbol{I}_1^H + \boldsymbol{R}_{xx} .$$

where, R_{xx} is the covariance matrix of noise plus rest of the interference, and is given by

$$\boldsymbol{R}_{xx} = \boldsymbol{xx}^H .$$

Therefore, the estimate of S is given by

$$\hat{\boldsymbol{s}} = \boldsymbol{B}_{eMMSE}\boldsymbol{r} . \quad (7)$$

The simulation results in this paper are obtained for the MMSE and eMMSE receiver by performing Monte Carlo simulations of the link and system level simulation assuming the channel state information (CSI) H known only at the UE for various antenna configurations. Also, for MMSE receiver the noise plus interference covariance R_{ww}, and for eMMSE receiver one dominant interference I_1 and noise plus rest of the interference covariance R_{ww} are assumed to be known. These assumptions will only help decouple the channel estimation problem, and will not in any way affect the conclusion drawn in this paper. Moreover, eMMSE receiver is used to obtain the post processing SINR [12] [14] [15] for making scheduling decisions for system simulations.

Figure 3 shows the link level performance comparison of MMSE and eMMSE receivers with two eNBs having one transmit antenna each, and a UE for two and three antennas. The UE is assumed to be located at equidistance from both eNBs so that the SIR is 0 dB, which reflects the worst case interference from one eNB when the desired and interfering eNB transmits at the same power. As expected, the eMMSE receiver is able to cancel the interference fully with knowledge of the one interfering channel for two UE antennas whereas the MMSE is flooring after 15 dB. Note that with three antennas MMSE receiver also suppress the interference completely and get an array gain of 3 dB when compared to the two antenna eMMSE receiver. However, with three UE antennas, the eMMSE receiver is also able to achieve diversity gain using the extra antenna, which is evident from the change in slope of the curve.

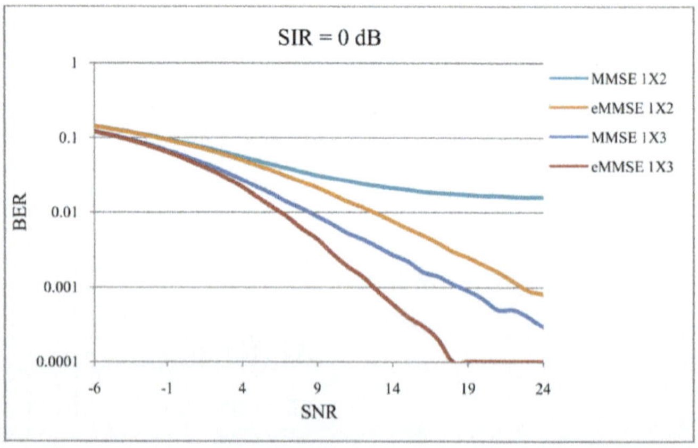

Figure 3. BER performance comparison of MMSE and eMMSE receiver with one dominant interferer.

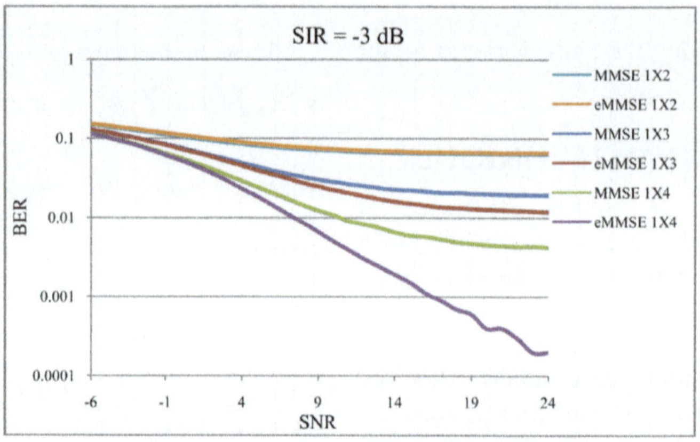

Figure 4. BER performance comparison of MMSE and eMMSE receiver with two dominant interferers.

Figure 4 shows the link level performance comparison of MMSE and eMMSE receivers with three eNBs having one transmit antenna each, and a UE for two, three and four antennas. The UE is assumed to be located at equidistance from all the three eNBs so that the SIR is −3 dB, which reflects the worst case interference from two eNB when the desired and interfering eNBs transmit at the same power. It is observed that both receivers are not able to suppress the interference fully with two and three antennas at the UE as it is evident due to the flooring after 15 dB. This is because either no or one interference channel is assumed to be known, and there are two dominant interferences. Note that with increase in number of UE antennas, interference suppression gain is increased for both the receivers. However, with four UE antennas, the eMMSE receiver is able to suppress the interference completely and achieve diversity gain using the extra antenna as evident from the slope of the curve.

SCHEDULING

The radio resource management (RRM) functions including packet scheduling are performed by the eNB in LTE. The scheduling is performed for a frequency resource corresponding to a time period of 1 ms, and is called as the transmit time interval (TTI) [11]. The number of UEs that can be scheduled in any given TTI is limited by the number of RBs in the frequency band and the scheduler [8] [10].

The functionalities of an eNB relevant to packet scheduling are illustrated in Figure 5. A buffer to store the incoming packets is assigned at the eNB for each UE. These packets that are arriving into the buffer are queued for transmission to the respective UEs. The arrival time of the packets are time stamped, and are taken out of the buffer for transmission on a first-in-first-out (FIFO) basis.

The UE computes the post processing SINR at the MMSE or eMMSE receiver, which are obtained from (4) to (7), respectively, and are given as

$$SINR_{post}^{MMSE} = \left[H^H \left(HH^H + R_{ww} \right)^{-1} H \right] / N_0 \quad (8)$$

and

$$SINR_{post}^{eMMSE} = \left[H^H \left(HH^H + I_1 I_1^H + R_{xx} \right)^{-1} H \right] / N_0 , \quad (9)$$

The post processing SINR is obtained using either (8) or (9) on every resource element (subcarrier) in the RB, and is converted in to a single effective SINR using mean mutual information bit (MMIB) [18] for a given bit error rate (BER). This value is mapped to a channel quality indicator (CQI), and is reported back to the eNB. The CQI is used by the eNB to choose appropriate modulation and coding scheme (MCS) to transmit the information bits in the physical channel [9]. This is called as link adaptation and a.k.a adaptive modulation and coding (AMC). The RBs with high post processing SINR is an indication of less interference in those RBs. When this is used as a metric for scheduling at the eNB, it will ensure maximum interference suppression at the allocated resources when either of the MMSE or eMMSE receivers is used at the UE for data detection.

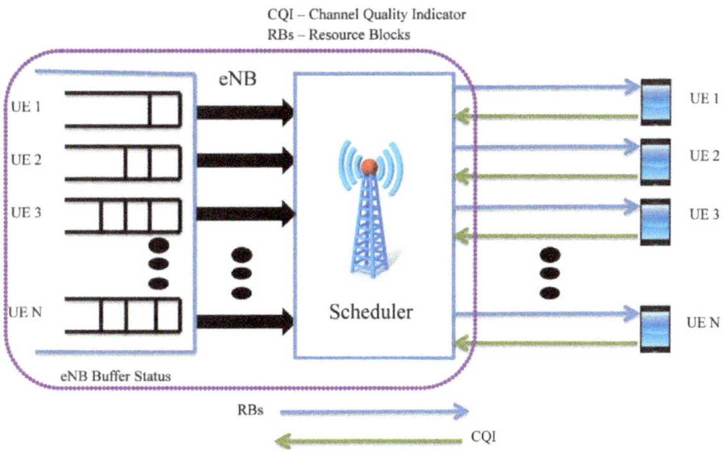

Figure 5. Communication system to illustrate the scheduling functionalities.

As the DL of LTE uses OFDMA as the multiple access scheme, it allows the eNB to perform opportunistic scheduling along both time and frequency to ensure that the allocations are given to an UE with the best post processing SINR in a particular portion of the band at a given point of time. The packet scheduler determines the UE to be scheduled on a RB based on its reported CQI, and other fairness or QoS metrics for PF and MLWDF schedulers, except in the case of RR scheduler that is independent of the channel conditions.

Round Robin Scheduler

The RR scheduler is simple resource allocation method that equally distributes the resources to all UEs [8] - [10]. The resources are allocated to available UEs in a cyclic fashion irrespective of their channel conditions. Hence, the overall system throughput will be low due to the possibility of channel conditions being poor at the time of actual resource allocation. The resource allocation can be done either in time domain, or frequency domain, or both of the time frequency resource in OFDMA systems. The RR scheduler is easy to implement, and it will provide the best fairness in resource allocation to UEs, however, at the cost of throughput.

Proportional Fair Scheduler

The PF scheduler aims at improving throughput without severely affecting the fairness by allocating resources to the UEs whenever their channel conditions are good with respect to their own average throughput of the past [9] [10]. It provides a trade-off between throughput and fairness, and is best suited for multiuser systems like OFDMA where the resources are shared in both time and frequency domain to exploit the multiuser diversity in the system. The PF scheduler is very suitable to support non real time traffic (NRT) traffic types as it ensures fairness among UEs over a long period of time. The metric to be computed on every RBs in each sub frame for all UEs to be scheduled is given by

$$P_{m1} = \left(\frac{r_q(t)}{R_q(t)} \right) . \quad (10)$$

where,

$r_q(t)$ is the instantaneous achievable data rate for a UE in a particular RB by its respective CQI obtained from the post processing SINR of the receiver computed using (8) or (9), and

$r_q(t)$ is the average data rate of the qth UE at time t obtained by averaging the data rate supported by this UE in the past over a long period of time.

The PF algorithm decides the UEs to be scheduled on a particular RB based on the priority function given below

$$P_f = \arg\max_{u_q} \left(P_{m1} \right) = \arg\max_{u_q} \left(\frac{r_q(t)}{R_q(t)} \right) . \quad (11)$$

where, u_q is the delay threshold of the qth UE.

Modified Largest Weighted Delay First Scheduler

The MLWDF scheduler aims to support real time traffic with additional performance metrics [10]. MLWDF when compared to the PF scheduler introduced additional two parameters, i.e. the head of line (HoL) packet delay and quality of service (QoS). For each packet in the queue, HoL packet delay, which is the time difference between the current time and the arrival time of the packet, is computed. Different delay deadlines are assigned to packets according to the QoS requirements for various services. The scheduler assigns RBs to different flows by considering the properties of the traditional PF rule and the HoL packet delay for the real time flows. Packets belonging to a real time service will be discarded when the target delay deadline assigned to that UE is exceeded while waiting at the MAC buffer. Therefore, the modified PF rule is appropriately weighted, and is given by

$$P_{m2} = W_q(t)\left(\frac{r_q(t)}{R_q(t)}\right). \tag{12}$$

where,

$r_q(t)$ is the instantaneous achievable data rate for a UE in a particular RB by its respective CQI obtained from the post processing SINR of the receiver computed using (8) or (9),

$R_q(t)$ is the average data rate of the qth UE at time t obtained by averaging the data rate supported by this UE in the past over a long period of time, and

$W_q(t)$ is the HoL packet delay of the qth UE at time t.

The MLWDF algorithm decides the UEs to be scheduled on a particular RB based when the delay exceeds its threshold t_q is decided based on the following priority function

$$P_M = \arg\max_{a_q}\left(P_{m2}\right) = \arg\max_{a_q} W_q(t)\left(\frac{r_q(t)}{R_q(t)}\right). \tag{13}$$

where,

$$a_q = -\left(\frac{\log p_q}{t_q}\right),$$

t_q is the delay threshold of the qth UE, and

p_q denotes the probability of HoL delay to exceed its threshold for the qth UE a_q.

SIMULATION RESULTS

The system level simulations to compare the performance of RR, PF, and MLWDF scheduler are performed using the network simulator tool NS2 for LTE-Advanced (LTE-A) system considering 5 MHz bandwidth with 25 RBs. The scheduling is done on a sub frame basis of 1 ms TTI as per LTE/LTE-A technologies and in one TTI all 25 RBs are scheduled to different UEs using one of the RR, PF, and MLWDF schedulers. After the users are scheduled for one TTI, the throughput is calculated as sum of all information bits transmitted successfully over all resource elements, which is obtained as the product of the number of bits in the modulation scheme with the code rate and (1 –*BER*).

A cell of radius 500 meters consisting of a macro eNB at the centre, and two pico eNBs are placed in the center of a non-overlapping random circular pico cell of radius 40 metres. A total of 40 UEs are considered with 24 UEs dropped uniformly in the entire cell, and 8 UEs each are dropped in the pico cell. The macro and pico eNBs transmit with a transmit power of 43 dBm and 30 dBm, respectively. Due to the difference in transmit power between the macro and the pico eNBs, more UEs tend to attach to the macro due to high SINR. To balance the load between the macro and the pico eNB, biasing is applied. i.e., a bias value of 0, 6, 9, and 15 dB is added to the preprocessing SINR given by (3). The UEs are attached to the strongest eNB depending on the modified preprocessing SINR after adding the bias value. Simulations are performed for multiple realizations by changing the bias values using RR, PF, and MLWDF schedulers for resource allocations. The post processing SINR is computed using (9) for all three schedulers in rate and throughput calculations, and applied in the priority function (11) and (13) for PF and MLWDF schedulers.

The performance of various schedulers in LTE system using interference mitigation receiver is obtained using system level simulations with the setup explained above. The throughput performance comparison of the various scheduling algorithm for different bias value is shown inFigure 6. It is evident from the figure that the throughput improves with the increase in bias values for all the three schedulers. This is due to the fact that the UEs are able to get more resources from moving to the pico eNB, and also thereby releasing more resources to the UEs attached to the macro eNB. The performance of RR scheduler is poor compared to the other two schedulers as it does not take into account of the CQI for scheduling. However, the MLWDF and PF scheduler perform better as it schedules UE with best channel and less interference in a particular resource block at a given time instant. Moreover, the best throughput performance is achieved by the MLWDF algorithm due to the fact that the fairness is determined by the QoS and delay permitted. This shows that it is advantageous to use MLWDF algorithm in real time systems.

The comparison of the average delay with increase in bias values for the PF, MLWDF and RR scheduling is shown in Figure 7. The increase in average delay with increase in the bias value is due to the fact that biasing results in increase in the interference when the UEs are moved from macro to pico eNB. This will reduce the post processing SINR, and hence less number of bits is transmitted in the allocated resources. It will in turn increase the delay in sending and receiving the packets. Here again the RR scheduler performs poor as the UEs are allocated with resources irrespective of the channel and interference conditions. However, the PF and MLWDF scheduler utilize the channel condition and the interference suppression capability of the receiver well, and hence can send more bits in their allocated resources, and hence reduce the average delay. It is found that the average delay is the least in the MLWDF algorithm, and this is due to the fact that QoS and delay metrics are used over PF during resource allocation.

The comparison of the packet delivery ratio with increase in bias values for the PF, MLWDF and RR scheduler is shown in Figure 8. The increase in the packet delivery ratio with increase in bias value is due to the fact that the UEs are able to get more resources from moving to the pico eNB, and also thereby releasing more resources to the UEs attached to the macro eNB. Moreover, the delivery ratio will increase with post processing SINR and the choice of the performance metric. It is evident that the MLWDF scheduler has the best performance with the additional QoS and delay metric compared to the PF scheduler. The RR scheduler performs poor for the similar reasons explained for the results in Figure 7.

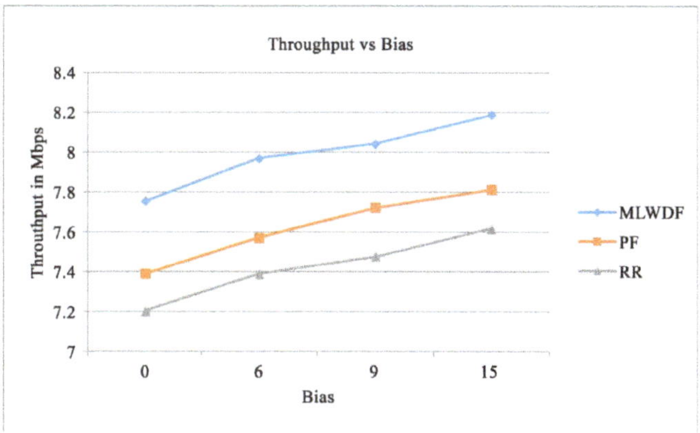

Figure 6. Throughput performance comparison for various bias values using MLWDF, PF and RR schedulers.

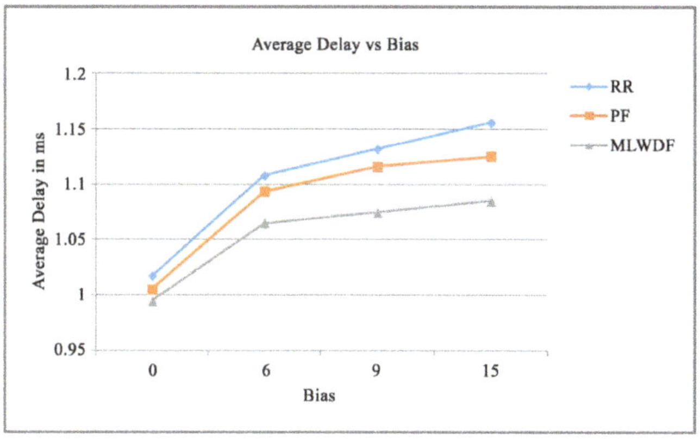

Figure 7. Average delay performance comparison for various bias values using MLWDF, PF and RR schedulers.

Figure 8. Packet delivery ratio performance comparison for various bias values using MLWDF, PF and RR schedulers.

CONCLUSION

The interference suppression capability of the MMSE and eMMSE receiver with a single dominant interference is analyzed, and their BER performance is studied. The link level performance of both receivers is compared with single eNB antenna and various UE antenna configurations. The performance of both receivers can be improved with increasing number of antennas at the UE. Moreover, it is found that both the receivers are able to completely suppress the interference when the number of antennas at the UE is sufficiently large. The system level performance of a system having one eNB and two pico eNB using MLWDF, PF, and RR schedulers is carried out by allocating the resources based on the post processing SINR of the eMMSE receiver. The throughput, average delay, and packet delivery ratio performances are obtained and compared for various biasing value of 3, 6, 9, and 15 dB. It is found that the MLWDF scheduler achieves higher throughput with the least average packet delay and the best packet delivery ratio compared to the RR and PF schedulers. The next best performance is obtained for a PF scheduler, and RR scheduler achieves the worst performance for all the three performance metrics. The performance of these schedulers along with further eMMSE receivers that can handle more number of eNB interferences using higher number of UE antenna configurations could be further explored.

REFERENCES

1. Feng, M.J., Jiang, T., Chen, D. and Mao, S.W. (2014) Cooperative Small Cell Networks: High Capacity for Hotspots with Interference Mitigation. IEEE Wireless Communications, 21, 108-116. http://dx.doi.org/10.1109/MWC.2014.7000978
2. Damnjanovic, A., et al. (2011) A Survey on 3GPP Heterogeneous Networks. IEEE Wireless Communications, 18, 10-21. http://dx.doi.org/10.1109/MWC.2011.5876496
3. Ghosh, A., Ratasuk, R., Mondal, B., Mangalvedhe, N. and Thomas, T. (2010) LTE-Advanced: Next Generation Wireless Broadband Technology. IEEE Wireless Communications, 17, 10-22. http://dx.doi.org/10.1109/MWC.2010.5490974
4. Pauli, V., Naranjo, J.D. and Seidel, E. (2010) Heterogeneous LTE Networks and Inter-Cell Interference Coordination. Nomor Research GmbH, Munich, 1-9.
5. Gerhard, W., et al. (2015) 5G Now Final MAC/Networking Concepts. 5GNow_D4.2_Final. http://www.5gnow.eu/wp-content/uploads/2015/04/5GNOW_D4.2_final1.pdf
6. Supratim, D., Pantelis, M., Jerzy, M. and James, P.S. (2014) Algorithms for Enhanced Inter-Cell Interference Coordination (eICIC) in LTE HetNets. IEEE/ACM Transactions on Networking, 22, 137-150. [Paper reference:3]
7. Andrews, J.G., Choi, W. and Heath Jr., R.W. (2007) Overcoming Interference in Spatial Multiplexing MIMO Cellular Networks. IEEE Wireless Communications, 14, 95-104. http://dx.doi.org/10.1109/MWC.2007.4407232
8. Balint, C., Budura, G. and Marza, E. (2010) Scheduling Techniques Evaluation in LTE systems with Mixed Data Traffic. Proceedings of the 9th International Symposium on Electronics and Telecommunications, Timisoara, 11-12 November 2010, 221-224.
9. Gomez, K., Goratti, L., Granelli, F. and Rasheed, T. (2014) A Comparative Study of Scheduling Disciplines in 5G Systems for Emerging Communications. Proceedings of the 1st International Conference on 5G for Ubiquitous Connectivity, Akaslompolo, December 2014, 1-6.

10. Capozzi, F., Piro, G., Grieco, L.A., Boggia, J., and Camarda, P. (2013) Downlink Packet Scheduling in LTE Cellular Networks: Key Design Issues and a Survey. IEEE Communications Surveys & Tutorials, 15, 678-700. http://dx.doi.org/10.1109/SURV.2012.060912.00100

11. 3rd Generation Partnership Project, Technical Specification Group, Radio Access Network (2013-02) Evolved Universal Terrestrial Radio Access (E-UTRA), Physical Channels and Modulation. Release 10, 3GPP TS 36.211 V10.7.0. http://www.3gpp.org/ftp/Specs/archive/36_series/36.211/36211-a70.zip

12. Ikuno, J.C., Wrulich, M. and Rupp, M. (2010) System Level Simulation of LTE Networks. Proceedings of the 71st IEEE Vehicular Technology Conference, Taipei, 16-19 May 2010, 1-5.

13. 3rd Generation Partnership Project, Technical Specification Group, Radio Access Network (2011-03) Spatial Channel Model for Multiple Input Multiple Output (MIMO) Simulations. Release 10, 3GPP TR 25.996. http://www.3gpp.org/ftp/Specs/archive/36_series/36.211/36211-a70.zip

14. Tetsushi, A. and Gerhard, B. (2010) Effective SINR Computation for Maximum Likelihood Detector in MIMO Spatial Multiplexing System. Proceedings of the IEEE Global Telecommunication Conference, Honolulu, 30 November-4 December 2009, 1-5.

15. Eraslan, E., Daneshrad, B. and Lou, C.-Y. (2012) Performance Indicator for MIMO MMSE Receivers in the Presence of Channel Estimation Errors. IEEE Wireless Communication Letters, 2, 211-214. http://dx.doi.org/10.1109/WCL.2013.012513.120824

16. Bai, Z.J., et al. (2011) On the Equivalence of MMSE and IRC Receiver in MU-MIMO Systems. IEEE Communication Letters, 15, 1288-1291. http://dx.doi.org/10.1109/LCOMM.2011.093011.110826

17. Tavares, F.M.L., Berardinelli, G., Mahmood, N.H., Sorensen, T.B. and Mogensen, P. (2014) On the Impact of Receiver Imperfections on the MMSE-IRC Receiver Performance in 5G Networks. Proceedings of the 79th IEEE Vehicular Technology Conference, Seoul, 18-21 May 2014, 1-6.

18. Tang, T.W., Doostnejad, R. and Lim, T.J. (2010) Mean Mutual Information Per Coded Bit Based Precoding in MIMO- OFDM Systems. Proceedings of the 72nd IEEE Vehicular Technology Conference, Ottawa, 6-9 September 2010, 1-5.

CHAPTER 19

APPLICATION OF EXTREME LEARNING MACHINE IN FAULT CLASSIFICATION OF POWER TRANSFORMER

Athikkan Venkatasami and Pitchai Latha

Keywords: Transformer, Dissolved Gas Analysis, Machine Learning

INTRODUCTION

Power transformer is one of the critical equipments in a power system. Hence, it is very essential to keep the power transformer in good health. Generally gases are produced during its operation and they get dissolved in transformer oil. These gases are extracted from oil and measured using gas chromatograph. Alternatively they are measured through gas sensors fitted in the power transformer. Dissolved gas analysis (DGA) is one of the oldest and reliable methods to evaluate whether there is any abnormality inside a power transformer or not [1] [2]. A number of methods are in use for the interpretation of dissolved gas analysis. Key Gas method, Roger's ratio method, IEC ratio method, Dual triangle etc. are the methods used. The individual concentrations of six fault gases are used in key gas method. Four common fault conditions are distinguished by the quoted percentage concentrations of the six gases. The Roger's ratio method use four gas ratios resulting into 12 proposed diagnosis. In the revised standard, only six diagnoses are used. In the IEC ratio method, the same three ratios as per the revised Roger's ratio method is used but it suggests different ranges and interpretations. However, there is no universally accepted interpretation method for the DGA. To overcome the limitations of the conventional diagnostic methods, soft computing tools are used to develop automated transformer fault classification.

Artificial neural network (ANN) can be trained to acquire knowledge. However, lots of training data are required for proper training. ANN also has universal approximation property and it enables to approximate any given continuous function with required level of accuracy. These networks are considered as a "black box" models and hence it is very difficult to interpret in their trained state [3]. Among various learning algorithms, the support vector machine (SVM) is one of the most important and widely used algorithms for transformer fault classifications. Yan Zhang [4] has reported the use of online DGA data using SVM. Four types of SVM models are used to classify transformer faults. Out of 60 dataset, 35 datasets are used for training and 25 for testing. Cross-validation method is used to calculate the free parameters of the SVM.

The state of art Extreme Learning Machine method offers a competitively good solution for complex tasks. The Extreme learning Machine is a recent second generation neural network algorithm. It is identified as the one which can give high performance in multifaceted problems and takes less computation time compared with other machine learning algorithms [5]. The ELM based model proposed by Huang et al. [6] performs quite fast with high learning phase for problems of relatively large dataset which does not require iterative tuning used in other neural networks. ELM based model is used for classification of images in medical applications [7]. The paper focuses on classification of tumour using three segment classifications of MR brain images in 3 dimensions. Using 60 neurons in the hidden layer, ELM model achieves 93.2% accuracy. The ELM is successfully applied to face recognition using Yale face database and Carnegie Mellon University face images database [8]. It is shown that the state preserving extreme learning machine used for face recognition gives best performance than the other known classifiers.

Citation: Venkatasami, A. and Latha, P. (2016). Application of Extreme Learning Machine in Fault Classification of Power Transformer. *Circuits and Systems*, 7, 2837-2845. DOI: 10.4236/cs.2016.710242.

Single layer feed forward Neural Network is modeled with hidden neurons and activation function by Guang- Bin et al. [9]. Extreme learning machine for RBF (Radial Basis Function) network produces faster results. Huang et al. [10] proved that single layer feed-forward neural network with randomly assigned input weights and hidden layer biases and with any nonzero activation function can approximate any continuous functions. Thermodynamic estimation of fault severity is suggested in [11] for transformer fault energy dissipated within the transformer. It is based on the enthalpy of forming one substance from another substance. When a thermodynamic system undergoes transformation, change in enthalpy takes place. The enthalpy [12] for forming one substance from another is the amount of energy required to produce from one mole of one substance to form another expressed in grams per mole.

This research paper focuses on six segment classifications model with dissolved gas analysis data of DB1 database and four segment classifications model with DB2 database. One method based on the conventional approach using gas concentration and the other improved method based on the enthalpy of the gas concentration is used. Section 2 covers the various ELM models for transformer fault classification. Application of ELM for fault classification in transformer using DB1 database is elaborated in Section 3. Section 4 details the application of ELM for fault classification in transformer using DB2 database. Conclusions based on the present study are given in Section 5.

ELM MODELS

ELM is a single layer with feed forward neural network (SLFNN). It randomly selects input weights and hidden layer biases without training. The output weights are obtained analytically using the norm least square solution and Moore-Penrose inverse of a general linear system [7]. Activation functions used are unipolar, bipolar and radial basis function (RBF) for the output neurons. Two databases (DB1 and DB2) are used. DB1 database contain 151 records published by IEC TC10 working group, whereas DB2 has 219 records based on data collected from utilities in INDIA. Four models namely ELM-I, ELM-II, ELM-III and ELM-IV are developed. ELM-I and ELM-III models are based on conventional approach using gas concentrations as input features. In ELM-II and ELM-IV, instead of gas concentrations as input features, enthalpy of the gas concentrations are used. Enthalpy is calculated from the gas concentrations as per the Table 1.

ELM Models Using DB1 Database

The database DB1 contains 151 records of IEC TC10 database. It contains methane, ethane, ethylene, hydrogen and acetylene carbon monoxide and carbon dioxide. The gas data are obtained from gas sensors fitted in the Power transformer or from gas chromatograph. Out of seven gases as shown in Figure 1, five gases namely acetylene, ethylene, ethane, methane and hydrogen are selected for ELM-I and ELM-II models. ELM-I model uses gas concentrations in parts per million (ppm) as input features.

Typical dataset of DB1 database showing input and output features are shown in Table 2. Partial discharge, discharges of low and high energy (D1 and D2), thermal faults (T1, T2 and T3) and normal (NF) form the output features.

For ELM-II model, the gas concentrations are converted into enthalpy and used as input feature.Table 3 shows the training and testing dataset used for each output feature. Partial discharge has only 9 dataset whereas discharge of high energy has total 48 dataset.

Both ELM-I and ELM-II models use total 151 dataset. Out of 151 dataset, 106 dataset is used for training and 45 dataset is used for testing.

ELM Models Using DB2 Database

Figure 2 shows the methodology of using DB2 database. DB2 has seven extracted features.

Table 1. Enthalpy of Transformer fault gases.

Sr. No.	Fault Gas	Chemical formula	Enthalpy (kJ/mol)
1	Methane	CH_4	77.70
2	Ethane	C_2H_4	93.50
3	Ethylene	C_2H_6	104.1
4	Hydrogen	H_2	128.5
5	Acetylene	C_2H_2	278.3

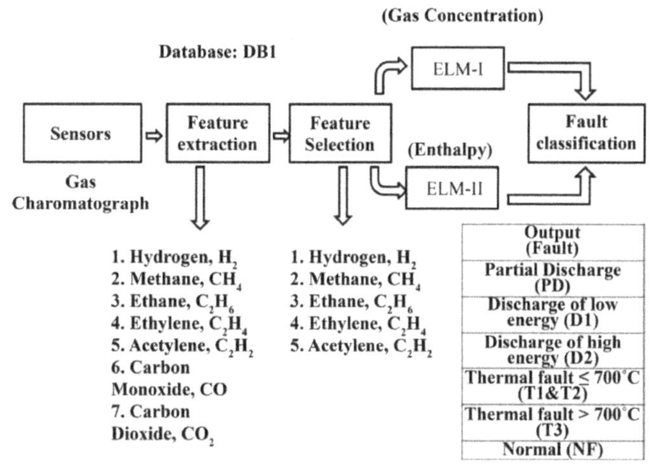

Figure 1. Structure of ELM Models with DB1 database.

Table 2. Typical DB1 database with input and output features.

Input Features Gas concentrations (parts per million, ppm)					Output features (Fault types)
H_2	CH_4	C_2H_4	C_2H_6	C_2H_2	
37,800	1740	8	249	8	Partial Discharge (PD)
305	100	161	33	541	Discharge of low energy (D1)
440	89	304	19	757	Discharge of high energy (D2)
3420	7870	6990	1500	33	Thermal fault ≤ 700°C (T1&T2)
6709	10,500	17,700	1400	750	Thermal fault > 700°C (T3)
100	200	200	200	20	Normal (NF)

Table 3. DB1 dataset with output features.

Sr. No.	Output features (Fault types)	Training Dataset	Testing Dataset	Total Dataset
1	Partial Discharge (PD)	6	3	9
2	Discharge of low energy (D1)	18	8	26
3	Discharge of High Energy (D2)	33	15	48
4	Thermal faults ≤ 700°C (T1 and T2)	12	4	16
5	Thermal faults > 700°C (T3)	13	5	18
6	Normal (NF)	24	10	34
Total dataset		106	45	151

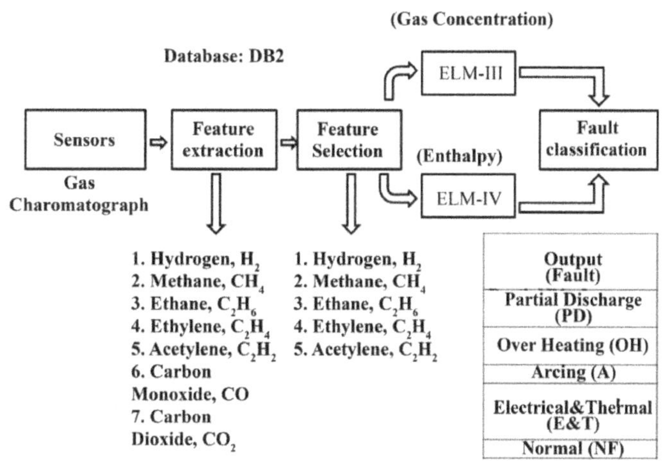

Figure 2. Structure of ELM Models with DB2 database.

Five features are selected for ELM-III and ELM-IV model. Model ELM-III is based on the gas concentrations in ppm and Model-IV is based on the enthalpy of gases. Four types of faults namely partial discharge, overheating, arcing and electrical and thermal faults are taken as output features as in Table 4.

Out of 219 dataset in DB2 database, 179 dataset is used for training and 40 dataset is used for testing. Total dataset used for partial discharge, overheating, arcing and Electrical & thermal (E&T) classes are 29, 120, 58 and 12 respectively.

Table 5 shows the training and testing dataset used for each output feature. Electrical and thermal fault as output feature has only 12 dataset.

APPLICATION OF ELM ON DB1 DATABASE

The transformer fault classification of DB1 database is analyzed using MATLAB™ software on ELM-I and ELM-II models. In ELM-I model, transformer fault classification is based on the gas concentration present in the transformer from DB1 database. Five gases namely hydrogen, methane, ethane, ethylene and acetylene forms the input features for the model. The performance of the model is analyzed with unipolar, bipolar and radial basis function (RBF) as activation functions. The training and testing efficiencies are calculated as shown in Table 6.

ELM classifier displays a confusion matrix showing the actual and predicted class. In the confusion matrix shown in Table 7, of the 3 cases of actual PD class, the system predicted 2 cases as PD and one case as class T1&T2. The system predicted all D1 class correctly. Of the 15 D2 class, 1 is classified as (T1&T2) class and 1 is classified as NF. With total 4 actual class of T1&T2, the system predicted 1 case as class of T1&T2 and 2 case as class T3 and 1 case as class NF. Out of five actual class of T3, 1 case is wrongly classified as class D2. All 10 actual class of NF is classified correctly. The overall testing efficiency is 84.4% using ELM with DB1 database and gas content as input.

The energy dissipated within the transformer is directly related to the enthalpy of hydrocarbon gases. The gas concentrations are converted to the enthalpy. The enthalpy of hydrogen, methane, ethane, ethylene and acetylene are input features for the ELM-II model. The model is evaluated with different activation functions and number of neurons in the hidden layer and the results are tabulated in Table 8.

Table 4. Typical DB2 database with input and output features.

Input Features Gas concentrations (parts per million, ppm)					Output features (Fault types)
H_2	CH_4	C_2H_4	C_2H_6	C_2H_2	
169	16	6	2	0	Partial Discharge (PD)
577	3541	2928	521	7	Over Heating (OH)
116	50	50	2	284	Arcing (A)
100	722	450	31	360	Electrical &Thermal (E&T)

Table 5. DB2 dataset with output features.

Sr. No.	Output features (Fault types)	Training Dataset	Testing Dataset	Total Dataset
1	Partial Discharge (PD)	24	5	29
2	Over Heating (OH)	99	21	120
3	Arcing (A)	47	11	58
4	Electrical &Thermal (E&T)	9	3	12
Total dataset		179	40	219

Table 6. Effect of ELM parameters on DB1 database with gas concentration.

Sr. No.	Activation Function	No. of hidden Neurons	Training efficiency (%)	Testing efficiency (%)
1	Unipolar	35	92	93
2	Bipolar	35	91	91
3	RBF	35	94	93

Table 7. Confusion matrix of DB1 database with gas concentration.

Actual Class	Predicted Class						Total Testing Dataset	Total correct prediction	Testing efficiency %
	PD	D1	D2	T1&T2	T3	NF			
PD	2	0	0	1	0	0	3	2	84.4
D1	0	8	0	0	0	0	8	8	
D2	0	0	13	1	0	1	15	13	
T1&T2	0	0	0	1	2	1	4	1	
T3	0	0	1	0	4	0	5	4	
NF	0	0	0	0	0	10	10	10	

Table 8. Effect of ELM parameters on DB1 database with enthalpy.

Sr. No.	Activation Function	No. of hidden Neurons	Training efficiency (%)	Testing efficiency (%)
1	Unipolar	35	92	93
2	Bipolar	35	91	91
3	RBF	35	94	93

Table 9. Confusion matrix of DB1 database with enthalpy.

Actual Class	Predicted Class						Total Testing Dataset	Total correct prediction	Testing efficiency %
	PD	D1	D2	T1&T2	T3	NF			
PD	3	0	0	0	0	0	3	3	93.3
D1	0	8	0	0	0	0	8	8	
D2	0	2	13	0	0	0	15	13	
T1&T2	0	0	0	4	0	0	4	4	
T3	0	0	0	0	5	0	5	5	
NF	0	0	1	0	0	9	10	9	

The simulation results are analyzed based on activation function, number of hidden neurons and gas concentration with enthalpy of gas. With gas concentration as input feature, the activation function using unipolar and radial basis function perform better than the bipolar activation function. Similar results are obtained with enthalpy of gas as input features. Optimum results are obtained with 35 neurons in the hidden layer.

In the example confusion matrix shown in Table 9, of the 3 cases of actual class of PD, the system predicted all correctly as class PD. Similarly, the system predicted all cases of class D1 also correctly. However for 15 cases of actual D2 class, the system predicted 13 cases as D2 class and 2 cases are wrongly classified as class D1. Similarly for 10 cases of actual class NF, 9 cases are predicted correctly as class NF and one wrongly predicted as class D2. The overall testing efficiency is 93.3% using DB1 database and enthalpy of gas as input feature.

With gas concentration as input feature, the activation function using unipolar and radial basis function perform better than the bipolar activation function. Similar results are obtained with enthalpy of gas as input features. Optimum results are obtained with 35 neurons in the hidden layer. The simulation results obtained on DB1 database using enthalpy of gas as input feature is compared with conventional gas concentration as input feature and the results are shown in Figure 3. The overall testing efficiency is 84.4% using ELM with DB1 database and gas content as input feature. The testing efficiency improves to 93.3% using DB1 database and enthalpy of gas as input feature.

APPLICATION OF ELM ON DB2 DATABASE

The DB2 database contains 219 records. Transformer fault classification of DB2 database is analyzed using MATLAB™ software on ELM-III and ELM-IV models. In the ELM-III model, gas concentration (ppm) is taken as input features and partial discharge, over hating, arcing and thermal and electrical faults are taken as output features.

The performance of the ELM model is evaluated using different activation functions and number of neurons in the hidden layer. The results showing training and testing efficiencies are shown in Table 10. Confusion matrix constructed for ELM-III model is given in Table 11.

The model predicted PD, OH, E&T class correctly whereas with 11 cases of actual class A (Arcing), the system predicted 10 cases as class A and 1 case wrongly predicted as class E&T.

Model ELM-IV uses DB2 database and the enthalpy of gases are taken as input features. Partial discharge, overheating, arcing and electrical and thermal faults are treated as output features. The performance of ELM-IV is analyzed by varying the activation function and the number of neurons in the hidden layer and the results are shown in Table 12.

With 35 neurons in the hidden layer, the training efficiency is 95%. The testing efficiency achieved is 100% with activation function of unipolar and bipolar whereas it is 97% with RBF activation function. In the example confusion matrix in Table 13, the system predicted all actual classes of PD, OH, A, E&T correctly.

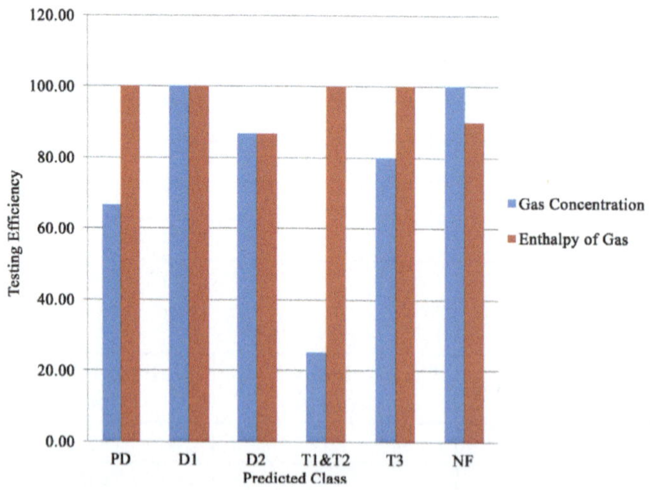

Figure 3. Performance comparison on DB1 database.

Table 10. Effect of ELM parameters on DB2 database with gas concentration.

Sr. No.	Activation Function	No. of hidden Neurons	Training efficiency (%)	Testing efficiency (%)
1	Unipolar	35	95	97
2	Bipolar	35	95	97
3	RBF	35	94	95

Table 11. Confusion matrix of DB2 database with gas concentration.

Actual Class	Predicted Class				Total Testing Dataset	Total correct prediction	Testing efficiency %
	PD	OH	A	E&T			
PD	5	0	0	0	5	5	97.5
OH	0	21	0	0	21	21	
A	0	0	10	1	11	10	
E&T	0	0	0	3	3	3	

Table 12. Effect of ELM parameters on DB2 database with enthalpy.

Sr. No.	Activation Function	No. of hidden Neurons	Training efficiency (%)	Testing efficiency (%)
1	Unipolar	35	95	100
2	Bipolar	35	95	100
3	RBF	35	95	97

Table 13. Confusion matrix of DB2 database with enthalpy.

Actual Class	Predicted Class				Total Testing Dataset	Total correct prediction	Testing efficiency %
	PD	OH	A	E&T			
PD	5	0	0	0	5	5	100
OH	0	21	0	0	21	21	
A	0	0	11	0	11	11	
E&T	0	0	0	3	3	3	

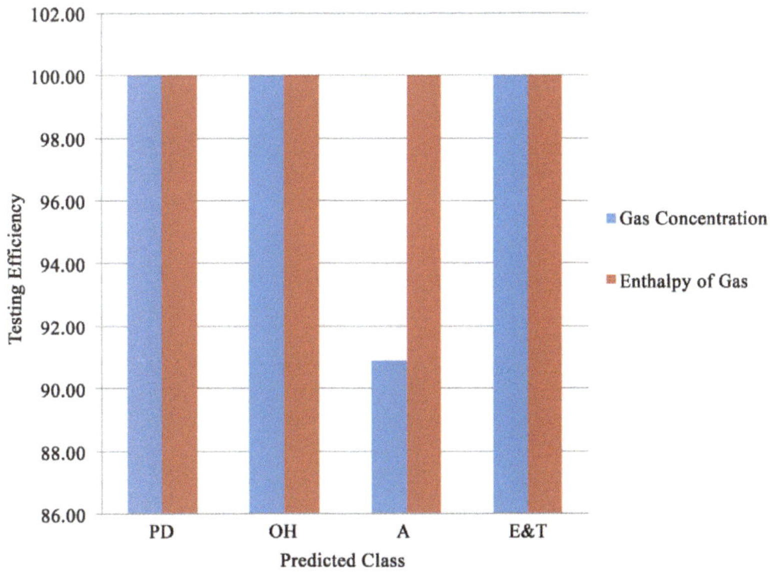

Figure 4. Performance comparison on DB2 database.

The simulation results are analyzed based on the activation function, number of hidden neurons, gas concentration and enthalpy of gas. With gas concentration as input features, the unipolar as well as bipolar activation function perform better than the radial basis function activation function. Also, when enthalpy of gas is used as input features, the unipolar, bipolar activation function perform better than radial basis activation function. Optimum results are obtained with 35 neurons in the hidden layer. The simulation results obtained on DB2 database using enthalpy of gas as input feature is compared with conventional gas concentration as input feature and the results are shown in Figure 4.

ELM testing efficiency of 97.5% is achieved with gas concentration as input features and 100% with enthalpy of gas concentration as input features.

CONCLUSION

This research paper presented the application of ELM approach to transformer fault classification. The proposed method is simulated in MATLAB™ Software. ELM method is validated on DB1 and DB2 database using four models. Conventional transformer fault classification based on gas concentration as input features is compared with fault classification using enthalpy of gas as input feature. The simulation carried out using DB1 database shows 84.4% testing efficiency with gas concentration as input feature and 93.3% with enthalpy as input feature. Results on DB2 database show 97.5% testing efficiency with gas concentration and 100% with enthalpy as input feature. The results show that ELM with enthalpy as input feature in fault classification of power transformer outperform over the conventional approach. It is proposed to use IEC gas ratios using enthalpy of gas in place of gas concentration as input feature in the future research work.

ACKNOWLEDGEMENTS

The authors would like to acknowledge the Principal, Dr. K. Ramar, Einstein College of Engineering, for supporting this research. The authors are also thankful to Mr. A. Amudhavanan, Managing Trustee, Einstein College of Engineering for granting permission to publish this paper.

REFERENCES

1. (2008) IEEE Std.C57.104, IEEE Guide for the Interpretation of Gases Generated in Oil-Filled Transformers.
2. (2007) IEC Publication 60599, Mineral Oil-Impregnated Electrical Equipment in Service: Guide to the Interpretation of Dissolved and Free Gas Analysis.
3. Kloppel, S., Stonnington, C.M., Dragnasski, C.C. and Scahill, R.I. (2008) Automatic Classification of MR Scans in Alzheimer's Disease. Brain. http://dx.doi.org/10.1093/brain/awm319
4. Yan, H. and Zhang, B. (2010) Transformer Fault Diagnosis Based on Support Vector Machine. IEEE International Conference on Computer Science and Information Technology, 2, 681-689.
5. Huang, G.B., Zhu, Q.Y. and Siew, C.K. (2006) Extreme Learning Machine Theory and Applications. Neuro Computing, 70, 489-501. http://dx.doi.org/10.1016/j.neucom.2005.12.126
6. Gulloway, M.M. Texture Analysis Using Grey Level Run Lengths. Elsevier Inc.
7. Deepa, S.N. and Arunadevi, B. (2013) Extreme Learning Machine for Classification of Brain Tumor in 3D MR Images. Informatol, 46, 111-121. [Paper reference:2]
8. Zahangir Alam, M., Sidike, V.K. and Taka, T.M. (2015) State Preserving Extreme Learning Machine for Face Recognition. International Joint Conference on Neural Networks, 1-7. http://dx.doi.org/10.1109/ijcnn.2015.7280788
9. Guang, B., Huang, A. and Siew, C.K. (2004) Extreme Learning Machine with Randomly Assigned RBF Kernels. The Proceedings of the Eighth International Conference on Control, Automation, Robotics and Vision.
10. Mustapha, B., Abdelhakim, H. and Abderrahim, B. (2009) High Accuracy Localization Method Using AOA in Sensor Networks. Computer Networks, 53, 3076-3088. http://dx.doi.org/10.1016/j.comnet.2009.07.015

11. Jakob, F. and James, J.D. (2015) Thermodynamic Estimation of Transformer Fault Severity. IEEE Transactions on Power Delivery, 39, 1941-1948.

12. Jacob, F., Noble, P. and Dukarm, J. (2012) A Thermodynamic Approach to Evaluation of the Severity of Transformer Faults. IEEE Transactions on Power Delivery, 27, 554-560. http://dx.doi.org/10.1109/TPWRD.2011.2175950

INDEX